ULTRA WIDEBAND WIRELESS COMMUNICATION

ULTRA WIDEBAND WIRELESS COMMUNICATION

Edited by

Hüseyin Arslan
University of South Florida, Tampa, Florida

Zhi Ning Chen
Institute for Infocomm Research, Singapore

Maria-Gabriella Di Benedetto
University of Rome La Sapienza, Italy

A JOHN WILEY & SONS, INC., PUBLICATION

This book is printed on acid-free paper. ∞

Copyright © 2006 by John Wiley & Sons, Inc. All rights reserved.

Published by John Wiley & Sons, Inc., Hoboken, New Jersey.
Published simultaneously in Canada.

No part of this publication may be reproduced, stored in a retrieval system, or transmitted in any form or by any means, electronic, mechanical, photocopying, recording, scanning, or otherwise, except as permitted under Sections 107 or 108 of the 1976 United States Copyright Act, without either the prior written permission of the Publisher, or authorization through payment of the appropriate per-copy fee to the Copyright Clearance Center, Inc., 222 Rosewood Drive, Danvers, MA 01923, 978-750-8400, fax 978-646-8600, or on the web at www.copyright.com. Requests to the Publisher for permission should be addressed to the Permissions Department, John Wiley & Sons, Inc., 111 River Street, Hoboken, NJ 07030, (201) 748-6011, fax (201) 748-6008.

Limit of Liability/Disclaimer of Warranty: While the publisher and author have used their best efforts in preparing this book, they make no representations or warranties with respect to the accuracy or completeness of the contents of this book and specifically disclaim any implied warranties of merchantability or fitness for a particular purpose. No warranty may be created or extended by sales representatives or written sales materials. The advice and strategies contained herein may not be suitable for your situation. You should consult with a professional where appropriate. Neither the publisher nor author shall be liable for any loss of profit or any other commercial damages, including but not limited to special, incidental, consequential, or other damages.

For general information on our other products and services please contact our Customer Care Department within the U.S. at 877-762-2974, outside the U.S. at 317-572-3993 or fax 317-572-4002.

Wiley also publishes its books in a variety of electronic formats. Some content that appears in print, however, may not be available in electronic format.

Library of Congress Cataloging-in-Publication Data:

Ultra wideband wireless communication / edited by Huseyin Arslan,
    Zhi Ning Chen, Maria-Gabriella Di Benedetto.
      p. cm.
    Includes index.
    ISBN 0-471-71521-2 (cloth)
      1. Broadband communication systems. 2. Ultra-wideband devices.
  3. Wireless communication systems. I. Arslan, Huseyin, 1968–
  II. Chen, Zhi Ning. III. Di Benedetto, Maria-Gabriella.
  TK5103.4.U44 2006
  621.384 - - dc22

2006008457

Printed in the United States of America

10 9 8 7 6 5 4 3 2 1

CONTENTS

Preface — xv

Contributors — xix

Chapter 1 Introduction to Ultra Wideband — 1
Hüseyin Arslan and Maria-Gabriella Di Benedetto

1.1 Introduction — 1
    1.1.1 Benefits of UWB — 2
    1.1.2 Applications — 3
    1.1.3 Challenges — 3
1.2 Scope of the Book — 4

Chapter 2 UWB Channel Estimation and Synchronization — 11
Irena Maravic and Martin Vetterli

2.1 Introduction — 11
2.2 Channel Estimation at SubNyquist Sampling Rate — 14
    2.2.1 UWB Channel Model — 14
    2.2.2 Frequency-Domain Channel Estimation — 15
    2.2.3 Polynomial Realization of the Model-Based Methods — 16
    2.2.4 Subspace-Based Approach — 20
    2.2.5 Estimation of Closely Spaced Paths — 24
2.3 Performance Evaluation — 25
    2.3.1 Analysis of Noise Sensitivity — 25
    2.3.2 Computational Complexity and Alternative Solutions — 27
    2.3.3 Numerical Example — 28
2.4 Estimating UWB Channels with Frequency-Dependent Distortion — 29
    2.4.1 Algorithm Outline — 31
2.5 Channel Estimation from Multiple Bands — 32
    2.5.1 Filter Bank Approach — 32
    2.5.2 Estimation from Nonadjacent Bands — 32
2.6 Low-Complexity Rapid Acquisition in UWB Localizers — 34

	2.6.1 Two-Step Estimation	36
2.7	Conclusions	39

Chapter 3 Ultra Wideband Geolocation   43
Sinan Gezici, Zafer Sahinoglu, Hisashi Kobayashi, and H. Vincent Poor

3.1	Introduction	43
3.2	Signal Model	44
3.3	Positioning Techniques	44
	3.3.1 Angle of Arrival	45
	3.3.2 Received Signal Strength	49
	3.3.3 Time-Based Approaches	51
3.4	Main Sources of Error in Time-Based Positioning	52
	3.4.1 Multipath Propagation	52
	3.4.2 Multiple Access Interference	53
	3.4.3 Nonline-of-Sight Propagation	53
	3.4.4 High Time Resolution of UWB Signals	54
3.5	Ranging and Positioning	55
	3.5.1 Relationship Between Ranging and Optimal Positioning Algorithms	55
	3.5.2 ToA Estimation Algorithms	58
	3.5.3 Two-Way Ranging Protocols	69
3.6	Location-Aware Applications	70
3.7	Conclusions	71

Chapter 4 UWB Modulation Options   77
Hüseyin Arslan, İsmail Güenç, and Sadia Ahmed

4.1	Introduction	77
4.2	UWB Signaling Techniques	78
	4.2.1 UWB-IR Signaling	79
	4.2.2 Multiband UWB	83
	4.2.3 Multicarrier UWB	85
	4.2.4 OFDM	85
4.3	Data Mapping	87
	4.3.1 Binary Data Mapping Schemes	87
	4.3.2 M-ary Data Mapping Schemes	89
4.4	Spectral Characteristics	91
4.5	Data Mapping and Transceiver Complexity	92
4.6	Modulation Performances in Practical Conditions	93

		4.6.1	Effects of Multipath	93
		4.6.2	Effects of Multiple Access Interference	95
		4.6.3	Effects of Timing Jitter and Finger Estimation Error	96
	4.7	Conclusion		99

Chapter 5 Ultra Wideband Pulse Shaper Design 103
Zhi Tian, Timothy N. Davidson, Xiliang Luo, Xianren Wu, and Georgios B. Giannakis

	5.1	Introduction		103
	5.2	Transmit Spectrum and Pulse Shaper		105
	5.3	FIR Digital Pulse Design		108
	5.4	Optimal UWB Single Pulse Design		110
		5.4.1	Parks–McClellan Algorithm	110
		5.4.2	Optimal UWB Pulse Design via Direct Maximization of NESP	111
		5.4.3	Constrained Frequency Response Approximation	113
		5.4.4	Constrained Frequency Response Design with Linear Phase Filters	114
	5.5	Optimal UWB Orthogonal Pulse Design		115
		5.5.1	Orthogonality Formulation	115
		5.5.2	Sequential UWB Pulse Design	117
		5.5.3	Sequential UWB Pulse Design with Linear Phase Filters	118
	5.6	Design Examples and Comparisons		120
		5.6.1	Single-Pulse Designs and their Spectral Utilization Efficiency	120
		5.6.2	Multiband Pulse Design	122
		5.6.3	Multiple Orthogonal Pulse Design	123
		5.6.4	Pulse Designs for Narrowband Interference Avoidance	125
		5.6.5	Impact of Pulse Designs on Transceiver Power Efficiency	126
	5.7	Conclusions		128

Chapter 6 Antenna Issues 131
Zhi Ning Chen

	6.1	Introduction	131
	6.2	Design Considerations	132
		6.2.1 Description of Antenna Systems	132

	6.2.2	Single-Band and Multiband Schemes	134
	6.2.3	Source Pulses	136
	6.2.4	Transmit Antenna and PDS	136
	6.2.5	Transmit–Receive Antenna System	141
6.3	Antenna and Pulse versus BER Performance		148
	6.3.1	Pulsed UWB System	148
	6.3.2	Effects of Antennas and Pulses	151

Chapter 7 Ultra Wideband Receiver Architectures   157
Hüseyin Arslan

7.1	Introduction	157
7.2	System Model	158
7.3	UWB Receiver Related Issues	160
	7.3.1 Sampling	160
	7.3.2 UWB Channel and Channel Parameters Estimation	161
	7.3.3 Interference in UWB	164
	7.3.4 Other Receiver-Related Issues	165
7.4	TH IR-UWB Receiver Options	165
	7.4.1 Optimal Matched Filter	167
	7.4.2 TR-Based Scheme	171
	7.4.3 Differential Detector	175
	7.4.4 Energy Detector	176
7.5	Conclusion	178

Chapter 8 Ultra Wideband Channel Modeling and Its Impact on System Design   183
Chia-Chin Chong

8.1	Introduction	183
8.2	Principles and Background of UWB Multipath Propagation Channel Modeling	184
	8.2.1 Basic Multipath Propagation Mechanisms	184
	8.2.2 Classification of UWB Channel Models	185
8.3	Channel Sounding Techniques	187
	8.3.1 Time-Domain Technique	187
	8.3.2 Frequency-Domain Technique	188
8.4	UWB Statistical-Based Channel Modeling	189
	8.4.1 Modeling Philosophy and Mathematical Framework	189
	8.4.2 Large-Scale Channel Characterization	190
	8.4.3 Small-Scale Channel Characterization	193

	8.4.4	Temporal Dispersion and Correlation Properties	197
8.5		Impact of UWB Channel on System Design	199
8.6		Conclusion	200

Chapter 9 MIMO and UWB — 205
Thomas Kaiser

9.1	Introduction		205
9.2	Potential Benefits of MIMO and UWB		206
9.3	Literature Review of UWB Multiantenna Techniques		208
	9.3.1	Spatial Multiplexing	208
	9.3.2	Spatial Diversity	209
	9.3.3	Beamforming	209
	9.3.4	Related Topics	210
9.4	Spatial Channel Measurements and Modeling		211
	9.4.1	Spatial Channel Measurements	211
	9.4.2	Spatial Channel Modeling	213
9.5	Spatial Multiplexing		215
9.6	Spatial Diversity		216
9.7	Beamforming		220
9.8	Conclusion and Outlook		223

Chapter 10 Multiple-Access Interference Mitigation in Ultra Wideband Systems — 227
Sinan Gezici, Hisashi Kobayashi, and H. Vincent Poor

10.1	Introduction		227
10.2	Signal Model		228
	10.2.1	Transmitted Signal	228
	10.2.2	Received Signal	229
10.3	Multiple-Access Interference Mitigation at the Receiver Side		231
	10.3.1	Maximum-Likelihood Sequence Detection	232
	10.3.2	Linear Receivers	232
	10.3.3	Iterative (Turbo) Algorithms	240
	10.3.4	Other Receiver Structures	243
10.4	Multiple-Access Interference Mitigation at the Transmitter Side		244
	10.4.1	Time-Hopping Sequence Design for MAI Mitigation	245
	10.4.2	Pseudochaotic Time Hopping	246
	10.4.3	Multistage Block-Spreading UWB Access	247
10.5	Concluding Remarks		248

Chapter 11 Narrowband Interference Issues in Ultra Wideband Systems 255
Hüseyin Arslan and Mustafa E. Şahin

- 11.1 Introduction 255
- 11.2 Effect of NBI in UWB Systems 258
- 11.3 Avoiding NBI 261
  - 11.3.1 Multicarrier Approach 261
  - 11.3.2 Multiband Schemes 263
  - 11.3.3 Pulse Shaping 264
  - 11.3.4 Other NBI Avoidance Methods 266
- 11.4 Canceling NBI 267
  - 11.4.1 MMSE Combining 268
  - 11.4.2 Frequency Domain Techniques 268
  - 11.4.3 Time–Frequency Domain Techniques 269
  - 11.4.4 Time Domain Techniques 270
- 11.5 Conclusion and Future Research 271

Chapter 12 Orthogonal Frequency Division Multiplexing for Ultra Wideband Communications 277
Ebrahim Saberinia and Ahmed H. Tewfik

- 12.1 Introduction 277
- 12.2 Multiband OFDM System 278
  - 12.2.1 Band Planning 278
  - 12.2.2 Sub-Band Hopping 278
  - 12.2.3 OFDM Modulation 280
  - 12.2.4 Frequency Repetition Spreading 280
  - 12.2.5 Time Repetition Spreading 280
  - 12.2.6 Coding 281
  - 12.2.7 Supported Bit Rates 281
  - 12.2.8 MB-OFDM Transceiver 282
  - 12.2.9 Improvement to MB-OFDM 283
- 12.3 Multiband Pulsed-OFDM UWB system 284
  - 12.3.1 Pulsed-OFDM Transmitter 284
  - 12.3.2 Pulsed-OFDM Signal Spectrum 284
  - 12.3.3 Digital Equivalent Model and Diversity of Pulsed-OFDM 286
  - 12.3.4 Pulsed-OFDM Receiver 288
  - 12.3.5 Selecting the Up-sampling Factor 289
- 12.4 Comparing MB-OFDM and MB-Pulsed-OFDM systems 290

	12.4.1	System Parameters	290
	12.4.2	Complexity Comparision	290
	12.4.3	Power Consumption Comparison	290
	12.4.4	Chip Area Comparison	291
	12.4.5	Performance Comparison	293
12.5	Conclusion		295

Chapter 13   UWB Networks and Applications — 297
Krishna M. Sivalingam and Aniruddha Rangnekar

13.1	Introduction		297
13.2	Background		298
	13.2.1	UWB Physical Layer	298
	13.2.2	IEEE 802.15.3 Standards	299
13.3	Medium Access Protocols		300
	13.3.1	IEEE 802.15.3 MAC Protocol	300
	13.3.2	Impact of UWB Channel Acquisition Time	303
	13.3.3	Multiple Channels	305
13.4	Network Applications		310
13.5	Summary and Discussion		311
Acknowledgments			311

Chapter 14   Low-Bit-Rate UWB Networks — 315
Luca DeNardis and Gian Mario Maggio

14.1	Low Data-Rate UWB Network Applications		315
	14.1.1	802.15.4a: A Short History	315
	14.1.2	The 802.15.4a PHY	316
	14.1.3	PHY: 802.15.4a versus 802.15.4	316
	14.1.4	Technical Requirements	317
	14.1.5	Applications	319
14.2	The 802.15.4 MAC Standard		321
	14.2.1	Network Devices and Topologies	321
	14.2.2	Medium Access Strategy	322
	14.2.3	From 802.15.4 to 802.15.4a	324
14.3	Advanced MAC Design for Low-Bit-Rate UWB Networks		324
	14.3.1	$(UWB)^2$: Uncoordinated, Wireless, Baseborn Medium Access for UWB Communication Networks	325
	14.3.2	Transmission Procedure	328
	14.3.3	Reception Procedure	331
	14.3.4	Simulation Results	333

Chapter 15 An Overview of Routing Protocols for Mobile Ad Hoc Networks   341
David A. Sumy, Branimir Vojcic, and Jinghao Xu

- 15.1 Introduction   341
- 15.2 Ad Hoc Networks   343
- 15.3 Routing in MANETs   345
- 15.4 Proactive Routing   345
  - 15.4.1 DSDV   346
  - 15.4.2 WRP   348
  - 15.4.3 CGSR   350
  - 15.4.4 STAR   351
  - 15.4.5 HSR   352
  - 15.4.6 OLSR   355
  - 15.4.7 TBRPF   356
  - 15.4.8 DREAM   358
  - 15.4.9 GSR   360
  - 15.4.10 FSR   360
  - 15.4.11 HR   362
  - 15.4.12 HSLS and A-HSLS   363
- 15.5 Reactive Routing   364
  - 15.5.1 DSR   365
  - 15.5.2 ARA   367
  - 15.5.3 ABR   369
  - 15.5.4 AODV   372
  - 15.5.5 BSR   374
  - 15.5.6 CHAMP   376
  - 15.5.7 DYMO   377
  - 15.5.8 DNVR   378
  - 15.5.9 LAR   380
  - 15.5.10 LBR   381
  - 15.5.11 MPABR   383
  - 15.5.12 NDMR   384
  - 15.5.13 PLBM   385
  - 15.5.14 RDMAR   387
  - 15.5.15 SOAR   388
  - 15.5.16 TORA   391
- 15.6 Power-Aware Routing   393
  - 15.6.1 BEE   394

	15.6.2	EADSR	395
	15.6.3	MTPR/MBCR/MMBCR/CMMBCR	395
	15.6.4	PARO	396
	15.6.5	PAWF	398
	15.6.6	$MFP/MIP/MFP_{energy}/MIP_{energy}$	400
15.7	Hybrid Routing		400
	15.7.1	MultiWARP	401
	15.7.2	SHARP	402
	15.7.3	SLURP	403
	15.7.4	ZRP	406
	15.7.5	AZRP	408
	15.7.6	IZR	408
	15.7.7	TZRP	408
15.8	Other		410
15.9	Conclusion		411
	Appendix		418

Chapter 16 Adaptive UWB Systems   429
Francesca Cuomo and Crishna Martello

16.1	Introduction		429
	16.1.1	Related Work on Adaptive UWB Systems	431
16.2	A Distributed Power-Regulated Admission Control Scheme for UWB		432
	16.2.1	Problem Formalization	434
	16.2.2	Power Selection in UWB	435
	16.2.3	Steps of the Access Scheme	438
16.3	Performance Analysis		439
	16.3.1	Impact of the Initial MEI on Performance of MEI-Based Power Regulation Schemes	442
	16.3.2	Performance Behavior as a Function of the Offered Load	445
16.4	Summary		449

Chapter 17 UWB Location and Tracking—A Practical Example of an UWB-Based Sensor Network   451
*Ian Oppermann, Kegen Yu, Alberto Rabbachin,
Lucian Stoica, Paul Cheong, Jean-Philippe Montillet,
and Sakari Tiuraniemi*

17.1	Introduction	451
17.2	Multiple Access in UWB Sensor Systems	452

	17.2.1	Location/Ranging Support	453
	17.2.2	Constraints and Implications of UWB Technologies on MAC Design	453
17.3	UWB Sensor Network Case Study	454
17.4	System Description—UWEN	456
	17.4.1	Communications System	456
	17.4.2	Transmitted Signal	456
	17.4.3	Framing Structure	458
	17.4.4	Location Approach	458
17.5	System Implementation	459
	17.5.1	Transceiver Overview	459
	17.5.2	Transmitter	460
	17.5.3	UWB Pulse Generator	462
17.6	Location System	463
17.7	Position Calculation Methods	468
17.8	Tracking Moving Objects	473
	17.8.1	Simulation Results	474
17.9	Conclusion	476
Acknowledgments	477

Index 481

PREFACE

Ultra wideband (UWB) radio has gained popularity worldwide thanks to its promise of providing very high bit rates at low cost. The interest in UWB led in 2001 to the creation of the IEEE 802.15.3a Study Group, with the aim of defining a novel standard for wireless personal area networks (WPANs) based on a UWB physical layer capable of bit rates on the order of 500 Mbps. The research and development efforts on UWB further intensified after the release of the first world-wide official UWB emission masks by the US Federal Communication Commission in February 2002. This release officially opened the way, at least in the USA, to the development of commercial UWB products. The stringent power limitations set by the FCC naturally determined the application scenarios suitable for UWB communication, that is either high bit rates over short ranges, dealt within the IEEE 802.15.3aTG, or low bit rates over medium-to-long ranges, dealt within the recently formed IEEE 802.15.4aTG.

The several different UWB PHY proposals originally submitted to IEEE 802.15.3aTG converged into two main proposals: multiband (MB-OFDM), based on the transmission of continuous OFDM signals combined with frequency hopping (FH) over instantaneous frequency bandwidths of 528 MHz, and direct-sequence (DS) UWB, based on impulse radio transmission of UWB DS-coded pulses. Although not specifically designed for ranging support, both MB-OFDM and DS-UWB foresee UWB emissions with bandwidths exceeding 500 MHz in order to comply with the FCC definition of UWB, and can thus potentially provide high ranging accuracy. The UWB ranging capability is a particularly attractive feature for location-aware applications, in particular in ad hoc and sensor networks, and introducing positioning in low data rate networks has recently become the main goal of the IEEE 802.15.4aTG, where impulse radio ultra wideband (IR-UWB) radio emerges as a most appealing principle.

The above application scenarios are typical for self-organizing and distributed networks, such as ad hoc and sensor networks, in which groups of wireless terminals located in a limited-size geographical area, communicate in an infrastructure-free multihop fashion, and without any central coordinating unit. Ultra wideband's special features like the need for operating at low power and accurate ranging capability bring about significant impacts on the design of the MAC and routing algorithms, and hence new strategies for algorithm and protocol development.

The objective of this book is to provide an introduction to the above major research issues in UWB communication that are currently occupying research attention worldwide. As such, the book is primarily intended to serve as a reference for comprehensive understanding of recent advances in both theory and practical design of UWB communication networks.

BROAD TOPICAL COVERAGE

The book covers issues related to physical layer, medium-access layer, networking layer, and also applications.

Following the Introduction (Chapter 1), the structure of the book consists of basically three parts:

- Analysis of physical layer and technology related issues (Chapters 2–6);
- Introduction to system design aspects including channel modeling, coexistence, and interference mitigation and control (Chapters 7–11);
- Review of MAC and network layer related issues, up to the application (Chapters 12–16).

A detailed description of how the book is organized and introduction to the different chapters of the book can be found in the Introduction (Chapter 1).

AUDIENCE

Our intention is that the book could serve as an introductory survey of important topics related to UWB, as well as providing an advanced mathematical treatise intended for technical professionals in the communications industry, technical managers, and researchers in both academia and industry. A basic background of wireless communications is preferable for a full understanding of the topics covered by the book.

COURSE USE

The book provides an organic and harmonized coverage of UWB communication, from radio to application. Within this framework, the book chapters are quite independent from one another. Therefore, different options are possible according to different course structures and lengths, as well as targeted audience background.

For each chapter we expect that a reader may skip the advanced mathematical description and still greatly benefit from the book. The topics are covered in fact in both descriptive and mathematical manners, and can therefore cater to different readers needs.

ACKNOWLEDGMENTS

We wish to thank all our colleagues who are authors of this book for working with us and giving their time and effort in making and supporting this book project.

We would also like to particularly thank our editor Paul Petralia, and former editor Valérie Molière, as well as all the entire editorial staff at John Wiley & Sons.

HÜSEYIN ARSLAN
ZHI NING CHEN
MARIA-GABRIELLA DI BENEDETTO

CONTRIBUTORS

Hüseyin Arslan, Ismail Güvenç, Mustafa E. Şahin, and Sadia Ahmed, Electrical Engineering Department, University of South Florida, Tampa, Florida

Maria-Gabriella Di Benedetto, University of Rome La Sapienza, Italy

Irena Maravić, European Molecular Biology Laboratory, Heidelberg, Germany

Martin Vetterli, IC, Swiss Federal Institute of Technology, Lausanne, Switzerland and EECS Department, University of California at Berkeley, Berkeley, California

Sinan Gezici, Hisashi Kobayashi, and H. Vincent Poor, Department of Electrical Engineering, Princeton University, Princeton, New Jersey

Zafer Sahinoglu, Mitsubishi Electric Research Laboratories, Cambridge, Massachusetts

Zhi Tian, Michigan Technological University, Houghton, Michigan

Timothy N. Davidson, McMaster University, Hamilton, Ontario, Canada

Xiliang Luo, University of Minnesota, Minneapolis, Minnesota

Xianren Wu, Michigan Technological University, Houghton, Michigan

Georgios B. Giannakis, University of Minnesota, Minneapolis, Minnesota

Zhi Ning Chen, Institute for Infocomm Research, Singapore

Chia-Chin Chong, DoCoMo USA Labs, San Jose, California

Thomas Kaiser, Smart Antenna Research Team, University of Duisburg-Essen, Duisburg, Germany

Ebrahim Saberinia, University of Nevada, Las Vegas, Nevada

Ahmed H. Tewfik, University of Minnesota, Minneapolis, Minnesota

Krishna M. Sivalingam and Aniruddha Rangnekar, Department of CSEE, University of Maryland, Baltimore, Maryland

Luca De Nardis and Gian Mario Maggio, Berkeley Wireless Research Center, Berkeley, California

David A. Sumy, Branimir Vojcic, and Jinghao Xu, The George Washington University, Washington, DC

Francesca Cuomo and Cristina Martello, Department Info-Com, University of Rome La Sapienza, Rome, Italy

Ian Oppermann, Kegen Yu, Alberto Rabbachin, Lucian Stoica, Paul Cheong, Jean-Philippe Montillet, and Sakari Tiuraniemi, Centre for Wireless Communications, University of Oulu, Finland

CHAPTER 1

Introduction to Ultra Wideband

HÜSEYIN ARSLAN and MARIA-GABRIELLA DI BENEDETTO

1.1 INTRODUCTION

Wireless communication systems have evolved substantially over the last two decades. The explosive growth of the wireless communication market is expected to continue in the future, as the demand for all types of wireless services is increasing. New generations of wireless mobile radio systems aim to provide flexible data rates (including high, medium, and low data rates) and a wide variety of applications (like video, data, ranging, etc.) to the mobile users while serving as many users as possible. This goal, however, must be achieved under the constraint of the limited available resources like spectrum and power. As more and more devices go wireless, future technologies will face spectral crowding, and coexistence of wireless devices will be a major issue. Therefore, considering the limited bandwidth availability, accommodating the demand for higher capacity and data rates is a challenging task, requiring innovative technologies that can coexist with devices operating at various frequency bands.

Ultra wideband (UWB), which is an underlay (or sometimes referred as shared unlicensed) system, coexists with other licensed and unlicensed narrowband systems. The transmitted power of UWB devices is controlled by the regulatory agencies [such as the Federal Communications Commission (FCC) in the United States], so that narrowband systems are affected from UWB signals only at a negligible level. UWB systems, therefore, are allowed to coexist with other technologies only under stringent power constraints. In spite of this, UWB offers attractive solutions for many wireless communication areas, including wireless personal area networks (WPANs), wireless telemetry and telemedecine, and wireless sensors networks. With its wide bandwidth, UWB has a potential to offer a capacity much higher than the current narrowband systems for short-range applications.

According to the modern definition, any wireless communication technology that produces signals with a bandwidth wider than 500 MHz or a fractional

---

Ultra Wideband Wireless Communication. Edited by Arslan, Chen, and Di Benedetto
Copyright © 2006 John Wiley & Sons, Inc.

bandwidth[1] greater than 0.2 can be considered as UWB. A possible technique for implementing UWB is impulse radio (IR), which is based on transmitting extremely short (in the order of nanoseconds) and low power pulses. Rather than sending a single pulse per symbol, a number of pulses determined by the processing gain of the system are transmitted per symbol. The processing gain serves as a parameter to flexibly adjust data rate, bit error rate (BER), and coverage area of transmission. Pulses can occupy a location in the frame based on the specific pseudo random (PN) code assigned for each user (as in the case of time-hopping UWB). Other implementations, such as direct sequence spreading, are also popularly used with impulse radio-based implementations. Impulse radio is advantageous in that it eliminates the need for up- and down-conversion and allows low-complexity transceivers. It also enables various types of modulations to be employed, including on–off keying (OOK), pulse-amplitude-modulation (PAM), pulse-position-modulation (PPM), phase-shift-keying (PSK), as well as different receiver types such as the energy detector, rake, and transmitted reference receivers.

Another strong candidate for UWB is multicarrier modulation, which can be realized using orthogonal frequency division multiplexing (OFDM). OFDM has become a very popular technology due to its special features such as robustness against multipath interference, ability to allow frequency diversity with the use of efficient forward error correction (FEC) coding, capability of capturing the multipath energy efficiently, and ability to provide high bandwidth efficiency through the use of sub-band adaptive modulation and coding techniques. OFDM can overcome many problems that arise with high bit rate communication, the most serious of which is time dispersion. In OFDM, the data-bearing symbol stream is split into several lower rate streams, and these sub-streams are transmitted on different carriers. Since this increases the symbol period by the number of nonoverlapping carriers (sub-carriers), multipath echoes affect only a small portion of neighboring symbols. Remaining intersymbol interference (ISI) can be removed by cyclically extending the OFDM symbol.

1.1.1 Benefits of UWB

The unique advantages of UWB systems are numerous. First of all, it introduces unlicensed usage of an extremely wideband spectrum, as mentioned above. The underlay usage of spectrum greatly increases spectral efficiency and opens new doors for wireless applications. The introduction of cognitive features along with opportunistic spectrum usage will further enhance current UWB applications.

UWB (both impulse radio and multicarrier) also offers great flexibility of spectrum usage. This system is characterized in fact by a variety of parameters that can enable the design of adaptive transceivers and that can be used for optimizing system performance as a function of the required data rate, range, power, quality-of-service, and user preference. UWB technology is likely to provide high data

---

[1]Fractional bandwidth $= 2 \cdot (F_H - F_L)/(F_H + F_L)$, where F_H and F_L are the upper and lower edge frequencies, respectively.

rates (on the order of 1 Gbps) over very short range (less than 1 m). The data rate can, however, be easily traded-off for extension in range by designing appropriate adaptive transceivers. Similarly, data rate and range can be traded-off for power, especially for low data rate and short range applications. Most importantly, *the same device* can be designed to provide service for multiple applications with a variety of requirements without the need for additional hardware.

The high temporal resolution of UWB signals results in low fading margins, implying robustness against multipath. Since UWB signals span a very wide frequency range (down to very low frequencies), they show relatively low material penetration losses, giving rise to better link margins. Moreover, often many distinct multipath components can be observed at the receiver (due to the large number of resolvable paths), and the system, therefore, has an excellent energy capturing capability. For example, rake receivers (with coherent combining) can be implemented to lock into multipath echoes, collect energy, and hence improve performance.

Excellent time resolution is another key benefit of UWB signals for ranging applications. Due to the extremely short duration of transmitted pulses, sub-decimeter ranging is possible. In IR-UWB systems, no up/down-conversion is required at the transceivers, with the potential benefit of reducing the cost and size of the devices. Other benefits of UWB include low power transmission and robustness against eavesdropping (since UWB signals look like noise).

1.1.2 Applications

UWB has several applications all the way from wireless communications to radar imaging, and vehicular radar. The ultra wide bandwidth and hence the wide variety of material penetration capabilities allows UWB to be used for radar imaging systems, including ground penetration radars, wall radar imaging, through-wall radar imaging, surveillance systems, and medical imaging. Images within or behind obstructed objects can be obtained with a high resolution using UWB.

Similarly, the excellent time resolution and accurate ranging capability of UWB can be used for vehicular radar systems for collision avoidance, guided parking, etc. Positioning location and relative positioning capabilities of UWB systems are other great applications that have recently received significant attention.

Last but not least is the wireless communication application, which is arguably the reason why UWB became part of the wireless world, including wireless home networking, high-density use in office buildings and business cores, UWB wireless mouse, keyboard, wireless speakers, wireless USB, high-speed WPAN/WBAN, wireless sensors networks, wireless telemetry, and telemedicine.

1.1.3 Challenges

In spite of all the advantages of UWB, there are several fundamental and practical issues that need to be carefully addressed to ensure the success of this technology in the wireless communication market. Multiaccess code design, multiple access interference (MAI) cancellation, narrowband interference (NBI) detection and cancellation, synchronization of the receiver to extremely narrow pulses, accurate modeling of UWB channels, estimation of multipath channel delays and coefficients, and adaptive transceiver design are some of the issues that still require a great deal of investigation. In addition to the above physical layer issues, the fundamental role of UWB technology in wireless networks is still open, and a wide range of research questions continue to present challenges, such as the particular role of UWB in wireless ad-hoc and sensors networks.

Among the challenges of UWB, a limited list can be given as follows:

- Coexistence with other services and handling strong narrowband interference;
- Shaping (adapting) spectrum of transmitted signals (multiband, OFDM-based UWB, etc.);
- Practical, simple, and low-power transceiver design;
- Accurate synchronization and channel parameter estimation;
- High sampling rate for digital implementations;
- Powerful processing capabilities for high performance and coherent digital receiver structures;
- Wideband RF component designs (such as antennas, low noise amplifiers, etc.);
- Multiple accessing, multiple access code designs, and multiuser interference;
- Accurate modeling of the ultra wideband channel in various environments;
- Adaptive system design and cross-layer adaptation for UWB;
- UWB tailored network design.

1.2 SCOPE OF THE BOOK

This book covers several aspects of the UWB technology, starting from the radio aspects all the way to UWB networking and UWB applications with the aim of shedding light on the UWB challenges listed at the end of the previous section. Although more emphasis is given to impulse radio UWB, OFDM-based UWB is also discussed throughly.

In UWB, the transmission bandwidth is extremely large, leading to multiple resolvable paths. At a given total transmitted power, power is distributed over an extremely large bandwidth. In the time domain, the high resolvability due to ultra wide bandwidth can affect the receiver performance. Since the total power is distributed over many multipath components, the power on each path might be very low [1]. Also, due to the broadband nature of UWB signals, the components propagating along different paths may undergo different frequency selective distortions. As a result, a received signal is made up of pulses with different pulse shapes, which makes synchronization, channel estimation, and optimal receiver design more challenging than in other wideband systems. In addition, implementation of standard techniques in digital UWB receivers would require very fast analog-to-digital (A/D) converters, operating in the gigahertz range, and thus high power consumption. As a result, synchronization and channel estimation are two of the most important issues in UWB. Therefore, one whole chapter will be devoted to discussion of synchronization and channel estimation issues. The problem of low-complexity channel estimation and synchronization issues in digital UWB receivers will be considered in detail in Chapter 2, "UWB Channel Estimation and Synchronization."

A very close subject to UWB synchronization is the accurate estimation of time of arrival of UWB signals. Accurate synchronization and fine resolution in time of arrival are not only important for reception and detection, but also for accurate ranging. Locationing and ranging applications can be developed on the basis of proper and low complex synchronization algorithms. Hence, Chapter 3, "Ultra Wideband Geolocation," covers this aspect. An overview of conventional ranging and positioning techniques, as well as the study of their performance for range estimation, is provided in this chapter.

Selecting the appropriate modulation technique for UWB still remains a major challenge. There are various possible modulation options depending on the application, design specifications and constraints, range, transmission and reception power, quality of service requirements, regulatory requirements, hardware complexity, data rate, reliability of channel, and capacity. Therefore, it is crucial to select the appropriate modulation according to purpose. Possible choices for UWB are binary phase shift keying (BPSK), quadrature phase shift keying (QPSK), PAM, OOK, PPM, pulse interval modulation (PIM), and pulse shape modulation (PSM) [2]. Among these options, BPSK is the most popular in UWB applications due to its smooth power spectrum and low BER. However, accurate phase detection of the modulated signal in BPSK requires complex channel estimation algorithms at the receiver. Compared with BPSK, OOK and PPM only require the knowledge of the presence or absence of energy and therefore channel estimation is not necessary for noncoherent reception. However, it is also possible to employ coherent receivers for these modulations for improved performance. Noise levels over the wireless channel also influence the choice of modulation. Higher-order modulation ensures high data rate at the cost of poor BER over noisy channels. Therefore, lower order modulation for low data rate applications is desirable under poor channel conditions. Transmission over multiple frequency bands or over multiple carriers, and various multiple accessing options such as time hopping (TH) and direct sequence (DS) could also be considered under the umbrella of UWB modulations. These issues will be covered in Chapter 4, "UWB Modulation Options," where several modulation options will be compared.

Similar to modulation options, there are also various ways to control the UWB spectrum shape by pulse shaping. As mentioned in the previous section, for appropriate spectrum overlay, the local regulators impose spectral masks that strictly constrain the transmission power of a UWB signal. Spectral masks are often not uniform, that is, there are stronger restrictions in some parts of the spectrum compared with others. The spectrum of a transmitted signal is influenced by the modulation format, the multiple access scheme, and most critically by the spectral shape of the underlying UWB pulse. The choice of the pulse shape is thus a key design decision in UWB systems. Chapter 5, "Ultra Wideband Pulse Shaper Design," will discuss the UWB pulse design issues.

Another important challenge in UWB wireless systems is the design of antennas. Most difficult issues include broadband response of impedance matching, gain, phase, radiation patterns, and polarization. Therefore, Chapter 6, "Antenna Issues," discusses antenna design in UWB systems along with the effects of antenna design on the transmission of UWB signals. Also, antenna design and pulse shaping issues are related in this chapter, and special considerations are given for UWB antenna design by taking pulse sources into account.

Many of the current applications of UWB require power efficient, low cost, and small-sized UWB transceivers. Therefore, practical and low complexity implementation of transceivers is of vital importance for the successful penetration of the UWB technology. UWB transceiver requirements and related trade-offs regarding practical designs will be discussed in Chapter 7, "Ultra Wideband Receiver Architectures." Different receiver structures will be discussed and these various approaches will be compared in terms of their ability to exploit *a priori* information (side information). The robustness of these various receivers depending on the availability and accuracy of the side information will also be investigated.

In order to be able to develop efficient and high performance transceiver algorithms and to design reliable radio systems, accurate and realistic modeling of the radio channel is needed. Unfortunately, the mechanisms that govern radio propagation in a wireless communication channel are complex and diverse. Consequently, channel modeling has been a subject of intense research for a long time. UWB channel modeling presents many differences compared with the well-known narrowband channel models. Therefore, Chapter 8, "Ultra Wideband Channel Modeling and its Impact on System Design," will provide an overview of the UWB propagation channel modeling work and its impact on the UWB communication system design. Establishment of the fundamental concepts and background for modeling the UWB multipath propagation channel, discussion of the two commonly used channel sounding techniques, description of the UWB statistical-based channel modeling work, and discussion of the impact of UWB channel on the system design are some of the important aspects that will be discussed in this chapter.

Exploiting the radio channel properties for improving the transceiver performance has a rich and long history in the wireless communication literature. Multiple antenna systems is one of these techniques that has been used for different purposes including diversity combining, interference cancellation, and data rate increase. Multi-input multi-output (MIMO) antenna systems is a major topic that has received significant interest in the wireless community over recent years. MIMO, which is often interpreted as an add-on technology, can be incorporated in any type of wireless technology, one of which is UWB. Therefore, in Chapter 9, "MIMO and UWB," the potential benefits of MIMO and UWB in terms of range extension, data rate improvement, interference rejection, and potential technological simplifications are introduced. Also, in the same chapter, a literature review on UWB multiantenna techniques, subdivided in spatial multiplexing, spatial diversity, beam-forming, and related topics, is provided. Complementing the channel models of Chapter 8, spatial UWB channel measurements and modeling will be highlighted to provide a solid basis for algorithmic design of MIMO and UWB transceivers.

In order to effectively share the available spectrum between different users, multiple accessing is of fundamental importance in wireless communication systems. Time division multiple access (TDMA), frequency division multiple access (FDMA), and code division multiple access (CDMA) are the most popular multi-access techniques for wireless systems. As in any communication system, multiple access is a key issue in UWB networks. In an ideal scenario, the system should be designed in such a way that there will be no interference from other users on a desired user. In reality this is not the case, as the systems are trying to provide access to more users so that the spectrum can be exploited more efficiently. As a result, multiple-access interference (such as co-channel interference, adjacent channel interference, and correlation of the other users code with the desired user code) becomes a tricky issue in wireless communications. Chapter 10, "Multiple-access Interference Mitigation in Ultra Wideband systems," covers the issues related to multiple-access IR-UWB, and explains signal processing techniques for combating the effects of interfering users on the detection of information symbols.

Another major interference source, specifically in UWB systems, is narrowband interference. The influence of narrowband technologies on UWB system can be significant, and in the extreme case, these signals may completely jam the UWB receiver. Even though narrowband signals interfere with only a small fraction of the UWB spectrum, due to their relatively high power with respect to the UWB signal, the performance and capacity of UWB systems can be affected considerably [3]. Recent studies show that the BER of UWB receivers is greatly degraded due to the impact of narrowband interference [4–8]. The high processing gain of the UWB signal can cope with the narrowband interferers to some extent. In many cases, however, even the large processing gain alone is not sufficient to suppress the effect of the high power interferers. Therefore, either the UWB system needs to avoid transmission over frequencies of strong narrowband interferers, or UWB receivers need to employ NBI suppression techniques to improve performance, capacity, and range. Narrowband interference issues will be discussed in detail in Chapter 11, "Narrowband Interference Issues in Ultra Wideband Systems."

Several of the above issues affect both impulse radio and multicarrier-based implementations of UWB. There are some specific issues and advantages, however, related to the OFDM based approach that deserved at least one whole chapter, considering also that the multiband OFDM system is currently one of the leading proposals for the IEEE 802.15.3a standard and is supported by more than 100 large companies and universities. For this purpose, Chapter 12, "Orthogonal Frequency Division Multiplexing for Ultra Wideband Communications," discusses in detail the OFDM based UWB approach.

The physical (PHY) and multiple access issues do not constitute the only research and development challenges and opportunities for UWB. Many other aspects are related to networking, adaptation, and crosslayer optimization. UWB networks have the potential to offer high bandwidth rates with low spectral energy, besides other features such as accurate localization and lower probability of jamming and detection. This has led to an increased interest in building UWB-based data networks. For instance, the IEEE TG802.15.3a standards group is in the process of developing an alternative high-speed link layer design conformable with the IEEE 802.15.3 wireless personal area network (WPAN) multiple-access protocol, operating at a few tens of meters and speeds of the order of several hundred megabits per second. UWB based networks are also being considered for wireless sensor networks and military applications. Chapter 13, "UWB Networks and Applications," contains a survey that will cover these issues.

Besides the strong push for high-data-rate UWB networks, there has also been a growing interest towards applying UWB to low-power and low-data-rate networks, such as in sensor networks [9]. The low bit rate applications and network issues of UWB will be discussed in Chapter 14, "Low-bit-rate UWB Networks."

Related to the UWB networking, one of the biggest challenge is to develop efficient routing protocols for mobile ad-hoc networks. The routing protocols in ad-hoc networks in general, and some specific aspects of these for UWB, will be discussed in detail in Chapter 15, "An Overview of Routing Protocols for Mobile Ad-hoc Networks." Power (or energy) aware routing protocols, which are described in this chapter, can be efficiently applied to ad-hoc networks with UWB.

As mentioned in the previous section, one of the great benefits of UWB is the flexibility for adaptive transceiver and network design. The adaptive network design and cross-layer optimization techniques are gaining significant interest in wireless communications. Therefore, Chapter 16, "Adaptive UWB Systems," focuses on adaptivity in UWB systems. In particular, it addresses the problem of how to exploit the UWB adaptability to support wireless links in ad-hoc networks as well as how to dynamically set up wireless communications among devices distributed in a given area, without the support of a centralized infrastructure.

Finally, a case study chapter on the application of UWB on wireless sensors network and for geolocationing is provided in Chapter 17, "UWB Location and Tracking—a Practical Example of an UWB-based Sensor Network." Impulse radio-based UWB technology has a number of inherent properties which are well suited to sensor network applications. In particular, impulse radio-based UWB systems (with potentially low complexity and low-cost designs and with noise-like signals) are resistant to severe multipath and have very good time domain resolution supporting location and tracking applications. In this chapter, an example architecture of a sensor system based on low-power, low-complexity UWB transceivers and a TDMA-based MAC will be provided.

REFERENCES

1. D. Cassioli, M. Z. Win, and A. F. Molisch, "Effects of spreading bandwidth on the performance of UWB RAKE receivers," in *Proc. IEEE Int. Conf. Commun. (ICC)*, vol. 5, May 2003, pp. 3545–3549.

2. I. Guvenc and H. Arslan, "On the modulation options for UWB systems," in *Proc. IEEE Military Commun. Conf. (MILCOM)*, vol. 2, Boston, MA, October 2003, pp. 892–897.
3. J. Foerster, "Ultra-wideband technology enabling low-power, high-rate connectivity (invited paper)," in *Proc. IEEE Workshop Wireless Commun. Networking*, Pasadena, CA, September 2002.
4. J. R. Foerster, "The performance of a direct-sequence spread ultra-wideband system in the presence of multipath, narrowband interference, and multiuser interference," in *Proc. IEEE Vehic. Technol. Conf.*, vol. 4, Birmingham, AL, May 2002, pp. 1931–1935.
5. J. Choi and W. Stark, "Performance of autocorrelation receivers for ultra-wideband communications with PPM in multipath channels," in *Proc. IEEE Ultrawideband Syst. and Technol. (UWBST)*, Baltimore, MD, May 2002, pp. 213–217.
6. L. Zhao and A. Haimovich, "Performance of ultra-wideband communications in the presence of interference," *IEEE J. Select. Areas Commun.*, vol. 20, pp. 1684–1691, December 2002.
7. G. Durisi, J. Romme, and S. Benedetto, "Performance of TH and DS UWB multiaccess systems in presence of multipath channel and narrowband interference," in *Proc. Int. Workshop Ultrawideband Systems*, Oulu, June 2003.
8. R. Tesi, M. Hamelainen, J. Iinatti, and V. Hovinen, "On the influence of pulsed jamming and coloured noise in UWB transmission," in *Proc. Finnish Wireless Commun. Workshop (FWCW)*, Espoo, May 2002.
9. M. G. Di Benedetto, L. De Nardis, M. Junk, and G. Giancola, "(UWB)^2: uncoordinated, wireless, baseborn medium access control for UWB communication networks," *Journal of Mobile Networks and Applications*, vol. 10, no. 5, pp. 663–674, October 2005.

CHAPTER 2

UWB Channel Estimation and Synchronization

IRENA MARAVIĆ and MARTIN VETTERLI

2.1 INTRODUCTION

Ultra wideband (UWB) systems are characterized as systems with instantaneous spectral occupancy larger than 500 MHz, or with a bandwidth greater than 20% of the center frequency. UWB radios can use frequencies from 3.1 GHz to 10.6 GHz—a band more than 7 GHz wide. In order to allow for such a large signal bandwidth, the FCC introduced severe broadcast power restrictions [11], meaning that UWB devices can make use of an extremely wide frequency band while not emitting enough energy to be noticed by other narrower band systems nearby, such as 802.11a/g radio. Such strict power limits, along with extreme bandwidths involved, bring about new challenges to both the analysis and practice of reliable systems. Namely, some of the critical issues include high sensitivity to synchronization errors, optimal exploitation of fading propagation effects in frequency-selective channels, low-power designs and co-existence with other wireless devices, as well as the development of novel signal processing techniques that are suitable for fully digital implementation.

Traditional UWB systems, often referred to as impulse radio, use trains of pulses of very short duration (on the order of a nanosecond), thus spreading the signal energy from near DC to several GHz [34, 45]. To maintain adequate signal energy for reliable detection, each symbol is made up of a sequence of pulses and transmitted over a large number of frames, with one pulse per frame. Such a signaling scheme is widely considered as a perfect candidate for a variety of bandwidth-demanding applications in wireless communications, including precise position location, ranging, and imaging through materials, among others. Yet, realizing the full potential of impulse radio communications depends critically on the success of timing synchronization, as its accuracy and complexity directly affect the system performance. Timing synchronization is required both at the frame level,

---

Ultra Wideband Wireless Communication. Edited by Arslan, Chen, and Di Benedetto
Copyright © 2006 John Wiley & Sons, Inc.

to determine when the first frame of each symbol starts, and at the pulse level, in order to find where a pulse is located within a frame. While synchronization represents the critical step in other wideband systems, such as DS-CDMA [2, 22], and methods developed for DS-CDMA can be adapted to UWB systems as well, the need for a much higher sampling rate in the latter makes this problem more challenging and calls for a different solution.

Recently, there has been increased interest in using other transmission techniques that would use multiple subbands rather than a single band to occupy such extremely wide bandwidths. In particular, baseband pulses can be modulated by several analog carriers to multiple frequency bands (typically 500–800 MHz wide). Such a transmission technique, usually referred to as multiband UWB [1], has several advantages over-pulse-based signaling scheme, including more efficient use of the FCC spectral mask [10], and reduced interference to/from coexisting systems by flexible selection of subbands [33]. As in pulse-based systems, timing acquisition and channel estimation also pose difficulties in multiband UWB systems; however, one of the major challenges in the system design is carrier frequency synchronization, especially if OFDM or fast frequency hopping is employed across multiple subbands.

In this chapter, we will consider the problem of timing synchronization and channel estimation in UWB systems, focusing mainly on the pulse-based signaling scheme. A vast amount of literature on this topic has appeared recently, with a common trend to minimize the number of analog components needed, and perform as much of the processing digitally as possible [9, 14, 17, 31]. Yet, given the extreme bandwidths involved, digital implementation may lead to prohibitively high costs in terms of power consumption and receiver complexity. For example, conventional techniques based on sliding correlators would require very fast and expensive A/D converters (operating in the gigahertz range) and therefore high power consumption. Furthermore, implementation of such techniques in digital systems would have almost unaffordable complexity in real systems as well as slow convergence time, since one has to perform an exhaustive search over thousands of fine bins, each at the nanosecond level. In order to improve the acquisition speed, several modified timing recovery schemes have been proposed, such as a bit reversal search [14], or the correlator-type approach which exploits properties of beacon sequences [12]. Even though some of these methods have already been in use in certain analog systems [11], the need for very high sampling rates, along with the search-based nature of these methods, makes them less attractive for digital implementation. Recently, a family of blind synchronization techniques was developed [31], which takes advantage of the so-called cyclostationarity of UWB signaling, that is, the fact that every information symbol is made up of UWB pulses that are periodically transmitted (one per frame) over multiple frames. While such an approach relies on frame-rate rather than Nyquist rate sampling, it requires relatively large data sets in order to achieve good synchronization performance.

Another challenge arises from the fact that the design of an optimal UWB receiver must take into account certain frequency-dependent effects on the received waveform. That is, due to the broadband nature of UWB signals, the components propagating along different paths typically undergo different frequency-selective distortions

[7, 36]. As a result, a received signal is made up of pulses with different pulse shapes, which makes the problem of optimal receiver design a much more delicate task than in other wideband systems [2, 22, 27]. In [7], an array of sensors is used to spatially separate the multipath components, which is then followed by identification of each path using an adaptive method, the so-called Sensor-CLEAN algorithm. However, due to the complexity of the method and the need for an antenna array, the method has been used primarily for UWB propagation experiments. In recent work [30], the authors present a data-aided maximum likelihood (ML) estimation approach, which uses symbol-rate samples of the correlator output, assuming that the received signal is correlated with a received *noisy template*. In particular, the term noisy template (or dirty template) comes from the fact that each received segment is noisy, distorted by the same, unknown channel and subject to the same time offset (corresponding to the time delay of an "aggregate" channel). A similar technique is also discussed in [32] where, at the receiver, integrate-and-dump operations are carried out on products of such segments, and the timing offset is found from symbol-rate samples. While such an approach significantly reduces the sampling rate compared with conventional techniques based on sliding correlators, it can be used primarily for timing acquisition in UWB impulse radios, but cannot be directly extended for estimating the channel impulse response.

In this chapter, we will mainly focus on a frequency-domain approach to channel estimation and synchronization in pulse-based ultra-wideband systems. Specifically, we will show how to extend some of our recent results on sampling certain classes of parametric nonband-limited signals [19, 29] to the problem of channel estimation in UWB systems, and estimate unknown channel parameters from a set of samples taken at a sub-Nyquist rate. The outline of the chapter is as follows. In Section 2.2, we introduce a model of a multipath fading channel and present a frequency domain framework for channel estimation. To provide a better insight into the basic principles behind the frequency domain approach, we first introduce a polynomial method for parameter estimation, which uses a concept of annihilating filters [30] and requires polynomial rooting to obtain parameters of interest. Next, we present a subspace-based method, which allows for more robust parameterization using the state-space approach [23]. In Section 2.3, we discuss the numerical performance and computational complexity of the presented algorithms and discuss alternative methods of lower computational requirements. In Section 2.4, we consider the case of more realistic UWB channel models, and present results on the problem of joint estimation of pulse shapes and time delays along different propagation paths. In Section 2.5, we discuss an extension of the frequency-domain framework to the problem of channel estimation from multiple (not necessarily adjacent) frequency bands. Namely, we present a more general solution that incorporates a filter bank at the receiver and allows for the estimation of the channel from multiple frequency bands with highest signal-to-noise ratio. Finally, in Section 2.6, we discuss the application of our framework to UWB systems for precise position location. In particular, we present a multiresolution or two-step approach to acquisition in such systems, which provides unique advantages over existing techniques in terms of acquisition speed and computational requirements.

2.2 CHANNEL ESTIMATION AT SUBNYQUIST SAMPLING RATE

2.2.1 UWB Channel Model

A number of propagation studies for ultra wideband signals have been done, taking into account temporal properties of a channel or characterizing a spatio-temporal channel response [7]. A typical model for the impulse response of a multipath fading channel is given by

$$h(t) = \sum_{l=1}^{L} a_l \delta(t - t_l), \qquad (2.1)$$

where t_l and a_l denote a signal delay and propagation coefficient along the lth path, respectively. Although this model does not adequately reflect specific frequency-dependent effects, it is commonly used for diversity reception schemes in conventional wideband receivers (e.g., RAKE receivers) [22]. Equation (2.1) can be interpreted as saying that a received signal $y(t)$ is made up of a weighted sum of attenuated and delayed replicas of the transmitted signal $s(t)$, that is,

$$y(t) = \sum_{l=1}^{L} a_l s(t - t_l) + \eta(t), \qquad (2.2)$$

where $\eta(t)$ denotes receiver noise.

In order to estimate the unknown delays and propagation coefficients, several classes of algorithms have been developed so far. In [4], the authors propose a least-squares (LS) procedure, taking into account the clustered structure of the channel. However, such an approach requires the Nyquist sampling rate (that is, a sub-pulse rate) and has prohibitively high computational requirements. Another class of algorithms is based on the maximum likelihood (ML) criterion for estimating the channel parameters [18, 36]. For example, in [18], the authors use the ML parameter estimation of UWB multipath channels in the presence of multiple-access interference (MUI). In particular, the impulse response estimates are formed using either training symbols or information-bearing symbols, while treating MUI as white Gaussian noise. Similarly to the LS approach, the computational complexity of the ML estimator increases rapidly as the number of multipath components increases, and becomes almost unaffordable in real-time applications. Besides, the sampling rate suggested in [18] is in the range $12.5/T_p - 25/T_p$ (where T_p denotes the pulse duration). With a typical value of $T_p = 0.7$ ns [38], the required sampling rate is prohibitively high and ranges between 17.9 GHz and 35.7 GHz.

To avoid the high sampling rate and reduce complexity, there has been a renewed interest in using the so-called transmitted-reference (TR) signaling; that is, rather than estimating the channel impulse response $h(t)$, one should estimate the aggregate analog channel $s(t) * h(t)$. Namely, the idea is to couple each information-bearing pulse $s(t)$ with an unmodulated (or pilot) pulse. For example, the transmitted pulse can be of the form $p(t) = s(t) + b \cdot s(t - T_f)$, $b = \{\pm 1\}$, where the frame duration T_f is chosen such that, after multipath propagation, the information and pilot pulses do not overlap. The receiver then correlates the received signal $y(t)$ with its delayed version $y(t - T_f)$ to yield the symbol estimate, assuming that the timing of each pulse is known. While such an approach requires only frame-rate samples, it results in 50% energy or rate loss, as half of the transmitted waveforms are used as pilot symbols. Recently, several modifications of the TR scheme have been also proposed in the literature. In [5], the so-called generalized likelihood ratio test (GLRT) schemes were investigated, whereas in [37] the authors propose a maximum-likelihood approach, which computes the autocorrelation of the channel impulse response at various delays. Yet, such techniques trade off computational requirements for performance, and have nearly the same complexity as the methods developed in [8, 36].

In the following, we will discuss a frequency-domain approach to channel estimation and timing in digital UWB receivers, which allows for sub-Nyquist sampling rates and reduced receiver complexity, while retaining a good performance. The idea is based on our recent results on sampling of certain classes of parametric nonband-limited signals that have a finite number of degrees of freedom per unit of time, or *finite rate of innovation* [19, 29]. Namely, the key is to note that the received signal $y(t)$ has only $2L$ degrees of freedom, time delays t_l and propagation coefficients a_l. Therefore, it seems intuitive that, when $s(t)$ is known *a priori* and there is no noise, it is possible to perfectly reconstruct the signal by taking only $2L$ samples of $y(t)$. That is, the minimum required sampling rate is, in general, determined by the number of degrees of freedom per unit of time, or the so-called rate of innovation [34]. While all the unknown parameters can be estimated using the time domain model (2.2), an efficient, closed-form solution is possible if we consider the problem in the frequency domain.

2.2.2 Frequency-Domain Channel Estimation

Assume that, during the channel estimation phase, the signal $s(t)$ is periodically transmitted, with a period T, to the receiver. For example, if $s(t)$ is made up of a modulated sequence of ultra short pulses, this would mean that the same sequence, or the same symbol, is periodically transmitted over the channel. Note that we made this assumption only to simplify the derivation, while it is generally not required (and sometimes cannot be even met in practice). If the channel is stationary,[1] the received noiseless signal $y(t)$ can be expressed in terms of its Fourier series as:

$$y(t) = \sum_{m=-\infty}^{\infty} Y[m] e^{jm\omega_0 t}, \quad (2.3)$$

---

[1] In fact, the only requirement is that the channel is quasi-stationary, namely, that its statistics do not change in a single burst, but can change from burst to burst.

where $\omega_0 = 2\pi/T$, while $Y[m]$ are the Fourier series coefficients of $y(t)$. If we denote by $S[m]$ the Fourier series coefficients of the transmitted signal $s(t)$, and using the channel model given by (2.1), the coefficients $Y[m]$ can be expressed as:

$$Y[m] = \frac{1}{T} \sum_{l=1}^{L} a_l S[m] e^{-jm\omega_0 t_l}. \qquad (2.4)$$

Clearly, the spectral components of the received signal are given by a sum of complex exponentials, where the unknown time delays t_l appear as complex frequencies while propagation coefficients a_l appear as unknown weights. Therefore, by considering the frequency domain representation of the received signal, one can convert the problem of estimating the unknown channel parameters $\{t_l\}_{l=1}^{L}$ and $\{a_l\}_{l=1}^{L}$ into the classical harmonic retrieval problem, commonly encountered in spectral estimation [25].

High-resolution harmonic retrieval is well-studied: there exists a rich body of literature on both theoretical limits and efficient algorithms for reliable estimation [15, 16, 21, 23]. There is a particularly attractive class of model-based algorithms, called super-resolution methods, which can resolve closely spaced sinusoids from a short record of noise-corrupted data. In [23, 25], a polynomial realization is discussed, where the parameters are estimated from zeros of the so-called prediction or annihilating filter. In [23], a state space method is proposed to estimate parameters of superimposed complex exponentials in noise, which provides an elegant and numerically robust tool for parameter estimation using a subspace-based approach. The ESPRIT algorithm is developed in [21], which can be viewed as a generalization of the state space method applicable to general antenna arrays. In [16], several subspace techniques for estimating generalized eigenvalues of matrix pencils are addressed, such as the Direct matrix pencil algorithm, Pro-ESPRIT, and its improved version TLS-ESPRIT. Another class of algorithms is based on the optimal maximum likelihood (ML) estimator [30], however, ML methods require L-dimensional search and are computationally more time-consuming than the subspace-based algorithms. Besides, in most cases encountered in practice, subspace methods can achieve performance close to that of the ML estimator [15], and are thus considered to be a viable alternative, provided that a low-rank system model is available.

In the following, we will adopt a model-based approach and show that it is possible to obtain high-resolution estimates of all the relevant parameters by sampling the received signal below the traditional Nyquist rate. The general setup we will be considering is shown in Figure 2.1. We will first discuss a polynomial realization of the estimator, which provides a good insight into fundamental principles behind high-resolution estimation from a subsampled version of the received signal. Later, we will present a more practical, subspace-based approach.

2.2.3 Polynomial Realization of the Model-Based Methods

Suppose that the received signal $y(t)$ is filtered with an ideal bandpass filter $H_b = [-N_1\omega_0, -M_1\omega_0] \cup [M_1\omega_0, N_1\omega_0]$ of bandwidth $B = (N_1 - M_1)\omega_0$, and assume for

2.2 CHANNEL ESTIMATION AT SUBNYQUIST SAMPLING RATE

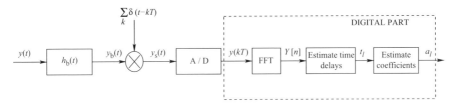

Figure 2.1 Receiver block diagram.

simplicity that $M_1\omega_0 = kB$, where k is a nonnegative integer number. At this point, we should note that the minimum size of the filter bandwidth is determined by the number of degrees of freedom of the received noiseless signal, that is, $2L\omega_0$.

Let $\{y_n\}_{n=1}^{N}$ denote the uniform samples taken from a filtered version of the received signal,

$$y_n = \langle h_b(t - nT_s) y(t) \rangle, \quad n = 0, 1, \ldots, N - 1,$$

where T_s is the sampling period, while $h_b(t)$ is the time domain representation of the filter H_b. Then the following relation holds:

$$y_n = \langle h_b(t - nT_s), \sum_m Y[m] e^{jm\omega_0 t} \rangle \tag{2.5}$$

$$= \sum_m Y[m] H_b(m\omega_0) e^{jm\omega_0 nT_s} \tag{2.6}$$

$$= \sum_{m=-N_1}^{-M_1} Y[m] e^{jm\omega_0 nT_s} + \sum_{m=M_1}^{N_1} Y[m] e^{jm\omega_0 nT_s}. \tag{2.7}$$

In fact, the above assumption on the position of the filter passband allows one to sample the signal at a rate determined by the bandwidth of the filter $R_s \geq 2(B/2\pi)$, which is commonly referred to as bandpass sampling [28]. Otherwise, one can use a more traditional approach of downconverting the filtered version prior to sampling, which also allows for sub-Nyquist sampling rates, but requires additional hardware stages in the analog front end. Under the above assumptions on the minimum sampling rate and the minimum bandwidth of the filter, the system of Equations (2.7) is invertible and will yield a unique solution for the coefficients $Y[m]$, $m \in [-N_1, -M_1] \cup [M_1, N_1]$ [19, 29]. In the following, we will consider only the coefficients $Y[m]$ with positive indices m, that is $m \in [M_1, N_1]$, while all the results can be also extended to the case when the coefficients with negative indices are included as well [19].

If we denote by $Y_s[n] = Y[M_1 + n]/S[M_1 + n]$, and assuming that in the considered frequency band the above division is not ill-conditioned, the samples $Y_s[n]$

can be expressed as a sum of complex exponentials, that is,

$$Y_s[n] = \sum_{l=1}^{L} a_l e^{-j(M_1\omega_0 + n\omega_0)t_l} = \sum_{l=1}^{L} \tilde{a}_l e^{-jn\omega_0 t_l}, \qquad (2.8)$$

where $\tilde{a}_l = a_l e^{-jM_1\omega_0 t_l}$. In practice, the discrete Fourier transform (DFT) will be used to determine $Y[n]$ and $S[n]$. Therefore, in the case of a nonperiodic transmission or when the channel is not stationary, Equation (2.8) will not hold exactly. When $y(t)$ is a periodic signal (e.g., as in the case discussed in Section 2.6), the DFT coefficients will exactly satisfy Equation (2.8).

The annihilating filter approach exploits the fact that in the absence of noise, each exponential $\{e^{-jn\omega_0 t_l}\}_{n \in \mathbb{Z}}$ can be "nulled out" or annihilated by a first-order finite impulse response (FIR) filter $H_l(z) = (1 - e^{-j\omega_0 t_l} z^{-1})$, that is,

$$e^{-jn\omega_0 t_l} * [1, -e^{-j\omega_0 t_l}] = 0.$$

Consider thus an Lth order FIR filter $H(z) = \sum_{m=0}^{L} H[m] z^{-m}$, having L zeros at $z_l = e^{-j\omega_0 t_l}$,

$$H(z) = \prod_{l=1}^{L} (1 - e^{-j\omega_0 t_l} z^{-1}). \qquad (2.9)$$

Note that $H[m]$ is the convolution of L elementary filters with coefficients $[1, -e^{-j\omega_0 t_l}]$, $l = 1, \ldots, L$. Since $Y_s[n]$ is the sum of complex exponentials, each will be annihilated by one of the roots of $H(z)$, thus we have

$$(H * Y_s)[n] = \sum_{k=0}^{L} H[k] Y_s[n-k] = 0, \quad \text{for} \quad n = L, \ldots, N-1. \qquad (2.10)$$

Therefore, the information about the time delays t_l can be extracted from the roots of the filter $H(z)$. The corresponding coefficients $\tilde{a}_l$ are then estimated by solving the system of linear Equations (2.8). In the following, we give an outline of the algorithm, while a more detailed discussion on the annihilating filter method can be found in [25] and [29].

Annihilating Filter Method

1. Find the coefficients $H[k]$ of the annihilating filter

$$H(z) = \prod_{l=1}^{L} (1 - e^{-j\omega_0 t_l} z^{-1}) = \sum_{k=0}^{L} H[k] z^{-k}, \qquad (2.11)$$

which satisfies Equation (2.10), that is,

$$(H * Y_s)[n] = 0, \quad \text{for} \quad n = L, \ldots, N-1.$$

2.2 CHANNEL ESTIMATION AT SUBNYQUIST SAMPLING RATE

By setting $H[0] = 1$, at critical sampling Equation (1.10) becomes

$$\begin{pmatrix} Y_s[L-1] & Y_s[L-2] & \cdots & Y_s[0] \\ Y_s[L] & Y_s[L-1] & \cdots & Y_s[1] \\ \vdots & \vdots & \ddots & \vdots \\ Y_s[2L-2] & Y_s[2L-3] & \cdots & Y_s[L-1] \end{pmatrix} \cdot \begin{pmatrix} H[1] \\ \vdots \\ H[L] \end{pmatrix}$$
$$= - \begin{pmatrix} Y_s[L] \\ Y_s[L+1] \\ \vdots \\ Y_s[2L-1] \end{pmatrix}. \qquad (2.12)$$

This system of equations is usually referred to as a high-order Yule–Walker system [30].

2. Find the values of t_l by finding the roots of $H(z)$.
3. Solve for the coefficients $\tilde{a}_l$ by solving the system of linear equations in Equation (2.8). This is a Vandermonde system, which has a unique solution since the t_ls are assumed to be distinct. The propagation coefficients a_l are then given by $a_l = \tilde{a}_l e^{jM_1 \omega_0 t_l}$.

The above result can be interpreted in the following way: the signal $y(t)$ is projected onto a low-dimensional subspace corresponding to its bandpass version. This projection is a unique representation of the signal as long as the dimension of the subspace is greater than or equal to the number of degrees of freedom. Specifically, since $y(t)$ has $2L$ degrees of freedom, $\{t_l\}_{l=0}^{L-1}$ and $\{a_l\}_{l=0}^{L-1}$, it suffices to use only $2L$ adjacent coefficients $Y_s[n]$, as can be seen from Equation (2.12). While in the noiseless case the critically sampled scheme leads to perfect estimates of all the parameters, in the presence of noise, such an approach suffers from poor numerical performance. In particular, any least-squares procedure that determines the filter coefficients directly from Equation (2.12) has poor numerical precision. In practice, this problem can be dealt with by oversampling and using standard techniques from noisy spectral estimation, such as the singular value decomposition (SVD). Namely, one should consider an extended system of Equations (2.12), that is,

$$\begin{pmatrix} Y_s[L-1] & Y_s[L-2] & \cdots & Y_s[0] \\ Y_s[L] & Y_s[L-1] & \cdots & Y_s[1] \\ \vdots & \vdots & \ddots & \vdots \\ Y_s[L_1-2] & Y_s[L_1-3] & \cdots & Y_s[L_1-L-1] \end{pmatrix} \cdot \begin{pmatrix} H[1] \\ \vdots \\ H[L] \end{pmatrix} = - \begin{pmatrix} Y_s[L] \\ Y_s[L+1] \\ \vdots \\ Y_s[L_1-1] \end{pmatrix}$$
$$\iff \mathbf{Y} \cdot \mathbf{h} = -\mathbf{y_s}, \qquad (2.13)$$

with $L_1 > 2L$, and decompose the matrix $\mathbf{Y}$ as

$$\mathbf{Y} = \mathbf{U}_s \mathbf{\Lambda}_s \mathbf{V}_s^H + \mathbf{U}_n \mathbf{\Lambda}_n \mathbf{V}_n^H, \qquad (2.14)$$

with the first term corresponding to the best (in the Frobenius-norm sense) rank L approximation of the matrix $\mathbf{Y}$. The filter coefficients $\mathbf{h}$ are then computed as

$$\mathbf{h} = -\mathbf{V}_s \mathbf{\Lambda}_s^{-1} \mathbf{U}_s^H \cdot \mathbf{y}_s. \qquad (2.15)$$

Although this modification considerably improves numerical accuracy on the estimates of filter coefficients, it is not sufficient for a good overall performance of the algorithm. In particular, in order to reduce sensitivity of the frequency estimates to noise, typically a high-order polynomial must be used [15], which imposes a significant computational burden since it is necessary to find roots of a large size polynomial in order to extract a small number of signal poles. In the following section, we will present an alternative subspace approach, based on state space modeling [23], which avoids root finding and relies only on a correct deployment of matrix manipulations. It leads to robust parameter estimates without overmodeling, by appropriately exploiting the algebraic structure of the signal subspace.

2.2.4 Subspace-Based Approach

The use of subspace techniques for channel estimation in wideband systems, such as DS-CDMA, is not new in the literature [2, 22, 27]. Yet almost all existing methods solve for the desired parameters from a sample estimate of the covariance matrix and resort to the Nyquist sampling rate (or even use fractional sampling). Clearly, applying such techniques to UWB systems would require sampling rates on the order of GHz and computational requirements not affordable in most of the UWB applications. We will show that it is possible to estimate all the parameters from a low-dimensional signal subspace, and this by avoiding explicit computation of the covariance matrix.

The main idea behind the state space approach is the following. Given the set of coefficients $Y_s[n]$, Equation (2.8), construct a Hankel[2] data matrix $\mathbf{Y}_s$ of size $P \times Q$, where $P, Q > L$,

$$\mathbf{Y}_s = \begin{pmatrix} Y_s[0] & Y_s[1] & \ldots & Y_s[Q-1] \\ Y_s[1] & Y_s[2] & \ldots & Y_s[Q] \\ \vdots & & & \\ Y_s[P-1] & Y_s[P] & \ldots & Y_s[P+Q-2] \end{pmatrix} \qquad (2.16)$$

---
[2] A Hankel matrix is a matrix in which the (i,j)th entry depends only on the sum $i+j$.

2.2 CHANNEL ESTIMATION AT SUBNYQUIST SAMPLING RATE

Consider first the simple case of a channel with only $L = 1$ propagation path. In the absence of noise, the elements of the matrix $\mathbf{Y}_s$ are given by

$$\mathbf{Y}_s[p, q] = \tilde{a}_1 z_1^{p+q}, \quad 0 \leq p \leq P-1, \quad 0 \leq q \leq Q-1, \quad (2.17)$$

where $z_1 = e^{-j\omega_0 t_1}$ denotes the signal pole. Therefore, $\mathbf{Y}_s$ can be written as $\mathbf{Y}_s = \mathbf{U}\mathbf{\Lambda}\mathbf{V}^H$, where the matrices $\mathbf{U}$, $\mathbf{\Lambda}$ and $\mathbf{V}$ are given by

$$\begin{aligned} \mathbf{U} &= (1 \quad z_1 \quad z_1^2 \quad \ldots \quad z_1^{P-1})^T \quad \mathbf{\Lambda} = (\tilde{a}_1) \\ \mathbf{V} &= (1 \quad z_1^* \quad z_1^{*2} \quad \ldots \quad z_1^{*Q-1})^T. \end{aligned} \quad (2.18)$$

The state space method is based on two properties of the data matrix $\mathbf{Y}_s$. The first one is that in the case of noiseless data, $\mathbf{Y}_s$ has rank $L = 1$. The second one is a Vandermonde structure of $\mathbf{U}$ and $\mathbf{V}$, that is, they both satisfy the so-called shift-invariant subspace property,

$$\overline{\mathbf{U}} = \underline{\mathbf{U}} \cdot z_1 \quad \text{and} \quad \overline{\mathbf{V}} = \underline{\mathbf{V}} \cdot z_1^*, \quad (2.19)$$

where $\overline{(\cdot)}$ and $\underline{(\cdot)}$ denote the operations of omitting the first and the last row of $(\cdot)$ respectively. Obviously, in the absence of noise, the signal pole z_1 (or its conjugate z_1^*) can be perfectly estimated from only two adjacent elements of $\mathbf{U}$, or alternatively, $\mathbf{V}$. In practice, z_1 should be fitted to a larger data set, using any of the two relations in Equation (2.19), specifically,

$$z_1 = \underline{\mathbf{U}}^+ \cdot \overline{\mathbf{U}} \quad \text{or} \quad z_1^* = \underline{\mathbf{V}}^+ \cdot \overline{\mathbf{V}}, \quad (2.20)$$

where $(\cdot)^+$ denotes the pseudoinverse of $(\cdot)$. Once the signal pole has been estimated, the time delay t_1 can be determined from its complex frequency, namely, $z_1 = e^{-j\omega_0 t_1}$, while $\tilde{a}_1$ can be found as a least-squares solution to Equation (2.8).

Let us next show how the same approach can be used to estimate the channel with $L > 1$ paths. As in the previous case, given the set of coefficients $Y_s[n] = \sum_{l=1}^{L} \tilde{a}_l z_l^n$, one should first construct the data matrix $\mathbf{Y}_s$ as in Equation (2.16). In the absence of noise, $\mathbf{Y}_s$ can be decomposed as $\mathbf{Y}_s = \mathbf{U}\mathbf{\Lambda}\mathbf{V}^H$, where $\mathbf{U}$, $\mathbf{\Lambda}$ and $\mathbf{V}$ are now given by

$$\mathbf{U} = \begin{pmatrix} 1 & 1 & 1 & \ldots & 1 \\ z_1 & z_2 & z_3 & \ldots & z_L \\ \vdots & & & & \\ z_1^{P-1} & z_2^{P-1} & z_3^{P-1} & \ldots & z_L^{P-1} \end{pmatrix} \quad (2.21)$$

$$\mathbf{\Lambda} = \text{diag}(\tilde{a}_1 \quad \tilde{a}_2 \quad \tilde{a}_3 \quad \ldots \quad \tilde{a}_L) \quad (2.22)$$

$$\mathbf{V} = \begin{pmatrix} 1 & 1 & 1 & \cdots & 1 \\ z_1^* & z_2^* & z_3^* & \cdots & z_L^* \\ \vdots & & & & \\ z_1^{*Q-1} & z_2^{*Q-1} & z_3^{*Q-1} & \cdots & z_L^{*Q-1} \end{pmatrix}. \qquad (2.23)$$

Clearly, we can again exploit the Vandermonde structure of $\mathbf{U}$ and $\mathbf{V}$, with the following shift-invariant subspace property:

$$\overline{\mathbf{U}} = \underline{\mathbf{U}} \cdot \mathbf{\Phi} \quad \text{and} \quad \overline{\mathbf{V}} = \underline{\mathbf{V}} \cdot \mathbf{\Phi}^H, \qquad (2.24)$$

where in this case, $\mathbf{\Phi}$ is a diagonal matrix having z_ls along the main diagonal. At this point, we should note that the above factorization is not unique. That is, if $\mathbf{Y_s} = \mathbf{U}\mathbf{S}\mathbf{V}^H$, then $\mathbf{Y_s} = \mathbf{U}\mathbf{A} \cdot \mathbf{A}^{-1}\mathbf{S}\mathbf{B} \cdot \mathbf{B}^{-1}\mathbf{V}^H$ is another possible factorization, for every choice of $L \times L$ nonsingular matrices $\mathbf{A}$ and $\mathbf{B}$. However, as we will show in the following, any such factorization can be used to estimate the signal poles.

The second key property is that, in the absence of noise, $\mathbf{Y}_s$ has rank L. This will allow us to reduce the noise level by approximating a noisy data matrix with a rank L matrix, and this by computing its singular value decomposition (SVD). Note that when $\mathbf{Y}_s$ is decomposed using the SVD, one would not obtain the same matrices $\mathbf{U}$, $\mathbf{\Lambda}$ and $\mathbf{V}$ as in Equations (2.21)–(2.23); however, the shift-invariance property would hold as well. In order to prove this, assume that the SVD of $\mathbf{Y_s}$ is given by

$$\mathbf{Y_s} = \mathbf{U_s}\mathbf{\Lambda_s}\mathbf{V_s}^H + \mathbf{U_n}\mathbf{\Lambda_n}\mathbf{V_n}^H, \qquad (2.25)$$

where the columns of $\mathbf{U_s}$ and $\mathbf{V_s}$ are L principal left and right singular vectors of $\mathbf{Y_s}$, respectively, while the second term contains remaining nonprincipals. Since both $\mathbf{U_s}$ and $\mathbf{V_s}$ are matrices of rank L [as well as $\mathbf{U}$ and $\mathbf{V}$ in Equations (2.21) and (2.23)], there will exist $L \times L$ nonsingular matrices $\mathbf{A}$ and $\mathbf{B}$ such that $\mathbf{U_s} = \mathbf{U} \cdot \mathbf{A}$ and $\mathbf{V_s} = \mathbf{V} \cdot \mathbf{B}$. Consider next the Vandermonde matrix $\mathbf{U}$, given by Equation (2.21), which can be written in the following, more compact form:

$$\mathbf{U} = \begin{pmatrix} \mathbf{b} \\ \mathbf{b} \cdot \mathbf{\Phi} \\ \mathbf{b} \cdot \mathbf{\Phi}^2 \\ \vdots \\ \mathbf{b} \cdot \mathbf{\Phi}^{P-1} \end{pmatrix}, \qquad (2.26)$$

where $\mathbf{\Phi}$ is the same diagonal matrix as before, $\mathbf{\Phi} = \text{diag}(z_l)$, while $\mathbf{b}$ is a row vector of length L, given by $\mathbf{b} = [1 \; 1 \; \ldots \; 1]_{L \times 1}$. The key is to observe that the matrix

$\mathbf{U}_s = \mathbf{U}\mathbf{A}$ can be expressed as

$$\mathbf{U}\mathbf{A} = \begin{pmatrix} \mathbf{b}\mathbf{A} \\ \mathbf{b}\mathbf{A} \cdot \mathbf{A}^{-1}\mathbf{\Phi}\mathbf{A} \\ \mathbf{b}\mathbf{A} \cdot \mathbf{A}^{-1}\mathbf{\Phi}^2\mathbf{A} \\ \vdots \\ \mathbf{b}\mathbf{A} \cdot \mathbf{A}^{-1}\mathbf{\Phi}^{P-1}\mathbf{A} \end{pmatrix}, \qquad (2.27)$$

where we have inserted $\mathbf{A}\mathbf{A}^{-1}$ between $\mathbf{b}$ and $\mathbf{\Phi}^k$. Given that $(\mathbf{A}^{-1}\mathbf{\Phi}\mathbf{A})^k = \mathbf{A}^{-1}\mathbf{\Phi}^k\mathbf{A}$, it becomes obvious that $\mathbf{U}\mathbf{A}$ satisfies the shift-invariance property as well, specifically,

$$\overline{\mathbf{U}\mathbf{A}} = \underline{\mathbf{U}\mathbf{A}} \cdot \mathbf{A}^{-1}\mathbf{\Phi}\mathbf{A} \qquad (2.28)$$

In this case, the matrix $\mathbf{A}^{-1}\mathbf{\Phi}\mathbf{A}$ in Equation (2.28) will be similar to $\mathbf{\Phi}$; therefore, it will have the same eigenvalues as $\mathbf{\Phi}$, that is, $\{z_k\}_{k=0}^{K-1}$. The same will be true in the case when the signal poles are estimated from a matrix $\mathbf{V}$, or another matrix $\mathbf{V}\mathbf{B}$, where $\mathbf{B}$ is any $K \times K$ nonsingular matrix. In practice, when the matrix $\mathbf{Y}_s$ is decomposed using the SVD, the signal poles, and thus the time delays t_l, can be estimated from any of the two matrices $\mathbf{U}_s$ or $\mathbf{V}_s$, by finding the eigenvalues of the operator that maps $\underline{\mathbf{U}_s}$ onto $\overline{\mathbf{U}_s}$ (or $\underline{\mathbf{V}_s}$ onto $\overline{\mathbf{V}_s}$). Once the signal poles have been estimated, the propagation coefficients a_k can be found as a least-squares solution to Equation (2.8). The algorithm can be thus summarized as follows.

Subspace-Based Algorithm

1. Given the set of the coefficients $Y_s[n]$, construct a $P \times Q$ matrix $\mathbf{Y}_s$ as in Equation (2.16), where $P, Q > L$.
2. Compute the singular value decomposition of $\mathbf{Y}_s$ as in Equation (2.25), and approximate the noiseless data matrix with a rank L matrix, using only L principal components, that is,

$$\mathbf{Y}_s \approx \mathbf{U}_s \mathbf{\Lambda}_s \mathbf{V}_s^H. \qquad (2.29)$$

3. Estimate the signal poles $z_l = e^{-j\omega_0 t_l}$ by computing the eigenvalues of a matrix $\mathbf{Z}$, defined as

$$\mathbf{Z} = \underline{\mathbf{U}_s}^+ \cdot \overline{\mathbf{U}_s}. \qquad (2.30)$$

Alternatively, if $\mathbf{V}_s$ is used in Equation (2.30) instead of $\mathbf{U}_s$, one would estimate complex conjugates of z_ls.

4. Find the coefficients $\tilde{a}_l$ from the Vandermonde system (2.8), that is,

$$Y_s[n] = \sum_{l=1}^{L} \tilde{a}_l e^{-jn\omega_0 t_l},$$

by fitting L exponentials $e^{-jn\omega_0 t_l}$ to the data set $Y_s[n]$.

Following the above discussion, it is obvious that we have converted the nonlinear estimation problem into the simpler problem of estimating the parameters of a linear model. Nonlinearity is postponed for the step where the information about the time delays is extracted from the estimated signal poles [23]. However, by considering the estimation problem in the frequency domain, we have avoided the estimation of the signal covariance matrix, which generally requires a larger data set and represents a computationally demanding part in other methods [2, 21].

Finally, we should note that, since we are estimating the signal parameters from the coefficients $Y_s[n] = Y[M_1 + n]/S[M_1 + n]$ (where $S[n]$ and $Y[n]$ are the DFT coefficients of the transmitted and received signal, respectively), in general, when noise is present, it will no longer remain white. However, since we are using a portion of the signal bandwidth, we can estimate the parameters from the frequency band where the power spectral density of the transmitted signal is nearly flat,[3] thus having the assumption on white noise can be still considered valid. Otherwise, one should use a noise-whitening transformation prior to estimating the channel parameters [19]. For example, one can compute a Cholesky decomposition [13] of the noise covariance matrix, that is, $\mathbf{R_w} = \mathbf{C^T C}$ (assuming that $\mathbf{R_w}$ is a positive definite matrix), and multiply the data matrix by $\mathbf{C^{-T}}$ prior to computing its singular value decomposition.

2.2.5 Estimation of Closely Spaced Paths

In most practical cases, the model-based methods provide an attractive alternative to more complex maximum likelihood techniques [27]. Yet the problem encountered in all parametric methods for harmonic retrieval is that their performance typically degrades if there are closely spaced sinusoidal frequencies, which in our case corresponds to the problem of estimating the parameters of closely spaced paths. While this can be avoided by assuming a low-rank channel model and estimating the parameters of only dominant components (provided that there is sufficient separation among them) [25], it is interesting to note that a relatively simple modification of the subspace-based method from Section 2.2.4, can significantly improve its resolution capabilities, which we present in the following.

Consider the data matrix $\mathbf{Y_s}$, defined in Equation (2.16). In order to estimate the signal poles z_ls, we have exploited the shift-invariant subspace property

---
[3]In systems that are properly designed, this is always the case.

(2.24), that is, $\overline{\mathbf{U}} = \underline{\mathbf{U}} \cdot \mathbf{\Phi}$, or alternatively, $\overline{\mathbf{V}} = \underline{\mathbf{V}} \cdot \mathbf{\Phi}$, where $\mathbf{\Phi}$ is a diagonal matrix with z_ls along the main diagonal. However, the Vandermonde structure of $\mathbf{U}$ and $\mathbf{V}$ allows for a more general version of Equation (2.24), specifically,

$$\overline{\mathbf{U}}^d = \underline{\mathbf{U}}_d \cdot \mathbf{\Phi}^d \quad \text{and} \quad \overline{\mathbf{V}}^d = \underline{\mathbf{V}}_d \cdot \mathbf{\Phi}^d, \qquad (2.31)$$

where $\overline{(\cdot)}^d$ and $(\cdot)_d$ denote the operations of omitting the first d rows and last d rows of $(\cdot)$, respectively. In this case, the matrix $\mathbf{\Phi}^d$ has elements $z_l^d = e^{-j\omega_0 dt_l}$ on its main diagonal, meaning that the effective separation among the estimated time delays is increased d times. This can improve the resolution performance of the method considerably, in particular for low values of SNR [19].

2.3 PERFORMANCE EVALUATION

2.3.1 Analysis of Noise Sensitivity

The statistical properties of the estimates obtained using high-resolution methods have been studied extensively, primarily in the context of estimating the frequencies of superimposed complex sinusoids from noisy measurements [15, 16, 24]. Expressions for the mean square error (MSE) of the frequency estimates suggest that the numerical performance of such methods is very close to the Cramer–Rao bound [26], which represents the lowest achievable MSE by any unbiased estimator, such as an ML estimator. A detailed presentation of the statistical properties is fairly involved and we will not pursue any such analysis here. However, we give simplified expressions for the MSE of the frequency estimate in the case of a single exponential (with amplitude a_1), which, in our framework, corresponds to the estimate of the time delay t_1 of the dominant path.

Consider first the subspace-based approach from Section 2.2.4. Let the data matrix $\mathbf{Y}_s$ be of size $P \times Q$, and let $N = P + Q - 1$ be the total number of DFT coefficients $Y_s[n]$ used for estimation. Recall that the coefficients $Y_s[n]$ are obtained from the bandpass version of the received signal $y(t)$, that is, $Y_s[n] = Y[n]/S[n]$, $n \in [N_1, N_1 + N - 1]$, Equation (2.8). If we define $\omega_1 = \omega_0 t_1$, and assume that the signal and noise are uncorrelated, the MSE of the state space approach can be expressed as [15]

$$E\{\Delta\omega_1^2\} \approx \begin{cases} \dfrac{1}{Q(N-Q)^2} \dfrac{\sigma_n^2}{|a_1|^2} \dfrac{N}{\sum_n |S[n]|^2}, & \text{for } Q \leq N/2 \\[2ex] \dfrac{1}{Q^2(N-Q)} \dfrac{\sigma_n^2}{|a_1|^2} \dfrac{N}{\sum_n |S[n]|^2}, & \text{for } Q > N/2 \end{cases} \qquad (2.32)$$

where σ_n^2 is noise variance. Note that the error is inversely proportional to the SNR at the output of the bandpass filter, defined as

$$\text{SNR} = \frac{|a_1|^2}{\sigma_n^2} \frac{\sum_n |S[n]|^2}{N}.$$

Therefore, for a given bandwidth of the filter, it is desirable to estimate the channel from a frequency band where the SNR is highest. The optimum performance is then achieved when $Q = N/3$ or $Q = 2N/3$, resulting in the MSE of time delay estimation

$$E\{\Delta t_1^2\} \approx \frac{1}{\omega_0^2} \frac{27}{4N^3} \frac{1}{\text{SNR}}. \tag{2.33}$$

This is very close to the Cramer–Rao bound (CRB) [31], given by

$$\text{CRB} = \frac{1}{\omega_0^2} \frac{6}{N^3} \frac{1}{\text{SNR}}, \tag{2.34}$$

which indicates desirable numerical performance of the state space approach. Similar performance can be achieved with the annihilating filter method [17], with an MSE of the form

$$E\{\Delta \omega_1^2\} \approx \begin{cases} \dfrac{2(2Q+1)}{3(N-Q)^2 Q(Q+1)} \dfrac{1}{\text{SNR}}, & \text{for } Q \leq N/2 \\[2mm] \dfrac{2[-(N-Q)^2 + 3Q^2 + 3Q + 1]}{3(N-Q)Q^2(Q+1)^2} \dfrac{1}{\text{SNR}}, & \text{for } Q > N/2 \end{cases} \tag{2.35}$$

where, in this case, Q represents the polynomial degree. As mentioned earlier, a choice of the polynomial degree directly affects the estimation performance, and the minimum MSE is achieved for $Q = N/3$ or $Q = 2N/3$, leading to

$$E\{\Delta t_1^2\} \approx \frac{1}{\omega_0^2} \frac{9}{N^3} \frac{1}{\text{SNR}}. \tag{2.36}$$

At this point, we should note that expressions for performance bounds (2.33), (2.34), and (2.36) are obtained using the first-order perturbation analysis and are generally valid only for medium to high signal-to-noise ratios. Still, these results give us a good indication as to the performance of the proposed methods at different sampling rates. That is, since the root mean square error (RMSE) for the time delay estimation is on the order of $\mathcal{O}(1/N^{3/2})$, by decreasing the sampling rate K times, RMSE increases by a factor of (approximately) $K^{3/2}$. Specifically, the following general relation between the RMSE of a subsampled estimator (RMSE_{ss}) and the RMSE of a Nyquist-sampled estimator (RMSE_{nq}) holds for all the considered methods,

$$\text{RMSE}_{ss} \approx \text{RMSE}_{nq} K^{3/2} \left(\frac{\text{SNR}_{nq}}{\text{SNR}_{ss}} \right)^{1/2}, \qquad (2.37)$$

where SNR_{nq} denotes the overall signal-to-noise ratio, while SNR_{ss} is the signal-to-noise ratio at the output of the corresponding bandpass filter. Clearly, even though the SNR after filtering may increase, the performance of a subsampled estimator is expected to degrade, due to a smaller data set used for estimation. Finally, note that Equation (2.37) implies that the performance bounds of subsampled state space or annihilating filter methods are again very close to the CRB of a subsampled ML estimator.

2.3.2 Computational Complexity and Alternative Solutions

A major computational requirement for the presented algorithms is associated with the singular value decomposition step, which is an iterative algorithm with computational order $\mathcal{O}(N^3)$ per iteration. In some cases, however, we are interested in estimating the parameters of only a few strongest paths; therefore, computing the full SVD of the data matrix $\mathbf{Y}_s$ is not necessary. In such a case, one can use simpler methods to find principal singular vectors, which have lower computational requirements and converge very fast to the desired solution [8, 13]. We first give an outline of the *Power method* [13], that can be used to compute only one dominant right (or left) singular vector of $\mathbf{Y}_s$. This can be of interest for initial synchronization or in applications such as ranging or positioning. Later, we present its extended version applicable to the general case of estimating $M_d > 1$ principal singular vectors.

Power Method Consider a matrix $\mathbf{F} = \mathbf{Y}_s \mathbf{Y}_s^H$ of size $P \times P$, and suppose that $\mathbf{F}$ is diagonalizable, that is, $\mathbf{\Lambda}^{-1} \mathbf{F} \mathbf{\Lambda} = \text{diag}(\lambda_1, \ldots, \lambda_P)$ with $\mathbf{\Lambda} = [\mathbf{y}_1, \ldots, \mathbf{y}_P]$ and $|\lambda_1| > |\lambda_2| \geq \cdots |\lambda_P|$. Given $\mathbf{y}^{(0)}$, the Power method produces a sequence of vectors $\mathbf{y}^{(k)}$ in the following way:

$$\begin{aligned} \mathbf{z}^{(k)} &= \mathbf{F} \mathbf{y}^{(k-1)}, \\ \mathbf{y}^{(k)} &= \mathbf{z}^{(k)} / \|\mathbf{z}^{(k)}\|_2. \end{aligned} \qquad (2.38)$$

The method converges if λ_1 is dominant and if $\mathbf{y}^{(0)}$ has a component in the direction of the corresponding dominant eigenvector $\mathbf{y}_1$. It is easily verified that $\mathbf{y}_1, \ldots, \mathbf{y}_P$ are the left singular vectors of $\mathbf{Y}_s$, therefore, once the principal singular vector $\mathbf{y}_1$ has been estimated, the signal pole z_1 (corresponding to the strongest signal component) is given by $z_1 = \mathbf{y}_1^+ \overline{\mathbf{y}_1}$. A potential problem with this method is that its convergence rate depends on $|\lambda_2/\lambda_1|$, a quantity which may be close to 1 and thus cause slow convergence. Improved versions of the algorithm which overcome this problem are discussed in [8]. Note that the power method involves only simple matrix multiplications and has a computational order of $\mathcal{O}(P^2)$ per iteration.

Orthogonal Iteration A straightforward generalization of the power method can be used to compute higher-dimensional invariant subspaces, that is, to find $M_d > 1$ dominant singular vectors. The method is typically referred to as *orthogonal iteration* or *subspace iteration* and can be summarized as follows.

Given a $P \times M_d$ matrix $\mathbf{W}^{(0)}$, the method generates a sequence of matrices $\mathbf{W}^{(k)}$ through the iteration

$$\mathbf{Z}^{(k)} = \mathbf{F}\mathbf{W}^{(k-1)}, \qquad (2.39)$$

$$\mathbf{W}^{(k)}\mathbf{R}^{(k)} = \mathbf{Z}^{(k)} \ (Q-R \text{ factorization}). \qquad (2.40)$$

The computational complexity of the method is on the order of $\mathcal{O}(P^2 M_d)$ per iteration, and clearly when $M_d = 1$ the algorithm is equivalent to the power method. In practice, $\mathbf{F}$ is first reduced to upper Hessenberg form (that is, $\mathbf{F}$ is zero below the first subdiagonal) and the method is implemented in a simpler way, avoiding explicit $Q-R$ factorization in each iteration. A more detailed discussion on this topic can be found in [8].

2.3.3 Numerical Example

To illustrate the numerical performance of the subspace algorithm, we consider the case of a channel model given by Equation (2.1), assuming $L = 70$ propagation paths with $M_d = 8$ dominant paths (containing 85% of the total power), as illustrated in Figure 2.2(a). We assumed that a symbol is made up of a sequence of 127 coded impulses, periodically transmitted over multiple cycles, and that pulse-amplitude modulation with a pseudo-random sequence of length 127 is used at the transmitter. Since we are considering discrete time signals, time is expressed in terms of samples, where one sample corresponds to the period of Nyquist-rate sampling. In Figure 2.2(b), we show the RMSE of delay estimation for the dominant components vs SNR. We used the approach presented in Section 2.2.5, where the parameter d is chosen to be $d = 30$. The method yields highly accurate estimates for a wide range of SNRs. For example, when the sampling rate N_s is one-quarter of the Nyquist rate N_n, and $SNR = -5$dB, the delay of the dominant components can be estimated with an RMSE of approximately 1 sample.

The effects of quantization on the estimation performance are shown in Figure 2.2(c). In particular, we considered 4–7 bit architectures and for each case we plotted the RMSE vs received SNR. The results are also compared to the "ideal" case when $n_b = 32$ bits are used for quantization. Clearly, as the number of bits increases, the overall performance improves. Generally, the 5-bit architecture already yields a very good performance. Also note that when $n_b \geq 5$ and the value of SNR is low (e.g., SNR < 0 dB), quantization has almost no impact on the estimation

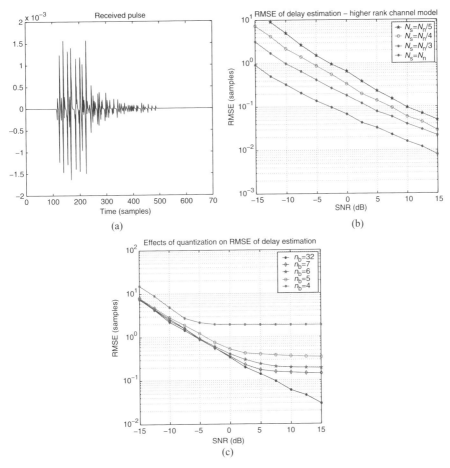

Figure 2.2 Higher-rank channel models. (a) Received UWB signal made up of 70 pulses, with eight dominant components (containing approximately 85% of the total power). (b) RMSE of delay estimation of the dominant components vs SNR. (c) Effects of quantization on the RMSE of delay estimation for 4–7 bit receiver architectures. The results are compared with the case when the number of bits is $n_b = 32$. The sampling rate is one-quarter the Nyquist rate ($N_s = N_n/4$).

performance. However, as the value of SNR increases, quantization noise becomes dominant and determines the overall numerical performance.

2.4 ESTIMATING UWB CHANNELS WITH FREQUENCY-DEPENDENT DISTORTION

In this section, we will touch upon the problem of estimating a channel that takes into account certain frequency-dependent properties. Namely, as a result of the very large bandwidth of UWB signals, components propagating along different propagation paths undergo different frequency selective distortion and a more realistic channel model for UWB systems is of the form [7]:

$$h(t) = \sum_{l=1}^{L} a_l p_l(t - t_l), \qquad (2.41)$$

where $p_l(t)$ are different pulse shapes that correspond to different propagation paths. In this case, the DFT coefficients computed from a bandpass version of the received signal can be expressed as

$$Y[n] = S[n] \sum_{l=1}^{L} P_l[n] \tilde{a}_l e^{-jn\omega_0 t_l}, \qquad (2.42)$$

where $P_l[n]$ are now unknown coefficients. Recall that $\tilde{a}_l = a_l e^{-jM_1 \omega_0 t_l}$. Clearly, in order to completely characterize the channel, we need to estimate the a_ls and t_ls as well as all the coefficients $P_l[n]$, which, in general, requires a nonlinear estimation procedure. However, one possible way to obtain a closed form solution is to approximate the coefficients $P_l[n]$ in the considered frequency band with polynomials of maximum degree $R - 1$, that is,

$$P_l[n] = \sum_{r=0}^{R-1} p_{l,r} n^r. \qquad (2.43)$$

Equation (2.42) now becomes

$$Y[n] = S[n] \sum_{l=1}^{L} \tilde{a}_l \sum_{r=0}^{R-1} p_{l,r} n^r e^{-jn\omega_0 t_l}. \qquad (2.44)$$

By denoting $c_{l,r} = \tilde{a}_l p_{l,r}$ and $Y_s[n] = Y[n]/S[n]$, we obtain

$$Y_s[n] = \sum_{l=1}^{L} \sum_{r=0}^{R-1} c_{l,r} n^r e^{-jn\omega_0 t_l}. \qquad (2.45)$$

Therefore, by using the polynomial approximation of the coefficients $P_l[n]$, the channel estimation problem can be reduced to the one of estimating the unknown parameters $c_{l,r}$ and t_l, from the coefficients $Y_s[n]$. In [19], we proved that such an estimation problem allows for a closed-form solution. In fact, both the annihilating filter method and the subspace approach can be easily extended to handle this type of nonlinear estimation problems. In the former case, the key is to observe that the annihilating filter will have multiple roots at $e^{-j\omega_0 t_l}$, that is,

$$H(z) = \prod_{l=1}^{L} H_{l,R-1}(z) = \prod_{l=1}^{L} (1 - e^{-j\omega_0 t_l} z^{-1})^R. \qquad (2.46)$$

Therefore, the information about the time delays t_l can be extracted from the roots of the filter $H(z)$, while the corresponding pulse shapes are then estimated by solving for the coefficients $c_{l,r}$ in Equation (2.45).

In order to derive the subspace solution, one can use the result on equivalence between the annihilating filter and the subspace estimator, and show that the estimated eigenvalues in the subspace approach will be given by $e^{-j\omega_0 t_l}$, each of algebraic multiplicity R. In the following, we briefly summarize the subspace approach, while for proofs and derivations we refer to [19].

2.4.1 Algorithm Outline

1. Given a set of coefficients $Y_s[m]$, construct an $M \times N$ matrix data $\mathbf{Y}_s$ as in Equation (2.16), where $M, N \geq RL$.
2. Compute the singular value decomposition of $\mathbf{Y}_s$, that is, $\mathbf{Y}_s = \mathbf{U}\mathbf{S}\mathbf{V}^H$. Find RL principal left and right singular vectors, $\mathbf{U}_s$ and $\mathbf{V}_s$, as the singular vectors corresponding to the K largest singular values of $\mathbf{Y}_s$.
3. Estimate the signal poles $z_l = e^{-j\omega_0 t_l}$ by computing the eigenvalues of a matrix $\mathbf{H}$, defined as

$$\mathbf{H} = \underline{\mathbf{U}}_s^+ \cdot \overline{\mathbf{U}}_s. \qquad (2.47)$$

Alternatively, if $\mathbf{V}_s$ is used in Equation (2.47), one would estimate complex conjugates of z_ks. While in the noiseless case one should find RL eigenvalues, each of multiplicity L, in the presence of noise, it is more desirable to approximate the signal poles with the eigenvalues of $\mathbf{H}$ that are closest to the unit circle.

4. Find the coefficients c_k as a least-squares solution to the Vandermonde system (2.8), that is,

$$Y_s[n] = \sum_{l=1}^{L} \sum_{r=0}^{R-1} c_{l,r} n^r e^{-jn\omega_0 t_l}.$$

At this point, it is important to note that the reconstruction of the pulse shapes from the set of estimated coefficient $c_{l,r}$ must be done carefully. If the pulse shapes are reconstructed from the estimated lowpass version of the signal, using the polynomial approximation (2.43), one can create ripples in the reconstructed signal due to the Gibbs phenomenon. Similarly, reconstructing the signal from a larger set of DFT coefficients, obtained by spectral extrapolation from Equation (2.43), is often numerically unstable. A conventional way to treat this problem is to use a less abrupt truncation of the DFT coefficients by appropriate windowing [20]. One possible solution is to do weighting of the extrapolated DFT coefficients with an exponentially decaying function, which can significantly improve the accuracy of reconstruction.

2.5 CHANNEL ESTIMATION FROM MULTIPLE BANDS

2.5.1 Filter Bank Approach

So far, we have considered only a low-dimensional subspace of the received signal and all the methods were developed under the assumption that one has access to consecutive DFT coefficients of the signal. While in the noiseless case it would be possible to estimate the parameters from any subspace of appropriate dimension, in the presence of noise the best performance of the algorithm is expected when the channel is estimated from a frequency band with highest signal-to-noise ratio. An alternative approach would be to estimate the channel from a larger subspace, using a filter bank at the receiver, where each subband is sampled at a rate determined by the filter bandwidth. The set of coefficients $Y_s[n]$ is then computed separately for each subband and combined to form the matrix $\mathbf{Y_s}$ in Equation (2.16), or to compute the annihilating filter coefficients in Equation (2.12). An obvious advantage of this approach is that a larger data set is used for estimation, which results in improved numerical performance, yet at the expense of increased computational and power requirements.

In the case when the channel parameters are estimated from adjacent subbands, the algorithm presented in Section 2.2.4 remains essentially the same, since we have access to consecutive coefficients $Y_s[n]$. A more interesting case is when the parameters are estimated from bands that are not necessarily adjacent. For example, if the noise level in certain bands is relatively high, or if some bands are subject to strong interference (e.g., interference from coexisting systems, such as GPS), it is desirable to estimate the channel by considering only those bands where SNIR (signal-to-noise-plus-interference ratio) is sufficiently high. To date, several solutions have been proposed to mitigate the problem of strong interference. For example, in ref [33], the authors discuss techniques for pulse shaping at the transmitter, where the idea is to design pulses with desirable spectral properties, using either carrier-modulation or baseband filtering of the pulse. Another approach requires filtering the signal at the receiver; however, this requires building high-Q notch filters on chip, which is technically not easy to achieve. We will show that the above algorithm can be adapted rather simply to handle this case, and this without introducing any additional stages at the transmitter or receiver.

2.5.2 Estimation from Nonadjacent Bands

Consider the channel model given by Equation (2.1). For simplicity, we will analyze the case when the channel parameters are estimated by sampling only two nonadjacent bands $B_1 = (M_1\omega_0, N_1\omega_0)$ and $B_2 = (M_2\omega_0, N_2\omega_0)$, while the same approach can be generalized to the case with multiple frequency bands. Let $Y[n]$ be the DFT coefficients of the received signal corresponding to the bands B_1 and B_2, and let $Y_s[n] = Y[n]/S[n]$ (assuming again that this division is well-conditioned). Under the above assumptions, the noiseless coefficients $Y_s[n]$ are given by $Y_s[n] = \sum_{l=1}^{L} a_l z_l^n$, where

$n \in [M_1, N_1] \cup [M_2, N_2]$. Next define a block-Hankel data matrix $\mathbf{Y_s}$ as

$$\mathbf{Y_s} = \begin{pmatrix} Y_s[M_1] & Y_s[M_1+1] & \cdots & Y_s[M_1+Q-1] \\ \vdots & & & \\ Y_s[M_1+P_1-1] & Y_s[M_1+P_1] & \cdots & Y_s[M_1+P_1+Q-2] \\ Y_s[M_2] & Y_s[M_2+1] & \cdots & Y_s[M_2+Q-1] \\ \vdots & & & \\ Y_s[M_2+P_2-1] & Y_s[M_2+P_2] & \cdots & Y_s[M_2+P_2+Q-2] \end{pmatrix}. \quad (2.48)$$

In the noiseless case, the matrix $\mathbf{Y_s}$ can be written as $\mathbf{Y_s} = \mathbf{U \Lambda V}^H$, where $\mathbf{U}$, $\mathbf{\Lambda}$, and $\mathbf{V}$ are now given by

$$\mathbf{U} = \begin{pmatrix} z_1^{M_1} & z_2^{M_1} & z_3^{M_1} & \cdots & z_L^{M_1} \\ \vdots & & & & \\ z_1^{M_1+P_1-1} & z_2^{M_1+P_1-1} & z_3^{M_1+P_1-1} & \cdots & z_L^{M_1+P_1-1} \\ z_1^{M_2} & z_2^{M_2} & z_3^{M_2} & \cdots & z_L^{M_2} \\ \vdots & & & & \\ z_1^{M_2+P_2-1} & z_2^{M_2+P_2-1} & z_3^{M_2+P_2-1} & \cdots & z_L^{M_2+P_2-1} \end{pmatrix}, \quad (2.49)$$

$$\mathbf{\Lambda} = \text{diag}(a_1 \quad a_2 \quad a_3 \quad \cdots \quad a_L), \quad (2.50)$$

$$\mathbf{V} = \begin{pmatrix} 1 & 1 & 1 & \cdots & 1 \\ z_1^* & z_2^* & z_3^* & \cdots & z_L^* \\ \vdots & & & & \\ z_1^{*Q-1} & z_2^{*Q-1} & z_3^{*Q-1} & \cdots & z_L^{*Q-1} \end{pmatrix}. \quad (2.51)$$

Clearly, the matrix $\mathbf{V}$ has the same Vandermonde structure as in Equation (2.23), meaning that the shift-invariance property (2.24) holds in this case as well, that is, $\overline{\mathbf{V}} = \underline{\mathbf{V}} \cdot \mathbf{\Phi}$, where $\mathbf{\Phi}$ is the diagonal matrix having z_ls along the main diagonal. Therefore, one can use the algorithm described in Section 2.2.4 to estimate the signal poles z_ls from $\mathbf{V}$, or alternatively, from the right singular vectors of $\mathbf{Y_s}$. However, a similar approach can also be used to estimate the poles from the left singular vectors. This is the case of interest when the number of rows in the data matrix $\mathbf{Y_s}$ is larger than the number of columns, which may come about as a result of sampling multiple frequency bands that are relatively narrow compared with the signal bandwidth. Namely, the key is to observe the following property of the matrix $\mathbf{U}$

$$\overline{\mathbf{U}} = \underline{\mathbf{U}} \cdot \mathbf{\Phi}, \quad (2.52)$$

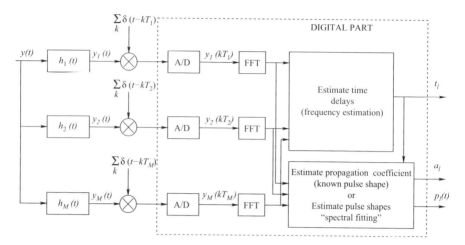

Figure 2.3 Estimation from multiple bands: receiver block diagram.

where $\overline{\overline{(\cdot)}}$ stands for the operation of omitting the rows 1 and $P_1 + 1$ of $(\cdot)$, and similarly, $\underline{\underline{(\cdot)}}$ denotes the operation of omitting the rows P_1 and $P_1 + P_2$ of $(\cdot)$. That is, the shift-invariance property can be exploited in this case as well, while the only modification in the developed algorithm is that the matrices $\overline{\overline{\mathbf{U}}}$ and $\underline{\underline{\mathbf{U}}}$ are constructed by removing the first and the last row respectively from each block of $\mathbf{U}$ (Figure 2.3).

When there is additive noise, we should first extract the principal components by computing the singular value decomposition of $\mathbf{Y}_s$ (2.25), and then estimate the signal poles $z_l = e^{-j\omega_0 t_l}$ as eigenvalues of a matrix $\mathbf{Z}$, defined as

$$\mathbf{Z} = \underline{\underline{\mathbf{U}}}_s^+ \cdot \overline{\overline{\mathbf{U}}}_s. \tag{2.53}$$

Alternatively, we could also define $\mathbf{Z}$ as

$$\mathbf{Z} = \underline{\underline{\mathbf{V}}}_s^+ \cdot \overline{\overline{\mathbf{V}}}_s, \tag{2.54}$$

in which case the eigenvalues of $\mathbf{Z}$ are complex conjugates of z_ls.

2.6 LOW-COMPLEXITY RAPID ACQUISITION IN UWB LOCALIZERS

One of the most interesting applications of pulse-based signaling scheme can be found in ultra wideband transceivers intended for precise position location. Such UWB transceivers, called localizers, have already been developed [11] and they use low duty-cycle periodic transmission of a coded sequence of impulses to ensure low-power operation and good performance in a multipath environment, as illustrated in Figure 2.4. Yet, rapid synchronization still presents the most

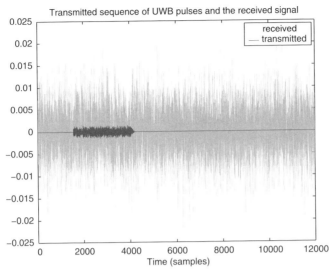

Figure 2.4 Signaling scheme in UWB localizers. A coded sequence of 127 UWB impulses (red) is periodically transmitted over multiple cycles, while the sequence duration spans approximately 20% of the cycle time T_c. Coding is achieved with a PN sequence of length 127, and the relative delay between the transmitted pulses is 20 samples. The received signal (blue) is dominated by noise. In this case, the received signal-to-noise ratio is SNR = −15 dB.

challenging part in the transceiver design. Current solutions are still analog and use a cascade of correlators to perform exhaustive search through all possible code positions [12], which is inherently time-consuming. A similar architecture, based on a "mostly digital" conception, is proposed in [9], where sampling is achieved using an A/D converter designed to run at 2 GHz. In addition to the high sampling rates, implementation of the cascade of correlators can take up to 30% of the circuit area and tends to consume a major amount of the total power. Thus developing alternative methods that would allow for faster acquisition and lower power consumption is still an open problem.

One possibility is to use the subspace approach presented in Section 2.2.4, by modeling the received noiseless signal $y(t)$ as a convolution of L delayed impulses $p(t)$ with a known coding sequence $g(t)$, that is,

$$y(t) = \sum_{l=1}^{L} a_l p(t - t_l) * g(t). \qquad (2.55)$$

As $y(t)$ is a periodic signal, its spectral coefficients are given by

$$Y[n] = \sum_{l=1}^{L} a_l P[n] G[n] e^{-jn\omega_c t_l}, \qquad (2.56)$$

where $\omega_c = 2\pi/T_c$, while T_c denotes a cycle time. Note that the spectral coefficients $G[n]$ corresponding to the coding sequence $g(t)$ are assumed to be known at the receiver, thus the total number of degrees of freedom per cycle is $2L$. Therefore, the signal parameters can be estimated using the method presented in Section 2.2.

At this point, we should mention that in a very general signaling scenario, such an approach can be numerically unstable. Specifically, in order to take advantage of the shift-invariance property, one has to consider the coefficients $Y_s[n] = Y[n]/P[n]G[n]$ (where $P[n]$ are the DFT coefficients of the transmitted pulse). This division is well-conditioned only in the case when both the pulse spectrum $P[n]$ and the spectrum of the coding sequence $G[n]$ has no zeros in the considered frequency band. While the first requirement can be satisfied by a proper design of the transmitted pulse [14], the additional modulation with a coding sequence can create spectral zeros. If one uses amplitude modulation with a pseudo-random (or pseudo-noise, PN), as in current systems [11, 12], the problem of dealing with spectral zeros in the PN sequence does not appear. However, in the case of time-hopping systems, this problem would arise. Thus, it is conceivable that in such a case it would be better to use the known spectrum as a sort of constraint in estimation, yet one will no longer be able to exploit the shift-invariance property directly.

In Figure 2.5, we show the delay estimation performance obtained with the subspace, assuming one dominant component, and this in the case when received pulses are distorted versions of a transmitted impulse. We assumed that a sequence of 127 coded impulses is periodically transmitted, where a transmitted impulse and a measured received waveform are illustrated in Figure 2.5(a). The propagation experiment was performed at the Berkeley Wireless Research Center [3]. The normalized power spectral density and the bands used for estimation are shown in Figure 2.5(b). We used the *Power method* to estimate the time delay of the dominant component, and this for three different values of the sampling rate: $N_s = 0.1N_n$, $N_s = 0.2N_n$ and $N_s = 0.3N_n$. The results are compared with those obtained using the matched filter approach (at Nyquist rate sampling N_n) [9], indicating that the presented method is more robust to waveform mismatch. For example, with the sampling rate of $N_s = 0.2N_n$, the timing performance is very similar to that of the matched filter, while for $N_s = 0.3N_n$, the subspace framework yields better performance.

2.6.1 Two-Step Estimation

Another improvement of the presented method in ranging/positioning applications is to use a "multiresolution" approach. That is, one can first obtain a rough estimate of the sequence timing, by taking uniform samples at a low rate over the entire cycle. Later, precise delay estimation can be carried out by increasing the sampling rate, yet sampling the received signal only within a narrow time window where the signal is present. The rationale for using the two-step approach is that in such systems a sequence duration T_s typically spans a small fraction of the cycle time T_c (e.g., less than 20%). As a result, all search-based methods [9, 12, 14], require a very long acquisition time and apparently "waste" power in sampling and processing time slots where the signal is not present.

2.6 LOW-COMPLEXITY RAPID ACQUISITION IN UWB LOCALIZERS

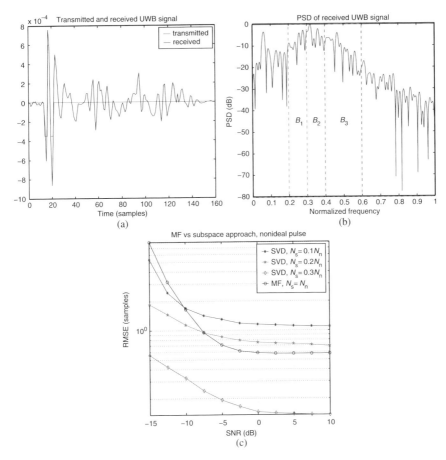

Figure 2.5 Timing recovery in non-ideal channels. (a) Received waveform (single pulse, including multipaths) and transmitted pulse. (b) Normalized power spectral density (PSD) of the received pulse and frequency bands used for estimation. (c) Timing estimation performances of the SVD-based method and the matched filter approach. The sampling rate for the SVD approach is $N_s = 0.1 N_n$ (the band B_1 is sampled), $N_s = 0.2 N_n$ (B_1 and B_2 are sampled) and $N_s = 0.3 N_n$ (B_2 and B_3 are sampled), while for the matched filter $N_s = N_n$.

A natural question arising from our discussion is how much one can reduce computational and power requirements using the two-step approach. In order to answer this question, consider the following scenario. Assume that the signal is first sampled at a low rate N_l over the entire cycle, and the Power method is used to achieve coarse synchronization. Assume next that the signal is sampled at a higher rate N_h (N_h is still below the Nyquist rate N_n) over a narrow time window of duration (roughly) T_s, and that M_d dominant signal components are estimated using the method of orthogonal iteration. Since we are mostly interested in the low SNR regime (SNR < 0 dB), a typical range for N_l is between $N_n/40$ and $N_n/20$, while N_h takes on values between $N_n/10$ and $N_n/2$.

TABLE 2.1 Comparison of Different Acquisition Algorithms: Computational Complexity, Power Consumption and the Number of Sampling Cycles

Method	Two-Step Approach		Subspace Method	Matched Filter
	Coarse Synchronization	Fine Synchronization		
Complexity	$\mathcal{O}[(N_l T_c)^2]$	$\mathcal{O}[M_d(N_h T_s)^2]$	$\mathcal{O}[M_d(N_h T_c)^2]$	$\mathcal{O}[(N_n T_c)^2]$
Power consumption	$\sim N_l T_c$	$\sim N_h T_s$	$\sim N_h T_c$	$\sim N_n T_c$
Number of cycles	$N_c = 2$		$N_c = 1$	$N_c \approx N_n T_c / K_{cr}$

In Table 2.1, we list the computational complexity, power consumption of A/D converters and the number of sampling cycles required to acquire the signal, for the following methods: the two-step algorithm, the subspace-based approach from Section 2.2.4, assuming uniform sampling at the rate N_h during the entire cycle, and the matched filter approach [9], with a cascade of K_{cr} correlators working at the Nyquist sampling rate N_n. Note that we have considered only the power consumption associated with A/D conversion, assuming a linear dependence on the sampling frequency [6], while a more precise analysis should also take into account the power consumption due to processing.

The benefits of the two-step approach are obvious: as the ratio T_c/T_s increases, the computational and power requirements can be reduced significantly. For example, when $N_l = N_n/40$, $N_h = N_n/4$, $M_d = 1$ and $T_c/T_s = 10$, the two-step approach reduces the complexity of the original subspace method approximately by a factor of 50, while power consumption is reduced by a factor of 5. Similarly, as N_h decreases, the advantages of the subspace method over the matched filter approach become more evident. Also note that due to the search-based nature of the matched filter method, it requires a much longer acquisition time compared with the other two approaches, where it suffices to sample at most two signal cycles. In practice, in the low-SNR regime, it is desirable to average the samples from multiple cycles in order to increase the effective SNR and, therefore, improve the numerical performance. While this does not have a major effect on the computational requirements, power consumption increases linearly with the number of averaging cycles. Thus, a good choice of the number of cycles depends on power constraints, a desirable estimation precision and acquisition time. Note that for the two-step approach, the overall performance improves by averaging the samples during the second phase only, when the fine synchronization takes place. During the first phase, it is useful to average the samples only if the processing gain is not sufficiently high to allow for coarse acquisition from a subsampled signal, while it does not affect the overall performance.

In Figure 2.6, we show the performance of the multiresolution or two-step delay estimation in the case of one dominant component (containing 70% of the signal energy), and the channel model given by Equation (2.1). The RMSE of the two-step approach for $N_l = 0.05\ N_n$ and $N_h = 0.5 N_n$ is shown in Figure 2.6(a). As the

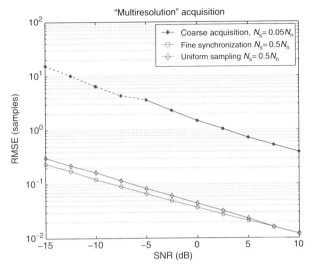

Figure 2.6 Two-step delay estimation. Coarse synchronization is obtained by sampling the received signal uniformly (over the entire cycle) at a low rate $N_l = N_n/20$. For low SNRs (less than -5 dB), the samples are averaged over multiple cycles (dashed line). Once a rough estimate of the sequence timing has been obtained, fine synchronization follows: the signal is sampled only within a narrow window, yet at a higher rate $N_h = N_n/2$. The RMSE of time delay estimation is compared with the RMSE obtained with high-rate uniform sampling over the entire cycle.

subsampling factor during the first phase is 20, for low values of SNR (that is, less than -5 dB), the samples are averaged over multiple cycles in order to increase the effective SNR. The error is compared with the RMSE obtained when the signal is sampled uniformly at a rate $N_h = 0.5\, N_n$ over the entire cycle. The results indicate that the two methods yield a very similar performance; however, in this case, the two-step approach reduces the computational requirements by a factor of 20, and the power consumption by a factor of 3.3.

2.7 CONCLUSIONS

In this chapter, we have discussed the problem of channel estimation and timing synchronization in digital UWB receivers. While there is a rich body of literature addressing this problem, most of which has appeared recently, this topic is far from being mature. In this context, developing novel signal processing techniques that could handle realistic channel models and this with relatively low complexity and in real time, still represents crucial task in meeting the challenges of UWB communications. To this end, there have been various approaches toward developing low-complexity solutions for digital UWB receivers. We first gave an outline of techniques presented in the literature and then discussed a subspace framework for channel estimation and timing in pulse-based UWB systems that yields estimates of unknown parameters from a subsampled version of the received UWB signal. Such an approach allows for fast algorithmic solutions, requires lower sampling rate and, therefore, lower complexity and power consumption compared to existing digital techniques. We specifically considered the application to UWB systems for precise position location; however, the algorithms we presented can also be used in other UWB applications as well as in other wideband systems, such as wideband CDMA.

REFERENCES

1. J. Balakrishnan, A. Batra, and A. Dabak, "A multi-band OFDM system for UWB communication," in *Proc. Conf. on Ultra-Wideband Systems and Technologies*, Reston, VA, 2003, pp. 354–358.
2. S. E. Bensley and B. Aazhang, "Subspace-based channel estimation for code division multiple access communication systems," *IEEE Transactions on Communications*, vol. 44, no. 8, pp. 1009–1020, August 1996.
3. http://bwrc.eecs.berkeley.edu
4. C. Carbonelli, U. Mengali, and U. Mitra, "Synchronization and channel estimation for UWB signals," in *Proc. of Global Telecommunications Conf.*, pp. 764–768, 2003.
5. Y. Chao and R. A. Scholtz, "Optimal and suboptimal receivers for ultra-wideband transmitted reference systems," in *Proc. Global Telecommunications Conf.*, San Francisco, CA, 2003, pp. 744–748.
6. T. Cho, D. Cline, C. Conroy, and P. Gray, "Design considerations for high-speed low-power low-voltage CMOS analog-to-digital converters," *Digest of Technical papers, Advanced Analog Integrated Circuit Symp.*, March 1994.
7. R. J. Cramer, R. A. Scholtz, and M. Z. Win, "Evaluation of an ultra-wideband propagation channel," *IEEE Transactions on Antennas and Propagation*, vol. 50, no. 5, pp. 561–570, May 2002.
8. J. W. Demmel, *Applied Numerical Linear Algebra*, SIAM, Philadelphia, PA, 1997.
9. I. O'Donnell, M. Chen, S. Wang, and R. Brodersen, "An integrated, low-power, ultra-wideband transceiver architecture for low-rate indoor wireless system," *IEEE CAS Workshop on Wireless Communications and Networking*, September 2002.
10. FCC First Report, and Order: In the Matter of Revision of Part 15 of the Commission's Rules Regarding Ultra-Wideband Transmission Systems, *FCC 02–48*, April 2002.
11. R. Fleming and C. Kushner, "Spread Spectrum Localizers," U.S. Patent 5,748,891, 5 May, 1998.
12. R. Fleming, C. Kushner, G. Roberts, and U. Nandiwada, "Rapid acquisition for ultra-wideband localizers," in *Proc. IEEE Conf. on UWB Systems and Technologies*, May 2002.
13. G. H. Golub and C. F. Van Loan, *Matrix Computations*, The Johns Hopkins University Press, Baltimore, 1989.
14. E. Homier and R. Scholtz, "Rapid acquisition of UWB signals in a dense multipath channel," in *Proc. IEEE Conf. on UWB Systems and Technologies*, May 2002.
15. Y. Hua and T. Sarkar, "Matrix pencil method for estimating parameters of exponentially damped/undamped sinusoids in noise," *IEEE Transactions on Acoustics, Speech and Signal Processing*, vol. 38, no. 5, pp. 814–824, May 1990.

16. Y. Hua and T. Sarkar, "On SVD for estimating generalized eigenvalues of singular matrix pencil in noise," *IEEE Transactions on Signal Processing*, vol. 39, no. 4, pp. 892–900, April 1991.
17. J. Y. Lee and R. Scholtz, "Ranging in a dense multipath environment using an UWB radio link," *IEEE Journal on Selected Areas in Communications*, vol. 20, no. 9, pp. 1677–1683, December 2002.
18. V. Lottici, A. D'Andrea, and U. Mengali, "Channel estimation for ultra-wideband communications," *IEEE Journal on Selected Areas in Communications*, vol. 20, no. 9, pp. 1638–1645, 2002.
19. I. Maravić and M. Vetterli, "Sampling and reconstruction methods for signals of finite rate of innovation in the presence of noise," *IEEE Transactions on Signal Processing*, vol. 53, no. 8, pp. 2788–2805, August 2005.
20. A. V. Oppenheim and R. W. Schafer, *Disrete-Time Signal Processing*, Prentice Hall, Englewood Cliffs, NJ, 1989.
21. R. Roy and T. Kailath, "ESPRIT estimation of signal parameters via rotational invariance techniques," *IEEE Transactions on Acoustics, Speech and Signal Processing*, vol. 37, no.7, pp. 984–995, July 1989.
22. A. Paulraj, B. Khalaj, and T. Kailath, "2-D RAKE receivers for CDMA cellular systems," in *Proc. IEEE GLOBECOM*, vol. 1, San Francisco, CA, December 1994, pp. 400–404.
23. B. D. Rao and K. S. Arun, "Model based processing of signals: a state space approach," *Proceedings of the IEEE*, vol. 80, no. 2, pp. 283–309, February 1992.
24. B. Rao, "Sensitivity analysis of state space methods in spectral estimation," in *Proc. IEEE ICASSP*, April 1987.
25. P. Stoica and R. Moses, *Introduction to Spectral Analysis*, Prentice Hall, Englewood Cliffs, NJ, 2000.
26. P. Stoica and A. Nehorai, "MUSIC, maximum likelihood and Cramer–Rao bound," *IEEE Transactions on Acoustics, Speech and Signal Processing*, vol. 37, no. 5, pp. 720–741, May 1989.
27. A. L. Swindlehurst, "Time delay and spatial signature estimation using known asynchronous signals," *IEEE Transactions on Signal Processing*, vol. 46, no. 2, pp. 449–462, February 1998.
28. R. G. Vaughan, N. L. Scott, and D. R. White, "Theory of bandpass sampling," *IEEE Transactions on Signal Processing*, vol. 39, no. 9, pp. 1973–1984, September 1991.
29. M. Vetterli, P. Marziliano, and T. Blu, "Sampling signals with finite rate of innovation," *IEEE Transactions on Signal Processing*, vol. 50, no. 6, pp. 1417–1428, June 2002.
30. Z. Tian and G. Giannakis, "Data-aided ML timing acquisition in ultra-wideband radios," in *Proc. Conf. on Ultra-Wideband Systems and Technologies*, pp. 142–146, Reston, VA, 2003.
31. L. Yang, Z. Tian, and G. Giannakis, "Non-data aided timing acquisition of ultra-wideband transmissions using cyclostationarity," in *Proc. of ICASSP*, April 2003.
32. L. Yang and G. Giannakis, "Blind UWB timing with a dirty template," in *Proc. of ICASSP*, Montreal, Canada, May 2004.
33. L. Yang and G. Giannakis, "Ultra-wideband communications: The idea whose time has come," *IEEE Signal Processing Magazine*, pp. 26–54, November 2004.
34. M. Z. Win and R. A. Scholtz, "Impulse radio: how it works", *IEEE Communications Letters*, vol. 2, pp. 36–38, February 1998.

35. M. Z. Win and R. A. Scholtz, "On the robustness of ultra-wide bandwidth signals in dense multipath environments", *IEEE Communications Letters*, vol. 2, pp. 51–53, February 1998.
36. M. Z. Win and R. A. Scholtz, "Characterization of ultra-wide bandwidth wireless indoor communication channel: a communication theoretic view," *IEEE J. Selected Areas in Communication*, vol. 20, pp. 1613–1627, December 2002.
37. H. Zhang and D. L. Goeckel, "Generalized transmitted-reference UWB systems," in *Proc. Conf. on Ultra-Wideband Systems and Technologies*, Reston, VA, 2003, pp. 147–151.

CHAPTER 3

Ultra Wideband Geolocation

SINAN GEZICI, ZAFER SAHINOGLU, HISASHI KOBAYASHI,
and H. VINCENT POOR

3.1 INTRODUCTION

A UWB signal is defined as one that possesses an absolute bandwidth larger than 500 MHz or a relative bandwidth larger than 20%. UWB systems offer many advantages for communications, such as high data rate transmission, robustness against small-scale fading, and low probability of interception [1–5]. Moreover, the inherent high time resolution of UWB signals facilitates very precise positioning, which is the subject of this chapter.

Since the US FCC approved the limited use of UWB technology [6], communication systems that employ UWB signals have drawn considerable attention. The initial standardization efforts of IEEE focused on high-data-rate applications of UWB for personal area networks (PANs) [7–9]. Currently, IEEE also focuses on another standard, IEEE 802.15.4a, for low-data-rate communications with precise ranging capabilities. UWB is the leading candidate for this standard, since it can provide high ranging accuracy and facilitates low-power and low-cost transceiver designs.

This decade will see a rise in sensor network applications and their widespread use. The diverse applications of sensor networks include home automation, industrial and environmental monitoring, asset management, security, surveillance, and many others. Since network densities and network sizes are expected to be quite high and the nodes operating on battery power are required to last for years, low power and low-cost technology is called for. Due to these requirements and its precision ranging capability, UWB impulse radio (IR) technology [1–4] is a strong candidate for emerging short-range and low-rate communication networks.

In this chapter, we discuss UWB geolocation. After introducing the signal model, we give an overview of conventional ranging and positioning techniques, and study their Cramer–Rao lower bounds (CRLBs) for range estimation. The following

---

Ultra Wideband Wireless Communication. Edited by Arslan, Chen, and Di Benedetto
Copyright © 2006 John Wiley & Sons, Inc.

sections focus on time of arrival (ToA) based range estimation techniques, because they benefit from the sharp time resolution of UWB signals. First, general causes of errors in ToA based range estimation techniques are discussed. Then, several specific ToA estimation schemes are presented: conventional correlation-based approaches, two-step estimation using low-rate sampling, simplified generalized maximum likelihood estimation and low-complexity timing offset estimation with dirty templates. We also present a two-way ranging protocol for accurate ranging in unsynchronized networks to deal with a clock offset between terminals. The chapter concludes with a brief look at emerging location aware applications and their market requirements.

3.2 SIGNAL MODEL

The received signal from an IR-UWB system over a multipath channel can be expressed as

$$r(t) = \sum_{l=1}^{L} \alpha_l s(t - \tau_l) + n(t), \qquad (3.1)$$

where L is the number of multipath components, α_l and τ_l are, respectively, the fading coefficient and the delay of the lth path, $n(t)$ is white Gaussian noise with zero mean and double-sided power spectral density $\mathcal{N}_0/2$, and $s(t)$ is given by

$$s(t) = \sqrt{E} \sum_{j=-\infty}^{\infty} d_j b_{\lfloor j/N_f \rfloor} w(t - jT_f - c_j T_c - a_{\lfloor j/N_f \rfloor}\Delta), \qquad (3.2)$$

with $w(t)$ denoting the received UWB pulse, E a constant that scales the transmitted pulse energy, T_f the "frame" time, and N_f the number of pulses representing one information symbol. For binary PAM, $b_{\lfloor j/N_f \rfloor} \in \{+1, -1\}$ and $a_{\lfloor j/N_f \rfloor} = 0 \,\forall j$, and for M-ary PPM, $b_{\lfloor j/N_f \rfloor} = 1 \,\forall j$ and $a_{\lfloor j/N_f \rfloor} \in \{0, 1, \ldots, M-1\}$ with Δ denoting the modulation index [5, 10]. In order to smooth the power spectrum of the transmitted signal and allow the channel to be shared by many nodes without causing catastrophic collisions, a TH sequence, $c_j \in \{0, 1, \ldots, N_c - 1\}$, is assigned to each node, where N_c is the number of chips in a frame, that is, $N_c = T_f/T_c$.

Additionally, random polarity codes, the d_js in Equation (3.2), can be employed, which are binary random variables taking ± 1 with equal probability, and are known to the receiver. Use of random polarity codes helps reduce the spectral lines in the power spectral density of the transmitted signal [11] and mitigate the effects of MAI [12].

An example PAM IR-UWB signal is shown in Figure 3.1, where six pulses are transmitted for each information symbol ($N_f = 6$) with the TH sequence $\{2, 1, 2, 3, 1, 0\}$.

3.3 POSITIONING TECHNIQUES

Conventional positioning techniques rely the on angle of arrival (AoA), received signal strength (RSS), ToA and the time difference of arrival (TDoA) measurements

Figure 3.1 A PAM TH-IR signal with pulse-based polarity randomization where $N_f = 6$, $N_c = 4$ and the TH sequence is $\{2, 1, 2, 3, 1, 0\}$. Assuming that $+1$ is currently being transmitted, the polarity codes for the pulses are $\{+1, +1, -1, +1, -1, +1\}$.

[13]. Each technique has its own merits and drawbacks under given cost and complexity constraints. Especially in sensor networks, low cost, low power and low complexity become important design considerations. In this section, we describe AoA, RSS, and time-based positioning techniques, present measurement models and their CRLBs, and discuss their feasibility for UWB applications.

3.3.1 Angle of Arrival

An AOA-based positioning technique involves measuring angles of the node seen by reference nodes. In order to determine the location of a node in a two-dimensional space, it is sufficient to measure the angles of the straight lines that connect the node and two reference nodes, as shown in Figure 3.2.

In order to provide a high resolution AoA measurement, both directional antennas and phased arrays can be applied to dither about the exact direction of the peak incident signal energy. Three types of array geometry have received considerable attention in this context: uniform linear array (ULA), rectangular lattice, and uniform circular array (UCA) (Figure 3.3). The ULA is simple to analyze, but it can provide only one-dimensional (1-D) information on wave arrivals and has a poor AoA estimation performance near end-fire. Therefore, a two-dimensional (2-D) array geometry is required to achieve 2-D signal arrival information. The study in [14] shows by comparison of the CRLB that an L-shaped array of sensors has 37% higher accuracy than a conventional cross array, which is

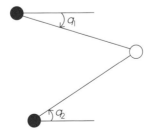

Figure 3.2 Positioning via AoA measurements. The black nodes are the reference nodes.

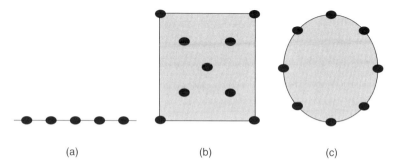

Figure 3.3 Illustration of antenna array geometries. (a) ULA, (b) rectangular lattice, and (c) UCA.

formed by two intersecting ULAs. It is important that the array geometry should result in uniform performance throughout the view of interest. It is noted in [15] that, because of the fact that the angular estimation accuracy and resolution capability decrease with decreasing array dimensions, the radius of the UCA should be selected sufficiently large. This puts practical limitations on the UCA. In general, the antenna element spacing must be on the order of half the wavelength of the carrier signal frequency to be able to model arrival time of the received signal at the antenna elements as a phase shift [16].

AoA Modeling A planar wave-front can be used to model the incoming signal in the far field, and the AoA can be determined by measuring the phase (time) difference of the wave-front at different antenna elements as shown in Figure 3.4. Assume that $s_n(t)$ is a single narrowband signal incident on a ULA antenna array with N equally spaced elements (with spacing ε) from angle α at time t and that the array output vector is $\mathbf{y}(t) = [y_1(t)\, y_2(t) \cdots y_N(t)]^T$, which can be expressed as

$$\mathbf{y}(t) = \mathbf{z}(\alpha)s_n(t) + \mathbf{n}(t), \tag{3.3}$$

where the so-called steering vector $\mathbf{z}(\alpha)$ is an $N \times 1$ vector, and describes the voltages induced on each array element by the incident signal when we can neglect the coupling between the array elements. Its ith element is given by

$$[z(\alpha)]_i = \exp\left[j\frac{w_0 \varepsilon}{c} \left(i - \frac{N-1}{2} \right) \cos \alpha \right], \tag{3.4}$$

for $i = 0, 1, \ldots, N-1$, where w_0 is the center frequency of $s_n(t)$ and c is the speed of light. The noise $\mathbf{n}(t)$ in Equation (3.3) is also an $N \times 1$ vector, whose components are assumed to be independent complex white Gaussian processes, with independent real and imaginary components each with variance σ^2 [14, 17]. In order to factor in the coupling between the array elements, a dimensionless

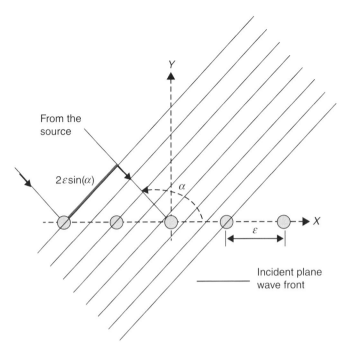

Figure 3.4 Illustration of a plane wave incident on a ULA antenna array. The antenna array elements are equidistantly placed along the x-axis. The red lines indicate the phase fronts of the incident wave. ε denotes the spacing between array elements and α is the arrival angle with respect to the x-axis.

symmetric impedance matrix $\mathbf{C}$ of size $N \times N$ can be defined. Then, the revised output vector becomes [17]

$$\mathbf{y}(t) = \mathbf{C}^{-1}\mathbf{z}(\alpha)s_n(t) + \mathbf{n}(t). \quad (3.5)$$

The Cramer–Rao Lower Bound To gain some insight into the derivation of the CRLB for the AoA measurements, imagine that we take a snapshot of the output vector by sampling it at $t = 0$, and neglect the coupling between array elements. Then, Equation (3.3) can be rewritten as $\mathbf{y} = \mathbf{z}(\alpha)s_n + \mathbf{n}$, where s_n now indicates a complex amplitude. In this model, there are three unknowns: the angle α, and the magnitude A and the phase ϕ of s_n. N antenna elements yield N observations, which can be treated as a signal plus complex white Gaussian noise, where the noise samples are assumed to be uncorrelated, hence independent:

$$y_i = Ae^{j\phi}e^{j[(w_0\varepsilon)/c]x(i)\cos\alpha} + n_i = z_i + n_i, \quad i = 0, 1, \ldots, N-1, \quad (3.6)$$

where $x(i)$ indicates the position of an array element on the x-axis and z_i is a complex signal. z_i and n_i can be decomposed into real and imaginary components

$z_i = z_i^R + jz_i^I$ and $n_i = n_i^R + jn_i^I$, where $z_i^R = A\cos[(w_0\varepsilon)/c\ x(i)\cos\alpha + \phi]$ and $z_i^I = A\sin[(w_0\varepsilon)/c\ x(i)\cos\alpha + \phi]$, respectively.

Then, the log-likelihood function can be written as [18]

$$\log p(\mathbf{y}|A,\phi,\alpha) = \ln\left(\frac{1}{(2\pi)^N \sigma^{2N}}\right) - \frac{1}{2\sigma^2}(\mathbf{y}-\mathbf{z})^H(\mathbf{y}-\mathbf{z}), \quad (3.7)$$

where the superscript H denotes the complex conjugate transpose, $\mathbf{y} = [y_0\ y_1\ \cdots\ y_{N-1}]^T$, and $\mathbf{z} = [z_0\ z_1\ \cdots\ z_{N-1}]^T$.

The computation of the CRLB requires the inverse of the Fisher information matrix (FIM) $\mathbf{J}$. The dimension of $\mathbf{J}$ is equal to the number of unknowns, and in the above example, it is 3×3:

$$\mathbf{J} = \begin{pmatrix} \frac{\partial^2 \log[p(\mathbf{y}|A,\phi,\alpha)]}{\partial\alpha\partial\alpha} & \frac{\partial^2 \log[p(\mathbf{y}|A,\phi,\alpha)]}{\partial\alpha\partial A} & \frac{\partial^2 \log[p(\mathbf{y}|A,\phi,\alpha)]}{\partial\alpha\partial\phi} \\ \frac{\partial^2 \log[p(\mathbf{y}|A,\phi,\alpha)]}{\partial A\partial\alpha} & \frac{\partial^2 \log[p(\mathbf{y}|A,\phi,\alpha)]}{\partial A\partial A} & \frac{\partial^2 \log[p(\mathbf{y}|A,\phi,\alpha)]}{\partial A\partial\phi} \\ \frac{\partial^2 \log[p(\mathbf{y}|A,\phi,\alpha)]}{\partial\phi\partial\alpha} & \frac{\partial^2 \log[p(\mathbf{y}|A,\phi,\alpha)]}{\partial\phi\partial A} & \frac{\partial^2 \log[p(\mathbf{y}|A,\phi,\alpha)]}{\partial\phi\partial\phi} \end{pmatrix}. \quad (3.8)$$

The CRLB for α can be expressed as $\text{Var}(\hat{\alpha}) \geq [\mathbf{J}^{-1}]_{1,1}$. The computation of this quantity is straightforward, and for a ULA with N elements can be expressed as [19, 20]

$$\text{CRLB}(\alpha) = \frac{1}{\frac{A^2}{\sigma^2}\left(\frac{w_0\varepsilon}{c}\right)^2 (\sin\alpha)^2 \left[\frac{N(N^2-1)}{12}\right]}. \quad (3.9)$$

As can be seen from this expression, the CRLB will decrease as the spacing between elements is increased or as more array elements are deployed. It also depends on the signal angle; for example, at higher obtuse angles the estimation accuracy is degraded.

AoA Approach for UWB Systems The AoA approach is not well suited to UWB positioning for several reasons. Because of the large bandwidth of a UWB signal, there is significant multipath time dispersion due to reflections from and diffraction around surrounding objects; hence the number of paths becomes very large, especially in indoor environments. An implementation of the maximum likelihood approach to estimate the AoA of each path requires a computationally expensive multidimensional search with the dimension determined by the number of signal paths [16, 17]. Therefore, accurate angle estimation becomes very challenging with the existence of scattering from objects in the environment. Furthermore, the use of antenna arrays makes the system costly, annulling the main advantage of a

UWB radio equipped with low-cost transceivers. As we will see later in this section, time-based approaches can provide very precise range estimates since the bandwidth is very large, and therefore they are better suited for UWB than are the costly AoA-based techniques.

3.3.2 Received Signal Strength

Relying on a path-loss model, the distance between two nodes can be estimated by measuring the energy of the received signal at one end. This distance-based technique requires at least three reference nodes to determine the two-dimensional location of a given node, using the well-known triangulation approach depicted in Figure 3.5 [13].

In order to determine the distance from RSS measurements, the characteristics of the channel must be known. The RSS is mainly determined by three propagation effects of the channel: inverse-power-law power decay with distance, slowly varying shadowing caused by obstructions, and multipath fading. Since the RSS measurements depend on the channel characteristics, RSS-based positioning algorithms are very sensitive to the estimation of channel parameters.

RSS Modeling The ambiguity in RSS measurements is caused by both small-scale and large-scale fading, propagation model parameters, antenna characteristics, and temperature and frequency dependency of radio components. In terrestrial settings, radio wave propagation and RSS are affected by reflections from large smooth surfaces and diffractions and scattering. Therefore, in order to have accurate range estimation from RSS measurements, the characteristics of the channel must be

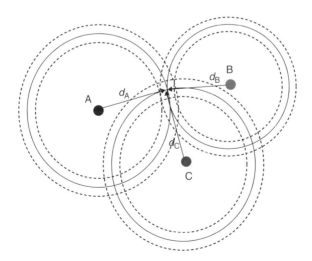

Figure 3.5 Illustration of positioning via the triangulation method based on three RSS observations from nodes A, B and C. The dashed circles indicate errors in RSS measurements, projected onto the range from each node.

known, and random small-scale attenuation should be mitigated and large-scale attenuation distilled. In wideband systems, mean received power can be calculated by summing the powers of the multipath in the power delay profile, whereas in narrowband systems averaging must be used for power estimation, because the receiver power experiences large fluctuations over a local area.

The measured received power R_{ij} at device i transmitted by device j is conventionally modeled as a log–normal variable, $R_{ij} \approx \mathcal{N}(P_r^i, \sigma_{sh}^2)$ [21], with $P_r^i = P_t^j - 10n_p \log_{10}(d_{ij})$, where P_r^i and P_t^j are, respectively, the decibel values of the mean received power at node i, and the transmitted power from node j, n_p is the propagation path loss exponent, d_{ij} is the distance between nodes i and j, and σ_{sh}^2 is the variance of the log–normal shadowing.

In UWB channels, frequency dependence of the path loss has been reported, and the frequency dependent and distance dependent losses can be modeled as being independent [22]. Hence, the received power at node i from node j can be expressed as

$$P_r^i = P_t^j - P_L^{ij}(d_{ij}) - P_L^{ij}(f), \qquad (3.10)$$

where $P_L^{ij}(d_{ij}) = 10n_p \log_{10}(d_{ij})$ denotes the decibel value of the distance dependent power loss and $P_L^{ij}(f)$ denotes that of the frequency dependent power loss. Some UWB channel measurement campaigns [23] find that the frequency dependent power loss is proportional to $a^2 f^{-2m}$, where the decay exponent m varies between 0.8 and 1.4, and a is an amplitude factor. The total loss can be computed by integrating the path loss over the entire frequency range. This frequency dependence will decrease the mean of the distribution of the R_{ij}s.

The Cramer–Rao Lower Bound The RSS-based positioning algorithm is very sensitive to the channel parameters (e.g., the path loss exponent and the variance of shadowing). This is clearly seen from the model in the previous section. Assume that the only parameter to estimate is $\theta = d_{ij}$, then the CRLB for the variance of an unbiased estimator $\hat{\theta}$ of θ can be found, using the log-likelihood function of the RSS observation, $\ln p(P_r^i | \theta) = -\ln(2\pi\sigma_{sh}^2) - [1/(2\sigma_{sh}^2)][P_r^i - 10n_p \log_{10}(d_{ij}) - P_L^{ij}(f)]^2$, to be

$$\operatorname{Var}(\hat{\theta}) \geq \frac{1}{-E\left\{\dfrac{\partial^2 \log[p(P_r^i | \theta)]}{\partial \theta^2}\right\}} = \left(\frac{d_{ij}\sigma_{sh}^2 \ln 10}{10n_p}\right)^2. \qquad (3.11)$$

From Equation (3.11), it is observed that the best achievable ranging limit depends on the channel parameters and the distance between the two nodes. For a more realistic bound, n_p and σ_{sh} can be assumed to be unknown nuisance parameters.

RSS Approach for UWB Systems It is clear from Equation (3.11) that the unique characteristic of a UWB signal, namely the enormous bandwidth, is not exploited to improve the best achievable accuracy in RSS-based schemes.

However, in some cases, the target node can be very close to some reference nodes, such as relay nodes in a sensor network, which can take RSS measurements only [24]. In such cases, RSS measurements can be used in conjunction with time-delay measurements of other reference nodes to improve the location estimation accuracy. The fundamental limits for such a hybrid positioning scheme are investigated in [24, 25].

3.3.3 Time-Based Approaches

Time-based positioning techniques rely on measurements of travel times of signals between nodes. If two nodes have a common clock, the node receiving the signal can estimate the ToA of the incoming signal that is time-stamped by the reference node.

Techniques for ToA estimation will be discussed in Section 3.5.2. After obtaining the estimates of ToA from a set of N reference nodes, the conventional ToA-based scheme estimates the position of the node using a least square (LS) approach [13]:

$$\hat{p} = \arg\min_{p} \sum_{i=1}^{N} w_i [\tau_i - d_i(p)/c]^2, \qquad (3.12)$$

where τ_i is the ith ToA measurement, $d_i(p) := \|p - p_i\|$ is the distance between the given node and the ith reference node, with p and p_i denoting their positions respectively, and w_i is a weight for the ith measurement that reflects the reliability of the ith ToA estimate.

The LS technique becomes optimal if the ToA measurements can be modeled as the summation of the true ToAs and independent Gaussian random variables with zero means and known variances. An asymptotically optimal ToA-based positioning algorithm will be mentioned in Section 3.5.

When there is no synchronization between a given node and the reference nodes, but there is synchronization among the reference nodes, the TDoA technique can be employed [13]. In this case, the TDoA of two signals traveling between the given node and two reference nodes is estimated, which determines the location of the node on a hyperbola with foci at the two reference nodes. Again a third reference node is needed for localization. In the absence of a common clock between the nodes, round-trip time between two transceiver nodes can be measured to estimate the distance between two nodes [26, 27].

The Cramer–Rao Lower Bound For a single-path additive white Gaussian noise (AWGN) channel,[1] it can be shown that the best achievable accuracy of a distance estimate $\hat{d}$ derived from ToA estimation satisfies the following inequality [28, 29]:

$$\sqrt{\text{Var}(\hat{d})} \geq \frac{c}{2\sqrt{2}\pi\sqrt{\text{SNR}}\beta}, \qquad (3.13)$$

---

[1] See [31] for the CRLB for ToA esimation in multipath channels.

where c is the speed of light, SNR is the signal-to-noise ratio and β is the effective (or RMS) signal bandwidth defined by

$$\beta \Delta \left[\int_{-\infty}^{\infty} f^2 |S(f)|^2 \, df \bigg/ \int_{-\infty}^{\infty} |S(f)|^2 \, df \right]^{1/2}, \qquad (3.14)$$

where $S(f)$ is the Fourier transform of the transmitted signal.

Note that unlike RSS-based techniques, the accuracy of a time-based approach can be improved by increasing the SNR or the effective signal bandwidth. Since UWB signals have very large bandwidths, this provides extremely accurate location estimates using time-based techniques.

Since the achievable accuracy under ideal conditions is very high, clock synchronization between the nodes becomes an important factor affecting ToA estimation accuracy. Hence, clock jitter must be considered in evaluating the accuracy of a UWB positioning system [30].

Time-Based Approaches for UWB Systems Time-based schemes provide very good accuracy due to the high time resolution (large bandwidth) of UWB signals. Moreover, they are less costly than the AoA-based schemes. Although it is easier to estimate RSS than ToA, the range information obtained from RSS measurements is very coarse compared with that obtained from the ToA measurements.

Due to the inherent suitability and accuracy of time-based approaches for UWB systems, we will focus our discussion on time-based UWB positioning in the rest of this chapter.

3.4 MAIN SOURCES OF ERROR IN TIME-BASED POSITIONING

Extremely accurate ToA and position estimation is possible in a single user, line-of-sight (LOS) and single-path environment. However, in a practical setting, multipath propagation, MAI and nonline-of-sight (NLOS) propagation make accurate positioning challenging. Moreover, due to the high resolution of UWB signals, the effects of clock inaccuracies, large number of bins (chips) to search, and limitations on sampling rates impose additional constraints on the positioning system.

3.4.1 Multipath Propagation

In conventional correlation-based ToA estimation algorithms, the time shift of a template signal that produces the maximum correlation with the received signal is used as the ToA estimate [32]. In other words, correlations of the received signal with shifted versions of a template signal are considered. In a single path channel, the transmitted waveform can be used as the optimal template signal, and conventional correlation-based estimation can be employed.[2] However, in the presence of an unknown multipath channel, the optimal template signal becomes the received

---

[2]In fact, even in a single-path environment, the received UWB pulse can have a different shape than the transmitted pulse due to the effects of the antennas.

waveform, which is the convolution of the transmitted waveform and the channel impulse response. Therefore, the correlation of the received signal with the transmit-waveform template is suboptimal in a multipath channel. If this suboptimal technique is employed in a narrowband system, the correlation peak[3] may not give the true ToA since multiple replicas of the transmitted signal partially overlap due to multipath propagation. In order to prevent this effect, super-resolution time delay estimation techniques, such as that described in [33], have been proposed. However, these techniques are more complex than the correlation-based algorithms. Fortunately, due to the large bandwidth of a UWB signal, multipath components are usually resolvable without the use of complex algorithms. However, the correlation peak will still not necessarily give the true ToA since the first multipath component is not always the strongest one. Therefore, first-path detection algorithms need to be considered, instead of simply choosing the delay of the signal path with the maximum correlation output (see Section 3.5.2).

3.4.2 Multiple Access Interference

In a multiuser environment, signals from other nodes can interfere with the signal of a given node and degrade the performance of ToA, and hence position, estimation algorithms. A technique for reducing the effects of MAI is to use different time slots for transmissions from different nodes. For example, in the IEEE 802.15.3 PAN standard [34], the transmissions from different nodes are time division multiplexed so that no two nodes in a given piconet transmit at the same time. However, even with such time multiplexing, there can still be MAI from neighboring piconets and MAI is still an issue.

In order to reduce the effects of MAI, TH codes with low cross-correlation properties can be employed [35], and pulse-based polarity randomization can be introduced [12, 36]. However, in order to be able to utilize coding at the timing stage, training codes should be predetermined; that is, they should be known to both the transmitter and the receiver. Otherwise, they would add additional uncertainty to ToA estimation. With known training patterns, template signals consisting of a number of pulses matched to both TH and polarity codes can be used to mitigate the effects of MAI.

In addition to TH and polarity codes, training sequences can be designed in order to facilitate ToA estimation in the presence of MAI [37].

3.4.3 Nonline-of-Sight Propagation

When the direct LOS between the target node and the reference node is blocked, only the reflections of the UWB signal from scatterers reach the receiving node. Therefore, the delay of the first arriving signal path does not represent the true ToA. Since the pulse travels an extra distance, a positive bias called the NLOS error is included in the measured time delay. In this case, using the conventional

---

[3]By selection of the correlation peak in a correlator receiver, we mean the selection of "delay parameter in the correlator" that gives the largest correlation output.

LS technique in Equation (3.12) would cause large errors in position estimation since the LS solution is optimal (maximum likelihood) only when each measurement error is a zero mean Gaussian random variable with known variance.

In the absence of any information about NLOS error, accurate positioning is not possible. In such a case, some nonparametric (pattern recognition) techniques, such as those described in [38, 39], can be employed. The main idea behind nonparametric positioning algorithms is to gather a set of ToA measurements from all the reference nodes at *known* locations beforehand and use this set as a reference to estimate the position when new measurements from a node are given.

In practical systems, it is usually possible to obtain some statistical information about the NLOS error. Wylie and Holtzman [40] observed that the variance of the ToA measurements in the NLOS case is usually much larger than that in the LOS case. They rely on this difference in the variance to identify NLOS situations and then use a simple LOS reconstruction algorithm to reduce the location estimation error. Also, by assuming a scattering model for the environment, the statistics of ToA measurements can be obtained, and then well-known techniques, such as maximum *a posteriori* probability (MAP) and ML, can be employed to mitigate the effects of NLOS errors [41, 42]. In the case of tracking a mobile user in a wireless system, biased and unbiased Kalman filters can be employed in order to estimate the location accurately [39, 43].

In addition to introducing a positive bias, NLOS propagation can also cause a situation where the first arriving signal path is not the strongest one. Therefore, the conventional ToA estimation method that chooses the strongest path would introduce another positive bias to the estimated ToA. Therefore, in UWB positioning systems, first-path detection algorithms are considered in order to mitigate the effects of the NLOS error, as we will discuss in Section 3.5.2.

3.4.4 High Time Resolution of UWB Signals

As we have noted above, the extremely large bandwidth of UWB signals results in very high time resolution, which enables very accurate ToA estimation. However, it also poses some challenges in practical systems.

First, clock jitter becomes an important factor in evaluating the accuracy of UWB positioning systems [30]. Since UWB pulses have very short (subnanosecond) duration, clock accuracies and drifts in the target and the reference nodes affect the ToA estimates.

Another consequence of high time resolution inherent in UWB signals is that the uncertainty region for ToA; that is, the set of delay positions that includes ToA, is usually very large compared with the chip duration. In other words, there is a large number of chips that need to be searched for ToA. This makes conventional correlation-based serial search approaches impractical, and calls for fast ToA estimation schemes (see Section 3.5.2).

Finally, high time resolution, or equivalently large bandwidth, of UWB signals makes it very impractical to sample the received signal at or above the Nyquist rate, which is typically on the order of tens of GHz. Therefore, ToA estimation schemes should make use of frame-rate or symbol-rate samples, which facilitate low-power designs.

3.5 RANGING AND POSITIONING

In this section, we first consider the relationship between ToA estimation (equivalently, ranging) and asymptotically optimal positioning algorithms. We present an asymptotically optimal scheme that obtains the ToA estimates by correlation techniques as its first step. Therefore, we can consider ToA estimation, or equivalently ranging, algorithms separately from the positioning algorithms without loss of optimality. In other words, ToA statistics based on correlator outputs provide sufficient statistics for final geolocation. Then, we consider some of the algorithms for ToA estimation in UWB systems. Finally, we describe a two-way ranging protocol that enables ToA estimation in the absence of synchronization between the nodes.

3.5.1 Relationship Between Ranging and Optimal Positioning Algorithms

The ranging problem is to estimate the distance between a given node and a reference node, as shown in Figure 3.6. Alternatively, it can be considered as a ToA estimation problem for the signal traveling between the two nodes. Positioning, however, refers to estimating the position of a node in a network (Figure 3.5). At least three reference nodes are required in order to determine the position of the target node in a two-dimensional space.

Given all the signals traveling between the target node and a set of reference nodes, the optimal positioning scheme is not necessarily the one that estimates the ToA's first and then uses them to estimate the position of the target node. However, under certain conditions, such a scheme can be shown to be asymptotically optimal [44].

In order to examine this issue, we first need to obtain the theoretical limiting accuracy for the positioning problem. Consider a synchronous UWB system with a target node and N reference nodes.[4] Let M of those reference nodes have NLOS to the target node, while the remaining nodes have LOS. The identities on NLOS or LOS reference nodes are assumed to be known; this information can be obtained by NLOS identification techniques [45–47]. Without this information, all first arrivals can be considered to be NLOS signals.

In the positioning problem, the aim is to estimate the position of the target node, given the signals traveling between the target node and the N reference nodes. The received signal related to the ith reference node can be expressed as

$$r_i(t) = \sum_{j=1}^{L_i} \alpha_{ij} s(t - \tau_{ij}) + n_i(t), \qquad (3.15)$$

---

[4]Note that practical UWB systems are not synchronous. However, two-way ranging protocols [26] are usually employed in order to compensate for the timing offset.

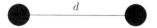

Figure 3.6 Ranging problem, where the aim is to estimate the distance between the two nodes.

for $i = 1, \ldots, N$, where L_i is the number of multipath components at the ith node, α_{ij} and τ_{ij} are, respectively, the fading coefficient and the delay of the jth path of the ith node, $s(t)$ is the UWB signal, and $n_i(t)$ is a zero mean AWGN process with spectral density $\mathcal{N}_0/2$. The delay of the jth path component at node i can be expressed, for two-dimensional positioning, as

$$\tau_{ij} = \frac{1}{c}\left[\sqrt{(x_i - x)^2 + (y_i - y)^2} + l_{ij}\right], \qquad (3.16)$$

for $i = 1, \ldots, N, j = 1, \ldots, L_i$, where $c = 3 \times 10^8$ m/s is the speed of light, $[x_i\, y_i]$ is the location of the ith node, l_{ij} is the NLOS propagation induced pathlength, and $[x\, y]$ is the location of the target node.

We assume, without loss of generality, that the first M nodes ($i = 1, \ldots, M$) have NLOS, and the remaining $N - M$ have LOS. Then, $l_{i1} = 0$ for $i = M + 1, \ldots, N$ since the signal directly reaches the related node in an LOS situation. Hence, the parameters to be estimated are the NLOS delays and the location of the node, $[x\, y]$, which can be expressed as $\boldsymbol{\theta} = [x\, y\, \boldsymbol{l}_{M+1} \cdots \boldsymbol{l}_N\, \boldsymbol{l}_1 \cdots \boldsymbol{l}_M]$, where

$$\boldsymbol{l}_i = \begin{cases} (l_{i1}\, l_{i2} \cdots l_{iL_i}) & \text{for } i = 1, \ldots, M, \\ (l_{i2}\, l_{i3} \cdots l_{iL_i}) & \text{for } i = M + 1, \ldots, N, \end{cases} \qquad (3.17)$$

with $0 < l_{i1} < l_{i2} \cdots < l_{iN_i}$ [44]. Note that for LOS signals the first delay is excluded from the parameter set, since these are known to be zero.

From Equation (3.15), the joint probability density function (p.d.f.) of the received signals from the N reference nodes, $\{r_i(t)\}_{i=1}^N$, can be expressed, conditioned on $\boldsymbol{\theta}$, as follows:

$$p_{\boldsymbol{\theta}}(\boldsymbol{r}) \propto \prod_{i=1}^N \exp\left\{-\frac{1}{\mathcal{N}_0}\int\left|r_i(t) - \sum_{j=1}^{L_i}\alpha_{ij}\, s(t - \tau_{ij})\right|^2 dt\right\}. \qquad (3.18)$$

From the expression in Equation (3.18), the lower bound on the variance of any unbiased estimator for the unknown parameter $\boldsymbol{\theta}$ can be obtained; that is, $E_{\boldsymbol{\theta}}\{(\hat{\boldsymbol{\theta}} - \boldsymbol{\theta})(\hat{\boldsymbol{\theta}} - \boldsymbol{\theta})^{\text{T}}\} \geq \mathbf{J}_{\boldsymbol{\theta}}^{-1}$, where $\mathbf{J}_{\boldsymbol{\theta}}$ is the FIM. It can be shown that the inverse of the FIM does not depend on the signals from the nodes that have NLOS to the target node [44]. In other words, the best accuracy can be achieved by using the signals only from the nodes with LOS. Moreover, the numerical examples in [44] show that, in most cases, the CRLB is almost the same whether all the multipath components from the LOS nodes, or just the first arriving paths of the LOS nodes are employed. Therefore, processing of the multipath components other than the first path does not increase the accuracy but increases the computational load.

Furthermore, the ML estimate of the node position based on the delays of the first incoming paths from LOS nodes achieves the CRLB as the SNR and/or the effective bandwidth increase to infinity [44]. This result implies that for UWB systems, the first arriving signal paths from the LOS nodes are sufficient for an approximately optimal positioning receiver design.

The asymptotically optimal receiver, shown in Figure 3.7, can be implemented by the following steps:

- Estimate the delays of the first multipath components by correlation techniques. In other words, for each reference node, choose the delay corresponding to the maximum correlation between the received signal and a receive-waveform template [44].
- Obtain the ML estimate for the position of the target node using the delays of the first multipath components of the LOS nodes.

In other words, the first step of the optimal receiver, the estimation of the first signal path, can be considered separately from the overall positioning algorithm without any loss in optimality.

Note that, in the previous scenario, no information on the statistics of the NLOS delays is assumed. When the p.d.f. of NLOS delays is available, it is shown in [44]

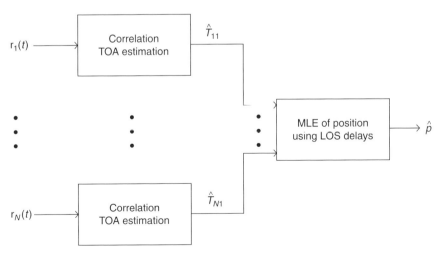

Figure 3.7 An asymptotically optimum receiver structure for positioning. No information about the statistics of the NLOS delays is assumed.

that the MAP estimate of the node position using the estimates of the delays of all the multipath components from all the nodes achieves asymptotic optimality. However, in practice, the distribution of the NLOS delays is usually not available. Also estimation of more multipath delays increases the computational complexity of the positioning algorithm. Therefore, only the ToA of the first signal path will be considered in the rest of this chapter.

3.5.2 ToA Estimation Algorithms

As considered in the previous subsection, the first step of the asymptotically optimal position estimator performs correlation-based ToA estimation. However, in a multipath environment, the correlation output needs to be maximized over a very large dimensional space due to a large number of multipath components, and hence unknown parameters, in a typical UWB system. Hence, the complexity of the ToA estimator in the optimal receiver is very high. Therefore, more practical ToA estimation algorithms have recently been proposed to estimate the arrival time of the first signal path. In this subsection, we will discuss several of these algorithms, which have low computational cost compared with the correlation-based estimation algorithm where the correlation between the received signal and a receive-waveform template is maximized.

Conventional Correlation-Based Approaches An optimal estimate of ToA can be obtained using a correlation receiver with the received waveform as the template signal (or, equivalently a matched filter matched to the received waveform), as shown in Figure 3.8, and choosing the time shift of the template signal that produces the maximum correlation with the received signal [32]. However, due to the multipath channel, the received waveform has many unknown parameters to be estimated. Hence, the optimal correlation-based ToA estimation, considered in Section 3.5, is impractical. Therefore, the transmitted waveform can be used in a conventional correlation-based receiver as the template signal. However, this is obviously

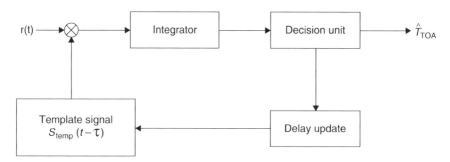

Figure 3.8 Correlation-based ToA estimation receiver.

suboptimal in a multipath environment. Also the peak picking operation does not necessarily give the true ToA in this case, since the first multipath component can be weaker than the others in some cases. Therefore, first path detection algorithms need to be considered with the suboptimal correlation-based schemes [48].

Moreover, due to the high time resolution of UWB signals, exhaustive search among thousands of bins (chips) is required for conventional correlation-based techniques, which results in very slow ToA estimation [49]. In order to speed up the process, different search strategies, such as random search or bit reversal search, can be employed [50].

Two-step ToA Estimation Using Low-Rate Samples One of the most challenging issues in UWB ToA estimation is to obtain a reliable estimate in a reasonable time interval under the sampling rate constraint. In order to have a low-power and low-complexity receiver, one should assume symbol-rate (or, sometimes frame-rate) sampling at the output of the correlators. However, when symbol-rate samples are employed, the ToA estimation can take a very long time. To address this problem, a two-step ToA estimation algorithm that can perform ToA estimation from symbol-rate samples in a reasonable time interval is proposed in [51]. In order to speed up the estimation process, the first step provides a rough estimate of ToA of the received signal based on RSS. Then, in the second step, the arrival time of the first signal path is estimated by employing a change detection approach [52].

Assume that the channel is bandlimited to $1/T_c$, and rewrite the received signal $r(t)$ of Equation (3.1) as follows:

$$r(t) = \sum_{l=1}^{\hat{L}} \hat{\alpha}_l s[t - (l-1)T_c - \tau_{\text{ToA}}] + n(t), \quad (3.19)$$

considering a tapped delay line version of the channel, where $\hat{\alpha}_l$ is the channel coefficient for the lth path, $\hat{L}$ is the number of multipath components, and τ_{ToA} is the ToA of the first-arriving incoming signal. Assume a training sequence for ToA estimation, and take $a_j = 0$ and $b_j = 1 \forall j$ in Equation (3.2); that is,

$$s(t) = \sqrt{E} \sum_{j=-\infty}^{\infty} d_j w(t - jT_f - c_j T_c), \quad (3.20)$$

where the UWB pulse $w(t)$ is assumed to have unit energy and duration T_c.

Also assume that the signal always arrives within one frame duration; that is, $\tau_{\text{ToA}} < T_f$, and there is no inter-frame interference (IFI); that is, $T_f \geq (\hat{L} + c_{\max})T_c$ (equivalently, $N_c \geq \hat{L} + c_{\max}$), where $c_{\max}$ is the maximum value of the TH sequence. As stated in [51], by means of predetermined TH codes, the algorithm can be extended to the case of $\tau_{\text{ToA}} > T_f$ as well; or a low-complexity algorithm, such as the dirty template approach [37], can be employed beforehand to reduce the uncertainty region to a frame interval.

Express the ToA as:[5]

$$\tau_{\text{ToA}} = kT_c = k_b T_b + k_c T_c, \qquad (3.21)$$

where $k \in [0, N_c - 1]$ is the ToA in terms of the chip interval T_c, T_b is the block interval consisting of B chips ($T_b = BT_c$), and $k_b \in [0, N_c/B - 1]$ and $k_c \in [0, B - 1]$ are the integers that determine, respectively, in which block and chip the first signal path arrives.

The two-step ToA algorithm first estimates the block in which the first signal path exists; then, it estimates the chip position in which the first path resides. In other words, this algorithm can be summarized as:

- Estimate k_b from RSS measurements.
- Estimate k_c (equivalently, k) from low-rate correlation outputs using a change detection approach.

First Step: Coarse ToA Estimation from RSS Measurements In the first step, the aim is to detect the *coarse* arrival time of the signal in the frame interval. Assume, without loss of generality, that the frame time T_f is an integer multiple of T_b; that is, $T_f = N_h T_b$.

In order to have reliable decision variables in this step, energy from N_1 different frames of the incoming signal is combined for each block. Hence, the decision variables are expressed as

$$Y_i = \sum_{j=0}^{N_1 - 1} Y_{i,j}, \qquad (3.22)$$

for $i = 0, \ldots, N_b - 1$, where

$$Y_{i,j} = \int_{jT_f + iT_b + c_j T_c}^{jT_f + (i+1)T_b + c_j T_c} |r(t)|^2 \, dt. \qquad (3.23)$$

Then, k_b in Equation (3.21) is estimated as

$$\hat{k}_b = \arg \max_{0 \le i \le N_b - 1} Y_i. \qquad (3.24)$$

In other words, the block with the largest signal energy is selected.

Second Step: Fine ToA Estimation from Low-Rate Correlation Outputs After determining the coarse estimation time from the first step, the second step tries to

---

[5]For simplicity, the ToA is assumed to be an integer multiple of the chip duration T_c. In a practical scenario, subchip synchronization can be obtained by employing a delay-lock-loop (DLL) after ToA estimation with chip-level uncertainty [30].

estimate k_c in Equation (3.21). Ideally, $k_c \in [0, B-1]$ needs to be searched for ToA estimation, which corresponds to searching $k \in [\hat{k}_b B, (\hat{k}_b + 1)B - 1]$, with $\hat{k}_b$ obtained from Equation (3.24). However, in some cases, the first signal path can reside in one of the blocks prior to the strongest one due to multipath effects. Therefore, instead of searching a single block, $k \in [\hat{k}_b B - M_1, (\hat{k}_b + 1)B - 1]$, with $M_1 \geq 0$, can be searched for the ToA in order to increase the probability of the detection of the first path. In other words, in addition to the block with the largest signal energy, an additional backwards search over M_1 chips can be performed. For notational simplicity, let $\mathcal{U} = \{n_s, n_s + 1, \ldots, n_e\}$ denote the uncertainty region for ToA, where $n_s = \hat{k}_b B - M_1$ and $n_e = (\hat{k}_b + 1)B - 1$ are the start and end points.

In order to estimate the ToA with chip-level resolution, we consider correlations of the received signal with shifted versions of a template signal. For delay iT_c, we obtain the following correlation output:

$$z_i = \int_{iT_c}^{iT_c + N_2 T_f} r(t) s_{\text{temp}}(t - iT_c) \, dt, \tag{3.25}$$

where N_2 is the number of frames over which the correlation output is obtained, and $s_{\text{temp}}(t)$ is the template signal given by

$$s_{\text{temp}}(t) = \sum_{j=0}^{N_2 - 1} d_j w(t - jT_f - c_j T_c). \tag{3.26}$$

From the correlation outputs for different delays, the aim is to determine the chip position, in which the first signal path has arrived. By appropriate choice of the block interval T_b and the number of chips M_1 for backwards search, and considering the large number of multipath components in a typical UWB environment, we can assume that the block starts with a number of chips with noise-only components and the remaining ones with signal plus noise components, as shown in Figure 3.9. Assuming that the statistics of the signal paths do not change significantly in the uncertainty region, we can express the different hypotheses approximately as follows [51]:

$$\begin{aligned} \mathcal{H}_0: & \quad z_i = \eta_i, & i = n_s, \ldots, n_e, \\ \mathcal{H}_k: & \quad z_i = \eta_i, & i = n_s, \ldots, k-1, \\ & \quad z_i = N_2 \sqrt{E} \hat{\alpha}_{i-k+1} + \eta_i, & i = k, \ldots, n_e, \end{aligned} \tag{3.27}$$

for $k \in \mathcal{U}$, where η_is denote the independent and identically distributed (i.i.d.) output noise distributed as $\mathcal{N}(0, N_2 \mathcal{N}_0/2)$, and $\hat{\alpha}_1, \ldots, \hat{\alpha}_{n_e - k + 1}$ are i.i.d. channel coefficients, assuming $n_e - n_s + 1 \leq \hat{L}$.

From the formulation in Equation (3.27), it is observed that the ToA estimation problem can be considered as a change detection problem [52]. Let $\boldsymbol{\theta}$ denote the unknown parameters of the distribution of $\hat{\alpha}$. Then, the log-likelihood ratio (LLR)

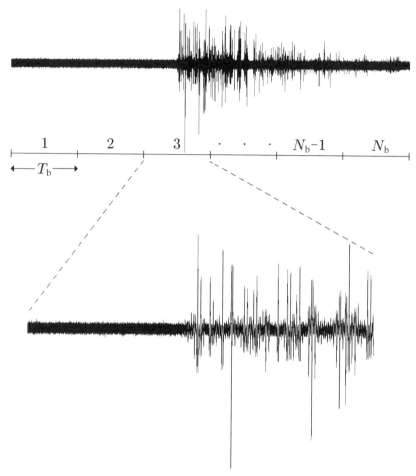

Figure 3.9 Illustration of the two-step ToA estimation algorithm. The signal on the top is the received signal in one frame. The first step checks the signal energy in N_b blocks and chooses the one with the highest energy (although one frame is shown in the figure, energy from different frames can be collected for reliable decisions). Assuming that the third block has the highest energy, the second step focuses on this block to estimate the ToA. The zoomed version of the signal in the third block is shown on the bottom.

can be calculated as

$$S_k^{n_e}(\boldsymbol{\theta}) = \sum_{i=k}^{n_e} \log \frac{p_{\boldsymbol{\theta}}(z_i|\mathcal{H}_k)}{p(z_i|\mathcal{H}_0)}, \qquad (3.28)$$

where $p_{\boldsymbol{\theta}}(z_i|\mathcal{H}_k)$ denotes the p.d.f. of the correlation output under hypothesis $\mathcal{H}_k$ and with unknown parameters given by $\boldsymbol{\theta}$, and $p(z_i|\mathcal{H}_0)$ denotes the p.d.f. of the correlation output under hypothesis $\mathcal{H}_0$.

Since $\boldsymbol{\theta}$ is unknown, its ML estimate can be obtained for a given hypothesis $\mathcal{H}_k$ and then that estimate can be used in the LLR expression. In other words, the generalized LLR approach [52] can be taken, where the ToA estimate is expressed as

$$\hat{k} = \arg\max_{k \in \mathcal{U}} S_k^{n_c}[\hat{\boldsymbol{\theta}}_{\mathrm{ML}}(k)], \qquad (3.29)$$

where

$$\hat{\boldsymbol{\theta}}_{\mathrm{ML}}(k) = \arg\sup_{\boldsymbol{\theta}} S_k^{n_c}(\boldsymbol{\theta}). \qquad (3.30)$$

For realistic UWB channel models [22], the ML estimate is very complicated to obtain. Hence, simpler estimators such as the method of moments (MM) estimator can be employed to obtain those parameters [51]. Then, the index of the chip having the first signal path can be obtained as

$$\hat{k} = \arg\max_{k \in \mathcal{U}} S_k^{n_c}[\hat{\boldsymbol{\theta}}_{\mathrm{MM}}(k)], \qquad (3.31)$$

where $\hat{\boldsymbol{\theta}}_{\mathrm{MM}}(k)$ denotes the MM estimate of $\boldsymbol{\theta}$ under hypothesis $\mathcal{H}_k$.

Note that the estimators in Equations (3.29) and (3.31) always choose one of the delays in the uncertainty region as the ToA, which means that the ToA is assumed to be in the uncertainty region $\mathcal{U}$. In order to prevent erroneous ToA estimation when the ToA is not an element of $\mathcal{U}$, additional tests can be performed [51].

The advantage of the change detection approach with MM estimation is that it can provide precise ToA estimation with reasonable complexity. However, in order for the estimators in Equations (3.29) and (3.31) to work efficiently, there should be sufficiently many multipath components, and their statistics should be almost the same in the uncertainty region determined by the second step.

Simplified Generalized Maximum Likelihood Scheme for First-Path Detection The approach in the previous subsection considers the ToA estimation as a change detection problem and estimates the delay of the first signal path using an approximate generalized LLR test. Another approach for the first path detection is to use the generalized maximum likelihood (GML) estimation principle and to obtain iterative solutions after some simplifications [26, 53].

Assume that a single UWB pulse is transmitted over a multipath channel. Then, the received signal in Equation (3.1) can be expressed as the sum of the first signal path, other multipath components and noise:

$$r(t) = \alpha_1 w(t - \tau_1) + \sum_{l=2}^{L} \alpha_l w(t - \tau_l) + n(t), \qquad (3.32)$$

where $\tau_1 < \tau_2 \cdots < \tau_L$, with the number of multipaths, L, being unknown. Note that, in practice, a number of UWB pulses is employed, which are combined to have sufficient SNR before the ToA estimation algorithm is applied. Here we consider a single UWB pulse without loss of generality by assuming that the signal in Equation (3.32) is obtained by combining a number of received pulses.

Assume that the delay and channel coefficient for the strongest multipath component, τ_{peak} and α_{peak}, are determined beforehand by means of a correlation technique. For example, they can be estimated by exhaustively searching possible delay positions, and choosing the one with maximum correlation output. Since the exhaustive or serial search takes a very long time for a UWB system, a two-step algorithm, as in the previous section or as in [54], can be employed first to obtain a rough estimate of the signal delays and then to seek the strongest component in the block. Using τ_{peak} and α_{peak}, we first obtain the following normalized signal [26]:

$$\tilde{r}(t) = \frac{r(t + \tau_{\text{peak}})}{|\alpha_{\text{peak}}|}$$
$$= \tilde{\alpha}_1 w(t + \tilde{\tau}_1) + \sum_{\tilde{\tau}_l \geq 0} \tilde{\alpha}_l w(t + \tilde{\tau}_l) + \sum_{\tilde{\tau}_l < 0} \tilde{\alpha}_l w(t + \tilde{\tau}_l) + \tilde{n}(t), \quad (3.33)$$

where $\tilde{\tau}_l = \tau_{\text{peak}} - \tau_l$, and $\tilde{\alpha}_l = \alpha_l/|\alpha_{\text{peak}}|$ for $l = 1, 2, \ldots, L$. Note that in Equation (3.33), the second term represents the multipath components before the strongest path, and the third term represents the paths after that.

Then, consider the signal components including and prior to the strongest path:

$$\tilde{r}(t) = \bar{r}(t), \quad t \leq T_p/2 \quad (3.34)$$

$$= \tilde{\alpha}_1 w(t + \tilde{\tau}_1) + \sum_{l=2}^{M} \tilde{\alpha}_l w(t + \tilde{\tau}_l) + \tilde{n}(t), \quad (3.35)$$

where T_p is the width of the UWB pulse $w(t)$, $M - 1$ is the number of multipath components before the the strongest component, and $\tilde{n}(t)$ is white Gaussian noise truncated to the time interval $(-\infty, T_p/2]$ [26].

After wide-band filtering and sampling, $\tilde{r}(t)$ can be expressed as a vector

$$\tilde{r} = \tilde{\alpha}_1 w_{\tilde{\tau}_1} + \sum_{l=2}^{M} \tilde{\alpha}_l w_{\tilde{\tau}_l} + \tilde{n}, \quad (3.36)$$

where $w_{\tilde{\tau}_l}$ consists of the samples from $w(t + \tilde{\tau}_l)$.

Since $\tilde{n}$ is a white Gaussian vector, the GML estimate for $\tilde{\tau}_1$ is given by the following [26]:

$$\hat{\tilde{\tau}}_1 = \arg\max_{\tilde{\tau}_1} \left[\min_{\tilde{\alpha}_1, M, \tilde{\alpha}, \tilde{\tau}} \left\| \tilde{r} - \tilde{\alpha}_1 w_{\tilde{\tau}_1} - \sum_{l=2}^{M} \tilde{\alpha}_l w_{\tilde{\tau}_l} \right\|^2 \right], \quad (3.37)$$

where $\tilde{\alpha} = [\tilde{\alpha}_2 \cdots \tilde{\alpha}_M]$ and $\tilde{\tau} = [\tilde{\tau}_2 \cdots \tilde{\tau}_M]$. Note that computational complexity of ToA estimation by Equation (3.37) is very high. Therefore, the following modifications are employed to obtain a simpler scheme, where $\Delta_{\tilde{\tau}}$ and $\Delta_{\tilde{\alpha}}$ are the thresholds of the algorithm [26]:

- Search the ToA over the portion of $r(t)$ satisfying $t \geq -\Delta_{\tilde{\tau}}$ so that the false alarm probability in the noise only portion of the signal is restricted;
- Stop the search when $|\tilde{\alpha}_1| < \Delta_{\tilde{\alpha}}$; and
- Ignore the multipath components that arrive after already detected paths.

Then, the following ToA estimation algorithm is obtained [26]:

(1) Set $n = 1$, $\delta_1 = 0$, and $\mu_{11} = 1$.
(2) Increase n by 1.
(3) Find δ_n that satisfies

$$\delta_n = \arg \max_{\delta_{n-1} < \delta < \Delta_{\tilde{\tau}}} \left(\tilde{r} - \sum_{i=1}^{n-1} \mu_{(n-1)i} w_{\delta_i} \right)^T w_\delta. \quad (3.38)$$

(4) Find $[\mu_{n1} \cdots \mu_{nn}]$ such that

$$[\mu_{n1} \cdots \mu_{nn}] = \arg \min_{\mu'_1, \ldots, \mu'_n} \left\| \tilde{r} - \sum_{i=1}^{n} \mu'_i w_{\delta_i} \right\|^2. \quad (3.39)$$

(5) If $\mu_{nn} \geq \Delta_{\tilde{\alpha}}$, go to step 2. Otherwise, proceed to the next step.
(6) $\tilde{\tau}_1$ is estimated as $\widehat{\tilde{\tau}_1} = \delta_{n-1}$.

The thresholds of the algorithm, $\Delta_{\tilde{\tau}}$ and $\Delta_{\tilde{\alpha}}$ are important parameters that determine the performance of the estimator. Therefore, those critical parameters can be selected based on some statistical information obtained from an experiment in the same environment [26].

The advantage of the GML-based algorithm is that it is a recursive algorithm and very accurate ToA estimation can be performed as reported in [26]. However, the main drawback is that it requires very high rate sampling, which is not practical in many applications.

Low-Complexity Timing Offset Estimation with Dirty Templates

An alternative to the GML-based approach is a low complexity timing offset estimation technique based on *symbol-rate* samples based on the novel idea of "dirty templates" [37, 55–57].

Due to the unknown multipath channel, the optimal template signal for correlation, which is the received waveform, is not available at the timing stage. Therefore, symbol-length portions of the received signal can be employed as noisy ("dirty") templates, the cross-correlations of which are used to estimate the ToA of the received signal.

In order to see how to estimate the timing offset using dirty templates, consider PAM with no polarity codes and express the received signal of Equations (3.1) and (3.2) as

$$r(t) = \sqrt{E} \sum_{k=-\infty}^{\infty} b_k w_R(t - kT_s - \tau_1) + n(t), \tag{3.40}$$

where $T_s = N_f T_f$ is the symbol interval and

$$w_R(t) = \sum_{l=1}^{L} \alpha_l w_T(t - \tau_{l,1}), \tag{3.41}$$

with

$$w_T(t) = \sum_{j=0}^{N_f - 1} w(t - jT_f - c_j T_c), \tag{3.42}$$

and $\tau_{l,1} = \tau_l - \tau_1$, for $l = 1, \ldots, L$. Note that $w_T(t)$ and $w_R(t)$ denote the transmitted symbol waveform and the received symbol waveform, respectively. Then, the ToA estimation problem amounts to estimating τ_1 of Equation (3.40), which is assumed to be confined to one symbol duration, $[0, T_s)$, without loss of generality. Also assume that $T_f \geq \tau_{L,1} + T_p$ and $c_0 \geq c_{N_f - 1}$ so that the nonzero support of $w_R(t)$ is confined to $[0, T_s)$, and no ISI is present.

Consider symbol-long segments of the received signal and calculate the cross-correlations between them as follows [37]:

$$x_k(\tau) = \int_0^{T_s} r(t + 2kT_s + \tau) r(t + (2k - 1)T_s + \tau) \, dt, \tag{3.43}$$

for $\tau \in [0, T_s)$. The signals $r(t + 2kT_s + \tau)$ and $r[t + (2k - 1)T_s + \tau]$, for $t \in [0, T_s)$ are the dirty templates since they are noisy and act as template signals for each other. Also they include the effects of the unknown multipath channel including the timing offset τ_1.

After some manipulation, Equation (3.43) can be expressed as [37]:

$$x_k(\tau) = b_{2k-1}[b_{2k-2} E_A(\tilde{\tau}_1) + b_{2k} E_B(\tilde{\tau}_1)] + \eta_k(\tau), \tag{3.44}$$

where

$$\tilde{\tau}_1 = (\tau_1 - \tau) \bmod T_s, \ E_A(\tau) = E \int_{T_s - \tau}^{T_s} w_R^2(t) \, dt, \ E_B(\tau) = E \int_0^{T_s - \tau} w_R^2(t) \, dt,$$

and $\eta_k(\tau)$ is the noise term.

For the nondata-aided (blind) case, the mean square of Equation (3.44) can be obtained as

$$E\{x_k^2(\tau)\} = \frac{1}{2}[E_A(\tilde{\tau}_1) + E_B(\tilde{\tau}_1)]^2 + \frac{1}{2}[E_A(\tilde{\tau}_1) - E_B(\tilde{\tau}_1)]^2 + \sigma_\eta^2, \tag{3.45}$$

where equiprobable information symbols are assumed. Since $E_A(\tilde{\tau}_1) + E_B(\tilde{\tau}_1) = E\int_0^{T_s} w_R^2(t)\,dt$ is constant, and $E_B(\tilde{\tau}_1) - E_A(\tilde{\tau}_1)$ is maximized at $\tau = \tau_1$, the time offset can be estimated as

$$\hat{\tau}_1 = \arg\max_{\tau \in [0, T_s)}\left\{\frac{1}{K}\sum_{k=1}^{K} x_k^2(\tau)\right\}, \qquad (3.46)$$

where the expected value is replaced by its sample mean estimate obtained from K symbol-long pairs of received segments.

For the data-aided case, use of special training sequences speeds up the time offset estimation process. For example, by using $b_k = (-1)^{\lfloor k/2 \rfloor}$, Equation (3.44) can be expressed as [37]:

$$x_k(\tau) = [E_A(\tilde{\tau}_1) - E_B(\tilde{\tau}_1)] + \eta_k(\tau), \qquad (3.47)$$

the mean square value of which is given by

$$E\{x_k^2(\tau)\} = [E_A(\tilde{\tau}_1) - E_B(\tilde{\tau}_1)]^2 + \sigma_\eta^2. \qquad (3.48)$$

Since Equation (3.48) is maximized at $\tau = \tau_1$, the estimator in Equation (3.46) can again be employed. The advantage of the data-aided scheme is that the sample mean converges faster to the mean square value in Equation (3.48); therefore, more accurate time offset estimates can be obtained with the same number of symbols.

Further improvements can be obtained over the estimator of Equation (3.46) in the data-aided case by different combinations of the correlator outputs or by analog implementation [37].

The main advantage of the dirty template approach is the rich multipath energy collection since the templates are circularly shifted versions of the received symbol waveform. However, the noise in those templates results in noise–noise cross-terms, which causes some performance loss. This effect can be mitigated to some extent by averaging [37].

Since the dirty template approach uses the received signal waveform, it does not need to estimate any parameters of the multipath components except the delay of the first path. This is not the case for the change detection and GML based approaches of the previous two subsections.

One main disadvantage of the timing with dirty templates (TDT) algorithm is that its ToA estimate will have an ambiguity equal to the extent of the noise-only region between consecutive symbols, because in such a case there can be a *set* of symbol-long signal segments pairs that are proportional to each other. Therefore, another algorithm needs to be implemented after the dirty template based algorithm obtains the timing offset estimate. In other words, this scheme can be used as the first step of a two-step ToA estimation algorithm.

Other ToA Estimation Algorithms, and Design Criteria In addition to the algorithms mentioned in the previous subsections, there are also other timing offset estimation algorithms. In [50], coded Beacon sequences are employed to speed up the acquisition process by enabling searches over larger intervals. This avoids exhaustive search of the whole uncertainty region on the chip level, and fast ToA estimation becomes possible.

A frequency domain approach based on sub-Nyquist uniform sampling is proposed in [58]. It uses low-rate samples and thus allows for slower A/D converters.

Similar to the scheme in Section 3.5.2, a ToA estimation technique that tries to estimate the breakpoint between the noise-only and signal part of the received signal process is proposed in [59]. This technique requires some knowledge of the power delay profile of the channel and considers Gaussian models for the received signal both before and after the signal arrival.

Based on the cyclostationarity of IR-UWB signals, blind timing offset estimation techniques are proposed in [60]. However, for an ambiguity up to a pulse width, pulse-rate samples are required by these algorithms.

Although each algorithm has its advantages and disadvantages, the main issues for ToA estimation schemes for UWB systems are the following:

- A low sampling rate is required in order to have a low power and practical design. Therefore, algorithms using symbol-rate or frame-rate samples are preferable to those that employ chip-rate samples.
- For a given accuracy, ToA estimation should be performed using as few training symbols as possible. In other words, the time it takes to estimate the ToA should not be very long.
- Related to the previous issue for a given time interval or a given number of training symbols, the ToA estimation should provide sufficient accuracy.

Considering these criteria, the TDT approach combined with a change detection or a conventional correlation-based first-path detection scheme is a reasonable scheme. This is because the TDT algorithm can reduce the uncertainty about the ToA to a small region quickly using symbol-rate samples. Then, a higher resolution algorithm based on correlation outputs of the received signal with a template signal matched to the transmitted symbol[6] can be used to estimate the ToA of the incoming signal. The algorithm to estimate the ToA from those correlation outputs can be a simple first-path detection algorithm or based on a change detection approach depending on the complexity constraints.

Design of UWB ToA estimators that provide a tradeoff between complexity and performance is still an active research area. Designing an optimal ToA estimator within the constraints discussed above, such as the maximum sampling rate and estimation interval, remains an open problem.

[6]Symbol-rate samples can be obtained by using a template signal matched to the transmitted UWB symbol.

3.5.3 Two-Way Ranging Protocols

In order to estimate the ToA of a signal from one node to another, the two nodes must have a common reference clock. In the absence of such a timing reference, two-way ranging protocols can be employed to determine the round-trip time.

In [26], a two-way ranging scheme, which is employed in a two-way remote synchronization technique in satellite systems [61], is used to estimate the round-trip time in UWB systems. In this scheme, each node switches between transmission and reception modes every T s. Let t_1 and t_2 denote the local clocks of node 1 and node 2, respectively. As shown in Figure 3.10, at $t_1 = 0$, node 1 starts transmitting signals to node 2, which sets its clock to $t_2 = 0$ at the *coarse* estimation of the signal arrival time. However, this estimation is different than the true arrival time by $t_{\text{off},2}$. Then, after T s node 2 starts transmitting signals to node 1. At $t_1 = 2t_{\text{prop}} + t_{\text{off},2} + T$, the signal arrives at node 1. However, the coarse estimation of the arrival time causes node 1 to detect the incoming signal with a difference of

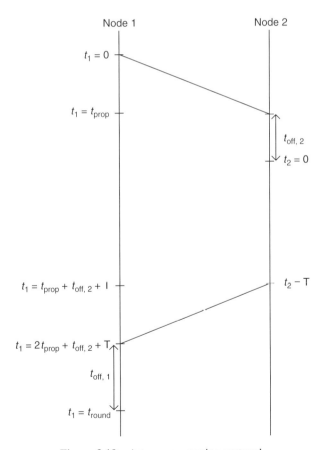

Figure 3.10 A two-way ranging protocol.

$t_{\text{off},1}$ from the true arrival time. By the second step of the ToA estimation algorithm, node 1 can estimate $t_{\text{off},1}$. However, it also needs $t_{\text{off},2}$ in order to be able to calculate the propagation time as

$$t_{\text{prop}} = 0.5(t_{\text{round}} - T - t_{\text{off},1} - t_{\text{off},2}). \quad (3.49)$$

For that reason, node 2 later sends node 1 a few bits to inform node 1 about $t_{\text{off},2}$. Then, the range can be estimated as $d = t_{\text{prop}}c$ using Equation (3.49), where c is the speed of light.

Note that this protocol considers a two-step ToA estimation algorithm at each node, where the first step performs a coarse estimation of the signal arrival time, and the second step tries to estimate the exact ToA (e.g., the algorithm in Section 3.5.2). Then, $t_{\text{off},1}$ and $t_{\text{off},2}$ denote the errors in the first step of the ToA estimation algorithm.

3.6 LOCATION-AWARE APPLICATIONS

Mobility makes location awareness a critical attribute for terminals in wireless networks. There is an increasing trend towards development of location-aware applications for both cellular and short-range communication networks. While location awareness in cellular networks is desired generally for multimedia services, content adaptation and distribution of personalized content [62], in short range networks, the main interest and market opportunities lie in tracking, control, monitoring and automation related applications. Therefore, positioning capability for both indoor and outdoor applications is crucial to penetrate a number of large markets.

The signals of conventional positioning systems such as the global positioning system (GPS) may not be available in indoor environments such as dwellings, warehouses and factories. Furthermore, in dense outdoor networks, adding a GPS functionality to every device becomes costly. Therefore, an alternative technology to GPS that would satisfy low cost and precision location would be viable.

One of the objectives of the emerging IEEE 802.15.4a standards is to modify and add precision ranging capability to the existing IEEE 802.15.4 physical layer. These standards will certainly spur development of various short-range and low-rate location-aware applications. Intuitively, each application will have varying requirements for communication range, data rate, ranging precision, mobility support and delay. In [63], location-aware applications are listed and their technical requirements are given. We provide a summary of some of those applications and their market requirements here.

One of the large application areas for location awareness is inventory tracking and pinpointing of goods in warehouses, and real-time tracking of shipments and valuable items in manufacturing plants. The desired ranging precision for such applications is approximately 3 cm to 3 feet according to the market analysis results presented in [63]. Another interesting field arises in civil government, safety and health care applications. Tracking people indoors, tracking fire-fighters, emergency responders, and miners, and finding avalanche victims, locating hospital staff, finding wandering patients, etc., may require ranging accuracy of only 50 cm. Security will certainly be another important area for location-aware applications. Daytime intruder detection, visitor management, enforcing restricted zones, and escort policies are some specific cases with a ranging accuracy requirement of 10–30 cm. Other interesting applications include locating cars in a parking lot and remote key unlocking, wireless body area networking for fitness and medical purposes, and real-time phone call forwarding.

Besides precise ranging requirements, battery lifetime of positioning capable devices is important, and it is generally expected to be longer than a year for UWB-based systems. Therefore, the UWB technology is a good candidate to satisfy both precision and longevity requirements.

3.7 CONCLUSIONS

In this chapter, we have considered positioning via UWB signals. Due to its high resolution capability in the time domain, time-based positioning algorithms are usually preferred over those involving AoA or RSS measurements. Although this high time resolution enables very accurate positioning, it also poses some challenges for practical systems, which may prevent the implementation of the optimal ToA estimation scheme in a positioning receiver. Therefore, some suboptimal ToA estimation algorithms have been considered, which offer varying tradeoffs between levels of accuracy and complexity.

The precise positioning capability of UWB systems facilitates many applications such as medical monitoring, security and asset tracking. Standardization efforts are underway in the IEEE 802.15.4a PAN standard, which will possibly make use of the unique features of the UWB technology.

REFERENCES

1. M. Z. Win and R. A. Scholtz, "Impulse radio: how it works," *IEEE Communications Letters*, vol. 2, no. 2, pp. 36–38, February 1998.
2. M. Z. Win and R. A. Scholtz, "On the energy capture of ultra-wide bandwidth signals in dense multipath environments," *IEEE Communications Letters*, vol. 2, pp. 245–247, September 1998.
3. M. L. Welborn, "System considerations for ultra-wideband wireless networks," *Proc. IEEE Radio and Wireless Conf.*, pp. 5–8, Boston, MA, August 2001.
4. R. A. Scholtz, "Multiple access with time-hopping impulse modulation," *Proc. IEEE Military Communications Conf., 1993 (MILCOM'93)*, vol. 2, pp. 447–450, Bedford, MA, October 1993.
5. M. Z. Win and R. A. Scholtz, "Ultra-wide bandwidth time-hopping spread-spectrum impulse radio for wireless multiple-access communications," *IEEE Transactions on Communications*, vol. 48, no. 4, pp. 679–691, April 2000.

6. US Federal Communications Commission, "First Report and Order 02-48," Washington, DC, 2002.
7. A. F. Molisch, Y. P. Nakache, P. Orlik, J. Zhang, Y. Wu, S. Gezici, S. Y. Kung, H. Kobayashi, H. V. Poor, Y. G. Li, H. Sheng and A. Haimovich, "An efficient low-cost time hopping impulse radio for high data rate transmission," *EURASIP Journal on Applied Signal Processing (Special Issue on UWB—State of the Art)*, vol. 2005, no. 3, pp. 397–412, March 2005.
8. J. Balakrishnan, A. Batra and A. Dabak, "A multi-band OFDM system for UWB communication," *IEEE Conf. on Ultra Wideband Systems and Technologies (UWBST'03)*, Reston, VA, November 2003.
9. P. Runkle, J. McCorkle, T. Miller and M. Welborn, "DS-CDMA: the modulation technology of choice for UWB communications," *IEEE Conf. on Ultra Wideband Systems and Technologies (UWBST'03)*, Reston, VA, November 2003.
10. C. J. Le Martret and G. B. Giannakis, "All-digital impulse radio for wireless cellular systems," *IEEE Transactions on Communications*, vol. 50, no. 9, pp. 1440–1450, September 2002.
11. Y.-P. Nakache and A. F. Molisch, "Spectral shape of UWB signals—influence of modulation format, multiple access scheme and pulse shape," *Proc. IEEE 57th Vehicular Technology Conf. (VTC 2003-Spring)*, vol. 4, pp. 2510–2514, Jeju, Korea, April 2003.
12. E. Fishler and H. V. Poor, "On the tradeoff between two types of processing gain," *Proc. 40th Annual Allerton Conf. on Communication, Control, and Computing*, Monticello, IL, October 2002.
13. J. Caffery, Jr., *Wireless Location in CDMA Cellular Radio Systems*, Kluwer Academic, Boston, MA, 2000.
14. Y. Hua, T. K. Sarkar, and D. D. Weiner, "An L-shaped array for estimating 2-D directions of wave arrival," *IEEE Transactions on Antennas and Propagation*, vol. 49, no. 2, pp. 143–146, February 1991.
15. Y. L. C. de Jong and M. H. A. J. Herben, "High-resolution angle of arrival measurement of the mobile radio channel," *IEEE Transactions on Antennas and Propagation*, vol. 47, no. 11, pp. 1677–1687, November 1999.
16. T. S. Rappaport, J. H. Reed, and B. D. Woerner, "Position location using wireless communications on highways of the future," *IEEE Communications Magazine*, pp. 33–41, October 1996.
17. S. W. Ellingston, "Design and evaluation of a novel antenna array for azimuthal angle-of-arrival measurement," *IEEE Transactions on Antennas and Propagation*, vol. 49, no. 6, pp. 971–978, June 1999.
18. S. M. Kay, *Fundamentals of Statistical Signal Processing: Detection Theory*, Prentice Hall, Englewood Cliffs, NJ, 1998.
19. R. O. Nielsen, "Estimation of azimuth and elevation angles for a plane wave sine wave with a 3-D array, *IEEE Transactions on Signal Processing*, vol. 42, no. 11, pp. 3274–3276, November 1994.
20. R. O. Nielsen, "Accuracy of angle estimation with monopulse processing using two beams," *IEEE Transactions on Aerospace and Electronic Systems*, vol. 37, no. 4, pp. 1419–1423, October 2001.
21. N. Patwari, A. O. Hero, M. Perkins, N. S. Correal, and R. J. O'Dea, "Relative location estimation in wireless sensor networks," *IEEE Transactions on Signal Processing*, vol. 51, no. 8, pp. 2137–2148, August 2003.

22. A. Molisch, "Status of models for UWB propagation channels," *IEEE 802 Interim Meeting*, IEEE P802.15-04/195r0, Orlando, FL, March 2004.
23. J. Kunisch and J. Pamp, "Measurement results and modeling aspects for the UWB radio channel," *Proc. IEEE Conf. on UWB Systems and Technologies*, pp. 19–23, Baltimore, MD, May 2002.
24. Z. Sahinoglu and A. Catovic, "A hybrid location estimation scheme (H-LES) for partially synchronized wireless sensor networks," *Proc. IEEE International Conf. on Communications (ICC'04)*, Paris, June 2004.
25. A. Catovic and Z. Sahinoglu, "The Cramer–Rao bounds of hybrid TOA/RSS and TDOA/RSS location estimation schemes," *IEEE Communications Letters*, vol. 8, no. 10, pp. 626–628, October 2004.
26. J.-Y. Lee and R. A. Scholtz, "Ranging in a dense multipath environment using an UWB radio link," *IEEE Transactions on Selected Areas in Communications*, vol. 20, no. 9, pp. 1677–1683, December 2002.
27. J. C. Adams, W. Gregorwich, L. Capots, and D. Liccardo, "Ultra-wideband for navigation and communications," *Proc. IEEE Aerospace Conf.*, vol. 2, pp. 785–792, Big Sky, MT, March 2001.
28. H. V. Poor, *An Introduction to Signal Detection and Estimation*, Springer-Verlag, New York, 1994.
29. C. E. Cook and M. Bernfeld, *Radar Signals: an Introduction to Theory and Applications*, Academic Press, New York, 1970.
30. Y. Shimizu and Y. Sanada, "Accuracy of relative distance measurement with ultra wideband system," *IEEE Conf. on Ultra Wideband Systems and Technologies (UWBST'03)*, Reston, VA, November 2003.
31. S. Gezici, Z. Tian, G. B. Giannakis, H. Kobayashi, A. F. Molisch, H. V. Poor, and Z. Sahinoglu, "Localization via ultra-wideband radios," *IEEE Signal Processing Magazine (Special Issue on Signal Processing for Positioning and Navigation with Applications to Communications)*, vol. 22, no. 4, pp. 70–84, July 2005.
32. G. L. Turin, "An introduction to matched filters," *IRE Transactions on Information Theory*, vol. IT-6, no. 3, pp. 311–329, June 1960.
33. M.-A. Pallas and G. Jourdain, "Active high resolution time delay estimation for large BT signals," *IEEE Transactions on Signal Processing*, vol. 39, no. 4, pp. 781–788, April 1991.
34. IEEE 802.15 WPAN Task Group 3 (TG3) [Online]. Available at: www.ieee802.org/15/pub/TG3.html
35. I. Guvenc and H. Arslan, "Design and performance analysis of TH sequences for UWB impulse radio," *Proc. IEEE Wireless Communications and Networking Conf. (WCNC'04)*, vol. 2, pp. 914–919, Atlanta, GA, March 2004.
36. S. Gezici, H. Kobayashi, H. V. Poor, and A. F. Molisch, "The trade-off between processing gains of an impulse radio system in the presence of timing jitter," *Proc. IEEE International Conf. on Communications (ICC 2004)*, vol. 6, pp. 3596–3600, Paris, June 2004.
37. L. Yang and G. B. Giannakis, "Timing ultra-wideband signals with dirty templates," *IEEE Transactions on Communications*, submitted.
38. M. McGuire, K. N. Plataniotis, and A. N. Venetsanopoulos, "Location of mobile terminals using time measurements and survey points," *IEEE Transactions on Vehicular Technology*, vol. 52, no. 4, pp. 999–1011, July 2003.

39. S. Gezici, H. Kobayashi, and H. V. Poor, "A new approach to mobile position tracking," *Proc. IEEE Sarnoff Symposium on Advances in Wired and Wireless Communications*, pp. 204–207, Ewing, NJ, March 2003.
40. M. P. Wylie and J. Holtzman, "The non-line of sight problem in mobile location estimation," *Proc. 5th IEEE International Conf. on Universal Personal Communications*, vol. 2, pp. 827–831, Cambridge, MA, September 1996.
41. S. Al-Jazzar and J. Caffery, Jr, "ML and Bayesian TOA location estimators for NLOS environments," *Proc. IEEE 56th Vehicular Technology Conf. (VTC) Fall*, vol. 2, pp. 1178–1181, Vancouver, BC, September 2002.
42. S. Al-Jazzar, J. Caffery, Jr, and H.-R. You, "A scattering model based approach to NLOS mitigation in TOA location systems," *Proc. IEEE 55th Vehicular Technology Conf. (VTC) Spring*, pp. 861–865, Birmingham, AL, May 2002.
43. B. L. Le, K. Ahmed, and H. Tsuji, "Mobile location estimator with NLOS mitigation using Kalman filtering," *Proc. IEEE Conf. on Wireless Communications and Networking (WCNC'03)*, vol. 3, pp. 1969–1973, New Orleans, LA, March 2003.
44. Y. Qi, H. Kobayashi and H. Suda, "On time-of arrival positioning in a multipath environment," *IEEE Transactions on Vehicular Technology*, March 2004 (downloadable from www.princeton.edu/~sgezici/Qi et al TOA Positioning in MP.pdf), submitted.
45. J. Borras, P. Hatrack, and N. B. Mandayam, "Decision theoretic framework for NLOS identification," *Proc. IEEE 48th Vehicular Technology Conf. (VTC'98 Spring)*, vol. 2, Ottawa, Canada, pp. 1583–1587, 18–21 May, 1998.
46. S. Gezici, H. Kobayashi and H. V. Poor, "Non-parametric non-line-of-sight identification," *Proc. IEEE 58th Vehicular Technology Conf. (VTC 2003 Fall)*, vol. 4, pp. 2544–2548, Orlando, FL, 6–9 October, 2003.
47. S. Venkatraman and J. Caffery, "A statistical approach to non-line-of-sight BS identification," *Proc. IEEE 5th International Symposium on Wireless Personal Multimedia Communications (WPMC 2002)*, Honolulu, Hawaii, October 2002.
48. W. C. Chung and D. S. Ha, "An accurate ultra wideband (UWB) ranging for precision asset location," *Proc. IEEE Conf. on Ultra Wideband Systems and Technologies (UWBST'03)*, pp. 389–393, Reston, VA, November 2003.
49. V. S. Somayazulu, J. R. Foerster, and S. Roy, "Design challenges for very high data rate UWB systems," *Conf. Record of the Thirty-Sixth Asilomar Conf. on Signals, Systems and Computers*, vol. 1, pp. 717–721, November 2002.
50. E. A. Homier and R. A. Scholtz, "Rapid acquisition of ultra-wideband signals in the dense multipath channel," *Proc. IEEE Conf. on Ultra Wideband Systems and Technologies (UWBST'02)*, pp. 105–109, Baltimore, MD, May 2002.
51. S. Gezici, Z. Sahinoglu, A. F. Molisch, H. Kobayashi, and H. V. Poor, "A two-step time of arrival estimation algorithm for impulse radio ultra wideband systems," *Proc. 13th European Signal Processing Conf.*, Antalya, 4–8 September, 2005.
52. M. Basseville and I. V. Nikiforov, *Detection of Abrupt Changes: Theory and Application*, Prentice-Hall, Englewood Cliffs, NJ, 1993.
53. M. Z. Win and R. A. Scholtz, "Energy capture vs. correlator resources in ultra-wide bandwidth indoor wireless communictions channels," *Proc. IEEE Military Communications Conf. (MILCOM'97)*, vol. 3, pp. 1277–1281, Monterey, CA, November 1997.
54. S. Gezici, E. Fishler, H. Kobayashi, H. V. Poor, and A. F. Molisch, "A rapid acquisition technique for impulse radio," *Proc. IEEE Pacific Rim Conf. on Communications,*

Computers and Signal Processing (PACRIM 2003), vol. 2, pp. 627–630, Victoria, 28–30 August 2003.
55. L. Yang and G. B. Giannakis, "Low-complexity training for rapid timing acquisition in ultra-wideband communications," *Proc. IEEE Global Telecommunications Conf. (GLOBECOM'03)*, vol. 2, pp. 769–773, San Francisco, CA, December 2003.
56. L. Yang and G. B. Giannakis, "Blind UWB timing with a dirty template," *Proc. IEEE International Conf. on Acoustics, Speech and Signal Processing*, Montreal, QB, 17–21 May, 2004.
57. L. Yang and G. B. Giannakis, "Ultra-wideband communications: an idea whose time has come," *IEEE Signal Processing Magazine*, vol. 21, no. 6, pp. 26–54, November 2004.
58. I. Maravic, J. Kusuma, and M. Vetterli, "Low-sampling rate UWB channel characterization and synchronization," *Journal of Communications and Networks*, vol. 5, no. 4, pp. 319–327, December 2003.
59. C. Mazzucco, U. Spagnolini, and G. Mulas, "A ranging technique for UWB indoor channel based on power delay profile analysis," *Proc. IEEE 59th Vehicular Technology Conf. (VTC'04 Spring)*, Milan, May 2004.
60. Z. Tian, L. Yang, and G. B. Giannakis, "Symbol timing estimation in ultra wideband communications," *Proc. IEEE Asilomar Conf. on Signals, Systems, and Computers*, Pacific Grove, CA, vol. 2, pp. 1924–1928, November 2002.
61. W. C. Lindsey and M. K. Simon, *Phase and Doppler Measurements in Two-Way Phase-Coherent Tracking Systems*. Dover, New York, 1991.
62. Z. Sahinoglu and A. Vetro, "Mobility characteristics for multimedia service adaptation," *Signal Processing: Image Communication (Special Issue on Multimedia Adaptation)*, vol. 18, no. 8, pp. 699–719, September 2003.
63. IEEE 15-03-0489-03-004a-application-requirement-analysis-031127 v0.4. Available at: www.ieee802.org/15/pub/TG4.html

CHAPTER 4

UWB Modulation Options

HÜSEYIN ARSLAN, İSMAIL GÜVENÇ, and SADIA AHMED

4.1 INTRODUCTION

In the world of wireless communication, UWB is considered as an attractive technology for its high capacity, high data rates, robustness against fading, low power consumption, low cost and low-complexity devices. In spite of all the benefits of UWB, the extremely wide frequency bands (greater than 500 MHz) and exceptionally narrow pulses (in the range of 10^2 ps) make it difficult to apply conventional narrowband modulation techniques into UWB systems. Therefore, a significant amount of research has been conducted to come up with the suitable modulation technique for UWB systems.

Selecting the appropriate modulation technique in the UWB systems still remains a major challenge. There are various possible modulation options that depend on the application, design specifications and constraints, range, transmission and reception power, quality of service requirements, regulatory requirements, hardware complexity, data rate, reliability of channel, and capacity. Therefore, it is crucial to choose the right modulation for the right purpose. Some of the well-studied modulation or mapping options in UWB are BPSK, QPSK, PAM, OOK, PPM, PIM, and PSM. Of these options, BPSK is one of the most popular modulation techniques in UWB applications due to its smooth power spectrum and low BER. However, accurate phase detection of the modulated signal in BPSK requires accurate channel estimation at the receiver. Compared with BPSK, OOK and PPM only require the knowledge of the presence or absence of signal and therefore channel estimation is not necessary. The noise level in wireless channels also influences the choice of modulation. Higher-order modulation ensures high data rate at the cost of poor BER in a noisy channel. Therefore, lower-order modulation for low-data-rate applications is desired in poor channel conditions. One can also consider transmission over multiple frequency bands or over multiple carriers, and various multiple accessing options such as TH and DS under the umbrella of UWB modulations.

Ultra Wideband Wireless Communication. Edited by Arslan, Chen, and Di Benedetto
Copyright © 2006 John Wiley & Sons, Inc.

78 UWB MODULATION OPTIONS

In this chapter, various multiple access schemes, carrier modulation, and data mapping will be studied for UWB communication systems. Digital communication requires *mapping* of the stream of bits into waveforms at the transmitter and conversely *demapping* is achieved at the receiver through various reception algorithms. These mapping and demapping operations are commonly referred as *modulation* and *demodulation*. In a more formal expression, modulation is defined as "the mapping of a sequence of binary digits into a set of corresponding waveforms" [1]. The mapped waveform can correspond to a single bit, or a sequence of bits. The term *modulation* is also often used to imply other concepts. Baseband signals are commonly mixed by carriers to up-convert them into intermediate frequencies (IF) and/or radio frequencies (RF). Carrier modulation, which is the process of embedding the information into a radio carrier [2], is also commonly referred as modulation. It is further possible to find certain multiple accessing and signaling techniques being referred as *modulation techniques* in the literature [3–5]. In the sequel, we will commonly refer the mapping of the information into waveforms as *data mapping*, which will be the main focus of our chapter. Different carrier-based and carrierless implementations of UWB systems as well as certain multiple accessing schemes that may be referred to in the literature as modulation approaches will be covered under *UWB signaling techniques*.

Within this context, in Section 4.2, TH and DS IR signaling will be presented using a unified signaling model. The same section will address two other popular UWB signaling approaches: multiband-UWB and multicarrier modulation schemes. In Section 4.3, several binary and higher-order data mapping formats for UWB will be analyzed in terms of their BER performances. Section 4.4 will address these data mapping or modulation techniques in terms of their spectral characteristics, which are important for compliance with regulatory requirements in UWB. Even though coherent modulation schemes have better power efficiencies, certain data mapping formats allow implementation of noncoherent reception and lower complexity hardware implementation, which will be discussed in Section 4.5. Finally, performances of certain data mapping formats in practical scenarios will be discussed in Section 4.6. Our analysis will cover the data modulation and inter-pulse interference effects of multipath, effects of multiple access interference, and effects of timing misalignment on the performance of various data mapping options.

4.2 UWB SIGNALING TECHNIQUES

Early implementation of UWB communication systems is based on transmission and reception of extremely short duration (typically subnanosecond) pulses, which are commonly referred as impulse radio. In the pioneering work by Scholtz [3] in 1993, time-hopping impulse radio was introduced as a carrierless modulation scheme, where no up/down conversion of the transmitted/received signal is required for the transceiver circuitry. Up until February 2002, the term UWB was tied solely to impulse radio modulation. In February 2002, the FCC released the

Part 15 amendment that allows (and specifies the rules for) the operation of UWB devices. The FCC definition for UWB in this document is that any signal having a fractional bandwidth[1] larger than 0.2, or a signal bandwidth[2] greater than 500 MHz is considered as UWB. These regulatory rules also specify indoor and outdoor *spectral masks*, which restricts transmission powers of UWB devices in order to minimize the interference with other narrowband technologies operating in the same frequency bands. The new FCC definition of UWB basically implies that *any* communication technology that has a bandwidth larger than 500 MHz is considered as UWB. This new ruling placed a variety of well-known and more established wireless communication technologies and applications under the umbrella of UWB systems. Multiband UWB is one of the possible examples where the entire UWB single band is split into several smaller bands, each sub-band satisfying the FCC bandwidth requirement. Multicarrier OFDM is another popular technology that can be applied to UWB systems under the same bandwidth ruling. Multiband and multicarrier concepts in UWB are studied in the subsections below. We will begin with a basic signal model for impulse radio UWB and move onto two possible multiple access schemes followed by the discussion on multiband and multicarrier.

4.2.1 UWB-IR Signaling

Let the generic transmitted and received UWB signals $s^{(\psi)}(t)$ and $r^{(\psi)}(t)$ by user ψ in a single-path, single-user environment be written as

$$s^{(\psi)}(t) = \sum_{j=-\infty}^{\infty} \sqrt{\frac{E_s}{N_s}} d_j^{(\psi)} \beta_{\lfloor j/N_s \rfloor}^{(\psi)} \omega_{tr}^{(\xi)}\left(t - jT_f - c_j^{(\psi)} T_h - \delta \alpha_{\lfloor j/N_s \rfloor}^{(\psi)}\right), \quad (4.1)$$

$$r^{(\psi)}(t) = \sum_{j=-\infty}^{\infty} \sqrt{\frac{E_s}{N_s}} d_j^{(\psi)} \beta_{\lfloor j/N_s \rfloor}^{(\psi)} \omega_{tr}^{(\xi)}\left(t - jT_f - c_j^{(\psi)} T_h - \delta \alpha_{\lfloor j/N_s \rfloor}^{(\psi)} + \epsilon_j\right) + n(t), \quad (4.2)$$

where j is the frame index, E_s is the energy per symbol, N_s is the number of pulses per symbol, T_f is the nominal interval between two pulses, δ is the modulation index if the modulation is PPM, and T_h is the chip duration which is larger than pulse width T_c for PPM. Decimal (time-hop) codes $c_j^{(\psi)}$ and binary (polarity) codes $d_j^{(\psi)}$ are pseudo-random codes unique to user ψ, and are used to employ TH-UWB (Figure 4.1) or DS-UWB (Figure 4.2) multiple access schemes, respectively, or their combination, as will be discussed later. The timing misalignment for the jth pulse (which may be due to timing jitter or finger estimation error), ϵ_j, is a zero-mean random variable, pulse amplitude is represented by $A = \sqrt{E_s/N_s}$, and $n(t)$ is the AWGN with a double-sided spectrum of $N_0/2$. With $\beta_{\lfloor j/N_s \rfloor}^{(\psi)}$ changing the amplitudes of the pulses (OOK, BPSK, positive PAM, M-ary PAM) or $\delta \alpha_{\lfloor j/N_s \rfloor}^{(\psi)}$ varying the

---
[1] Ratio of signal bandwidth to central frequency.
[2] Defined by the range within 10 dB of the peak transmission power.

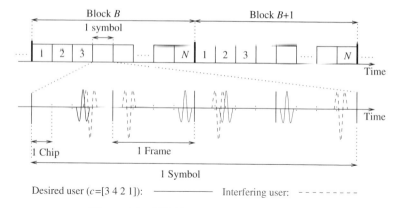

Figure 4.1 TH-UWB-IR signaling structure.

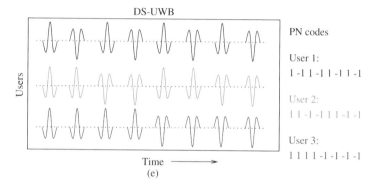

Figure 4.2 DS-UWB-IR signaling structure.

time positions of the pulses (PPM, M-ary PPM, PIM), or $\omega_{rx}^{(\xi)}$ modifying the shapes of the pulses (PSM), UWB-IR signals can be modulated in different ways as shown in Table 4.1 and in Figures 4.3 and 4.4, and as will be discussed in the next section.

The transmitted pulse shape is represented by $\omega_{tr}^{(\xi)}$, where the ξ term is used to refer to different pulse shapes for PSM, and can be dropped for simplicity of notation if identical pulse shapes are employed. In this chapter, for theoretical purposes, and unless otherwise stated, the transmitted pulse shape is modeled as the second derivative of the Gaussian pulse [6]:

$$\omega_{tr}(t) = \left[1 - 4\pi\left(\frac{t}{\tau}\right)^2\right] e^{-2\pi(\frac{t}{\tau})^2}, \quad (4.3)$$

where τ is used to adjust the pulse width, T_c, and it is assumed further that the pulse shape is normalized to have unit energy.

TABLE 4.1 Binary Data Mapping Formats and BER Performances

Binary	$\beta^{(k)}_{\lfloor j/N_s \rfloor}$	$\delta\alpha^{(k)}_{\lfloor j/N_s \rfloor}$	BER
Orthogonal PPM	1	$0, T_c$	$Q\left(\sqrt{\dfrac{N_s A E_p}{N_0}}\right)$
Optimum PPM	1	$0, \delta_{opt},$ $\delta_{opt} = \underset{\delta}{\operatorname{argmax}}\{R(0) - R(\delta)\}$	$Q\left(\sqrt{\dfrac{N_s A E_p}{N_0}[R(0) - R(\delta_{opt})]}\right)$
BPSK	± 1	0	$Q\left(\sqrt{\dfrac{2 N_s A E_p}{N_0}}\right)$
PAM	a_1, a_2	0	$Q\left(\sqrt{\dfrac{(a_2 - a_1)^2 N_s A E_p}{2 N_0}}\right)$
OOK	$0, a$	0	$Q\left(\sqrt{\dfrac{a^2 N_s A E_p}{2 N_0}}\right)$

Although other receiver techniques are possible, matched filtering (or correlator receiver) is the optimal reception approach as it maximizes the signal-to-noise ratio (SNR). In matched filtering, exact received pulse shape information is assumed to be available at the receiver to construct a local template. As shown in Figure 4.5, the received signal is correlated with this local template to calculate the decision statistics, which are then used for demapping (based on the employed data mapping format) to obtain the received bit stream. The normalized autocorrelation function (NACF), which depends on the received pulse shape, carries significant importance in a matched filter receiver for the calculation of the Euclidean distances between the

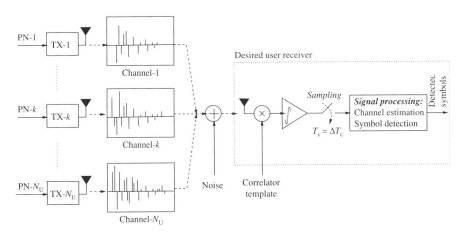

Figure 4.3 TH-UWB-IR signaling structure.

82 UWB MODULATION OPTIONS

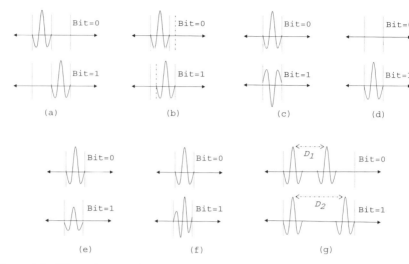

Figure 4.4 Binary data mapping schemes. (a) orthogonal PPM; (b) optimal PPM; (c) BPSK; (d) OOK; (e) PAM; (f) PSM; (g) PIM.

symbols for different modulation formats, and accounts for synchronization errors. For the pulse shape assumed in this chapter, the NACF is given by

$$R(\Delta t) = \frac{\int_{-\infty}^{\infty} \omega_{\text{rec}}(t) \omega_{\text{rec}}(t - \Delta t) \, dt}{\int_{-\infty}^{\infty} \omega_{\text{rec}}^2(t) \, dt}$$

$$= \left[1 - 4\pi \left(\frac{\Delta t}{\tau} \right)^2 + \frac{4\pi^2}{3} \left(\frac{\Delta t}{\tau} \right)^4 \right] e^{-\pi \left(\frac{\Delta t}{\tau} \right)^2}, \quad (4.4)$$

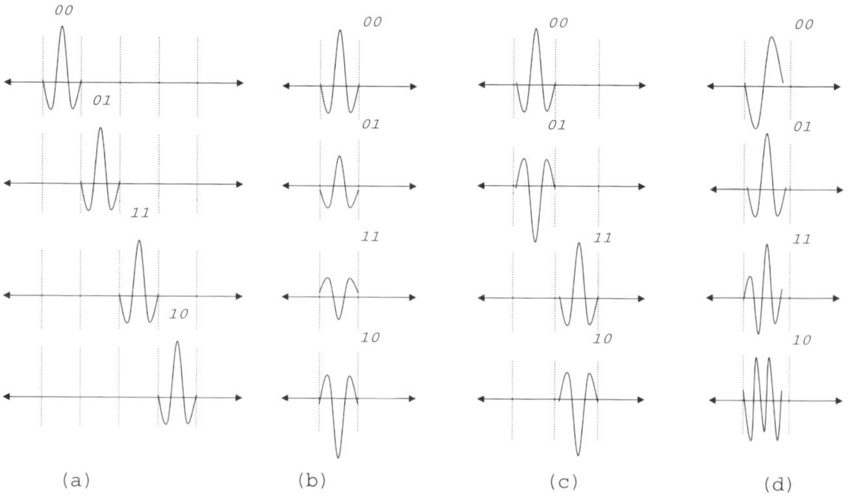

Figure 4.5 *M*-ary data mapping schemes. (a) *M*-ary PPM; (b) *M*-ary PAM; (c) biorthogonal signaling; (d) multilevel PSM.

where under perfect synchronization we have $R(0) = 1$, which implies the received pulse energy as

$$E_{\text{p}} = \frac{E_{\text{s}}}{N_{\text{s}}} = \int_{-\infty}^{\infty} \omega_{\text{rec}}^2(t)\, dt \qquad (4.5)$$

How the received signal energy affects the performances of different data mapping formats will be discussed in more detail in the next section.

Note that the system models given in Equations (4.1) and (4.2) are applicable for TH-UWB or DS-UWB, or their combination. For pure TH-UWB implementation, $d_j^{(\psi)}$ is taken to be unity for all j, and $c_j^{(\psi)}$ is used to hop the pulses in time [3, 6–11]. On the other hand, pure DS-UWB is achieved using $c_j^{(\psi)} = 0$ for all j, and using $T_{\text{f}} = T_{\text{h}}$ (i.e., decreasing the frame duration into a chip duration and transmitting the pulses consecutively), where this time the codes $d_j^{(\psi)}$ are used to change the polarities of the pulses as in DS code division multiple access (DS-CDMA) systems [12–14]. It is also possible to use TH and DS codes (also refered as polarity codes) simultaneously, which helps to smooth the spectrum further in a TH-UWB implementation [15–17], and yields a more robust system due to additional spreading gain. By using appropriate code designs, it is possible to have codes with better correlation properties for both TH-UWB [7, 8] and DS-UWB [12] implementations.

It is also possible to apply different multiple accessing schemes and mapping/demapping techniques into UWB systems where multiband or multiple carriers are utilized for efficient data modulation and transmission.

4.2.2 Multiband UWB

The basic idea behind multiband schemes is to split the total available bandwidth into multiple frequency bands for efficient utilization of the UWB spectrum by transmitting multiple UWB signals at different frequencies. Since the transmission is "almost" orthogonal over each of these bands (like frequency division multiplexing, FDM), the signals do not interfere with each other. By partitioning the spectrum into smaller chunks (each of which is still larger than 500 MHz to comply with FCC spectrum regulations), a better co-existence with other current and future wireless technologies can be achieved. This approach also enables worldwide interoperability of the UWB devices, as the spectral allocation for UWB could be different in various parts of the world. Another great benefit of multiband is the ability to avoid narrowband interference over the frequency spectrum where strong interferers exists. For example, transmission over the UNII band, where possible 802.11a-based WLAN devices pose a threat, can be avoided.

In spite of all the benefits given above, multiband system design might also give up some other benefits over the traditional single-band approaches. First of all, in multiband schemes the bandwidth adjustment is relatively coarse as the bandwidth in each subband should be at least 500 MHz wide. Hence, turning off the transmission over a big chunk of a frequency spectrum (like unlicensed UNII bands) all the time does not exploit the full bandwidth that the regulations allow. Ideally, a UWB solution should be robust against interference received from both licensed and unlicensed devices, and should provide spectral flexibility for current and future spectrum assignment worldwide, while providing the highest spectral efficiency possible. A fully adaptive solution that can take advantage of the available spectrum by dynamically adjusting the transmission depending on the measurement of the interference level over these frequencies is desirable. Note that multiband is not the only way of adjusting the transmitted power spectrum. There are other ways of controlling the spectrum. The analysis of these different approaches is beyond the scope of this chapter.

There has been a variety of multiband solutions for UWB communications. Even though the main concept is the same, there are some variations in each of these solutions. The bandwidth and number of available bands generate different performance tradeoffs and design challenges such as sampling rate, multipath, and multiple access interference. Some of the possible solutions to multiband are pulse-based, single-carrier-based and multiple-carrier-based (OFDM or other approaches) [19, 20]. Multiband can also be employed by modulating UWB pulses using direct sequence PN codes and then utilizing the resulting signal to modulate single carriers on sub-bands.

Figure 4.6 shows a pulse-based scheme where each unique pulse (top figure) transmits information over corresponding sub-band (bottom figure).

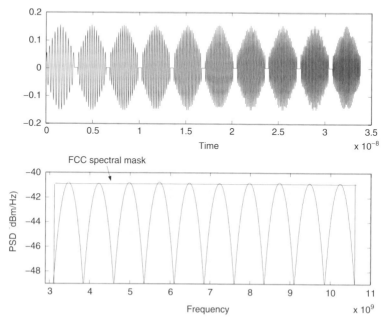

Figure 4.6 Multiband UWB signaling (pulse-based): top, pulses used to generate frequency bands 1–10; bottom, power spectral densities of individual pulse shapes in bands 1–10.

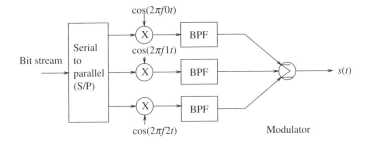

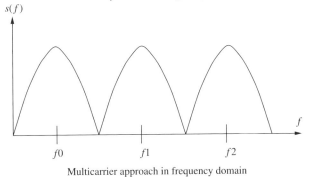

Figure 4.7 Multi-carrier UWB signaling: three carrier frequencies and power spectral densities of individual carriers.

4.2.3 Multicarrier UWB

The multicarrier approach is a strong candidate for UWB modulation where multiple carriers are modulated by UWB symbols to transmit information. There are several ways to implement multicarrier modulation. One such implementation involves multiple nonoverlapping orthogonal carriers. Another technique utilizes OFDM. Figure 4.7 describes the first approach where three nonoverlapping carriers are used to transmit information. Although multicarrier is a good solution against ISI, the use of separate carriers greatly reduces spectral efficiency and increases hardware complexity due to the increased number of mixers and filter banks at the transmitter and at the receiver. A popular way to employ the multicarrier approach is OFDM technology, which can efficiently tackle multicarrier-related issues.

4.2.4 OFDM

OFDM is a spectrally efficient way of implementing multicarrier modulation where multiple overlapping orthogonal carriers are used through fast Fourier transform (FFT) and inverse fast Fourier transform (IFFT) [5]. Although OFDM falls under the umbrella of multicarrier modulation, only a single carrier is required for the physical implementation, eliminating the need for mixers and filter banks at

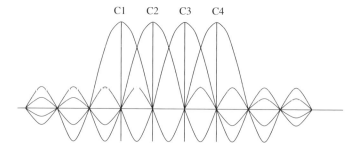

Figure 4.8 OFDM: four carriers and their power spectral densities.

transmitter and receiver. Figure 4.8 shows four orthogonal carriers of an OFDM symbol, where the sinc carrier functions overlap while maintaining orthogonality at the proper frequency sampling positions. The number of carriers within an OFDM symbol is a design issue and depends on the application.

Over the last few years OFDM has gained wide popularity in the wireless world for its robustness against multipath interference, ability to capture multipath energy efficiently, ability to allow frequency diversity with the use of efficient forward error correction (FEC) coding, and ability to provide high bandwidth efficiency through the use of sub-band adaptive modulation and coding techniques. In UWB, OFDM has an additional advantage, its ability to avoid transmission of narrowband interference in the carriers that are significantly corrupted. Also, timing mismatch is efficiently handled by OFDM through the use of a cyclic prefix. Since the carriers are modulated by the UWB symbols in OFDM, the pulses constructing the UWB symbol are also modulated by the data bits or symbols, known as data mapping (or *data modulation*). Some of the more popular data modulation options in OFDM are BPSK, QPSK, 16QAM and 64QAM [see Figure 4.9(a)].

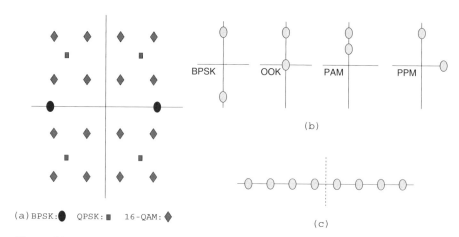

Figure 4.9 Signal constellation points for different modulation formats: (a) popular modulation schemes in OFDM; (b) binary modulations; (c) *M*-ary PAM.

The data rate and data bits per OFDM symbol vary depending on the respective data mapping or data modulation options. Another attractive feature of OFDM is the use of sub-band adaptive modulation where the data mapping can be different over different carriers depending on the SNR of the carriers.

Apart from all the benefits of employing OFDM modulation in UWB, the OFDM scheme comes with its own challenges like intercarrier interference (ICI) due to loss of orthogonality among carriers and nonlinearity in power amplifiers due to high peak-to-average-power-ratio (PAPR).

Carrier modulation includes modulating one or more carriers by UWB symbols or pulse streams. Each of these pulses is again modulated by data bits or symbols using data mapping and will be discussed in detail in the following section.

4.3 DATA MAPPING

Digital data is embedded in bit streams or symbols of 1s and 0s and requires transmission over UWB pulses. Data mapping or data modulation is essential to achieve mapping of binary bits onto these UWB pulse waveforms. The performances of different data mapping formats have been analyzed extensively in the past for other technologies [1, 21–23], including some recent work that discussed certain aspects of different UWB data mapping formats [24–27]. There are multiple possible solutions to data mapping and the choice depends on design, application, data-rate, BER requirements, complexity requirements, regulatory issues, and multiple accessing scheme. In this section, we will address appropriate binary and higher-order data mapping formats for UWB communications, followed by the discussion of other modulation-related issues in the subsequent sections.

4.3.1 Binary Data Mapping Schemes

Binary mapping of the digital data on UWB pulses means that only one bit is transmitted by a single pulse or a train of pulses (i.e., one bit per symbol). Some of the possible binary data modulation options for UWB IR are PPM, BPSK, OOK, binary PAM, PSM [28], and PIM [29], and can be implemented in TH-IR, DS-IR, and multiband IR. In PPM, the position of each pulse is modulated depending on the transmitted bit while the pulse phase and amplitude remain the same. PAM involves modulating the pulse amplitude according to data bits. OOK and BPSK are both special cases of PAM. In OOK, the presence or absence of transmitted signal determines a 1 or a 0, respectively. BPSK involves changing the pulse polarity according to binary bit information. In PIM, information is embedded within the pulse to pulse intervals. Due to the baseband nature of IR, all these modulation schemes do not use any phase information. Another modulation scheme is PSM where modified Hermite polynomials are used to construct orthogonal pulses to represent different symbols [28]. Mapping of the bits on a UWB pulse for these different binary modulation schemes is depicted in Figure 4.3(b), which also applies for multiple pulses per bit.

Based on how the bits are mapped onto the pulses, constellation for a specific data mapping format will have different forms, as depicted in Figure 4.9(b) for PAM, PPM, OOK, and BPSK modulations. The Euclidean distance is expressed as the distance between two adjacent symbols or signal points on a constellation. The BER performance of binary modulations in a AWGN channel using the Euclidean distance d_{12} between two different symbols is given as [1]:

$$P_b = Q\left(\sqrt{\frac{d_{12}^2}{2N_0}}\right) = \frac{1}{2}P(0|1) + \frac{1}{2}P(1|0), \qquad (4.6)$$

where $P(0|1)$ and $P(1|0)$ are the probabilities of detecting the opposites of the transmitted bits, and $Q(x) = \frac{1}{2}\text{erfc}(x/\sqrt{2})$. As shown in Figure 4.10 for BPSK modulation, the BER is calculated as the area under the tails of the Q-function for either possible data symbols, and depends on both the noise power and the signal power. Equation (4.6) dictates that BER performance is related to the Euclidean distance of a particular modulation scheme. For orthogonal PPM, the Euclidean distance is expressed as $\sqrt{2E_p}$; for BPSK it is expressed as $2\sqrt{E_p}$; for OOK it is $a\sqrt{E_p}$ and for PAM it is $(a_1 - a_2)\sqrt{E_p}$. Plugging these values in Equation (4.6) and considering the effects of processing gain N_s and pulse amplitude A, BER performances of binary modulations are obtained as summarized in Table 4.1. Maximizing Euclidean distance reduces the probability of detecting the wrong symbol. Optimum PPM is obtained by choosing the modulation index δ that maximizes the Euclidean distance.

Theoretical plots for binary modulations in AWGN and Rayleigh fading channels using $N_s = 1$ and $A = 1$ are depicted in Figure 4.11. It is seen that orthogonal PPM

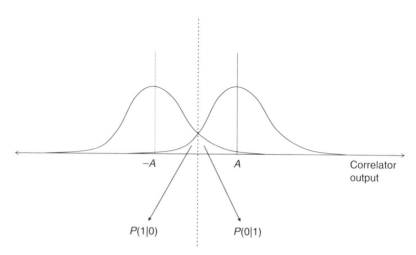

Figure 4.10 Probability density functions corresponding to bits 0 and 1, and calculation of the bit error probabilities.

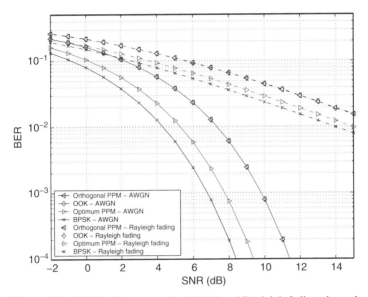

Figure 4.11 BER performances in AWGN and Rayleigh fading channels.

and OOK modulations have the same BER performance when the average transmitted pulse energies are the same ($a = \sqrt{2}$). BPSK is 3 dB more power-efficient than both OOK and orthogonal PPM. For the case of $T_c = 0.8$ ns, δ_{opt} is evaluated as 0.16 ns and $R(\delta_{opt})$ as -0.6, which makes optimum PPM 1 dB more power-inefficient than BPSK. Performance plots in Figure 4.11 for the Rayleigh fading channel show that similar degradation is observed in all modulation schemes. Since the Euclidean distance is reduced when using positive PAM due to the same polarity of the pulses, it has the worst power efficiency (and therefore is not preferred). For example when $a_1 = \sqrt{0.5}$ and $a_2 = \sqrt{1.5}$ are selected, positive PAM is 9 dB more power-inefficient than BPSK.

Some of the discussed binary data mapping schemes in this section are applicable to higher order M-ary mapping formats whenever there is a need for higher data rate and efficient data transmission, which will be discussed in the next section.

4.3.2 *M*-ary Data Mapping Schemes

In order to achieve increased data rate, multiple bits (rather than a single bit) can be mapped onto UWB pulses, which leads to M-ary data mapping options. Some of the possible M-ary data mapping or modulation options are M-ary PPM, M-ary PAM, and biorthogonal signaling, M-ary PSM, and M-ary PIM. Mapping of the bit streams onto the UWB pulses for different M-ary data mapping formats is depicted in Figure 4.4 for two bits per symbol (i.e., 4-ary modulation).

There exist trade-offs in the implementation of the different higher-order mapping schemes. M-ary PPM requires a bank of M correlators, whose outputs

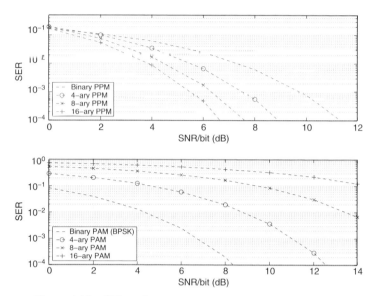

Figure 4.12 SER performances of M-ary PPM and M-ary PAM.

are compared and the highest correlator output yields the detected bits. A single correlator is sufficient for M-ary PAM; however, it requires accurate estimation of the channel in order to set optimal thresholds for various correlator output levels.[3] Biorthogonal signaling, which uses $M/2$ orthogonal waveforms and their negatives to construct a set of M waveforms, requires $M/2$ correlators. With appropriate design of templates, number of correlators in biorthogonal scheme can be decreased to $M/4+1$ and in PPM to $M/2$.

Theoretical BER peformances of M-ary modulations are shown in Table 4.2. Figure 4.12 compares the symbol error rate (SER) performances of M-ary PPM and M-ary PAM for various modulation orders, M, which are based on their theoretical performances in AWGN channels [1]. It is seen that, as M increases, the SER performance of M-ary PPM improves due to the increased dimensionality of the *Euclidean space*, while the SER performance of M-ary PAM declines because of the decreased Euclidean distance between the symbols. For M larger than 4, power efficiency of M-ary PPM is better than BPSK (binary PAM). The performance improvement in M-ary PPM comes at the expense of increased transceiver complexity and bounded data rate, since the larger symbol duration limits the pulse repetition frequency (PRF). For example, instead of using 8-ary PPM to transmit 3 bits, BPSK can be used in the same time interval to transmit 8 bits. Spectrally efficient M-ary PAM can be considered to achieve higher data rates with a moderate power efficiency. Compared with M-ary PPM, M-ary biorthogonal signaling has simpler receiver structure since half the number of correlators are employed [1]. Another advantage of this scheme is that it

---

[3]Note that, conversely, M-ary PPM can be implemented in a noncoherent manner, which does not require channel estimation (unless some kind of optimum rake combining is performed).

combines the power efficiency of *M*-ary PPM with the smooth power spectrum of *M*-ary PAM [25].

Application of various data modulation schemes (both binary and *M*-ary) also depend on some other factors, but not only their BER performances. In the rest of the chapter, we will cover the aspects of different data mapping formats related with the spectral characteristics of the resulting signal, transceiver complexity, and performances in practical scenarios (multipath, multiuser interference, and synchronization errors).

4.4 SPECTRAL CHARACTERISTICS

Spectral characteristics of different modulation schemes carry significant importance due to interference effects on the other technologies. Power spectral density (PSD) of the modulated UWB waveform must be within the spectral mask specified by the regulating agencies. For example, in the USA, the FCC allows -41 dBm/MHz power levels for the frequency band 3.1–10.6 GHz. Beyond this band, power levels are extremely low, except in the very low frequency band.

Spectra of transmitted signals when employing different UWB modulations are investigated in [15, 24]. The desired UWB spectrum should be smooth to avoid spectral peaks that violate the FCC declared UWB spectral mask. The application of different data mapping techniques results in different shapes of spectrum that need to be analyzed and compared. Due to the periodicity of the pulses, OOK, positive PAM, and PPM have discrete spectral lines in the signal PSD [30]. This forces a reduction in the overall transmitted power to fit within the required spectral mask. Methods such as pulse dithering [6], which randomly changes the pulse-to-pulse intervals, and polarity randomization [15], which randomly changes pulse polarities, are proposed to smooth the spectrum. On the other hand, antipodal modulation

TABLE 4.2 *M*-ary Data Mapping Formats and SER Performances

M-ary	$\beta^{(k)}_{\lfloor j/N_s \rfloor}$	$\delta\alpha^{(k)}_{\lfloor j/N_s \rfloor}$	SER
M-ary PPM	1	$mT_c, m = 0, 1, \ldots, (M-1)$	$\dfrac{1}{\sqrt{2\pi}} \displaystyle\int_{-\infty}^{+\infty} \{1 - [1 - Q(x)]^{M-1}\} \cdot e^{-\frac{(x - \sqrt{2\psi N_s A E_p / N_0})^2}{2}} dx,$ $\psi = \log_2 M$
M-ary PAM	$2m - 1 - M$ $m = 1, 2, \ldots, M$	0	$\dfrac{2(M-1)}{M} Q\left(\sqrt{\dfrac{6\psi N_s A E_{\text{pav}}}{(M^2 - 1)N_0}}\right),$ $E_{\text{pav}} = \dfrac{(M^2 - 1) E_p}{3},$ $\psi = \log_2 M$

TABLE 4.3 Power Spectral Densities of Various UWB Modulations

Modulation	Power Spectral Density
PAM	$\dfrac{\sigma_a^2}{T_\mathrm{f}}|\Omega(f)|^2 + \dfrac{\mu_a^2}{T_\mathrm{f}^2}\displaystyle\sum_{j=-\infty}^{\infty}\left|\Omega\left(\dfrac{j}{T_\mathrm{f}}\right)\right|^2 \delta_D\!\left(f - \dfrac{j}{T_\mathrm{f}}\right)$
OOK	$\dfrac{1}{T_\mathrm{f}}|\Omega(f)|^2 + \dfrac{1}{T_\mathrm{f}^2}\displaystyle\sum_{j=-\infty}^{\infty}\left|\Omega\left(\dfrac{j}{T_\mathrm{f}}\right)\right|^2 \delta_D\!\left(f - \dfrac{j}{T_\mathrm{f}}\right)$
BPSK	$\dfrac{1}{T_\mathrm{f}}|\Omega(f)|^2$

schemes (BPSK and M-ary PAM) inherently offer a smooth PSD due to random polarities of the modulated pulses. In Table 4.3, power spectral densities of different data mapping formats are summarized [24], where $\Omega(f)$ denotes the Fourier transform of $\omega_{\mathrm{tr}}(t)$, $\delta_D()$ is the Dirac delta function, and $\sigma_a^2, \tilde{}\ \mu_a^2$ are the variance and mean of the weight sequence, respectively (uniform pulse spacing in TH-UWB is assumed with no polarity randomization).

Pulse shaping and design of the TH sequences also plays a crucial role in meeting with regulatory requirements, and avoiding interference from other narrowband technologies. Although other appropriate pulse shaping approaches that fully exploit the FCC spectral mask are possible [31], derivatives of Gaussian pulse are commonly used due to their practicality, where the central UWB frequency increases with derivative order [32, 33]. In order to meet with FCC requirements, the fifth derivative of the Gaussian pulse is commonly reported in the USA. By appropriately shaping the pulses, or appropriately designing the TH sequences [34, 35], one can also place a notch in the interferer's spectrum to minimize the degradation on the UWB devices.

Data mapping or modulation takes place at transmitter at the expense of certain hardware components. Conversely, demodulation at the receiver requires additional components. Hardware complexity influences the choice of data modulation schemes and is discussed in the next section.

4.5 DATA MAPPING AND TRANSCEIVER COMPLEXITY

The performance of a modulation scheme also depends on implementation under practical conditions. The level of hardware implementation and computational complexity plays an important role in determining which modulation to use in what application. Noncoherent demodulation, such as envelope detection or square-law detection, is commonly used to decrease the complexity [24] and cost [36] of the receivers. Therefore, if the transceiver complexity and cost are the primary concerns, a scheme that enables noncoherent demodulation (OOK, positive PAM, PPM, and M-ary PPM) can be considered. On the other hand, BPSK and M-ary PAM require coherent demodulation since the information is embedded in the polarities of the pulses. The number of cross-correlators is another issue that increases the receiver complexity, and *M*-ary orthogonal schemes must be carefully designed considering the complexity/performance tradeoff.

Discussion of transceiver complexity opens a door for other practical conditions or parameters that play as essential role in wireless channel and are important determining factor for right modulation for right application.

4.6 MODULATION PERFORMANCES IN PRACTICAL CONDITIONS

The performance of different data mapping schemes can be affected differently in practical scenarios. In this section, we will address the effects of multipath and multiuser interference on various data modulation schemes. Also, the effect of timing jitter (or finger estimation error) on different data mapping formats will be discussed.

4.6.1 Effects of Multipath

Due to reflection, diffraction, and scattering effects, the transmitted signal arrives at the receiver through multiple paths with different delays. In narrowband systems, most of the multipath components arrive within the symbol duration, and therefore the receiver observes as if there was a single multipath component which has an extremely large fading margin. The time resolution of UWB systems allows the receiver to observe the individual multipath components, and makes it possible to collect the energy using rake receivers. A commonly used double-exponential channel model for UWB systems is developed in [37], and the channel impulse response is given by

$$h(t) = X \sum_{l \geq 0} \sum_{k \geq 0} \alpha_{l,k} \delta_D(t - T_l - \tau_{l,k}), \qquad (4.7)$$

where X is a log–normal random variable to model shadowing, and $\alpha_{k,l}$ are the tap gains for the kth tap in the lth cluster. The delays T_l and $\tau_{k,l}$ are the delays of the first multipath component of a cluster, and the relative delays of the multipath components within the cluster, respectively, both of which are conditionally exponentially distributed.

By using rake receivers, it is possible to collect the energy at the delayed taps. All-rake (*ARake*), selective-rake (*SRake*), or partial-rake (*PRake*) receivers are all feasible approaches to collect all, strongest, or first arriving resolvable multipath components, respectively [38]. Optimal combining of the multipath components in white noise is achieved by maximal ratio combining (MRC), where the finger weights are designed based on the channel tap weights to maximize the output SNR. Minimum mean square error (MMSE) combining of the taps yields optimal performance in correlated channels, however requires the computation/estimation of the correlation matrix over various multipath components. A reduced complexity combining technique that does not require either the estimates of the fading amplitudes or the correlation matrix is equal gain combining (EGC), where all the multipath components are weighted equally. Unlike MRC or MMSE, EGC can be used with noncoherent modulation schemes, such as OOK and orthogonal PPM with noncoherent detection.

If the spacing between certain multipath components in Equation (4.7) is smaller than the pulse duration T_c, this yields inter-pulse interference, and multipath components start acting as data modulation [39, 40]. This effect is more pronounced for PPM than BPSK, as the effective duration spanned by PPM signals is twice that of BPSK signals. Computer simulations are performed to observe the data modulation and inter-pulse interference effects in both data mapping formats. The channel model CM1 of [37] is used, 50000 realizations of the channel are generated, the channel impulse responses are truncated into 60 ns, and the average BER performance is evaluated. The pulse T_c is taken as 0.8 ns, yielding inter-pulse interference between various multipath components. An *ARake* receiver and *SRake* receivers with 1, 3, and 10 fingers are employed for both modulation formats to capture the energy. In Figures 4.13 and 4.14, it is observed that (as discussed in previous sections) there is a 3 dB performance difference between the ideal performance of BPSK and PPM. However, when a single correlator is used, this performance difference can be as large as 4 dB, due to the fact that the data modulation effect can be more catastrophic for PPM. Also note that in both data mapping formats, due to the inter-pulse interference effects, there exists a break point where increasing the SNR begins to yield somewhat linear decrease in the BER, rather than the waterfall-like

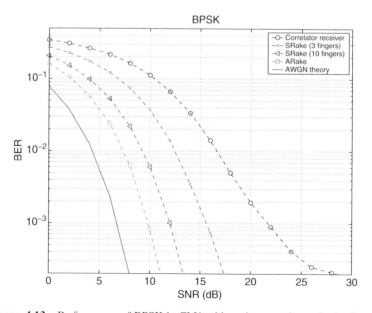

Figure 4.13 Performance of BPSK in CM1 with various numbers of rake fingers.

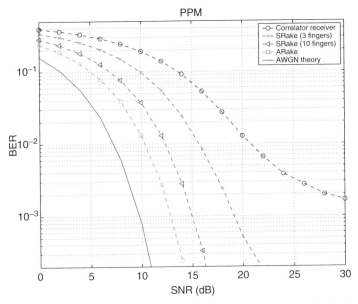

Figure 4.14 Performance of PPM in CM1 with various numbers of rake fingers.

decrease. Due to the same effect, in both figures, we notice that it is not possible to reach ideal performance even if *ARake* receivers are used.

4.6.2 Effects of Multiple Access Interference

The composite signal from N_u asynchronous users arriving at the receiver is given by

$$r_{\text{mai}}(t) = \sum_{\psi=1}^{N_u} s^{(\psi)}(t - \tau_\psi) + n(t), \qquad (4.8)$$

where τ_ψ is the random delay of the ψth user. Once this signal is sampled at the desired user's receiver, the matched filter output is given by (considering initially BPSK modulation)

$$Y = b^{(\xi)} \sqrt{E_b^{(\xi)} N_s} + M + N, \qquad (4.9)$$

where ξ refers to the desired user, $b^{(\xi)}$ is the desired users's symbol, $N \sim \mathcal{N}(0, \sigma^2)$ is the output noise and M is the total MAI, which is the sum of interference terms from the interfering users:

$$M = \sum_{k=1, k \neq \xi}^{N_u} M_k, \qquad (4.10)$$

where M_k is the MAI from user k and N_u is the total number of users. Similar to the approach in [41], when random polarity codes are used for each pulse, we can approximate the MAI from user k by the following Gaussian random variable, when the number of pulses per information symbol for user ξ, N_s, is large:

$$M_k \sim \mathcal{N}\left(0, \frac{E_{\text{rp}}^{(k)}}{N_h}\right), \qquad (4.11)$$

where $E_{\text{rp}}^{(k)}$ is the energy of a received pulse from user k. Then, we can express the *signal to interference plus noise ratio* (SINR) of the system for user ξ as

$$\text{SINR} \approx \frac{N_s E_{\text{rp}}^{(\xi)}}{\sigma_n^2 + \frac{1}{N_h} \sum_{\substack{k \neq \xi \\ k=1}}^{N_u} E_{\text{rp}}^{(k)}}, \qquad (4.12)$$

which can be directly inserted into $Q(\cdot)$ as $Q(\sqrt{\text{SINR}})$ to evaluate the BER for BPSK. Similar approaches can be repeated to calculate BER as $Q[\sqrt{(\text{SINR}/2)}]$ for PPM (due to doubled noise effects in both pulse positions [1]).

To demonstrate the performance differences of different modulation schemes (orthogonal PPM, BPSK) in such an asynchronous multiuser environment, we have done simple computer simulations, and compared them with the theoretical expressions. The parameters are selected as $N_s = 1$, $T_c = 0.8$ ns, SNR = 7 dB, $N_u = 100$, and $T_f = 20T_c$. Time delays τ_ψ of each user are selected randomly between 0 and $19T_c$ as multiples of pulse duration T_c. It is observed from Figure 4.15 that BPSK outperforms other data mapping formats for all numbers of users. Also, Gaussian approximation is seen to show some deviation for small numbers of users, while showing a good agreement when the number of users is large.

4.6.3 Effects of Timing Jitter and Finger Estimation Error

Accurate synchronization is extremely important for UWB communication systems due to extremely short duration pulses employed. The reasons for timing errors in UWB systems include timing jitter (which can be due to transceiver clock instabilities), and finger estimation error. While the timing jitter is typically on the order of 10 ps with the current transistor technology and is usually modeled by a Gaussian or uniform distribution [42, 43], distribution of finger estimation error depends on the pulse shape and the noise power [42], but may as well be modeled via uniform distribution for analytical tractability [44].

Timing mismatches between the correlator template and the received signal can result in serious degradations in the BER. Given a fixed amount of mismatch ϵ_j between the correlator template and the received signal, the SNR degradation is a function of the normalized pulse correlation function, and will be different for different data mapping formats. The BER performances of four different modulation formats for a fixed timing misalignment are presented in Table 4.4, and are depicted in Figure 4.16 for SNR = 10 dB. It is observed that the degradations in BPSK and PPM are similar, while OOK performs worse for large jitter. BER equation in

4.6 MODULATION PERFORMANCES IN PRACTICAL CONDITIONS

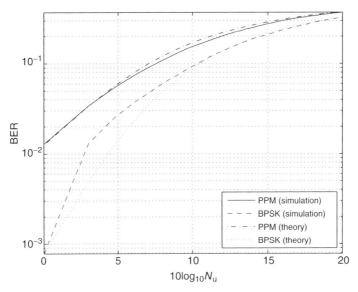

Figure 4.15 Theoretical and simulated performance of different modulation schemes in an asynchronous multiuser channel.

Table 4.4 for OOK implies that timing jitter does not affect the *false alarm rate*, but increases the *missed detection rate*, yielding biased decisions towards zero. The problem will be more pronounced in M-ary PAM, which also uses threshold detection. Note that, since the value of the normalized autocorrelation function $R(\epsilon_j)$ in Equation (4.4) will be unity at $\epsilon_j = 0$, the performances in Table 4.4 boil down to the ideal performances in Table 4.1 under perfect synchronization.

Timing jitter becomes a serious problem for multiband schemes as well, since the autocorrelation functions of the pulses used in higher-order bands decay much faster than the autocorrelation function of the *monopulse*. In order to see the effect of timing jitter, a sample multiband scheme is constructed by dividing the 3.1–10.6

TABLE 4.4 BER Performances Under Timing Misalignment

Modulation	BER
Orthogonal PPM	$Q\left(\sqrt{\dfrac{E_p}{N_0}R^2(\epsilon_j)}\right)$
Optimum PPM	$Q\left(\sqrt{\dfrac{E_p}{N_0}\dfrac{[R(\epsilon_j)-R(\delta-\epsilon_j)]^2}{R(0)-R(\delta)}}\right)$
BPSK	$Q\left(\sqrt{\dfrac{2E_p}{N_0}R^2(\epsilon_j)}\right)$
OOK	$\dfrac{1}{2}\left[Q\left(\sqrt{\dfrac{E_p}{N_0}}\right) + Q\left(\sqrt{\dfrac{E_p}{N_0}[2R(\epsilon_j)-1]^2}\right)\right]$

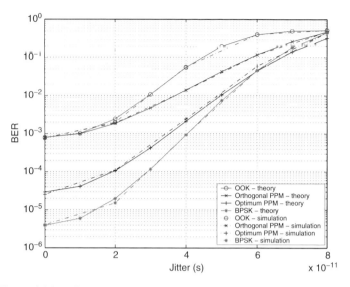

Figure 4.16 Effect of timing jitter on BER performances (SNR = 10 dB).

GHz band into 10 bands of 750 MHz each. The autocorrelation functions of the pulse used in band 1 (3.1–3.85 GHz) of the multiband scheme and the pulse in Equation (4.3), which is used in the standard UWB scheme, are compared in Figure 4.17. It is seen that, since the central frequencies of both systems are

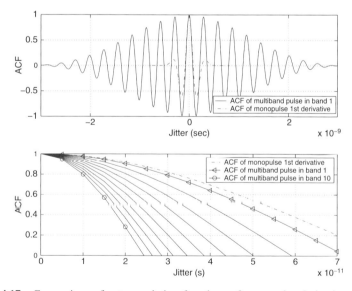

Figure 4.17 Comparison of autocorrelation functions of monopulse derivative and pulse used in band 1 of multiband, and comparison of the autocorrelations of the pulses used in all bands.

close, similar degradation will be observed. Figure 4.17 also compares the autocorrelation functions of the pulses used in different bands of the multiband scheme with respect to the fixed timing jitter value that ranges from 0 to 70 ps. It is observed that degradation due to timing jitter when using the multiband scheme is worse for higher order bands.

A similar analysis can be repeated for UWB schemes that employ modified Hermite polynomial-based pulses. Using these polynomials, it is possible to construct a new pulse that is orthogonal to the previous pulses by using certain transformations. Such higher-order pulses have larger number of zero crossings and their autocorrelation functions decay faster, implying less robustness against timing jitter [45].

4.7 CONCLUSION

UWB signal modulation remains a challenge in current and future wireless communication systems and is subject to ongoing research to achieve the best suitable modulation option for UWB systems. In this chapter, several popular UWB modulation options have been presented and compared in terms of the BER, spectral characteristics, and transceiver complexities. Effects of multipath, multiple access, narrowband interference, and timing jitter are analyzed and verified by simulation. It is shown that PPM performance degrades in multipath and multi-user environments since symbols occupy larger time durations. Compared with other modulations, OOK and M-ary PAM are more susceptible against timing jitter. Although a multiband scheme provides interference mitigation and flexibility of data rate, it is more susceptible to timing jitter, especially at the higher frequency bands. In summary, BPSK is preferred for its high power efficiency and smooth spectrum; OOK for its simple transceiver structure; M-ary PPM for its improved power efficiency; and M-ary PAM for higher data rates. In terms of coexistence with the other narrowband wireless technologies, BPSK modulation is the best choice since the spectral lines in the power spectral density of the signal are suppressed. If other modulation schemes are to be used, pulse dithering is required to smooth the spectrum. It is also possible to place a notch at the interferer frequency [34].

REFERENCES

1. J. G. Proakis, *Digital Communications*, 4th edn. New York: McGraw-Hill, 2001.
2. G. L. Stuber, *Principles of Mobile Communication*, 2nd edn. Boston, MA: Kluwer Academic, 2001.
3. R. A. Scholtz, "Multiple access with time-hopping impulse modulation," in *Proc. IEEE Mil. Commun. Conf. (MILCOM)*, vol. 2, Maryland, USA, October 1993, pp. 447–450.
4. P. Runkle, J. McCorkle, T. Miller, and M. Welborn, "DS-CDMA: the modulation technology of choice for UWB communications," in *Proc. Ultrawideband Syst. Technol. (UWBST)*, Reston, VA, November 2003, pp. 364–368.

5. A. Batra, J. Balakrishnan, G. R. Aiello, J. R. Foerster, and A. Dabak, "Design of a multiband OFDM system for realistic UWB channel environments," *IEEE Trans. Microwave Theory Techniques*, vol. 52, no. 9, pp. 2123–2138, September 2004.
6. M. Win and R. A. Scholtz, "Ultra-wide bandwidth time-hopping spread-spectrum impulse radio for wireless multiple-access communications," *IEEE Trans. Commun.*, vol. 48, no. 4, pp. 679–689, April 2000.
7. M. Iacobucci and M. G. Benedetto, "Multiple access design for impulse radio communication systems," in *Proc. IEEE Int. Conf. Commun. (ICC)*, vol. 2, New York, April 2002, pp. 817–820.
8. M. Z. Win, "A unified spectral analysis of generalized time-hopping spread-spectrum signals in the presence of timing jitter," *IEEE J. Select Areas Commun.*, vol. 20, no. 9, pp. 1664–1676, December 2002.
9. M. Z. Win and R. A. Scholtz, "Impulse radio: how it works," *IEEE Commun. Lett.*, vol. 2, no. 2, pp. 36–38, February 1998.
10. I. Guvenc, H. Arslan, S. Gezici, and H. Kobayashi, "Adaptation of multiple access parameters in time hopping UWB cluster based wireless sensor networks," in *Proc. IEEE Mobile Adhoc Sensor Systems Conf. (MASS)*, Fort Lauderdale, FL, October 2004, pp. 235–244.
11. I. Guvenc and H. Arslan, "TH sequence construction for centralised UWB-IR systems in dispersive channels," *IEE Electronics Letters*, vol. 6, pp. 491–492, April 2004.
12. J. R. Foerster, "The performance of a direct-sequence spread ultra-wideband system in the presence of multipath, narrowband interference, and multiuser interference," in *Proc. IEEE Conf. UWB Syst. Technol. (UWBST)*, vol. 3, Baltimore, MD, May 2002, pp. 87–91.
13. V. S. Somayazulu, "Multiple access performance in UWB systems using time hopping vs. direct sequence spreading," in *Proc. Wireless Commun. Networking Conf. (WCNC)*, vol. 2, Orlando, FL, March 2002, pp. 522–525.
14. J. R. Foerster, "Ultra-wideband technology enabling low-power, high-rate connectivity," in *Proc. IEEE CAS. Workshop Wireless Commun. Network*, California, USA, September 2002.
15. Y. P. Nakache and A. F. Molisch, "Spectral shape of UWB signals-influence of modulation format, multiple access scheme and pulse shape," Technical Report (TR2003-40), May 2003.
16. S. Gezici, H. Kobayashi, H. V. Poor, and A. F. Molisch, "Performance evaluation of impulse radio UWB systems with pulse-based polarity randomization," *Signal Processing, IEEE Trans. [see also Acoust., Speech, Signal Process., IEEE Trans.]*, vol. 53, no. 7, pp. 2537–2549, July 2005.
17. S. Gezici, H. Kobayashi, and H. V. Poor, "A comparative study of pulse combining schemes for impulse radio UWB systems," in *Proc. IEEE Sarnoff Symp.*, Princeton, NJ, April 2004, pp. 7–10.
18. I. Guvenc and H. Arslan, "Design and performance analysis of TH sequences for UWB-IR systems," in *Proc. Wireless Commun. Networking Conf. (WCNC)*, vol. 2, Atlanta, GA, April 2004, pp. 914–919.
19. M. O. Wessman and A. Svensson, "Comparison between DS-UWB, multiband UWB and multiband OFDM on IEEE UWB channels," in *Proc. Nordic Radio Symp. and Finnish Wireless Commun. Workshop*, Oulu, August 2004.

20. K. Mandke, H. Nam, L. Yerramneni, C. Zuniga, and T. Rappaport, "The evolution of ultra wide band radio for wireless personal area networks," *High Frequency Electron.*, pp. 22–32, September 2003.
21. K. C. Chen, "Direct detect modulations of high speed indoor diffused infrared wireless transmission," in *Proc. IEEE Int. Symp. Pers. Indoor Mob. Rad. Commun. (PIMRC)*, vol. 4, The Netherlands, September 1994, pp. 1096–1100.
22. J. Zhang, "Modulation analysis for outdoors applications of optical wireless communications," in *Proc. IEEE Int. Conf. Commun. Technol. (ICCT)*, vol. 2, Beijing, August 2000, pp. 1483–1487.
23. P. Okrah, Digital radio modulation: A wireless reference guide, *Commun. Syst. Des. Mag.*, March 2002.
24. M. Welborn, "System considerations for ultra-wideband wireless networks," in *Proc. IEEE Radio and Wireless Conf. (RAWCON)*, Boston, MA, August 2001, pp. 5–8.
25. I. Guvenc and H. Arslan, "On the modulation options for UWB systems," in *Proc. IEEE Mil. Commun. Conf. (MILCOM)*, vol. 2, Boston, MA, October 3003, pp. 892–897.
26. M. Hamalainen, R. Tesi, J. Iinatti, and V. Hovinen, "On the performance comparison of different UWB data modulation schemes in AWGN channel in the presence of jamming," in *Proc. IEEE Radio and Wireless Conf. (RAWCON)*, August 2002, pp. 83–86.
27. J. McCorkle, "Why such uproar over ultra-wideband?" *Commun. Syst. Des. Mag.*, March 2002.
28. L. B. Michael, M. Ghavami, and R. Kohno, "Effect of timing jitter on Hermite function based orthogonal pulses for ultra wideband communication," in *Proc. Int. Symp. on Wireless Pers. Multimedia Commun.*, Aalborg, Denmark, September 2001, pp. 441–444.
29. G. Cariolaro, T. Erseghe, and L. Vangelista, "Exact spectral evaluation of the family of digital pulse interval modulated signals," *IEEE Trans. Inform. Theory*, vol. 47, no. 7, pp. 2983–2992, November 2001.
30. C. Muller, S. Zeisberg, H. Seidel, and A. Finger, "Spreading properties of time hopping codes in ultra wideband systems," in *Proc. (IEEE) Int. Symp. Spread-Spectrum Tech. Appl.*, Prague, 2–5 September 2002, pp. 64–67.
31. X. Luo, L. Yang, and G. B. Giannakis, "Designing optimal pulse-shapers for ultra-wideband radios," *J. Commun. Networks (JCN)*, vol. 5, no. 4, pp. 344–353, December 2003.
32. M. G. D. Benedetto and B. R. Vojcic, "Ultra wide band wireless communications: A tutorial," *J. Commun. Networks (JCN)*, vol. 5, no. 4, pp. 290–302, December 2003.
33. M. G. D. Benedetto and G. Giancola, *Understanding Ultra Wide Band Radio Fundamentals*, 1st edn. Englewood Cliffs, NJ: Prentice Hall, 2004.
34. L. Piazzo and J. Romme, "Spectrum control by means of the TH code in UWB systems," in *Proc. (IEEE) Vehic. Technol. Conf. (VTC)*, vol. 3, Jeju, Korea, April 2003, pp. 1649–1653.
35. M. S. Iacobucci, M. G. D. Benedetto, and L. D. Nardis, "Radio frequency interference issues in impulse radio multiple access communications systems," in *Proc. IEEE Conf. UWB Syst. Technol. (UWBST)*, vol. 3, Baltimore, MD, May 2002, pp. 293–296.
36. G. R. Aiello, L. Taylor, and M. Ho, "A UWB architecture for wireless video networking," in *Proc. IEEE Int. Conf. Consumer Elect. (ICCE)*, June 2001, pp. 18–19.

37. J. Foerster, "IEEE P802.15 working group for wireless personal area networks (WPANs), channel modeling sub-committee report—final," March 2003. Available at: www.ieee802.org/15/pub/2003/Mar03/
38. D. Cassioli, M. Z. Win, F. Vatalaro, and A. F. Molisch, "Performance of low-complexity RAKE reception in a realistic UWB channel," in *Proc. IEEE Int. Conf. Commun (ICC)*, vol. 2, New York, April 2002, pp. 763–767.
39. L. Ge, G. Yue, and S. Affes, "On the BER performance of pulse-position-modulation UWB radio in multipath channels," in *Proc. IEEE Conf. UWB Syst. Technol. (UWBST)*, vol. 3, Baltimore, MD, May 2002, pp. 231–234.
40. F. E. Aranda, N. Brown, and H. Arslan, "Rake receiver finger assignment for Ultra-wideband radio," in *Proc. IEEE Workshop Sig. Processing Advances Wireless Commun. (SPAWC)*, Rome, June 2003, pp. 239–243.
41. S. Gezici, H. Kobayashi, H. V. Poor, and A. F. Molisch, "Performance evaluation of impulse radio uwb systems with pulse-based polarity randomization in asynchronous multiuser environments," in *Proc. Wireless Commun. Networking Conf. (WCNC)*, vol. 2, Atlanta, GA, March 2004, pp. 908–913.
42. I. Guvenc and H. Arslan, "Performance evaluation of UWB systems in the presence of timing jitter," in *Proc. IEEE Ultrawideband Syst. Technol. Conf. (UWBST)*, Reston, VA, November 2003, pp. 136–141.
43. W. M. Lovelace and J. K. Townsend, "The effects of timing jitter and tracking on the performance of impulse radio," *IEEE J. Select. Areas Commun.*, vol. 20, no. 9, pp. 1646–1651, December 2002.
44. H. Sheng, R. You, and A. M. Haimovich, "Performance analysis of ultra-wideband rake receivers with channel delay estimation errors," in *Proc. CISS*, Princeton, NJ, 2004, pp. 921–926.
45. L. B. Michael, M. Ghavami, and R. Kohno, "Effect of timing jitter on Hermite function based orthogonal pulses for ultra wideband communication," in *Proc. 4th Int. Symp. Wireless Personal Multimedia Commun.*, Aalborg, September 2001, pp. 441–444.

CHAPTER 5

Ultra Wideband Pulse Shaper Design

ZHI TIAN, TIMOTHY N. DAVIDSON, XILIANG LUO,
XIANREN WU and GEORGIOS B. GIANNAKIS

5.1 INTRODUCTION

With the release of the U.S. FCC spectral masks in 2002 [7], UWB technology has attracted great interest as a means of wresting additional capacity from the already heavily utilized store of wireless bandwidth. The scarcity of bandwidth resources coupled with the capability of UWB radios to overlay existing systems, welcomes UWB connectivity for short-range, high-data-rate wireless indoor pico-nets and potentially for low-power wireless sensor networks outdoors [21, 32]. However, the benefits of UWB signaling may be offset by the interference to and from existing systems operating over the same frequency bands. For spectrum overlay control, the FCC regulations imposed spectral masks that strictly constrain the transmission power of a UWB signal to be well below the noise floor in all bands. The spectrum of a transmitted signal is influenced by the modulation format, the multiple access scheme, and most critically by the spectral shape of the underlying UWB pulse. The choice of the pulse shape is thus a key design decision in UWB systems, with the following design objectives to fulfill:

1. *Efficient Spectral Utilization*—the transmission reliability of a UWB system is determined by the received signal-to-noise ratio (SNR). Given the stringent transmission power limitations, maximization of the received SNR requires efficient utilization of the bandwidth and power allowed by the FCC masks.
2. *Flexible Interference Avoidance*—to avoid interference to (and from) co-existing narrowband systems, the corresponding frequency bands must be avoided. The avoidance mechanism should allow for sufficient flexibility to adapt to the changing nature of co-existing services in terms of their number and center frequencies, and to accommodate the spectrum regulations in different countries as well.

Ultra Wideband Wireless Communication. Edited by Arslan, Chen, and Di Benedetto
Copyright © 2006 John Wiley & Sons, Inc.

3. *Multiple Orthogonal Pulses*—as an alternative to conventional time-hopping or direct-sequence techniques, UWB multiple access can be accomplished via multiple orthogonal pulse shapes. These orthogonal pulses can be nonoverlapping in frequency, as in multiband systems, or can even have overlapping spectra for high-rate multiple access with large diversity gain.

4. *Convenient Implementation*—the pulse shape design must consider the implementation challenges imposed by the ultrawide bandwidth, and the extent to which the design can be readily implemented using off-the-shelf hardware components. In addition, the design problems ought to be formulated in a manner that enables the application of reliable and efficient solution algorithms that require little or no interaction with the designer.

To design pulse shapers with desirable spectral properties, two approaches can be readily employed: carrier modulation and/or baseband analog/digital filtering of the baseband pulse shaper. The former relies on local sinusoidal oscillators at the UWB transmitter and receiver, which are prone to mismatch and can give rise to carrier frequency offset/jitter (CFO/CFJ). Current multiband UWB access proposals consider this approach to facilitate flexible and scalable spectrum use [15], but multiple CFO/CFJs emerge in the presence of frequency hopping (FH). UWB impulse radio, on the other hand, is a carrierless system that is built around baseband transceivers in order to reduce implementation cost and complexity. Unfortunately, the widely adopted baseband Gaussian monocycle pulse [22, 27] exhibits a poor fit to the FCC spectral masks and thus is not desirable for practical usage. This motivates alternative approaches for obtaining a pulse shape that satisfies the FCC masks. Although passing the (Gaussian) pulse through a baseband analog filter can re-shape the pulse without introducing CFO/CFJ, it is well known that analog filters of ultra wide bandwidth are quite expensive to produce and are not as flexible when compared with digital filters, which are accurate, highly linear, and perfectly repeatable. One approach to the design of digital pulse shapers that comply with the FCC spectral masks is to employ prolate spheroidal wave functions to generate pulses from the dominant eigenvectors of a channel matrix that is constructed by sampling the spectral mask [6, 18, 33]. Pulses generated from different eigenvectors are mutually orthogonal, but require a high sampling rate that could lead to implementation difficulties. Other pulse shaping methods include exploiting the properties of Hermite orthogonal polynomials [9], and fine-tuning higher-order derivatives of the Gaussian pulse [24], the latter of which is not flexible in fitting FCC spectral mask changes as well as other regional regulations. All these pulses do not achieve optimal spectral utilization. For flexible pulse shaping and convenient use of off-the-shelf hardware components, digital FIR filter design solutions may be more appropriate [12, 30, 32].

In this chapter, we will address the UWB pulse design issue by putting forward optimal design methodologies for waveforms synthesized by a digital FIR filter. A convenient basis pulse, such as the Gaussian monocycle, is used as the building block. Prior to modulating this basis pulse, the channel symbols are passed through a linear FIR prefilter, whose filter tap coefficients are carefully designed to generate the desirable synthesized pulse. Section 5.2 describes the spectral properties of the transmitted UWB signal and discusses the pulse design objectives. Section 5.3 introduces the FIR filter structure for UWB waveform synthesis. Section 5.4 focuses on single pulse designs, where the conventional Parks–McClellan algorithm [13] is briefly discussed, followed by derivations of optimal convex formulations of the pulse design problem that generate pulses with maximum spectral utilization under the spectral mask constraints. Multiple orthogonal pulse design is addressed via a sequential design strategy in Section 5.5. Design examples are provided in Section 5.6 for single-band, multiband and orthogonal pulses with overlapping spectra, along with comparisons of the spectral utilization efficiency of various pulses, and evaluation of the system-level impact of pulse shape design in terms of BER performance and robustness to narrowband interference. Concluding remarks are provided in Section 5.7.

5.2 TRANSMIT SPECTRUM AND PULSE SHAPER

In a UWB impulse radio system, each information symbol is conveyed over a train of N_f repeated basic pulses, with one pulse per frame of duration T_f corresponding to a pulse repetition frequency (PRF) of $1/T_f$. Each unit-energy pulse $p(t)$ has an ultra short duration $T_p (T_p \ll T_f)$ at the nanosecond scale, and hence occupies an ultra wide bandwidth. The equivalent symbol signature waveform is $p_s(t) := \sum_{n=0}^{N_f-1} p(t - c_n T_c - n T_f)$, and has symbol duration $T_s := N_f T_f$, where the sequence $\{c_n\}_{n=0}^{N_f-1}$ represents the user-specific pseudo-random time-hopping (TH) code with $c_n T_c < T_f, \forall n \in [0, N_f - 1]$. Let $b_k \in \{\pm 1\}$ be independent and identically distributed (i.i.d.) binary data symbols with energy $\mathcal{E}_s$ spread over N_f frames. When pulse amplitude modulation (PAM) is used, the transmitted PAM UWB waveform is given by:

$$u(t) = \sqrt{\mathcal{E}_s/N_f} \sum_k b_k p_s(t - k T_s). \quad (5.1)$$

The power spectral density (PSD) of $u(t)$ is then given by

$$\Phi_{uu}(f) = \frac{\mathcal{E}_s}{N_f} \cdot \frac{1}{T_s} |P_s(f)|^2, \quad (5.2)$$

where $P_s(f)$ is the Fourier transform (FT) of $p_s(t)$, and depends on both $p(t)$ and the TH code $\{c_n\}_{n=0}^{N_f-1}$. Specifically, $P_s(f)$ can be expressed as

$$P_s(f) = P(f) \sum_{n=0}^{N_f-1} e^{-j2\pi f n T_f} e^{-j2\pi f c_n T_c}, \quad (5.3)$$

where $P(f)$ is the FT of $p(t)$. When the TH code $\{c_n\}_{n=0}^{N_f-1}$ is independent and uniformly distributed over $[0, N_c - 1]$ with integer values, $\Phi_{uu}(f)$ can be approximated as [26]:

$$\Phi_{uu}(f) \approx \alpha |P(f)|^2, \tag{5.4}$$

where $\alpha = \mathcal{E}_s/T_f$ is a constant that depends on the frame interval and the energy per symbol. A similar result is also derived in [12] for pulse position modulation (PPM) UWB waveforms. In general, the spectral shape of a UWB signal is influenced by the modulation format, the multiple access scheme and the pulse shape. For radios operating above 960 MHz, there is a limit on the peak emission level contained within a 50 MHz bandwidth centered on the peak frequency f_M, at which the highest radiated emission occurs. UWB emissions are average-limited for PRFs greater than 1 MHz and peak-limited for PRFs below 1 MHz [17]. It is convenient to use a zero mean information stream (as in PAM) to control the spectral characteristics of the modulated signal. For combinations of nonequiprobable systems and nonantipodal modulation schemes such as PPM, spectral spikes may appear every $1/T_s$ Hz, which can be quite dense within the pulse bandwidth determined by $1/T_p$. Spectral lines can be reduced in number to every $1/T_c$ Hz when symbol-periodic random TH is employed, and eliminated by pulse polarity randomization strategies [16]. Even though spectral spikes are undesirable in military applications for low probability of detection/interception (LPD/LPI) concerns, the severity of interference from UWB transmissions to legacy systems depends on the average power, which is nevertheless small after averaging the power of spectral lines over the resolution bandwidth. In a nutshell, a UWB transmitter can be treated as a linear amplifier of the pulse shaper $p(t)$, as in Equation (5.4). Hence, the UWB pulse design problem is equivalent to designing the basic pulse $p(t)$ to meet the prescribed system specifications.

In order for $\Phi_{uu}(f)$ to satisfy the FCC regulatory requirements, the power spectrum of the UWB pulse $p(t)$ must satisfy $|P(f)|^2 \leq S_{FCC}(f)$, where $S_{FCC}(f)$ is a version of the regulatory mask normalized by a scalar $1/\alpha$; cf. Equation (5.4). The U.S. FCC First Order and Report (R&O) defined three spectral masks for imaging systems, communication and measurement systems, and vehicular radar systems, respectively. In particular, the bandwidth and spectral mask assigned for indoor communications is illustrated in Figure 5.1. We observe that most of the UWB signal power should be allocated to the band 3.1–10.6 GHz, while considerable attenuation is imposed in other regions of the spectrum to avoid interference to legacy narrowband systems, especially for frequencies up to 3.1 GHz. Accordingly, we define $\mathcal{F}_\mu := \{f | f \subset [3.1, 10.6] \text{GHz}\}$ as the UWB passband. In practice, one typically imposes a tighter mask, say $S(f)$, on $|P(f)|^2$ in the design phase, that is,

$$S_p(f) = |P(f)|^2 \leq S(f), \tag{5.5}$$

where $S(f) \leq S_{FCC}(f)$ for all f. A tighter mask is needed not only to provide stronger interference mitigation to a frequency band of concern, but also to ensure that the regulatory mask is satisfied in practical implementations. For example, the spectrum

5.2 TRANSMIT SPECTRUM AND PULSE SHAPER

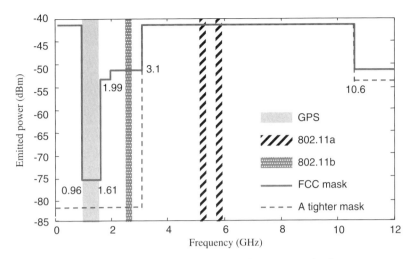

Figure 5.1 FCC spectral mask for indoor communications.

from 0.96 GHz to 3.1 GHz hosts GPS and 802.11.b/g bands, which prompts us to consider a tighter mask $S_T(f)$ over this spectrum, shown by the dotted line in Figure 5.1. Enforcing the tighter mask also allows some margin for "spectral re-growth" due to nonlinearities in the transmitter.

The goal of UWB pulse shape design is to find a waveform $p(t)$ that has high spectral utilization efficiency, while at the same time complying with the spectral mask $S(f)$. The spectral utilization efficiency can be measured in terms of the normalized effective signal power (NESP), which is the ratio of the power transmitted in the designated passband $\mathcal{F}_p$ of the spectral mask over the total power that is permissible under the given mask. For any $\mathcal{F}_p$, the NESP is defined as $\bar{\psi} = \int_{\mathcal{F}_p} S_p(f) \, df / \int_{\mathcal{F}_p} S(f) \, df$. Because $S(f)$ is independent of our design parameters, maximizing $\bar{\psi}$ is equivalent to maximizing

$$\psi = \int_{\mathcal{F}_p} S_p(f) \, df. \tag{5.6}$$

To motivate the need for efficient UWB pulse shaper design, we first consider the Gaussian monocycle, which is straightforward to produce over a large bandwidth by baseband antennas [11]. Due to the derivative characteristics of the antenna, when a Gaussian pulse $x(t) = [A/(\sqrt{\pi}\tau)]e^{-t^2/\tau^2}$ is transmitted, the output of the transmitter antenna can be modeled by the first derivative of the Gaussian pulse in the form $x^{(1)}(t) = -[2At/(\sqrt{\pi}\tau^3)]e^{-t^2/\tau^2}$, where the superscript $^{(n)}$ denotes the nth derivative. The amplitude spectrum of the nth derivative of a Gaussian pulse is [16]

$$|X_n(f)| = A(2\pi f)^n e^{-(\pi f \tau)^2}. \tag{5.7}$$

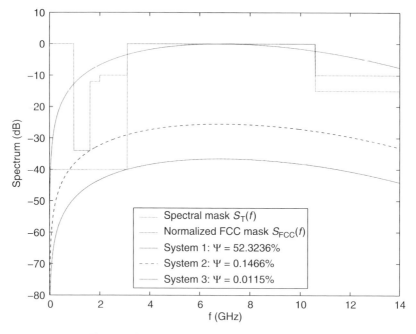

Figure 5.2 Power spectrum of the Gaussian pulse.

The peak of this spectrum is located at frequency $f_M = \sqrt{n}/(\sqrt{2}\pi\tau)$ and has a value $|X_n(f_M)| = A(\sqrt{2n}/\tau)^n e^{-n/2}$. For the first derivative of the Gaussian pulse ($n = 1$), the pulse width that contains 99.99% of the total pulse energy is well approximated by $T_x \approx 4.95\tau$.

Consider the Gaussian monocycle $x^{(1)}(t)$ of width $T_x = 0.1626\,\text{ns}$ ($\tau = 32.9\,\text{ps}$) at the transmitter antenna output, which corresponds to a peak frequency of $f_M = 6.85\,\text{GHz}$. We can scale the amplitude A to generate three system designs, as illustrated in Figure 5.2: System 1 complies with the FCC mask only in the UWB passband, system 2 has the largest possible amplitude under the FCC mask constraint, while system 3 complies with the tighter mask. It is clear that system 1 violates the FCC mask in the stopbands, whereas systems 2 and 3 comply with their respective spectral masks at the expense of very low spectrum utilization. As a result, the Gaussian monocycle might not be a wise choice for UWB systems and judicious pulse design is needed for better UWB spectrum utilization. In the remainder of this chapter, we discuss several candidate design techniques and the implementation of the resulting pulses.

5.3 FIR DIGITAL PULSE DESIGN

A convenient method for synthesizing a pulse shape is to take a linear combination of delayed versions of a "basis" pulse, $q(t)$. We will take this approach, and will

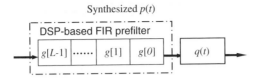

Figure 5.3 DSP-based UWB pulse design.

consider pulses of the form [12, 32]:

$$p(t) = \sum_{k=0}^{L-1} g[k]q(t - kT_0), \quad (5.8)$$

where T_0 is the delay interval, and the set $\{g[k]\}_{k=0}^{L-1}$ contains the L (real) coefficients that are to be designed. Natural choices for $q(t)$ include the widely promoted Gaussian monocycle $x^{(1)}(t)$ and the sinc pulse. If we view the design parameters $\{g[k]\}$ as the tap coefficients of an L-tap FIR filter, our approach to shaping the waveform $p(t)$ reduces to FIR linear prefiltering, as illustrated in Figure 5.3; see also [12].

In order to succinctly express our design strategies for the set $\{g[k]\}$, we let T_q denote the duration of $q(t)$ and let $Q(f)$ denote the Fourier transform of $q(t)$. Since T_0 corresponds to the sampling interval in the FIR filter, the clock rate of the transmitter is $F_0 := 1/T_0$. The duration of the synthesized pulse is $T_p = (L-1)T_0 + T_q$. For notational convenience, we will sometimes stack the impulse response of $g[k]$ into a vector $\mathbf{g} := [g[0], g[1], \ldots, g[L-1]]^T$, where $(\cdot)^T$ denotes the transpose. Similarly, the autocorrelation of $g[k]$, namely $r[m] = \sum_k g[k] \times g[k+m]$, will sometimes be represented by $\tilde{\mathbf{r}} = [r[0], r[1], \ldots, r[L-1]]^T$. To concisely describe the frequency components of an L-tap discrete-time FIR filter, we define

$$\mathbf{v}(f, L) := [1, e^{j2\pi f T_0}, e^{j2\pi f 2T_0}, \ldots, e^{j2\pi f(L-1)T_0}]^T, \quad (5.9)$$

$$\tilde{\mathbf{v}}(f, L) := [1, 2\cos(2\pi f T_0), 2\cos(2\pi f 2T_0), \ldots, 2\cos(2\pi f(L-1)T_0)]^T, \quad (5.10)$$

where $\mathbf{v}(f, L)$ and $\tilde{\mathbf{v}}(f, L)$ form the basis of complex-phase and linear-phase components, respectively. Consequently, the frequency response of $g[k]$ can be written as $G(e^{j2\pi f T_0}) = \sum_{k=0}^{L-1} g[k]e^{-j2\pi f kT_0} = \mathbf{v}^H(f, L)\mathbf{g}$, where $(\cdot)^H$ denotes the conjugate transpose. As a special case, when linear phase filters are of interest for their reduced storage requirements in implementation, $g[k]$ becomes symmetric, and the filter length is reduced to $\tilde{L} = (L-1)/2$ (assuming L is odd). In this case, $G(e^{j2\pi f T_0}) = e^{-j2\pi f T_0 \tilde{L}} \tilde{\mathbf{v}}^T(f, \tilde{L} + 1)\tilde{\mathbf{g}}$, where $\tilde{\mathbf{g}} := [g[\tilde{L}], g[\tilde{L} + 1], \ldots, g[L-1]]^T$.

The key property of the synthesized pulse is its power spectrum $S_p(f) := |P(f)|^2$. For pulses of the form in Equation (5.8), $S_p(f) = S_q(f)S_g(f)$, where $S_q(f) = |Q(f)|^2$ is the power spectrum of $q(t)$, and $S_g(f) = |G(e^{j2\pi f T_0})|^2$. The spectral utilization factor ψ defined in Equation (5.6) can therefore be written as a convex quadratic function of the filter coefficients $g[k]$:

$$\psi = \int_{\mathcal{F}_p} S_p(f)\,df = \mathbf{g}^T \mathbf{Q} \mathbf{g}, \tag{5.11}$$

where $\mathbf{Q} = \int_{\mathcal{F}_p} S_q(f)\mathbf{v}(f,L)\mathbf{v}^H(f,L)\,df$. If $S_q(f)$ has a sufficiently simple analytic form over the passband, then $\mathbf{Q}$ can be calculated analytically; for example, when $S_q(f)$ is constant over the passband. In other cases, $\mathbf{Q}$ can be computed numerically, even when we only know samples of $S_q(f)$ rather than its functional form.

As the design parameters in our approach are the filter coefficients, $g[k]$, it will be convenient to express the spectral mask constraint in Equation (5.5) as an explicit constraint on those coefficients. To do so, we define the scaled mask

$$\check{M}(f) = \begin{cases} S(f)/S_q(f) & \text{if } S_q(f) > 0, \\ +\infty & \text{if } S_q(f) = 0, \end{cases} \tag{5.12}$$

and observe that the spectral mask constraint $S_p(f) \leq S(f)$ is equivalent to

$$S_\mathbf{g}(f) \leq \check{M}(f) \quad \text{for all } f. \tag{5.13}$$

One can uniquely control $S_g(f)$ only over the domain $f \in [0, 1/(2T_0)]$, as outside this domain $G(e^{j\pi f T_0})$ is replicated periodically. By defining $M(f) = \min_{n \in \mathbb{Z}} \check{M}(f + n/T_0)$ over the domain $f \in [0, 1/(2T_0)]$,[1] the spectral mask constraint can be equivalently written as

$$\sqrt{S_\mathbf{g}(f)} = |G(e^{j2\pi f T_0})| \leq \sqrt{M(f)} \quad \text{for all } \in [0, 1/(2T_0)]. \tag{5.14}$$

Now that we have established the pulse shaping architecture, we will explore various methods for the design of the coefficients, $g[k]$, of the FIR filter. Instead of designing analog pulses with spectra $P(f)$ satisfying the prescribed mask $S(f)$, we will design digital filters with coefficients $g[k]$ that satisfy Equation (5.14).

5.4 OPTIMAL UWB SINGLE PULSE DESIGN

The goal of a single pulse design scheme is to find filter coefficients $g[k]$ such that the synthesized waveform $p(t)$ maximizes the spectral utilization efficiency ψ, while complying with the spectral mask $S(f)$. Before introducing the optimal designs, let us first consider a classical FIR filter design.

5.4.1 Parks–McClellan Algorithm

The Parks–McClellan (PM) algorithm [13], is one of the "workhorses" of digital filter design, and hence it is a natural candidate for the design of the pulse

---

[1]In many cases we will have $M(f) = \check{M}(f)$ for all $f \in [0, 1/(2T_0)]$.

shaping filter [12]. The objective of the resulting design method is to find an $\tilde{L}$-tap linear phase FIR filter with coefficients $\tilde{\mathbf{g}}$ so that the magnitude spectrum $|\tilde{G}(e^{j2\pi fT_0})|$ approximates the mask function $\sqrt{M(f)}, f \in [0, 1/(2T_0)]$. This design problem can be phrased as:

Given a set of frequency bands, $\{\mathcal{F}_{s_i}\}_{i=1}^{N_s}$, known as stop bands, which are separated from $\mathcal{F}_p$ by so-called transition bands, find a linear phase filter which achieves

$$\min_{\tilde{\mathbf{g}}} \cdot \max \left\{ \max_{f \in \mathcal{F}_p} \left| \sqrt{\check{M}(f)} - \delta - \tilde{G}(e^{j2\pi fT_0}) \right|, \eta_i \max_{f \in \mathcal{F}_{s_i}} |\tilde{G}(e^{j2\pi fT_0})| \right\}_{i=1}^{N_s} \qquad (5.15)$$

where δ is a small positive constant to account for "ripples" in the pass band, and $\{\eta_i\}$ is a set of weights.

This problem turns out to be a Chebyshev approximation problem with a desired frequency response $\sqrt{M(f)}$, and can be solved using numerical tools for the conventional PM algorithm [13].

The PM design method facilitates good approximations of the FCC spectral mask in a minimax sense, but poses several challenges for UWB pulse design. First, it does not directly optimize the spectral utilization of the pulse. [Spectral utilization involves approximation in an energy sense, rather than the minimax sense in Equation (5.15).] Second, the PM algorithm results in a filter with a power spectrum that approximates the spectral mask, but might not lie strictly below the mask. In other words, the equiripple nature of the PM filter may result in the mask being violated. In order to ensure mask compatibility, the stopbands $\{\mathcal{F}_{s_i}\}$, the ripple tolerance δ, and the weights $\{\eta_i\}$ must be carefully selected. The search for appropriate values of these parameters may require repeatedly solving Equation (5.15) with different parameter values in an interactive trial-and-error fashion, until an acceptable waveform is found. In most situations, the requirement of interaction is undesirable, and hence there is a need for optimal FIR filter design techniques that generate pulses with globally maximum spectral utilization and guaranteed compliance to the spectral mask, without having to interactively search for suitable design parameter values. One such method is provided in the next section.

5.4.2 Optimal UWB Pulse Design via Direct Maximization of NESP

The direct statement of our optimal pulse shaper design problem is as follows.

Problem 1: *Given L, T_0, $S_q(f)$ and $S(f)$, find a filter $g[k]$ of length L that maximizes ψ, subject to the spectral mask constraint $S_p(f) \leq S(f)$ for all f.*

In this section, we will analyze this problem and will provide a computationally efficient method for obtaining a globally optimal solution. We begin by establishing a more explicit formulation of the mask constraint in Equation (5.14). We define $\mathbf{A}(f) = [\text{Re}(\mathbf{v}(f, L)), -\text{Im}(\mathbf{v}(f, L))]^T$, where $\text{Re}(\cdot)$ and $\text{Im}(\cdot)$ denote the real and imaginary parts, respectively. By recognizing that $\sqrt{S_g(f)} = \|\mathbf{A}(f)\mathbf{g}\|_2$, Problem 1 can be formulated as

$$\max_{\mathbf{g}} \quad \psi = \mathbf{g}^T \mathbf{Q} \mathbf{g} \tag{5.16a}$$

$$\text{s.t.} \quad \|\mathbf{A}(f)\mathbf{g}\|_2 \leq \sqrt{M(f)} \quad \text{for all } f \in [0, 1/(2T_0)]. \tag{5.16b}$$

The feasible set in Equation (5.16) is defined by the intersection of an infinite number of second-order cone constraints on linear transformations of $\mathbf{g}$, one for each f; cf. Equation (5.16b). Hence, it is convex [4]. However, Equation (5.16b) defines an infinite number of constraints, that must be rendered finite. One way to approximate Equation (5.16b) is to sample it uniformly in frequency, and replace it by

$$\|\mathbf{A}(f_n)\mathbf{g}\|_2 \leq \sqrt{M(f_n)} - \epsilon_{N_d} \quad \text{for all } f_n \in \mathsf{F}^{N_d} := \{n/(2N_d T_0)\}_{n=0}^{N_d}, \tag{5.17}$$

where $\epsilon_{N_d} \geq 0$, and N_d is typically chosen to be in the order of 15 L [28].

Unfortunately, the objective in Equation (5.16) is a convex quadratic function of $\mathbf{g}$, and since it is to be maximized under cone constraints, Equation (5.16) is a nonconvex optimization problem. Therefore, any algorithm for the solution of Equation (5.16) must be able to deal with the intricacies of locally optimal solutions, including the choice of termination criteria, and the number and nature of the "starting points." As a result, the algorithms that are typically used to solve a problem of the form in Equation (5.16) involve a certain amount of interaction with the designer. (Alternative algorithms that do not require significant interaction tend to be computationally expensive.)

An observation that essentially removes the need for the designer to interact with the solution algorithm is that both the objective and the constraints in Equation (5.16) are linear functions of the autocorrelation of $\mathbf{g}$. In particular, $\psi = \mathbf{c}^T \tilde{\mathbf{r}}$, where $\mathbf{c} = \int_{\mathcal{F}_p} S_q(f) \tilde{\mathbf{v}}(f, L) \, df$ and $\|\mathbf{A}(f)\mathbf{g}\|_2^2 = \tilde{\mathbf{v}}^T(f, L) \tilde{\mathbf{r}}$. Therefore, Problem 1 can be reformulated as the following convex optimization problem in the autocorrelation of the filter:

$$\max_{\tilde{\mathbf{r}}} \quad \psi = \mathbf{c}^T \tilde{\mathbf{r}} \tag{5.18a}$$

$$\text{s.t.} \quad \tilde{\mathbf{v}}^T(f, L) \tilde{\mathbf{r}} \leq M(f) \quad \text{for all } f \in [0, 1/(2T_0)], \tag{5.18b}$$

$$\tilde{\mathbf{v}}^T(f, L) \tilde{\mathbf{r}} \geq 0 \quad \text{for all } f \in [0, 1/(2T_0)]. \tag{5.18c}$$

This problem is a semi-infinite linear program (SILP, e.g., [14]) in which there are two constraints for each f. The additional linear constraint set in Equation (5.18c) is a necessary and sufficient condition for $\tilde{\mathbf{r}}$ to represent a valid autocorrelation sequence. The semi-infinite constraints can be discretized to form a finite linear program, from which a globally optimal solution can be efficiently found [14, 17], without significant interaction with the designer. While that solution is an approximation to the true solution, discretization strategies exist that ensure that this approximation is a good one; for example, the direct analogy of Equation (5.17) above. (As an alternative to discretization, the constraints in Equation (5.18b) and (5.18c) can be precisely enforced using linear equality constraints on certain positive semi definite auxiliary matrix variables [2, 5].) Once the optimal autocorrelation has been found, an optimal filter $g[k]$ can be found via spectral factorization; for example, [10, 28].

5.4.3 Constrained Frequency Response Approximation

An alternative to directly maximizing the NESP is to take an indirect approach in which the objective is to keep the pulse frequency response $P(f) = Q(f)G(e^{j2\pi f T_0})$ as close to a desired response $e^{j\theta^{(d)}(f)}\sqrt{S(f)}$ as possible over the specified passband, where the power spectrum of the desired response is the spectral mask $S(f)$, and $\theta^{(d)}(f)$ is a phase component that can be chosen by the designer. Note that the phase component does not affect $S_p(f)$, but becomes relevant when designing $P(f)$. The indirect approach will be useful in extending the pulse design methodology to the case of multiple orthogonal pulses.

To formalize the notion of closeness between two frequency responses, we will use functional norms of the form

$$\phi_\ell := \left\| e^{j\theta^{(d)}(f)}\sqrt{S(f)} - Q(f)G(e^{j2\pi f T_0}) \right\|_{\mathcal{L}_\ell(\mathcal{F}_p)}, \qquad (5.19)$$

where

$$\|X(f)\|_{\mathcal{L}_\ell(\mathcal{F}_p)} = \begin{cases} \left(\int_{\mathcal{F}_p} |X(f)|^\ell \, df \right)^{1/\ell} & \text{for } 1 \leq \ell < \infty, \\ \max_{f \in \mathcal{F}_p} |X(f)| & \text{for } \ell = \infty. \end{cases}$$

The approximation error in Equation (5.19) involves both magnitude and phase components of the frequency response. We now formalize the problem.

Problem 2: *Given $\ell, L, T_0, S(f), \theta^{(d)}(f)$ and $Q(f)$, find a filter $g[k]$ of length L that achieves*

$$\min_{\mathbf{g}} \quad \phi_\ell = \left\| e^{j\theta^{(d)}(f)}\sqrt{S(f)} - Q(f)G(e^{j2\pi f T_0}) \right\|_{\mathcal{L}_\ell(\mathcal{F}_p)} \qquad (5.20a)$$

$$\text{s.t.} \quad S_p(f) \leq S(f) \quad \text{for all } f. \qquad (5.20b)$$

We will now show that this problem is convex in $\mathbf{g}$, and hence a globally optimal solution can be efficiently found without significant interaction with the designer.

The feasible set defined by Equation (5.20b) is the same as that in Problem 1. Depending on the choice of ℓ, different formulations of Problem 2 arise. In particular, since the NESP involves power spectra that are "squares" of the components of the arguments of the norm in Equation (1.19), a natural choice for ℓ is $\ell = 2$, which gives a good approximation to Problem 1. For this choice, the objective in Problem 2 can be written as

$$\phi_2 = \mathbf{g}^T \mathbf{Q} \mathbf{g} - \mathbf{b}^T \mathbf{g} + c_2, \qquad (5.21)$$

where $\mathbf{Q}$ was defined after Equation (1.11), $\mathbf{b} = 2\mathrm{Re}(\int_{\mathcal{F}_p} e^{j\theta^{(d)}(f)} \sqrt{S(f)}\, \mathbf{Q}^*(f) \mathbf{v}^*(f, L)\, df)$ with $*$ denoting conjugate, and $c_2 = \int_{\mathcal{F}_p} S(f)\, df$ a constant. If we choose a matrix $\mathbf{L}$ such that $\mathbf{L}^H \mathbf{L} = \mathbf{Q}$, then when $\ell = 2$, Problem 2 can be explicitly formulated as

$$\min_{\mathbf{g}, \mu} \quad \mu - \mathbf{b}^T \mathbf{g} \qquad (5.22\mathrm{a})$$

$$\text{s.t.} \quad \|\mathbf{L}\mathbf{g}\|_2^2 \le \mu \qquad (5.22\mathrm{b})$$

$$\|\mathbf{A}(f)\mathbf{g}\|_2 \le \sqrt{M(f)} \quad \text{for all } f \in [0, 1/(2T_0)]. \qquad (5.22\mathrm{c})$$

This is a convex optimization problem with a linear objective, a rotated second-order cone constraint (5.22b), and an infinite number of (conventional) convex quadratic constraints (5.22c). Discretization of Equation (5.22c) results in a formulation that can be efficiently solved for a (globally) optimal $\mathbf{g}$ using general purpose convex cone optimization tools [25]. Although solutions to Equation (5.22) do not generate waveforms that explicitly maximize the NESP, they efficiently generate pulses with large NESPs, without the need for the spectral factorization post-processing step that is required for the efficient solution of Problem 1 [via Equation (5.18)].

5.4.4 Constrained Frequency Response Design with Linear Phase Filters

The solution of Problem 2 can often be simplified if $g[k]$ is (further) constrained to have linear phase, at the expense of a performance penalty. For symmetric (linear phase) filters of odd length, the problem in Equation (1.22), which is a formulation of Problem 2 with $\ell = 2$, simplifies to

$$\min_{\tilde{\mathbf{g}}, \mu} \quad \mu - \tilde{\mathbf{b}}^T \tilde{\mathbf{g}} \qquad (5.23\mathrm{a})$$

$$\text{s.t.} \quad \|\tilde{\mathbf{L}}\tilde{\mathbf{g}}\|_2^2 \le \mu, \qquad (5.23\mathrm{b})$$

$$-\sqrt{M(f)} \le \tilde{\mathbf{v}}^T(f, \tilde{L} + 1)\tilde{\mathbf{g}} \le \sqrt{M(f)} \quad \text{for all } f \in [0, 1/(2T_0)], \qquad (5.23\mathrm{c})$$

where $\tilde{\mathbf{Q}} = \tilde{\mathbf{L}}^H \tilde{\mathbf{L}} = \int_{\mathcal{F}_p} S_q(f) \tilde{\mathbf{v}}(f, \tilde{L}+1) \tilde{\mathbf{v}}^T(f, \tilde{L}+1)\, df$, and $\tilde{\mathbf{b}} = 2\int_{\mathcal{F}_p} \cos[\theta^{(d)}(f) + 2\pi f T_0 \tilde{L} - \theta_q(f)]\sqrt{S(f) S_q(f)}\, \tilde{\mathbf{v}}(f, \tilde{L}+1)\, df$. It is clear from the expression for $\tilde{\mathbf{b}}$ that a natural choice for $\theta^{(d)}(f)$ in such a design is $\theta^{(d)}(f) = \theta_q(f) - 2\pi f T_0 \tilde{L}$. We also point out that Equation (5.23c) is a set of linear constraints (two for each f), whereas Equation (5.22c) was a set of convex quadratic constraints. Once again, Equation (5.23) can be efficiently solved using general purpose convex optimization tools [25]. The problem in Equation (5.23) is also particularly amenable to methods that employ multiple exchange techniques [1, 23].

The case of $\ell = \infty$ is also of interest because of its relationship with the PM designs. If we choose $\theta^{(d)}(f) = \theta_q(f) - 2\pi f T_0 \tilde{L}$, then for linear phase filters with a positive gain in the passband, we have the following formulation:

$$\min_{\tilde{g}} \max_{f \in \mathcal{F}_p} \sqrt{S_q(f)} \left| \sqrt{\check{M}(f)} - \tilde{G}(e^{j2\pi f T_0}) \right| \quad (5.24a)$$

$$\text{s.t.} \quad |\tilde{G}(e^{j2\pi f T_0})| \leq \sqrt{M(f)}, \quad (5.24b)$$

where $\tilde{G}(e^{j2\pi f T_0}) = \tilde{\mathbf{v}}^T(f, \tilde{L}+1)\tilde{\mathbf{g}}$ is the "phase centered" version of $G(e^{j2\pi f T_0})$. In contrast to the waveform generated by the solution to Equation (5.15) by the PM algorithm, which merely approximates the spectral mask, any solution to Equation (5.24) fully complies with the spectral mask. Furthermore, like each of the proposed designs in Sections 5.4.2–5.4.4, the determination of a globally optimal $\tilde{\mathbf{g}}$ requires the solution of only one optimization problem, and that problem can be efficiently solved.

The formulations of Problems 1 and 2 were based on precise knowledge of $S_q(f)$. In the case where $S_q(f)$ is not precisely known, we can replace $S_q(f)$ by an estimate $\hat{S}_q(f)$ that is also an upper bound, that is, $\hat{S}_q(f) \geq S_q(f)$ for all f. This replacement is well motivated for design convenience, because $S_q(f)$ is typically fairly flat over the band of interest, and because replacing $S_q(f)$ by a constant $\hat{S}_q \geq S_q(f)$, for all $f \in \mathcal{F}_p$, enables the integrals which constitute $\mathbf{Q}$, $\mathbf{c}$ and $\mathbf{b}$ to be analytically evaluated. Also, if $\hat{S}_q(f)$ is constant for all $f \in \mathcal{F}_p$, the precise transformation of some semi-infinite linear constraints into (finite) linear matrix inequalities [5] takes on a relatively simple form.

5.5 OPTIMAL UWB ORTHOGONAL PULSE DESIGN

Motivated by the demand for high data rates, orthogonal frequency multiplexing and high spectral efficiency multidimensional modulations [19], in this section we extend the FIR filter approach to design spectrally efficient orthogonal pulses. The goal is to design a set of mutually orthogonal pulses, each of which occupies the entire spectrum allowed by the spectral mask. This enables each pulse to benefit from the large multipath diversity provided by the ultra wide bandwidth.

5.5.1 Orthogonality Formulation

Let us consider two pulses $p_1(t)$ and $p_2(t)$ that are generated by two different sets of filter coefficients $\mathbf{g}_1$ and $\mathbf{g}_2$, each of length L. For these two pulses to be orthogonal, they have to satisfy the time-domain constraint $\int_{-\infty}^{+\infty} p_1(t) p_2(t) \, dt = 0$, which can be equivalently written in the frequency domain as $\int_{-\infty}^{+\infty} P_1(f) P_2^*(f) \, df = 0$. In a matrix-vector form, the orthogonality constraint can be written as

$$\int_{-\infty}^{+\infty} P_1(f) P_2^*(f)\,df = \mathbf{g}_1^H \mathbf{Q} \mathbf{g}_2 = 0, \qquad (5.25)$$

where $\mathbf{Q}$ was defined after Equation (5.11). Suppose that a filter $\mathbf{g}_1$ has been designed via a single pulse method, such as Problem 1. The problem of directly maximizing the NESP of $p_2(t)$ subject to the orthogonality constraint (5.25), can be formulated as follows.

Orthogonal pulse design 1: *Given L, T_0, $S_q(f)$, $M(f)$ and $\mathbf{g}_1$, find $\mathbf{g}_2$ that achieves*

$$\max_{\mathbf{g}_2}\ \psi = \mathbf{g}_2^T \mathbf{Q} \mathbf{g}_2 \qquad (5.26a)$$

$$\text{s.t.}\quad \mathbf{g}_1^T \mathbf{Q} \mathbf{g}_2 = 0, \qquad (5.26b)$$

$$\|\mathbf{A}(f)\mathbf{g}_2\|_2 \le \sqrt{M(f)}\quad \forall f \in [0, 1/(2T_0)]. \qquad (5.26c)$$

The linear equality constraint in Equation (5.26b) and the set of convex quadratic constraints in Equation (5.26c) describe a convex feasible set, but, as in Equation (5.16), the objective is to maximize a convex function of $\mathbf{g}_2$, and hence Equation (5.26) is a nonconvex optimization problem. Unfortunately, it is not possible to transform Equation (5.26b) into a function of the autocorrelation vector $\tilde{\mathbf{r}}_2$, and therefore, we cannot directly borrow the direct single pulse design techniques from Section 5.4.2. In order to avoid the intricacies of having to deal with the potential for locally optimal solutions in the solution of Equation (5.26), we now seek formulations of an indirect design problem that is easier to solve. Our indirect formulations are based on the frequency response approximation problem discussed in Section 5.4.3.

As shown in Section 5.4.3, a pulse with a large NESP can be efficiently obtained by making $P_k(f)$ close to a desired frequency response $P_k^{(d)}(f) := e^{j\theta_k^{(d)}(f)}\sqrt{S(f)}$. When designing multiple orthogonal pulses, the desired response is constructed so that each $P_k^{(d)}(f)$ has the same power spectrum $S(f)$, but has a distinct phase, $\theta_k^{(d)}(f)$. To impose orthogonality among different pulses, we can select the design parameters $\{\theta_k^{(d)}(f)\}$ such that the desired frequency responses $\{P_k^{(d)}(f)\}$ are mutually orthogonal. That is,

$$\int e^{j\left(\theta_k^{(d)}(f) - \theta_i^{(d)}(f)\right)} S(f)\,df = 0 \quad \text{for any } k \neq i. \qquad (5.27)$$

If the desired frequency responses are orthogonal, then the designed pulses will (essentially) inherit this property if the achieved approximation error is sufficiently small. We can formulate the resulting design problem as follows.

5.5 OPTIMAL UWB ORTHOGONAL PULSE DESIGN

Orthogonal pulse design 2: Given $L, T_0, \theta_q(f), M(f)$, a filter $\mathbf{g}_1$ designed via Equation (5.27), and $\theta_2^{(d)}(f)$ that is orthogonal to $\theta_1^{(d)}(f)$ according to Equation (5.27), find $\mathbf{g}_2$ that achieves

$$\min_{\mathbf{g}_2} \phi_\ell = \left\| e^{j\theta_2^{(d)}(f)}\sqrt{S(f)} - Q(f)G_2(e^{j2\pi f T_0}) \right\|_{\mathcal{L}_\ell(\mathcal{F}_p)} \quad (5.28a)$$

$$\text{s.t.} \quad \|A(f)\mathbf{g}_2\|_2 \leq \sqrt{M(f)} \quad \forall f \in [0, 1/(2T_0)]. \quad (5.28b)$$

An approximation of the orthogonality constraint in Equation (5.27) can significantly simplify the design of (essentially) orthogonal pulses. In particular, since $\mathcal{F}_p$ is the passband (or bands) of the spectral mask, then the constraint in Equation (5.27) can be approximated by taking the integral over $\mathcal{F}_p \cup \tilde{\mathcal{F}}_p$, where $\tilde{\mathcal{F}}_p$ is the mirror image of $\mathcal{F}_p$ in the negative frequencies. If $S(f)$ is constant over $\mathcal{F}_p$, then Equation (5.27) can be approximated by

$$\int_{\mathcal{F}_p \cup \tilde{\mathcal{F}}_p} e^{j(\theta_k^{(d)}(f) - \theta_i^{(d)}(f))} \, df = 0 \quad \text{for any } k \neq i. \quad (5.29)$$

If that integral is (further) approximated by its N-point centered Riemann sum, and N is large, then the desired responses will be (essentially) orthogonal if

$$\mathbf{w}_i^H \mathbf{w}_k + \mathbf{w}_k^H \mathbf{w}_i = 0 \quad \text{for any } k \neq i, \quad (5.30)$$

where the nth element of the phase vector $\mathbf{w}_k$ is $\mathbf{w}_k[n] = e^{j\theta_k^{(d)}(f_n)}$, $0 \leq n \leq N-1$, with $f_n = f_{\text{left}} + (n + 1/2)\Delta_f$, f_{left} being the lower band edge of $\mathcal{F}_p$ and Δ_f being the width of the Riemann rectangle, which is $(1/N)$th of the bandwidth of $\mathcal{F}_p$. For convenience, we will define F_p^N to be the set of such f_ns. A sufficient condition for Equation (5.30) is that $\mathbf{w}_k^H \mathbf{w}_i = 0$ for any $k \neq i$.

5.5.2 Sequential UWB Pulse Design

Having understood the orthogonality requirements, we now present a sequential (SEQ) strategy for (essentially) orthogonal pulse design. The procedure starts with the design of the first pulse $p_1(t)$ subject to the mask constraint only. Subsequent pulses $p_k(t)$, $k = 2, \ldots, K$, are then designed one by one to fit into the desired spectral mask, as well as to be (essentially) orthogonal to all previously designed pulses. To put this approach in a mathematical form, we suppose that $(k-1)$ (essentially) mutually orthogonal pulses $\{p_m(t)\}_{m=1}^{k-1}$ are already in place. Rather than directly minimizing the approximation error of the kth pulse, $\|P_k^{(d)}(f) - P_k(f)\|_{\mathcal{L}_\ell(\mathcal{F}_p)}$, we minimize the ℓ-norm of an N-element vector $\mathbf{x}$ whose nth element is $P_k^{(d)}(f_n) - P_k(f_n)$, $f_n \in \mathsf{F}_p^N$. In the case where $\ell < \infty$, this corresponds to minimizing the N-point centered Riemann sum approximation of $\|P_k^{(d)}(f) - P_k(f)\|_{\mathcal{L}_\ell(\mathcal{F}_p)}$. Since $P_k^{(d)}(f_n) = \mathbf{w}_k[n]\sqrt{S(f_n)}$, the subproblem in the sequential design strategy can be formulated as follows.

SEQ pulse design: *Given ℓ, L, T_0, $S_q(f)$, $S(f)$, N, and $k-1$ previously selected filter phase vectors $\{\mathbf{w}_m\}_{m=1}^{k-1}$ of mutually orthogonal pulses $\{p_m(t)\}_{m=1}^{k-1}$, find vectors $\mathbf{g}_k$ of length L and $\mathbf{w}_k$ of length N that satisfy*

$$\min_{\mathbf{g}_k, \mathbf{w}_k, \mathbf{x}} \quad \|\mathbf{x}\|_\ell \tag{5.31a}$$

$$\text{s.t.} \quad x[n] = \mathbf{w}_k[n]\sqrt{S(f_n)} - Q(f_n)\mathbf{v}^H(f_n, L)\mathbf{g}_k, \quad \forall f_n \in \mathsf{F}_p^N, \tag{5.31b}$$

$$\|\mathbf{A}(f)\mathbf{g}_k\|_2 \leq \sqrt{M(f)}, \quad \forall f \in [0, 1/(2T_0)], \tag{5.31c}$$

$$\mathbf{w}_k^H \mathbf{w}_m = 0, \quad \forall m = 1, \ldots, k-1. \tag{5.31d}$$

In this formulation, the objective (5.31a) [and (5.31b)] indirectly seeks a pulse with a large NESP, while Equations (5.31c) and (5.31d) ensure spectral mask compatibility and mutual orthogonality of the target spectra, respectively.

To reduce the complexity of searching for $\mathbf{w}_k$, a convenient approach is to select mutually orthogonal phase vectors prior to the SEQ procedure. Treating $\mathbf{w}_k$ as a length-N codeword with complex-valued elements, we can use the FFT matrix to design an example of K complex codewords as follows.

Complex orthogonal phase vectors via FFT: *A set of K mutually orthogonal phase vectors $\{\mathbf{w}_k\}_{k=1}^K$ can be obtained by setting $\mathbf{w}_k[n] = e^{j2\pi kn/N}$, $n = 0, \ldots, N-1$, for $k = 1, \ldots, K \leq N$.*

With the phase vectors $\{\mathbf{w}_k\}$ selected this way, we now mimic Equation (5.22) to propose the following convex formulation of the SEQ pulse design sub-problem with $\ell = 2$:

$$\min_{\mathbf{g}_k, \mu_k} \quad \mu_k - \hat{\mathbf{b}}_k^T \mathbf{g}_k \tag{5.32a}$$

$$\text{s.t.} \quad \|\mathbf{L}\mathbf{g}_k\|_2^2 \leq \mu_k, \tag{5.32b}$$

$$\|\mathbf{A}(f)\mathbf{g}_k\|_2 \leq \sqrt{M(f)} \quad \forall f \in [0, 1/(2T_0)], \tag{5.32c}$$

where $\hat{\mathbf{b}}_k = 2\Delta_f \text{Re}(\sum_{n=0}^{N-1} \mathbf{w}_k[n]\sqrt{S(f_n)}Q^*(f_n)\mathbf{v}^*(f_n, L))$.

5.5.3 Sequential UWB Pulse Design with Linear Phase Filters

The formulation in Equation (5.32) allows the FIR filters to have nonlinear phase characteristics. As was the case in Section 5.4.4, the design can be simplified if $g[k]$ is further constrained to have linear phase. With odd-length symmetric linear phase filters, the pulse frequency response becomes $P_k(f) = \sqrt{S_q(f)} e^{j[\theta_q(f) - 2\pi f_n T_0 \tilde{L}]} \tilde{\mathbf{v}}^T(f, \tilde{L}+1)\tilde{\mathbf{g}}_k$. Since $\tilde{\mathbf{v}}^T(f, \tilde{L}+1)\tilde{\mathbf{g}}$ is real, a natural choice for $\mathbf{w}_k$ [cf. Equation (5.31b)] is $\mathbf{w}_k = e^{j[\theta_q(f) - 2\pi f T_0 \tilde{L}]}\tilde{\mathbf{w}}_k$, $\forall f$, where $\tilde{\mathbf{w}}_k = \pm 1$. As such, the selection of the orthogonal phase vectors $\mathbf{w}_k$ is simplified to construction of binary orthogonal codewords, $\tilde{\mathbf{w}}_k$. If we choose N to be a power of 2, with $N \geq K$, such binary codewords can be selected from the set of length $N = 2^i$ Hadamard codewords [20, p. 424]. When $K < 2^i$, as will normally be the case, one must select elements from this set. For efficient spectral utilization, an appropriate method is to select Hadamard codewords in ascending order of the number of sign transitions in the codeword [30].

Binary orthogonal phase vectors via Hadamard partition: *In an SEQ pulse design problem with linear phase filters, the binary codewords $\{\tilde{\mathbf{w}}_k\}$ of length $N = 2^i$ should be selected from the set of length-2^i Hadamard codewords in the following way:*

1. *Arrange the 2^i Hadamard codewords in ascending order of the number of sign transitions in the codeword; and,*
2. *Set $\tilde{\mathbf{w}}_k$ to be the kth Hadamard codeword.*

This choice (essentially) minimizes the loss of NESP due to the "crossing bands" between bands of positive and negative gain in the passband of the filter. As an example, if we have $N = 8$ discretization points (i.e., eight Riemann rectangles), and we want to design $K = 3$ pulses, the Hadamard partition result suggests the following choices:

$$\tilde{\mathbf{w}}_1 = [1, \quad 1, \quad 1, \quad 1, \quad 1, \quad 1, \quad 1, \quad 1]^T, \quad (5.33a)$$

$$\tilde{\mathbf{w}}_2 = [1, \quad 1, \quad 1, \quad 1, \quad -1, \quad -1, \quad -1, \quad -1]^T, \quad (5.33b)$$

$$\tilde{\mathbf{w}}_3 = [1, \quad 1, \quad -1, \quad -1, \quad -1, \quad -1, \quad 1, \quad 1]^T. \quad (5.33c)$$

For the larger values of N that would be used in practice, the first three codewords have a similar sign structure.

With the phase vectors chosen as described above, for odd-length symmetric linear phase filters the formulation in Equation (5.31) simplifies to

$$\min_{\tilde{\mathbf{g}}_k, \mathbf{x}} \quad \|\mathbf{x}\|_\ell \quad (5.34a)$$

$$\text{s.t.} \quad x[n] = \sqrt{S_q(f_n)} \left[\tilde{\mathbf{w}}_k[n] \sqrt{\check{M}(f_n)} - \tilde{\mathbf{v}}^T(f_n, \tilde{L}+1)\tilde{\mathbf{g}}_k \right] \quad \forall f_n \in \mathsf{F}_p^N, \quad (5.34b)$$

$$-\sqrt{M(f)} \leq \tilde{\mathbf{v}}^T(f, \tilde{L}+1)\tilde{\mathbf{g}}_k \leq \sqrt{M(f)} \quad \forall f \in [0, 1/(2T_0)], \quad (5.34c)$$

from which explicit formulations for $\ell = 2$ and $\ell = \infty$ can be easily obtained.

5.6 DESIGN EXAMPLES AND COMPARISONS

We now present examples for different designs of UWB pulses which satisfy the US FCC mask $S_{FCC}(f)$ illustrated in Figure 5.1. We will compare their performance in terms of spectral utilization efficiency, system-level impact on bit error rates, as well as robustness to narrowband interference (NBI).

5.6.1 Single-Pulse Designs and their Spectral Utilization Efficiency

We first investigate various formulations for single pulse designs. For the digital FIR designs, the basis pulse $q(t)$ is chosen to be the Gaussian monocycle of pulse width 0.1626 ns, whose waveform parameters are specified in Section 5.2. The clock rate F_0 is set to be a relatively low frequency of $1/T_0 = 28$ GHz.

Design 1: Our first design is obtained by solving a version of Equation (5.18) in which Equation (5.18b) is discretized and Equation (5.18c) is precisely transformed into L linear constraints on an $L \times L$ positive semidefinite matrix [25]. This transformation leads to a semidefinite programming formulation that can be efficiently solved for the optimal $\tilde{r}$ using a general purpose solver [25]. Spectral factorization [10, 28] is then applied to extract the optimal pulse coefficients **g**.

Design 2: In this case we design an odd-length, symmetric, linear phase filter via Equation (5.23) in which Equation (5.23c) is discretized. The resulting optimization problem has a linear objective, a set of linear constraints and a single rotated second-order cone constraint.

The power spectra of the pulses emanating from Designs 1 and 2 for length $L = 33$ filters and the FCC mask are provided in Figure 5.4. (The power spectrum of a pulse obtained from Design 1 with the tighter mask and a different basis pulse appeared in [29].) For comparison, the power spectrum of the PM pulse

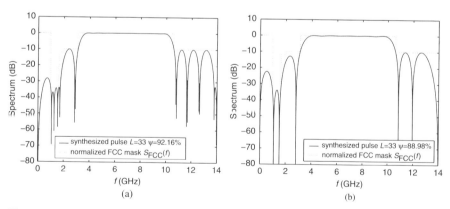

Figure 5.4 Power spectrum of the optimally synthesized pulse under the FCC mask $S_{FCC}(f)$: (a) Design 1; (b) Design 2.

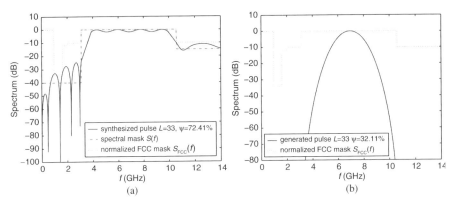

Figure 5.5 Power spectrum of pulses obtained using existing design approaches: (a) PM algorithm based pulse; (b) "prolate-spheroidal" pulse [18].

shaper in [12] is depicted in Figure 5.5(a), and that of the "prolate-spheroidal" (PS) pulse shaper in [18] is presented in Figure 5.5(b). The improved NESP of the proposed designs is immediately apparent from Figures 5.4 and 5.5. All these digitally designed pulses are configured to operate at the highest spectral utilization while conforming to the relevant spectral mask, using the same filter length, $L = 33$. In addition, analog pulses including the Gaussian monocycle and its variants are also investigated. For the basic Gaussian monocycle, we employ System 2 in Figure 5.2, as this system does not violate the FCC mask. High-order derivatives of the Gaussian pulse are considered for their resemblance of sinusoids modulated by a Gaussian pulse-shaped envelope [16]. As the derivative order increases, the number of zero crossings in the same pulse width increases, acting as if a higher "carrier" frequency sinusoid is modulated by an equivalent Gaussian envelope. By choosing the order of the derivatives of the Gaussian pulse n and a suitable pulse width, it is possible to find a pulse that satisfies the FCC mask. With reference to the amplitude spectrum expression in Equation (5.7), a good design choice is $n = 5$ and $\tau = 72$ ps [16], which yields a time-domain support of $T_p \approx 6\tau = 0.43$ ns and a 3 dB bandwidth of 3.46 GHz that falls within the range of $[5.25, 8.92]$ GHz and peaks at $f_M = 7.01$ GHz.

TABLE 5.1 NESPs of Various Pulses

Design Method	$S(f) = S_{FCC}(f)$
Design 1	$\bar{\psi} = 92.16\%$
Design 2	$\bar{\psi} = 88.98\%$
PM Pulse	$\bar{\psi} = 72.41\%$
PS Pulse	$\bar{\psi} = 32.11\%$
Gaussian ($n = 5$)	$\bar{\psi} = 50.76\%$
Gaussian ($n = 1$)	$\bar{\psi} = 0.15\%$

The NESPs of the different design methods are quantified and tabulated in Table 5.1. As expected, Design 1 provides the largest NESP, because it is based on direct maximization of the NESP (cf. Problem 1). However, it is interesting to note that both this design and Design 2 (which is based on the indirect method of Problem 2 with the linear phase constraint) provide considerably larger NESPs than the PM pulse, and substantially larger NESPs than the PS design and the Gaussian monocycle. These designs (and the PM-based pulses) also enjoy an implementation advantage in terms of their low sampling frequency requirement compared with that of the PS pulses (64 GHz) [18].

Achieving NESP optimality also offers the potential for using a lower sampling frequency or a shorter filter length than competing methods, which may simplify practical hardware realization. The filter length problem is formulated in [30] in terms of finding the shortest pulse duration that achieves a given NESP requirement, subject to satisfaction of the spectral mask. Under the FCC mask constraint, a pulse $p(t)$ with an NESP of greater than 80% can be synthesized by an FIR filter of length $L = 15$, corresponding to a pulse duration $T_p = 0.66$ ns. (Recall that when $L = 33$ the maximum achievable NESP is 92.16%.) Similarly, the minimum length filter required to achieve the NESP of 72.41% achieved by the length 33 Parks–McClellan filter is merely $L = 12$.

It is possible to further reduce the number of filter coefficients by generalizing the FIR-filter-based pulse synthesis architecture that we have considered [cf. Equation (5.8)] to include pulses of the form $p(t) = \sum_{k=0}^{L-1} g[k]q(t - \tau_k)$, where the delays τ_k are to be designed jointly with the coefficients $g[k]$ [31]. Unfortunately, that joint design problem is not convex, and hence joint design algorithms will require a considerable interaction and substantial computational resources. The computational cost of obtaining a good joint design can often be significantly reduced by using an iterative approach in which one cycles between optimizing $g[k]$ for the current estimate of the optimal delays and optimizing the delays for the current estimate of the optimal coefficients. Although the second step in the cycle remains computationally expensive, the methods proposed in this chapter can be used to efficiently solve the first step by simply generalizing the definition of $\mathbf{v}(f,L)$ in Equation (5.9) so that its kth element, $0 \leq k \leq L - 1$, is $e^{j2\pi f \tau_k}$. That said, we believe that the design and implementation efficiencies obtained by choosing $\tau_k = kT_0$, as we have in this chapter, outweigh the potential performance gains of the generalized scheme for all but the truly small values of L.

5.6.2 Multiband Pulse Design

An alternative signaling format to single-band UWB impulse radio is to use multiband (MB) UWB waveforms, which have been proposed for wireless personal area networks (WPANs) under IEEE 802.15 [3]. In a multiband system, multiple bands of bandwidth greater than or equal to 500 MHz are employed, with each band being occupied by a distinct pulse. With the entire bandwidth divided into several non-overlapping sub-bands, multiband UWB systems allow flexibility in efficiently "filling up" the spectral mask, and facilitate co-existence with legacy systems and worldwide deployment by enabling some sub-bands to be turned off in order to avoid interference and comply with different regulatory requirements. In addition, multiband systems provide another dimension for multiple access via frequency division. Different users can use different pulses for multiple access, and frequency hopping can also be easily implemented by switching among those baseband pulses to acquire greater frequency diversity.

Pulse design for a multiband system can directly borrow the results from single-pulse design. The only difference lies in the different specifications of passband and the corresponding spectral mask constraint. We assume that the passband is equally divided into K sub-bands, each of bandwidth of $B_s = (7.5/K)$ GHz. For the design of the pulse that occupies the kth sub-band, $1 \leq k \leq K$, we impose a spectral mask $S_k(f)$ that satisfies the global spectral mask and has low stopband levels in order to avoid significant interband interference. In particular, we impose

$$S_k(f) = \begin{cases} -60\text{dB} & [0, 3.1 + (k-1)B_s]\,\text{GHz}, \\ 0\text{dB} & [3.1 + (k-1)B_s, 3.1 + kB_s]\,\text{GHz}, \\ -60\text{dB} & [3.1 + kB_s, +\infty)\,\text{GHz} \end{cases} \quad (5.35)$$

The filter tap coefficients $\mathbf{g}_k$ for the kth pulse can be obtained by efficiently solving Problem 1 via the formulation in Equation (5.18). As a design example, we suppose that the entire UWB passband is equally divided into three sub-bands. Setting $L = 100$, we obtain the synthesized pulses $\{p_k(t)\}_{k=1}^3$ by replacing $S(f)$ in Problem 1 with $S_k(f)$ in Equation (5.35), $k = 1, 2, 3$, respectively. The power spectra and time-domain waveforms of the synthesized multiband pulses are shown in Figure 5.6.

The multiband configuration is convenient for fast FH UWB systems. To hop from one frequency sub-band to another, one can simply reset the memory of the shift register (SR) that stored the corresponding L-tap coefficient $\mathbf{g}$, or use a bank of SRs and switch among them to select the desired band. Figure 5.7 shows such a digital frequency hopping transmitter structure for UWB communication. The digital architecture implements linear combinations of the baseband Gaussian monocycle, and does not involve any analog carriers. This avoids the carrier frequency offset (CFO) effects, which are commonly encountered in analog FH implementations. This design has a relatively stringent requirement on the clock timing accuracy (down to several picoseconds), but the switching time between two SRs is faster than the switching time between two carriers.

5.6.3 Multiple Orthogonal Pulse Design

Although they are (almost) orthogonal to each other, each of the pulses in a multiband UWB system can only utilize a portion of the power allowed by the FCC mask, since each of them occupies only a portion of the entire bandwidth. Here we demonstrate orthogonal pulse sets with better spectral utilization. As an example, we have designed three orthogonal UWB pulses using the sequential design strategy with linear phase filters described in Section 5.5.3, including the proposed selection of

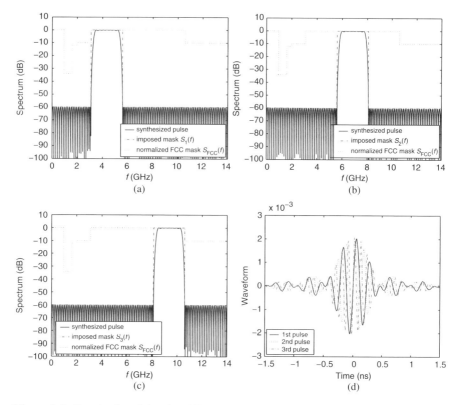

Figure 5.6 Results for a 3-band multiband UWB system. (a) first sub-band; (b) second sub-band; (c) third sub-band; and (d) time-domain wave forms.

the Hadamard codewords. We designed length $L = 33$ linear-phase filters, using a formulation of Equation (5.34) with $\ell = 2$ and length $N = 512$ codewords. Under the tighter mask constraint $S_T(f)$ (cf. Figure 5.1), the power spectra of the resulting (essentially) orthogonal pulses and their correlation properties are provided in Figure 5.8. The NESPs of these three pulses are 76.51%, 51.31% and 49.97%, respectively, and hence all three pulses provide higher spectral utilization efficiency

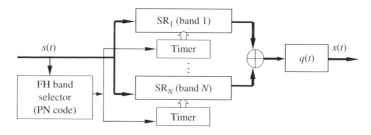

Figure 5.7 Implementation of multiband with FH.

5.6 DESIGN EXAMPLES AND COMPARISONS 125

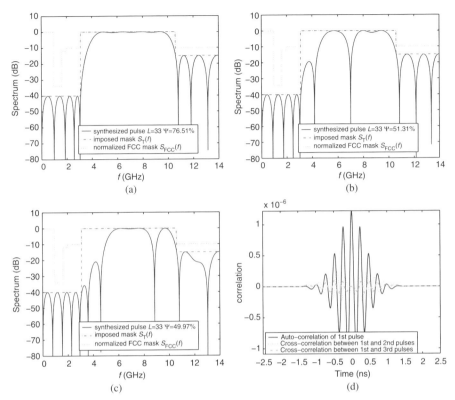

Figure 5.8 Three orthogonal pulses designed via the SEQ strategy. (a) first pulse; (b) second pulse; (c) third pulse; and (d) time-domain correlation functions.

than the pulse from the PS orthogonal set [18] with the largest spectral utilization efficiency. [The PS pulse has an NESP of 32.11% and its power spectrum is shown in Figure 5(b).]

5.6.4 Pulse Designs for Narrowband Interference Avoidance

Digital FIR filtering permits flexible pulse design to accommodate a range of system requirements, which is particularly attractive for UWB systems that need to avoid NBI to and from legacy wireless radios and services. A straightforward method to alleviate NBI is to impose stricter mask constraints over vulnerable frequency bands. As shown in Figure 5.1, we can impose a tighter mask $S_T(f)$ over the frequency range of 0.96–3.1 GHz, in order to protect the GPS and the 802.11 b/g WLAN bands located within this region. Depending on the locations of narrowband interferers, NBI avoidance may also involve imposing a spectral mask with notches. For example, 802.11a WLAN and Bluetooth systems operate in two narrow bands: [5.15, 5.35] and [5.725, 5.825] GHz. To avoid transmissions over these two bands,

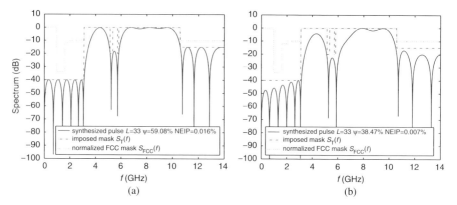

Figure 5.9 Pulses with robustness to NBI: (a) with notch constraints; (b) multiband approach.

we can modify the tight mask $S_T(f)$ by imposing two notches at the corresponding bands, each being −20 dB lower than the maximum allowable power spectrum.

An alternative method to enforce NBI avoidance is to employ the multiband approach, but maintain nearly complete spectrum utilization of the FCC mask for each designed pulse. As suggested in IEEE 802.15.3WPAN proposals, the allowable UWB spectrum can be partitioned into a lower passband over [3.1, 5.15] GHz and an upper passband over [5.825, 10.6] GHz. This multiband plan was originally suggested for carrier-based OFDM UWB systems to support variable data rates [15]. Here we design two baseband pulses, $p_1(t)$ and $p_2(t)$, to fit in the lower and upper bands respectively. The transmitted single-band pulse $p(t)$ is nothing but a summation of these two pulses, that is, $p(t) = p_1(t) + p_2(t)$. In contrast to its constituent sub-band pulses, $p(t)$ occupies almost the entire allowable spectrum to enable full diversity.

Figure 5.9 depicts pulses with NBI-avoidance capability. The NESPs of the pulses designed by the notch-based method and the multiband method, are 59.08% and 38.47%, respectively. The multiband method has a lower NESP, but is versatile in supporting flexible data rates by turning off one of the bands. On the other hand, the level of interference from UWB transmissions to WLAN users can be measured in terms of the normalized effective interference power (NEIP), which is the ratio of the power transmitted over the WLAN bands over the total permissible power under the given mask. The NEIPs of the above two designed pulses are 0.016% and 0.007% respectively, both of which are quite low. In contrast, the presence of NBI around 5–6 GHz imposes severe power limitations when the Gaussian pulse or its higher-order derivatives are used.

5.6.5 Impact of Pulse Designs on Transceiver Power Efficiency

Under the stringently constrained power spectral density, the spectral utilization efficiency (NESP) of the transmitted pulses directly affects the transmit SNR, which in

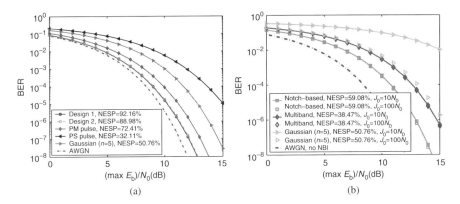

Figure 5.10 BER performance: (a) comparisons of different pulse shapers; (b) impact of NBI.

turn affects the reception quality. To quantify these effects, we now compare the BER performance of the pulses investigated in Section 5.6.1. We simulated a system in which binary PAM symbols are transmitted with transmitter parameters $N_f = 10$ and $T_f = 30$ ns. For each pulse shape, the system transmitted the maximum energy-per-bit allowed under the FCC mask. The multipath channel was generated according to the CM1 indoor channel model by IEEE 802.15.3a working group [8], with channel parameters with channel parameters $(\Lambda, \lambda, \Gamma, \gamma) = (0.0233$ ns$^{-1}, 2.5$ ns$^{-1}, 7.1$ ns, 4.3 ns$)$. The channel delay spread was truncated to 28 ns, and the total power gain of this channel was normalized to 1. An ideal receiver was employed to combine energy from all the channel taps. This provides a lower bound of the BER achievable in practice when using each pulse.

Figure 5.10(a) depicts the BER values for various transmit pulses with respect to (max $E_b)/N_0$, where max E_b is the maximum allowed transmit bit energy of the pulses under consideration, and N_0 is the PSD of the additive white Gaussian noise at the receiver. By employing one of the two pulses synthesized by the FIR DSP designs (which have NESPs around 90%), a UWB radio operating in the CM1 multipath channel would obtain BER performance that is less than 1 dB away from that of an ideal detector in an AWGN channel.

The BER performance of the UWB system in the presence of NBI is plotted in Figure 5.10(b), where three pulse shapers are considered: the NBI-resilient pulses designed using the notch-based and multiband design techniques described in Section 5.6.4, and the NBI-prone fifth-order derivative of a Gaussian pulse, that is, $x^{(n)}(t)$ with $n = 5$. The corresponding NESPs are 59.08%, 38.47% and 50.76%, respectively. In the simulated environment, there are two NBI sources. One occupies the band [0.96–3.1] GHz, and the other occupies the two 802.11a bands around 5.5 GHz, and both interferers have the same PSD, J_0. This PSD was set to $J_0 = 10 N_0$ or $J_0 = 100 N_0$. It can be observed that a transceiver employing either of the two NBI-resilient pulses has a BER performance that is quite robust to the NBI. This is due to the fact that the power spectra of both pulses have notches

20 dB lower than the mask in the WLAN bands, and hence their BERs are barely affected when the NBI power level is raised from $J_0 = 10\,N_0$ to $J_0 = 100\,N_0$. In contrast, the fifth-order Gaussian derivative pulse was not designed to avoid the NBI, and its BER performance suffers considerably when J_0 increases. It is also interesting to observe that, for a small level of NBI ($J_0 = 10\,N_0$), the BER performance of the multiband NBI-resilient pulse, which has an NESP $= 38.47\%$, is comparable to that of the NBI-prone fifth-order Gaussian derivative pulse, which has an NESP $= 50.76\%$. This illustrates the trade-off between NBI-avoidance and the NESP level. Since $J_0 = 10\,N_0$ is relatively small, the fifth-order Gaussian derivative pulse is able to make up for its NBI sensitivity through its larger NESP. It is only when J_0 becomes large that the NBI-avoidance capability of the multiband pulse out-weighs its loss in NESP. (As a reference, an FIR-prefiltered pulse without NBI-related constraints can reach a maximum NESP of 82% under the tight mask [30].) In a nutshell, the NBI-resilience of a pulse design needs to be judiciously traded off against the impact of the resulting NESP loss. This example illustrates that an adaptive NBI-avoidance strategy might be preferred, but the improved performance of such a scheme would have to be weighed against the consequent increase in implementation complexity.

5.7 CONCLUSIONS

Optimizing the spectral utilization of the transmitted signal is critical to UWB systems supporting high-data-rate wireless access. In this chapter, we presented waveform shaping methods that generate single pulses and sets of (essentially) orthogonal pulses for UWB systems that satisfy a spectral mask constraint, such as that imposed by the US FCC. In particular, the design of FIR prefiltering structures was described. This framework not only provides waveforms with high spectral utilization and guaranteed spectral mask compliance, but also permits simple modifications that can accommodate several other system objectives. Some algebraic transformations facilitated the formulation of various single pulse design problems as convex optimization problems, from which globally optimal solutions can be efficiently obtained. Related techniques were used to generate efficiently solvable formulations of a range of other UWB pulse design problems, including multiband pulse design and multiple orthogonal pulse design. These pulse shapers can support flexible avoidance of narrowband interference, as well as implementation of fast frequency hopping, free of analog carriers. Moreover, these designs have a lower sampling frequency requirement than competing digital methods, and can be implemented without putting excessive burden on ADC circuitry and without modifying the analog components of existing UWB antennas and transmitters.

REFERENCES

1. J. W. Adams and J. L. Sullivan, "Peak-constrained least-squares optimization", *IEEE Transactions on Signal Processing*, vol. 46, no. 2, pp. 306–321, February 1998.

2. B. Alkire and L. Vandenberghe, "Convex optimization problems involving finite autocorrelation sequences," *Mathematical Programming, Series A*, vol. 93, no. 3, pp. 331–359, December 2002.
3. A. Batra et al., "Multi-band OFDM physical layer proposal for IEEE 802.15 Task Group 3a," IEEE P802.15 Working Group for Wireless Personal Area Networks (WPANs) Publications Document: IEEE P802.15-03/268r3, March 2004.
4. J. O. Coleman and D. P. Scholnik, "Design of nonlinear-phase FIR filters with second-order cone programming," in *Proc. 1999 Midwest Symp. on Circuits and Systems*, Las Cruces, NM, USA, August 1999.
5. T. N. Davidson, Z.-Q. Luo, and J. F. Sturm, "Linear matrix inequality formulation of spectral mask constraints with applications to FIR filter design," *IEEE Transactions on Signal Processing*, vol. 50, no. 11, pp. 2702–2715, November 2002.
6. R. S. Dilmaghani, M. Ghavami, B. Allen and H. Aghvami, "Novel UWB pulse shaping using prolate spheroidal wave functions," in *Proc. IEEE Int. Symp. on Personal, Indoor and Mobile Radio Communications*, vol. 1, pp. 602–606, Beijing, September 2003.
7. FCC Report and Order, *In the matter of Revision of Part 15 of the Commission's Rules Regarding Ultra-Wideband Transmission Systems*, FCC 02-48, April 2002.
8. IEEE P802.15 Working Group for WPANs, *Channel Modeling Sub-Committee Report Final*, IEEE P802.15-02/368r5-SG3a, November 2002.
9. M. Ghavami, L. B. Michael and R. Kohno, "Hermite function based orthogonal pulses for UWB communication," in *Proc. Int. Symp. on Wireless Personal Multimedia Communications*, pp. 437–440, Aalborg, September 2001.
10. T. N. T. Goodman, C. A. Micchelli, G. Rodriguez, and S. Seatzu, "Spectral factorization of Laurent polynomials," *Advances in Computational Mathematics*, vol. 7, no. 4, pp. 429–454, 1997.
11. J. Han and C. Nguyen, "A new ultra-wideband, ultra-short monocycle pulse generator with reduced ringing," *IEEE Microwave and Wireless Components Letters*, vol. 12, no. 6, pp. 206–208, June 2000.
12. X. Luo, L. Yang, and G. B. Giannakis, "Designing optimal pulse-shapers for ultra-wideband radios," *Journal of Communications and Networks*, vol. 5, no. 4, pp. 344–353, December 2003.
13. J. H. McClellan and T. W. Parks, "A unified approach to the design of optimum FIR linear-phase digital filters," *IEEE Transactions on Circuit Theory*, vol. CT–20, no. 6, pp. 697–701, November 1973.
14. P. Moulin, M. Anitescu, K. O. Kortanek, and F. A. Potra, "The role of linear semi-infinite programming in signal-adapted QMF bank design," *IEEE Transactions on Signal Processing*, vol. 49, no. 9, pp. 2160–2174, September 1997.
15. Multiband OFDM Alliance, http://www.multibandofdm.org/
16. Y.-P. Nakache and A. F. Molisch, "Spectral shape of UWB signals—Influence of modulation format, multiple access scheme and pulse shape," in *Proc. IEEE Vehicular Technology Conf.*, vol. 4, pp. 2510–2514, April 2003.
17. S. Nordebo and Z. Zhang, "A unified approach to digital filter design with time and frequency-domain specifications," *IEEE Transactions on Circuits and Systems—II: Analog and Digital Signal Processing*, vol. 46, no. 6, pp. 765–775, June 1999.
18. B. Parr, B. Cho, K. Wallace, and Z. Ding, "A novel ultra-wideband pulse design algorithm," *IEEE Communications Letters*, vol. 7, no. 5, pp. 219–221, May 2003.

19. D. Porrat and D. Tse, "Bandwidth scaling in ultra wideband communication," in *Proc. Allerton Conf. on Communications, Control and Computing*, Monticello, IL, October 2003.
20. J. G. Proakis, *Digital Communications*, 4th edn, McGraw-Hill, New York, February 2001.
21. S. Roy, J. R. Foerster, V. S. Somayazulu, and D. G. Leeper, "Ultrawideband radio design: the promise of high-speed, short-range wireless connectivity," *Proceedings of the IEEE*, vol. 92, no. 2, pp. 295–311, February 2004.
22. R. A. Scholtz, "Multiple access with time hopping impulse modulation," in *Proc. Military Communications Conf.*, Boston, MA, pp. 447–450, October 1993.
23. I. W. Selesnick, M. Lang, and C. S. Burrus, "A modified algorithm for constrained least square design of multiband FIR filters without specified transition bands," *IEEE Transactions on Signal Processing*, vol. 46, no. 2, pp. 497–501, February 1998.
24. H. Sheng, P. Orlik, A. M. Haimovich, L. J. Cimini, and J. Zhang, "On the spectral and power requirements for ultra-wideband transmission," in *Proc. IEEE Int. Conf. on Communications*, pp. 738–742, Anchorage, AK, 2003.
25. J. F. Sturm, "Using SeDuMi 1.02, a MATLAB toolbox for optimization over symmetric cones," *Optimization Methods and Software*, vols. 11–12, pp. 625–653, 1999.
26. M. Z. Win, "Spectral density of random UWB signals," *IEEE Communication Letters*, vol. 6, no. 12, pp. 526–528, December 2002.
27. M. Z. Win and R. A. Scholtz, "Impulse radio: How it works," *IEEE Communication Letters*, vol. 2, pp. 36–38, February 1998.
28. S. Wu, S. Boyd, and L. Vandenberghe, "FIR filter design via spectral factorization and convex optimization," in *Applied and Computational Control, Signals and Circuits*, vol. 1, B. Datta, ed., Birkhauser, Boston, MA, 1999.
29. X. Wu, Z. Tian, T. N. Davidson, and G. B. Giannakis, "Optimal waveform design for UWB radios," in *Proc. IEEE Int. Conf. on Acoustics, Speech and Signal Processing*, Montreal, May 2004.
30. X. Wu, Z. Tian, T. N. Davidson, and G. B. Giannakis, "Optimal waveform design for UWB radios," *IEEE Transactions on Signal Processing* (in press).
31. Y. Wu, A. F. Molisch, S.-Y. Kung, and J. Zhang, "Impulse radio pulse shaping for ultra-wide bandwidth (UWB) systems," in *Proc. Int. Symp. on Personal, Indoor and Mobile Radio Communications*, vol. 1, pp. 877–881, Beijing, September 2003.
32. L. Yang and G. B. Giannakis, "Ultra-wideband communications: An idea whose time has come," *IEEE Signal Processing Magazine*, vol. 21, no. 6, pp. 26–54, November 2004.
33. H. Zhang and R. Kohno, "SSA realization in UWB multiple access systems based on prolate spheroidal wave functions," in *Proc. IEEE Wireless Communications and Networking Conf.*, pp. 1794–1799, Atlanta, GA, March 2004.

CHAPTER 6

Antenna Issues

ZHI NING CHEN

6.1 INTRODUCTION

For potential UWB devices, the design of antennas is a challenging issue. On the one hand, to achieve high data rates in wireless transmission, the UWB systems usually occupy extremely broad bandwidths, typically of a few gigahertz. Within the operating bandwidths, UWB antennas should have stable response in terms of impedance matching, gain, radiation patterns, phase, and polarization. Also, the demands for the antennas include small size, conformal design, low cost, easy integration into other RF circuits. Moreover, the requirements for broad bandwidths are associated with other crucial constraints such as small size and low cost because most promising UWB applications will be portable devices.

On the other hand, to avoid the possible inband/outband interference between the UWB systems and existing electronic systems, the frequency regulators must work out the emission limits for the applications. For example, the emission limits of the effective isotropic radiated power (EIRP) levels of -41.3 dBm/MHz for a frequency range of 3.1–10.6 GHz were released by FCC for the unlicensed use of commercial UWB communication systems. The emission limits will be determined by both the selection of source pulses and design of antennas in UWB systems.

In this chapter, the unique aspects of antenna designs in UWB systems will be highlighted first. Some typical antennas will be used to elucidate the effects of the antenna design on the transmission of UWB signals. After that, the special design considerations for UWB antennas and source pulses will be presented. Next, the design and evaluation of planar antennas used in UWB applications will be introduced. The last part discusses the effects of the antenna system and source/template pulses on the BER performance of the UWB communication systems.

Ultra Wideband Wireless Communication. Edited by Arslan, Chen, and Di Benedetto
Copyright © 2006 John Wiley & Sons, Inc.

6.2 DESIGN CONSIDERATIONS

As compared with conventional narrowband/broadband systems, two essential and special design considerations for antennas in UWB systems should be emphasized. One is that the power density spectrum (PDS) shaping of the radiated signals should conform to the emission limit masks for avoiding possible interference with other existing electronic systems. The other is that the source pulses and transmit–receive antennas should be evaluated in terms of the overall system performance, such as maximum S/N ratio or minimal BER. [1]

6.2.1 Description of Antenna Systems

Consider a typical transmit–receive antenna system, in the UWB system, as shown in Figure 6.1. The Friis transmission formula can be used to relate the output power of the receive antenna to the input power of the transmit antenna as given in Equation (6.1), where it is assumed that each antenna is in the far-field zone of the other. This formula is frequency-dependent for the general cases when the parameters in Equation (6.1) vary within operating frequency ranges.

$$\frac{P_r(\omega)}{P_t(\omega)} = [1 - |\Gamma_t(\omega)|^2][1 - |\Gamma_r(\omega)|^2]G_r(\omega)G_t(\omega)|\hat{\rho}_t(\omega) \cdot \hat{\rho}_r(\omega)|^2 \left(\frac{\lambda}{4\pi r}\right)^2 \quad (6.1)$$

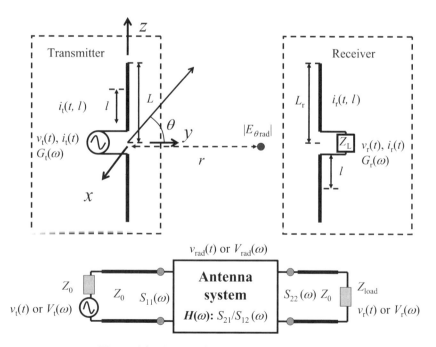

Figure 6.1 A transmit–receive antenna system.

where P_t, P_r are the time-average input power of the transmit antenna and output power of the receive antenna; Γ_t, Γ_r are the return losses at the input of the transmit antenna and the output of the receive antenna; G_t, G_r are the orientation-dependent (θ, ϕ) gains of the transmit and the receive antennas; $|\hat{\rho}_t \cdot \hat{\rho}_r|^2$ are the polarization matching factors between the transmit and receive antennas; λ is the operating wavelength at operating frequency ω; and r is the distance between the transmit and receive antennas. If a transfer function $H(\omega)$ is defined to describe the relation between the source and output signal (voltage) $[V_t(\omega)/2]^2/2 = [P_t(\omega)Z_0]$ and $V_r^2(\omega)/2 = [P_r(\omega)Z_{\text{load}}]$, Equation (6.1) can be simplified as Equation (6.2):

$$H(\omega) = \frac{V_r(\omega)}{V_t(\omega)} = \left|\sqrt{\frac{P_r(\omega)}{P_t(\omega)}\frac{Z_{\text{load}}}{4Z_0}}\right| e^{-j\phi(\omega)} = |H(\omega)|e^{-j\phi(\omega)};$$

$$\phi(\omega) = \phi_t(\omega) + \phi_r(\omega) + \omega r/c \quad (6.2)$$

where c denotes velocity of light, and $\phi_t(\omega)$ and $\phi_r(\omega)$ are, respectively, the phase variation caused by the transmit and receive antennas. Therefore, the transfer characteristics of the transmit–receive antenna system can be described by the transfer function $H(\omega)$ by means of the performance parameters of the transmit and receive antennas, such as impedance matching, gain, polarization matching, the distance between the antennas, operating frequency as well as the orientation of the antennas if the characteristics of an RF channel are completely ignored.

Also, the antenna system can be considered as a two-port network, as shown in Figure 6.1, and thus, the transfer function can be measured in terms of the S parameter S_{21} when the source impedance and loading are matching to the transmit and receive antennas, respectively. Therefore, the measurable parameter S_{21} or $H(\omega)$ can be used to evaluate the performance of the antenna systems. The setup for the measurement of the transfer function S_{21} or $H(\omega)$ in the frequency domain is shown in Figure 6.2. According to the Image Theory, the transmit and receive monopole antennas are installed above a ground plane which is big so that the antennas are in the far field zone of the other. A vector network analyzer is used to measure all S parameters through the two ports, namely the input of the transmit antenna and the output of the receive antenna. Therefore, the transfer function of the antenna system can be measued in the azimuth plane.

Furthermore, the relation between the radiated electric fields and source pulses at the transmit antenna can be expressed in Equation (6.3). The vector transfer function $\mathbf{E}_{\text{rad}}(\omega)$ is with the polarization direction $\hat{a}$ of the transmit antenna and determined by the characteristics of the transmit antenna, such as impedance matching, gain, and the orientation of observation point. $V_t(\omega)$ is the spectrum of a source signal (voltage). Therefore, the radiation transfer function can be used to describe the radiated PDS for the evaluation of the emission limits:

$$\mathbf{E}_{\text{rad}}(\omega) = \mathbf{H}_{\text{rad}}(\omega)V_t(\omega) = \hat{a}|H_{\text{rad}}(\omega)|e^{-j\phi_{\text{rad}}(\omega)}V_t(\omega);$$

$$\phi_{\text{rad}}(\omega) = \phi_t(\omega) + \omega r/c \quad (6.3)$$

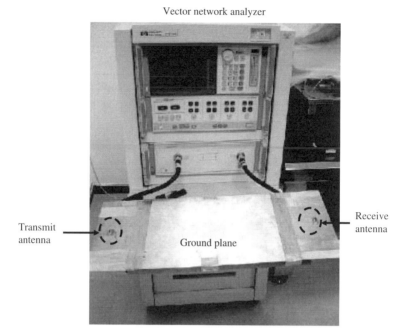

Figure 6.2 The setup of measurement of transfer functions of the transmit–receive antenna system with a vector network analyzer.

6.2.2 Single-Band and Multiband Schemes

The UWB systems can utilize the UWB band in a variety of ways. For instance, multiband (single-/multicarrier) and single-band schemes have been proposed for UWB systems. To comply with the emission limits, the design considerations for source pulses and transmit antennas are subject to the specific system schemes.

Under the multiband scheme, the available UWB band can be divided into several sub-bands. Each of the source pulses is shaped to occupy only one subband. For example, Figure 6.3(a) shows the scheme of 15 uniform sub-bands for 7.5 GHz UWB band, where the 10 dB bandwidths are of 500 MHz. Figure 6.3(b) displays a Gaussian pulse $v_0(t) = e^{-(t/\sigma)^2}$ with $\sigma = 1366$ ps, which is modulated by the sine signals with the frequencies of $(3.35 + n \times 0.5)$ GHz ($n = 0, 1, 2, \ldots, 14$). Under such a scheme, it is easy to control the PDS for the avoidance of possible inband/outband interference with other systems.

Alternatively, the single-band scheme was initially proposed for UWB technology. The single or few source pulses, which usually have a very short duration, are shaped so that their spectra occupy as wide as possible range within the UWB band for high data rates and S/N. The short pulses may be transmitted with a carrier.

From Equation (6.3), it can be seen that there are at least two ways to meet the emission limit masks. One is to optimize the spectra, $V_t(\omega)$ of source signals directly to make the 10 dB bandwidth of the source signal narrower than the UWB band

6.2 DESIGN CONSIDERATIONS   **135**

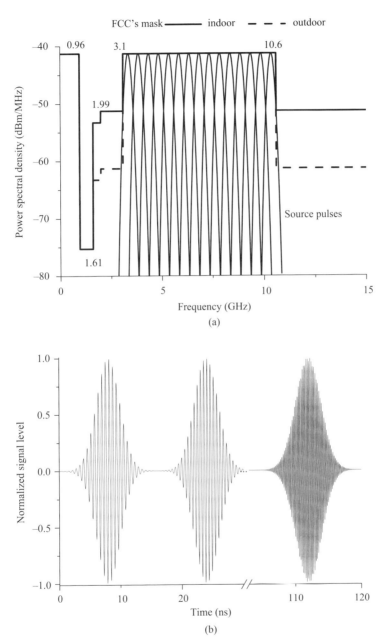

Figure 6.3 Pulses and spectra in multiband scheme: (a) frequency domain (FD); (b) time domain (TD).

when the antenna system has a constant unchanged radiation transfer function, $\mathbf{H}_{rad}(\omega)$ within the UWB band. This involves two scenarios. One is that the 10 dB bandwidth fully falls into the UWB band by properly selecting source pulses. Otherwise, the 10 dB bandwidth spectrum of the pulse can be shifted into the UWB band by modulating the pulse with a proper sine signal (carrier). Both cases will make antenna design easy.

The other one is to tailor the spectra, $V_t(\omega)$ of source signals by using the filtering function of $\mathbf{H}_{rad}(\omega)$, viz. to control $\mathbf{H}_{rad}(\omega) V_t(\omega)$ if $V_t(\omega)$ does not meet the emission limit mask itself. Using this method, the transmit antenna acts as not only a radiator but also a filter, which is designed to suppress the unwanted radiation outside the UWB band or in the specific band. This will make antenna designs complicated.

6.2.3 Source Pulses

In principle, all the impulses with the spectra (wider than 500 MHz stipulated by FCC) can be used as the signals. However, practically, only the pulses which can be easily generated and controlled and have low power-consumption [no direct current (DC) component] are selected to generate UWB signals. Owing to unique temporal and spectral properties, a family of Rayleigh (differentiated Gaussian) pulses, $v_n(t)$ or $\tilde{v}_n(\omega)$ is widely used as the source pulses in the UWB systems, as given in Equation (6.4).

$$v_n(t) = \frac{d^n}{d^n t}[e^{-(t/\sigma)^2}]; \quad \tilde{v}_n(\omega) = (j\omega)^n \sigma \sqrt{\pi} e^{-(\omega\sigma/2)^2} \quad (6.4)$$

where the pulse parameter σ stands for the time when $v_0(\sigma) = e^{-1}$. The pulse duration T is defined as the interval between the start and the end of the pulse where the values $|v_n(t = \pm T/2)|$ decreases from the normalized peak value to e^{-9} as shown in Figure 6.4, where the pulses $v_0(t)$ and $v_1(t)$ are shown. Obviously, only $v_1(t)$ is a monocycle pulse, which is easily generated by RF circuits and does not generate any DC component in the frequency domain (FD).

The calculation shows that the 10 dB bandwidths of the first-order Rayleigh pulses with $\sigma > 61$ ps ($T > 305$ ps) are <7.5 GHz, as shown in Figure 6.4(a). However, their spectra do not fully fall into the UWB band defined by the FCC. Some higher-order Rayleigh pulses can match the UWB band directly, such as the fourth-order Rayleigh pulses with $67 < \sigma < 76$ ps, the fifth-order Rayleigh pulses with $72 < \sigma < 91$ ps, and the sixth-order Rayleigh pulses with $76 < \sigma < 106$ ps.

6.2.4 Transmit Antenna and PDS

The effects of source pulses and transmit antennas on the radiated PDS shaping are taken into account. Two types of the antennas with narrow and broad impedance bandwidths are exemplified.

First, a thin-wire straight dipole of $L = 11$ mm in length and having a 0.3 mm radius was simulated. Figure 6.5(a)–(c) shows the simulated $|S_{11}|$ and the

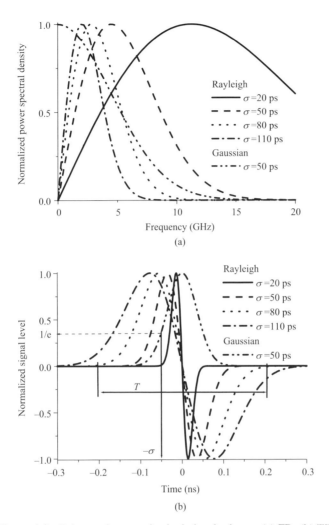

Figure 6.4 Pulses and spectra in single-band scheme: (a) FD; (b) TD.

normalized radiated transfer function $|\mathbf{H}_{rad}(\omega)|$. The 10 dB bandwidth is 25% with the well-matched frequency of 5.85 GHz. Three typical first-order Rayleigh pulses with σ = 30, 45, and 80 ps are used as source pulses. The radiated fields are the co-polarization components $|\mathbf{E}_{rad}(\omega)|$ in the direction of $\theta = 90°$ and at a distance of r = 1960 mm, as shown in Figure 6.1. Both inner resistance of the voltage source and resistive loading of the receive antenna are 100 Ω.

From the transfer function $|H(\omega)|$ shown in Figure 6.5(a)–(c), it is readily observed that the dipole acts as a high-pass filter within the UWB band. Thus, the tailored radiated spectrum of the short pulse cannot fully meet the emission limit mask when the pulse has the high emission levels at the frequencies higher than 10.6 GHz as shown in Figure 6.5(a). In contrast, Figure 6.5(c) displays that the

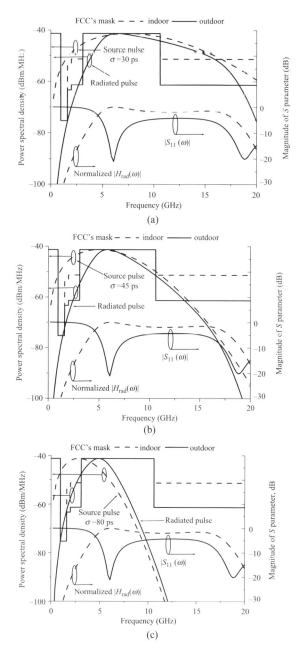

Figure 6.5 Comparison of spectral density shaping of radiated electrical fields by a narrow band thin-wire dipole and a broadband planar dipole, both driven by first-order Rayleigh source pulses: (a) $\sigma = 30$ ps, thin-wire dipole; (b) $\sigma = 45$ ps, thin-wire dipole; (c) $\sigma = 80$ ps, thin-wire dipole; (d) $\sigma = 30$ ps, planar dipole; (e) $\sigma = 45$ ps, planar dipole; (f) $\sigma = 80$ ps, planar dipole.

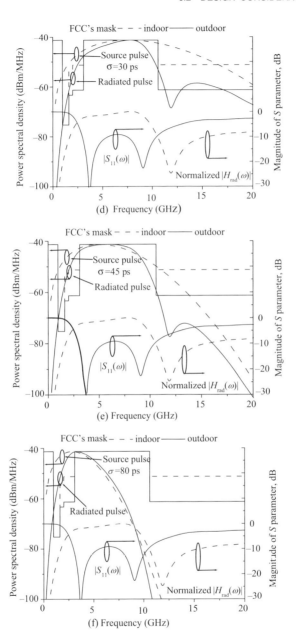

Figure 6.5 *Continued.*

radiated spectrum of the longest pulse also does not meet the emission limit masks because of its high emission levels at the frequencies lower than 3.1 GHz. Figure 6.5(b) evidently demonstrates that the radiated spectrum can completely comply with the specific emission constraints mask by properly selected source pulses for a given transmit antenna. Figure 6.5(d)–(f) shows the results for a planar square dipole of size of 18 mm × 18 mm. The results show that the planar dipole features better impedance matching within UWB band than the thin-wire dipole. The spectra of the received pulses by the planar dipole keep almost the same shape within the UWB band as those of source pulses, whereas the narrowband thin-wire dipole acts as a narrowband filter to tailor the spectra of the source pulses.

Another important parameter, the efficiency of the transmit antenna can be evaluated in Equation (6.5):

$$\eta_{rad} = \frac{\int_0^\infty P_t(\omega)[1 - |S_{11}(\omega)|^2]\,d\omega}{\int_0^\infty P_t(\omega)\,d\omega} \times 100\% \qquad (6.5)$$

Efficiency η_{rad} is determined by source pulse and transmit antenna. As an example, the calculated efficiency of the case discussed in Figure 6.5(b) is about 53%, where the spectrum conforms well to the emission limit mask but the antenna is a narrowband design with high return lose in the most of the UWB band.

Figure 6.6 compares the waveforms of the radiated electric fields with $\sigma = 30$, 45, and 80 ps or $\sigma_{ant}/\sigma = 1.22$, 0.82, and 0.46. The parameter $\sigma_{ant}/\sigma = L/c$

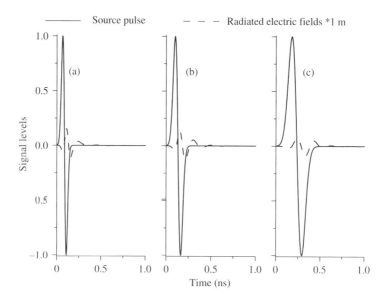

Figure 6.6 Comparison of waveforms of source pulses and the radiated electric fields (at a point 10 m from the transmit antenna): (a) $\sigma = 30$ ps; (b) $\sigma = 45$ ps; (c) $\sigma = 80$ ps.

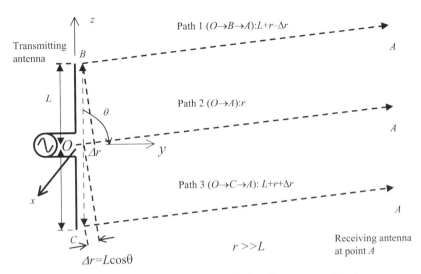

Figure 6.7 Multipath model of the radiation from a transmit antenna.

indicates the time for light to travel the length, L of the antenna arm at the velocity $c = 3 \times 10^8$ m/s. Owing to the highpass filtering of the antennas in the FD, the waveforms of the radiated pulses distorted [2]. In the Time Domain (TD), the distorted waveforms of the radiated pulses are basically attributed to the reflection appearing at the ends (including the input) of the dipole. The pulses radiated from the ends of the dipole, namely points, O, B, and C, arrive at the receiving antenna located at point A (in the far-field zone) through the paths of different lengths, namely paths 1, 2, and 3, as illustrated in Figure 6.7. The length difference between the paths also causes the time delay in the TD or the phase difference in the FD. Therefore, the length of the dipoles and the orientation of the observation points affect the waveforms of the radiated pulses significantly.

6.2.5 Transmit–Receive Antenna System

The other crucial criterion of the UWB antennas is associated with the performance of overall transmit–receive antenna systems. This consideration stems from the fact that, compared with the antennas in narrowband systems, the antennas or antenna systems in the UWB systems hardly maintain invariable performance across a range of a few gigahertz. The variation in the performance of the antennas or antenna systems significantly affects the waveforms and spectra of the radiated pulses, as discussed above. As a result, the distortion of the signals received by a receiver is usually severe. The transfer function can be used to assess the performance of the antenna systems and evaluate the distortion of the received signals. However, it is difficult to exactly formulate the transfer function between arbitrary transmit–receive antennas in a close form due to the complicated frequency-dependent features of the antennas [3–6]. Therefore, the effects of the transfer function $|H(\omega)|$ given in Equation (6.2) on the output signals are evaluated. Also, the performance of the antenna systems is measured by pulse fidelity for the single-band scheme and system transmission efficiency.

The received signals, which are transmitted through both narrowband and broadband antenna systems, are compared for the single-band and multiband schemes. The source pulses used in Figures 6.3 and 6.4 are adopted, respectively. The narrowband and broadband antenna systems comprise a pair of thin-wire dipoles and planar square dipoles, which have been used in Figure 6.5.

Figure 6.8 illustrates the system transfer function $|H(\omega)|$ and the radiation transfer function $|\mathbf{H}_{rad}(\omega)|$ for the narrowband and broadband antenna systems. The comparison shows that the broadband antenna system features a flatter $|H(\omega)|$ or $\mathbf{H}_{rad}(\omega)|$ than the narrowband antenna system within the UWB band.

Figure 6.9(a)–(c) shows the waveforms of the received signals in the single-band and multiband schemes, when the signals go through the narrowband thin-wire dipole system. The pulse waveforms illustrated in Figure 6.9(a) and (b) are not identical with the waveforms of source pulses, even the radiated pulses shown in Figure 6.6. In the single-band scheme, the severe distortion results mainly from the narrowband filtering of the antenna systems. In other words, the dispersion in magnitude and phase of $|H(\omega)|$ results in the distortion of the pulse waveforms. In a multiband scheme, the change in the magnitude of $|H(\omega)|$ causes the uneven envelope of the magnitudes of the received signals, which accords with the shape of $|H(\omega)|$ shown in Figure 6.8. Figure 6.9(c) displays the group delay and the variation in the group delay, which also distort the signals, especially when a large change in the group delay occurs. Due to narrowband operation, the

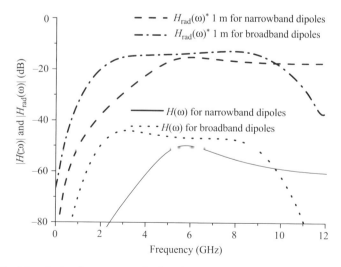

Figure 6.8 Magnitude of system transfer function $|H(\omega)|$ and the radiation transfer function $|\mathbf{H}_{rad}(\omega)| \times 1$ m for narrowband and broadband antenna systems.

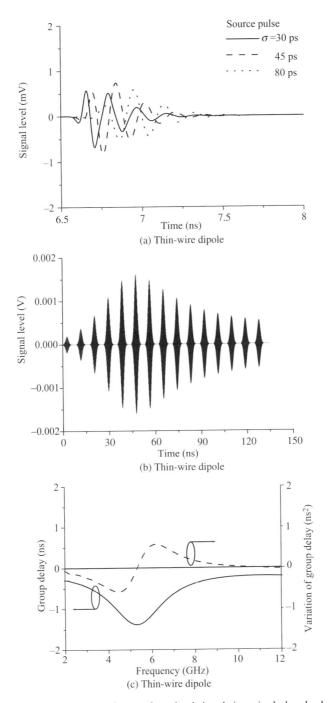

Figure 6.9 (a) and (d) The waveforms of received signals in a single-band scheme; (b) and (e) the waveforms of received signals in a multiband scheme; (c) and (f) the group delay and its variation of $|H(\omega)|$.

144 ANTENNA ISSUES

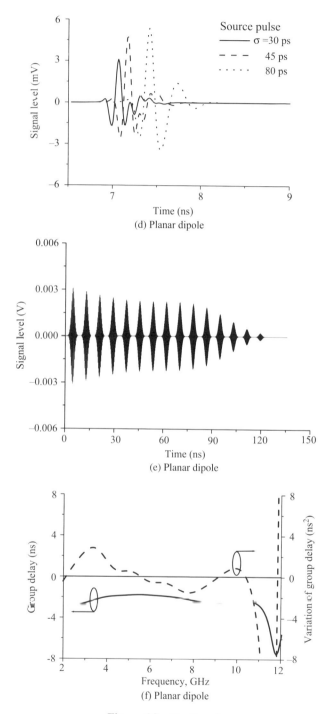

Figure 6.9 *Continued.*

latter usually has linear phase response across the operating band. The important influence of the group delay leads to the shift in carriers. That means that the maximum energy can be detected at different frequencies from the original carrier.

As a comparison, a broadband planar dipole system was used to transmit and receive the pulses. The waveforms of the received signals in the single-band and multiband schemes are illustrated in Figure 6.9(d)–(f). Compared with the pulse waveforms illustrated in Figure 6.9(a) and (b), the pulse waveforms shown in Figure 6.9(d) and (e) change less. The pulse waveforms in the single-band scheme are similar to the second Rayleigh pulses. The longer the pulse is, the longer the time delay is. This reveals that, in the TD, the reflection occurring at the ends of the dipoles essentially brings on the distortion of the waveforms, although the source is well matched to the input of the antenna within a broad bandwidth. Because of the flat magnitude response of the planar dipole, the envelope of the pulse magnitudes in the multiband scheme is more even than that shown in Figure 6.9(b), which also agrees with the shape of the magnitude planar dipole shown in Figure 6.8. However, it should be noted that the group delay and the variation of the group delay depicted in Figure 6.9(f) are greater than those shown in Figure 6.9(c) since the planar dipoles are larger than the thin-wire dipoles.

Figure 6.10 demonstrates the frequency shift of each modulated pulses with respect to the 15 carriers. The curves of the frequency shift are closely related to the system response shown in Figure 6.8. It suggests that, in the multiband scheme, the signals are primarily affected by the magnitude of the system response because the group delay varies slightly within one width-limited sub-band.

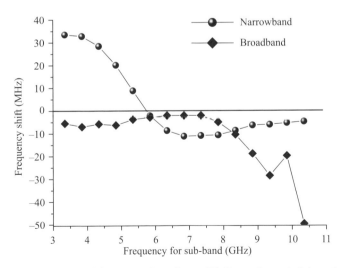

Figure 6.10 Frequency shift in terms of maximum fidelity at the varied detecting frequencies for a broadband antenna system.

To evaluate the transmit and receive capability of the antenna systems, Equation (6.1) is rewritten as Equation (6.6):

$$10 \log \left[\frac{P_r(\omega)}{P_t(\omega)}\right] = 10 \log \left\{ [1 - |\Gamma_t(\omega)|^2][1 - |\Gamma_r(\omega)|^2] \right.$$

$$\left. \times G_t(\omega) G_r(\omega) |\hat{\rho}_t(\omega) \cdot \hat{\rho}_r(\omega)|^2 \left(\frac{\lambda^2}{4\pi}\right) \right\} - 10 \log (4\pi r^2)$$

$$= \eta(\text{dB}) - 10 \log (4\pi r^2) \qquad (6.6)$$

where the term η is independent of the distance between the transmit and receive antennas and indicates the transmit and receive capability of the antenna systems.

Figure 6.11 shows that for both single-band and multiband schemes, the broadband antenna system always transmits and receives the pulses much more efficiently than the narrowband antenna system. Figure 6.11(a) further suggests that, for a given antenna system in a single-band scheme, the system efficiency be also dependent on the pulse widths. Moreover, Figure 6.11(b) points out that the efficiency of a given antenna system varies with the carriers applied to a multiband scheme.

In addition, the fidelity of the signal of an antenna system is used to assess the quality of a received pulse and select a proper detection template, particularly for the single-band scheme [7]. The definition of the fidelity is given in Equation (6.7):

$$F = \max_{\tau} \int_{-\infty}^{\infty} L[p_{\text{source}}(t)] p_{\text{output}}(t - \tau) \, dt \qquad (6.7)$$

where the source pulse $p_{\text{source}}(t)$ and output pulse $p_{\text{output}}(t)$ are normalized by their energy, respectively. The fidelity F is the maximum integration by varying time delay τ. The linear operator $L[\cdot]$ operates on the input pulse $p_{\text{source}}(t)$. Evidently, the template at the output of a receiving antenna may be $L[p_{\text{source}}(t)]$ not the simple $p_{\text{source}}(t)$ for maximum fidelity. The calculated fidelity F for the single band scheme and different operators $L[\cdot]$ is tabulated in Table 6.1.

From Table 6.1, it is seen that the waveforms of the received pulses are not identical to those of the source pulses, especially for narrowband antenna systems. For sinusoidal templates, the fidelity F for the narrowband antenna system is much higher than that for the broadband one with the narrow bandpass filtering function. By optimizing Rayleigh's templates, the fidelity F can be increased greatly up to more than 0.9. The order $n > 1$ of the Rayleigh pulse suggests the differential functions of the antenna system. For the broadband antenna system, the order is just 4. However, the order n for the narrowband antenna system is larger than that for the broadband antenna system and increases as the pulse becomes wider or the ratio of $L:c$ less.

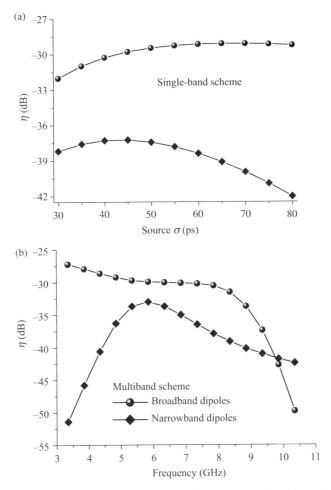

Figure 6.11 System transmission efficiency, η, for narrowband and broadband antenna systems: (a) single-band scheme; (b) multiband scheme.

TABLE 6.1 Calculated Fidelity for the Pulses in the Single Band Scheme

Antenna System	Source $p_{source}(t)$ with σ, ns	F for Template $P_{source}(t)$	F/ω, GHz, for Template $\sin(\omega t)$	Template, nth-order Rayleigh Pulse		
				F	Order, n	σ, ns
Narrowband	30	0.74	0.80/6.16	0.89	6	83
	45	0.77	0.83/5.82	0.95	7	99
	80	0.58	0.88/5.25	0.99	12	149
Broadband	30	0.70	0.62/3.86	0.94	4	78
	45	0.81	0.72/3.73	0.93	4	91
	80	0.87	0.87/3.21	0.95	4	129

6.3 ANTENNA AND PULSE VERSUS BER PERFORMANCE

In Section 6.2, the design considerations of the UWB antennas and source pulses have been discussed in terms of S parameters, transfer functions, systems efficiency, group delay and fidelity. These parameters have been used in the design of UWB antenna systems. However, in a systems point of view, the key performance parameter for wireless communications is BER. In this section, the effects of the antennas and source/template pulses on BER performance of pulsed UWB systems are evaluated. The pulsed UWB system is characterized in both TD and FD. The BER performance of the system is formulated in terms of antenna system transmission efficiency, fidelity between the received pulse and the template pulse, and incident power. As examples, the BER performance of the system with varying antenna systems and source/template pulses is discussed in the free-space channel with additive white Gaussian noise (AWGN).

6.3.1 Pulsed UWB System

Figure 6.12 shows a pulsed UWB system with a pair of transmit and receive antennas separated at a distance D in a free space, where the antennas in the far field region of each other in the UWB band. The transmit antenna is driven by a source with an internal impedance Z_0, and the receive antenna is terminated with a load impedance Z_{load} [7].

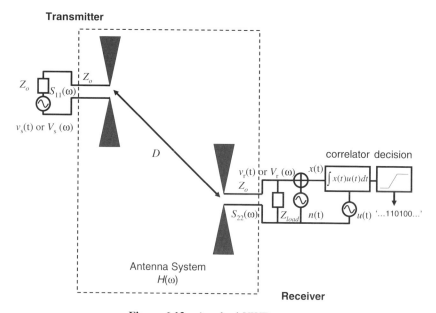

Figure 6.12 A pulsed UWB system.

6.3 ANTENNA AND PULSE VERSUS BER PERFORMANCE

In the TD, the source signal $v_s(t)$ and the received signal at the load $v_r(t)$ are given by Equations (6.8) and (6.9)

$$v_s(t) = \sum_{k=1}^{N} a_k w(t - kT_s) \qquad (6.8)$$

$$v_r(t) = \sum_{k=1}^{N} a_k r(t - kT_s) \qquad (6.9)$$

where, N is the transmitted symbol number, $w(t)$ and $r(t)$ are the source and received pulses, and $\{a_k\}$ are the transmitted symbols. Assume each a_k takes the value of either $+1$ or -1 with equal probability, and a_m is independent of a_n when $m \neq n$. T_s is the symbol duration, and the data rate is $R_b = 1/T_s$.

The received load voltage is contaminated with additive white Gaussian noise (AWGN) $n(t)$, which is a zero mean stationary stochastic process with the Gaussian distribution and a constant power spectral density (PSD). The noise PSD N_0 is related to the time autocorrelation of $n(t)$ by using Equation (6.10)

$$\frac{E(n(t)n(t+\alpha))}{Z_{\text{load}}} = \frac{1}{2\pi} \int_{-\infty}^{+\infty} N_0 e^{j\omega\alpha} d\omega = N_0 \delta(\alpha) \qquad (6.10)$$

where, α is a time delay, $E(X)$ takes the expectation of the random variable X, and $\delta(\alpha)$ is the Dirac function. The signal $x(t) = v_r(t) + n(t)$ mixed with noise is detected by the correlator with the local template $u(t)$ given by Equation (6.11)

$$u(t) = \sum_{k=1}^{N} p(t - kT_s - \tau) \qquad (6.11)$$

where, $p(t)$ is the template pulse, and τ is the time shift which can be adjusted by the correlator for the desired BER performance. For the k-th transmitted symbol, the correlation output C_k is given by Equation (6.12)

$$C_k = R_k + N_k \qquad (6.12a)$$

$$R_k = a_k \int_{-\infty}^{+\infty} r(t - kT_s) p(t - kT_s - \tau) dt \qquad (6.12b)$$

$$N_k = \int_{-\infty}^{+\infty} n(t) p(t - kT_s - \tau) dt. \qquad (6.12c)$$

where, R_k and N_k are the signal and noise components as in [8].

Then the correlation outputs are fed into the decision device, where the carried symbols are determined and output. Since the noise $n(t)$ is AWGN in nature, N_k is a zero mean random variable with the Gaussian distribution. As a result, the

SNR and the BER can be computed by Equations (6.13) and (6.14) [9]

$$SNR = \frac{|R_k|^2}{E(|N_k|^2)} = \frac{\int_{-\infty}^{+\infty} |r(t)|^2 dt}{N_0 Z_{\text{load}}} \left| \int_{-\infty}^{+\infty} \frac{r(t)}{\sqrt{\int_{-\infty}^{+\infty} |r(t)|^2 dt}} \frac{p(t-\tau)}{\sqrt{\int_{-\infty}^{+\infty} |p(t)|^2 dt}} dt \right|^2$$

(6.13)

$$BER = Q(\sqrt{SNR}) = \frac{1}{\sqrt{2\pi}} \int_{\sqrt{SNR}}^{+\infty} e^{-t^2/2} dt \qquad (6.14)$$

where the correlator usually chooses the maximum τ (τ_{max}) for the maximum SNR, namely

$$SNR = \frac{\int_{-\infty}^{+\infty} |r(t)|^2 dt}{N_0 Z_{\text{load}}} \left| \int_{-\infty}^{+\infty} \frac{r(t)}{\sqrt{\int_{-\infty}^{+\infty} |r(t)|^2 dt}} \frac{p(t-\tau_{\text{max}})}{\sqrt{\int_{-\infty}^{+\infty} |p(t)|^2 dt}} dt \right|^2$$

$$= \frac{\int_{-\infty}^{+\infty} |r(t)|^2 dt}{N_0 Z_{\text{load}}} (F(r(t), p(t)))^2 \qquad (6.15)$$

where, the fidelity $F(r(t), p(t))$, as a measure of the resemblance between $r(t)$ and $p(t)$, is defined by Equation (6.16) [1]

$$F(r(t), p(t)) = \max_{\tau} \left| \int_{-\infty}^{+\infty} \frac{r(t)}{\sqrt{\int_{-\infty}^{+\infty} |r(t)|^2 dt}} \frac{p(t-\tau)}{\sqrt{\int_{-\infty}^{+\infty} |p(t)|^2 dt}} dt \right| \qquad (6.16)$$

In the FD, the antenna system can be described by an orientation-dependent transfer function $H(\omega)$ which was defined by Equation (6.2). Using the incident wave power P_{inc} to the transmit antenna and the received power P_r at the load as well as the received and source pulses $R(\omega)$ and $W(\omega)$, the antenna system transmission efficiency η_{sys} can be expressed as Equation (6.17)

$$\eta_{\text{sys}} = \frac{P_r}{P_{\text{inc}}} = \frac{4 Z_0 \int_{\text{BW}} |H(\omega)|^2 |W(\omega)|^2 d\omega}{Z_{\text{load}} \int_{\text{BW}} |W(\omega)|^2 d\omega}. \qquad (6.17)$$

Therefore, the SNR in Equation (6.15) can be rewritten in terms of P_{inc}, η_{sys}, $F(r(t), p(t))$, R_b, and N_0 as

$$SNR = \frac{P_{\text{inc}} \eta_{\text{sys}}}{R_b N_0} (F(r(t), p(t)))^2. \qquad (6.18)$$

6.3 ANTENNA AND PULSE VERSUS BER PERFORMANCE

By using Equations (6.14) and (6.18), the performance of the antenna system and the selections of the source/template pulses are related to the BER performance of the pulsed UWB system. The most important parameter of the antenna system is the transfer function $H(\omega)$ for characterizing the pulsed UWB system in the FD because it integrates all of the important antenna parameters including gain, impedance matching, polarization matching, path loss, and phase delay as mentioned in the previous section.

From Equations (6.14) and (6.15), it can be seen that for certain incident power at a transmitter, the SNR is proportional to the system transmission efficiency and the fidelity between the received pulse and the template pulse. In the other words, with the fixed data rate and the noise environment, the higher transmission efficiency and fidelity which are determined by the selected antenna systems and source/template pulses will be capable of providing the higher SNR.

6.3.2 Effects of Antennas and Pulses

The effects of antenna systems and source/template pulses on the BER performance of UWB systems are discussed by selecting the antenna systems with varying source and templates. In the study, the Rayleigh pulses given in Equation (6.4) are used as the source pulses and templates. Two antenna systems are employed, and each of them consists of two identical dipoles placed face to face with the distance $D = 1$ m for the maximum system gain. Figure 6.13 shows a planar 9×2-mm center-fed strip and a 14×14-mm center-fed square dipole antenna systems [10].

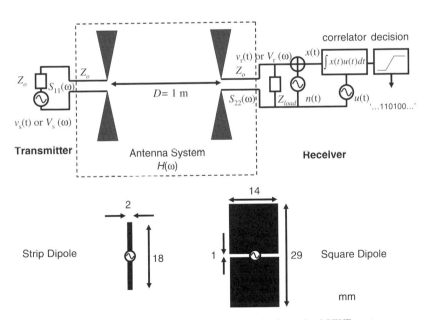

Figure 6.13 Three types of antenna systems in the pulsed UWB system.

152 ANTENNA ISSUES

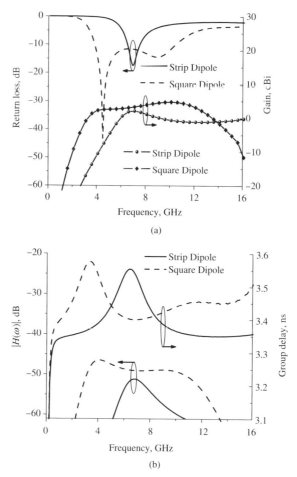

Figure 6.14 (a) The return loss and the gain and (b) the magnitude and group delay of $H(\omega)$ for the antenna systems.

Figure 6.14 shows the simulated results for the antenna systems, where the antenna systems have $Z_0 = Z_{\text{load}} = 100\,\Omega$, the transfer function $H(\omega) = |S_{21}(\omega)|$, the gain $G_t(\omega) = G_r(\omega)$, as well as the group delay of $H(\omega)$. From Figure 6.14(a), it can be found that the strip dipole antenna is a narrowband design with a well-matched frequency range of 6.5–7.5 GHz, whereas the broadband square dipole covers the whole UWB band well. The 10-dB transmission bandwidth are, respectively 5.6–10.8 GHz for the strip dipole antenna and 2.6–12.5 GHz for the square dipole. The group delay of two systems shown in Figure 6.14(b) is between 3.3 ns and 3.6 ns. The narrowband strip dipole system has the lower $|H(\omega)|$.

Figure 6.15 plots the system transmission efficiency η_{sys} for three antenna systems for varying source σ. The results are similar to that in Figure 6.11. As σ

6.3 ANTENNA AND PULSE VERSUS BER PERFORMANCE

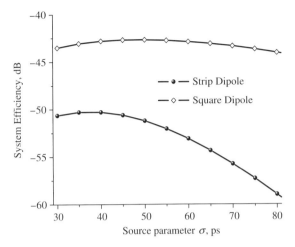

Figure 6.15 The system transmission efficiency η_{sys} for the antenna systems when source σ is between 35 ps and 80 ps.

increases from 35 ps to 80 ps, the η_{sys} of the strip dipole antenna system decreases, whereas the η_{sys} for the square dipole antenna system keeps almost consistent.

Then, the modulated Gaussian pulse $g(t)$ given in Equation (6.19a) with $\sigma = 45$ ps is selected as the source pulse, and the n-th order Rayleigh pulse $(n > 1)$ $v_n(t)$ Equation (6.19b) is chosen as the template to detect the received pulses.

$$g(t) = e^{-\left(\frac{t}{150 \times 10^{-12}}\right)^2} \cos(2\pi f_0 t) \tag{6.19a}$$

$$v_n(t) = \frac{d^n}{d^n t}\left[e^{-\left(\frac{t}{\sigma_n}\right)^2}\right] \tag{6.19b}$$

The parameters for the templates and the fidelity are tabulated in Table 6.2. The fidelity for the broadband square dipole antenna system is higher than that for the strip dipole antenna system if the templates are identical to the source pulses. However, using the optimal templates, the system can achieve high fidelity even

TABLE 6.2 The Parameters for the Template Pulses and the Corresponding Fidelity When Source $\sigma = 45$ ps

Antenna System	Modulated Gaussian Pulse		Source Pulse		The n-th Order Rayleigh Pulse		
	f_0, GHz	Fidelity	σ, ps	Fidelity	n	σ_n, ps	Fidelity
A	7.0	0.96	45	0.70	10	105	0.95
B	5.0	0.87	45	0.89	3	70	0.93

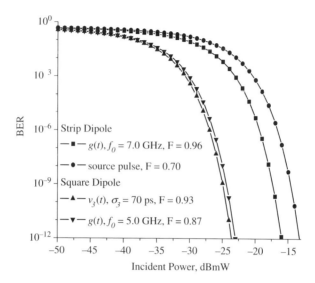

Figure 6.16 The BER with respect to the incident power and the template pulses for the source with $\sigma = 45$ ps.

for narrowband antenna systems. This suggests that the narrowband antenna system distorts the transmitted pulses more severely than the broadband antenna system. To simplify the receiver, a broadband antenna is desired.

Last, the effects of the antenna systems, the template pulses, and the source pulse with $\sigma = 45$ ps on the BER performance of the system are examined where the data rate is $R_b = 500$ Mbit/s and the noise PSD is $N_0 = kTF/2$, where k is the Boltzmann's constant 1.38×10^{-23} J/K, T is the room temperature 300K, and F is the noise figure 6 dB. Figure 6.16 shows the BER with respect to the incident power and the template pulses for the antenna systems. Two important observations have been found from the results. One is that the selection of the templates affects the BER dramatically. For the templates which have higher fidelity, the BER of the system is lower. The other one is that the antenna performance such as bandwidth or gain has a significant effect on the BER performance of the system. The narrowband antenna system has much higher BER than the broadband antenna systems due to its lower system transmission efficiency.

REFERENCES

1. Z. N. Chen, X. H. Wu, H. F. Li, N. Yang, and M. Y. W. Chia, "Considerations for source pulses and antennas in UWB radio systems," *IEEE Transaction on Antennas and Propagation*, vol. 52, no. 7, pp. 1739–1748, July 2004.
2. G. S. Smith, "On the interpretation for radiation from simple current distributions," *IEEE Antennas and Propagation Magazine*, vol. 40, no. 3, pp. 9–14, June 1998.

3. G. F. Ross, "A time domain criterion for the design of wideband radiating elements," *IEEE Transaction on Antennas and Propagation*, vol. 16, no. 3, p. 355, May 1968.
4. A. Shlivinski, E. Heyman, and R. Kastner, "Antenna characterization in the time domain," *IEEE Transaction on Antennas and Propagation*, vol. 45, no. 7, pp. 1140–1149, July 1997.
5. D. M. Pozar, "Waveform optimizations for ultra-wideband radio systems," *IEEE Transaction on Antennas and Propagation*, vol. 51, no. 9, pp. 2335–2345, September 2003.
6. D. Lamensdorf and L. Susman, "Baseband-pulse-antenna techniques," *IEEE Antennas and Propagation Magazine*, vol. 36, no. 1, pp. 20–30, February 1994.
7. T. Wang, Z. N. Chen, and K. S. Chen, "Effect of selecting antennas and templates on BER performance in pulsed UWB wireless communication systems," *Proceedings of IEEE International Workshop on Antenna Technology*, Singapore, pp. 446–449, March 2005.
8. T. Wang, Y. Wang, K. S. Chen, "Analyzing the interference power of narrowband jamming signal on UWB system," in *Proc. IEEE 14th Sympos. on Personal Indoor Mobile and Radio Commun.*, vol. 1, pp. 612–615, September 2003.
9. J. G. Proakis, *Digital Communications*, 4th edition, McGraw-Hill, New York.
10. M. J. Ammann and Z. N. Chen, "An asymmetrical feed arrangement for improved impedance bandwidth of planar monopole antennas," *Microwave Optical Technology Letters*, vol. 40, no. 2, pp. 156–158, 2004.

CHAPTER 7

Ultra Wideband Receiver Architectures

HÜSEYIN ARSLAN

7.1 INTRODUCTION

As wireless communication systems are making the transition from wireless telephony to interactive Internet data and multimedia types of applications, the desire for higher data rate transmission is increasing tremendously. As more and more devices go wireless, future technologies will face spectral crowding, and coexistence of wireless devices will be a major issue. Considering the limited bandwidth availability, accommodating the demand for higher capacity and data rates is a challenging task, requiring innovative technologies that can coexist with devices operating at various frequency bands. UWB, which has been considered primarily for use with radar applications in the past, offers attractive solutions for many wireless communication areas, including wireless personal area networks (WPANs), wireless telemetry and telemedicine, and wireless sensors networks. With its wide bandwidth, ultra wideband has the potential to offer much higher capacity than the current narrowband systems. Other benefits of UWB include low power transmission, potential for low-cost and simple transceiver design, immunity to multipath effects, high resolution (sub-decimeter range), robustness against eavesdropping, and easier material penetration.

High-capacity, high-data-rate, simple, power-efficient, low-cost, and small UWB transceiver design is a challenging task. There are several receivers proposed for UWB communication. Fully coherent receivers like optimal matched filtering, typically employed by rake reception, perform well but at the expense of extremely high computational and hardware complexity. In general, a coherent receiver requires several parameters (side information) concerned with the received signal, radio channel, and interference characteristics. Multipath delays, channel coefficients for each delayed multipath components, and distortion of the pulse shape need to be estimated for optimal coherent reception. Note that, in UWB, the number of multipath components is very large (can be a few hundred). Note also that, given

---

Ultra Wideband Wireless Communication. Edited by Arslan, Chen, and Di Benedetto
Copyright © 2006 John Wiley & Sons, Inc.

the total constant transmitted power, the power in each of these multipath components will be very low. Therefore, estimating the delays and coefficients from the received multipath components is an extremely challenging task. If care is not taken, fully coherent reception might lead to very poor overall system design. Therefore, receivers that relax these would also be preferable.

Noncoherent (or lightly coherent) receiver designs in UWB relax the amount of information that needs to be estimated accurately for the detection of the transmitted bits. In other words, the synchronization, channel estimation, and pulse shape estimation is not as stringent as in the case of the fully coherent receivers. Some of the noncoherent transceiver designs include transmitted reference (TR) based UWB, energy detector, and differential detector. Common to all these approaches is that the channel estimation and received pulse estimation are not necessary. Also, the timing estimation is easier and the receiver performance is more immune to the timing mismatch.

In this chapter, several transceiver designs for IR-based UWB signals will be studied. The transceiver requirements and the related trade-offs like performance, capacity, hardware, and computational complexity regarding practical designs will be discussed. The impact of UWB radio channel, self- and other user-interference, sampling rate, and modulation options on the receiver design will be explained. Various UWB receivers will be compared in terms of performance and other important criteria. First, an optimal matched filtering and its implementation using rake receivers will be provided. Then, various simpler and less coherent versions of the rake reception that trade off performance for complexity will be overviewed. Another popular receiver that employs reference pulses associated with the transmitted data pulses, referred to as TR-based UWB, will be examined and compared with the rake reception. It will be seen that the TR scheme has some similarities with the rake reception, and these common points will be exploited to unify these approaches. Techniques to enhance the performance of the TR scheme and various possible implementations will be discussed. Another technique which is similar to the TR scheme and avoids the transmission of the reference pulse by employing differential encoding in the transmitted symbols will be explained. The receivers that will be discussed in essence correlate the received signal with itself (auto-correlation) or with a local template (cross-correlation) at the receiver. Finally, receivers that only receive the energy of the received signal over transmission intervals and subsequently make decisions based on the received energy (energy detector) will be studied. The energy detector (ED) will be compared with the other receivers and possible improvements of the detection performance will be explained.

7.2 SYSTEM MODEL

Depending on the type of multiple accessing and other system parameters, various system models are used for UWB. In order to narrow the scope the focus is on TH-IR-based UWB. A simple TH multiaccess model of the UWB communication system is shown in Figure 7.1. The pulsed UWB approach often transmits many low-duty-cycle pulses to represent a bit, where the number of pulses in a symbol is a design criterion which determines the processing gain of the system. The *off-time* between two consecutive pulses implies a second type of processing gain which helps against multipath and multiuser interference, and other users may transmit in the gaps between these pulses. Note that, rather than a constant pulse-to-pulse interval, a user-specific TH code can be used to help the channelization of the system, while smoothing the power spectral density and allowing a secure transmission.

The information sequence corresponding to a single user is transmitted using the TH codes assigned for this user, and can be represented as

$$s(t) = \sum_{j=-\infty}^{\infty} A\beta_{\lfloor j/N_s \rfloor} \omega^{tx}(t - jT_f - c_j T_h - \delta\alpha_{\lfloor j/N_s \rfloor}), \quad (7.1)$$

where T_f is the nominal interval between two pulses, N_s is the number of pulses per symbol, and δ is the modulation index if the modulation is PPM. The pulse amplitude is represented by A and will be normalized to 1 for simplification. T_h represents

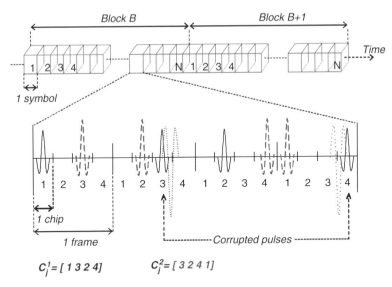

Figure 7.1 A simple TH-IR-UWB signal structure: each symbol carrying the information is transmitted with a number of pulses, where in this figure four pulses represent a symbol. Pulses occupy a location in the frame based on the specific pseudorandom (PN) code assigned for each user. Two different codes and the corresponding pulse locations are shown. Note that these two codes are orthogonal (do not interfere with each other). Another user's pulses that interferes with first user's code is also shown to demonstrate how interference from other users affects the system. A block in this figure represents a number of symbols where FEC coding, interleaving, and other MAC layer protocols might be applied.

the chip duration. Decimal codes $\{c_j\}$ represent a pseudo-random TH code for this transmission. UWB signals can be modulated in different ways; for example, by changing the amplitudes of the pulses with $\beta_{\lfloor j/N_s \rfloor}$, modulations like OOK, BPSK, positive PAM, and M-ary PAM can be obtained. Similarly, by varying the time positions of the pulses with $\delta\alpha_{\lfloor j/N_s \rfloor}$, modulations like PPM and M-ary PPM can be obtained.

The transmitted signal passes through a multipath radio propagation channel to reach the receiver. A single antenna UWB receiver is considered in this chapter. In addition to the desired signal, interfering signals and AWGN are present at the receiver input. Therefore, the received signal can be represented as

$$r(t) = \sum_{j=-\infty}^{\infty} \sum_{l=1}^{L} \gamma_l(t) \beta_{\lfloor j/N_s \rfloor} \omega_l^{rx}[t - jT_f - c_j T_h - \delta\alpha_{\lfloor j/N_s \rfloor} - \tau_l(t)] + n(t), \quad (7.2)$$

where $\gamma_l(t)$ is the time-varying tap coefficient for the lth tap arriving at time $\tau_l(t)$, L is the total number of paths, and $n(t)$ is the noise term which includes both AWGN and other interference sources. Note that the received pulse is shown to be different from the transmitted pulse and varies for each multipath component.

The composite noise $n(t)$ is often modeled as AWGN (folding the interfering sources to the AWGN), assuming that the number of interferers is large and the receiver is not hit by a dominant interfering source. This assumption simplifies the receiver design and analysis of many receivers. However, often the noise can be dominated by a strong interference and, as will be mentioned later, by designing receivers that take into account the statistics of the dominant interfering sources, the performance and capacity of the system can be enhanced. The specific characteristics of the interference sources and the appropriate models for various interferences will be described later.

7.3 UWB RECEIVER RELATED ISSUES

There are several issues that impact the design of a receiver in UWB. In this section, some of these issues will be discussed.

7.3.1 Sampling

The ability of sampling the received signal is one of the most critical issues in UWB receiver design. A full software-defined radio (SDR) based architecture requires sampling very close to the receiver antenna above the Nyquist rate. Nyquist sampling also allows implementation of many possible adaptive receiver implementation and baseband signal processing algorithms for performance improvement. However, sampling the received signal at (or above) the Nyquist rate requires extremely high sampling frequency, which is currently not feasible for practical implementation of UWB technology. In addition to the extremely high sampling frequency, the analog-to-digital-converter (ADC) must support a very large dynamic range to resolve the signal from the strong interferers like narrowband interference. A high sampling rate also requires very powerful signal processing capability. Field programmable gate arrays (FPGA) and very high speed digital-signal processors (DSP) are considered for handling the high processing requirements.

Pushing the sampling away from antenna and incorporating some analog front-end circuitry before the ADC relaxes sampling and ADC requirements. For example, analog multipliers (for correlating the signal with its delayed version) followed by analog integrate-and-dump circuitry are popularly used for TR-based and differential detectors. In an energy detector, an analog square device followed by analog integrate-and-dump circuitry is used. In more coherent rake-type receivers, reduced sampling rates after the correlator outputs are also possible. Depending on the implementation of the template and correlation, the output of the correlator can have samples chip-spaced, frame-spaced, and symbol-spaced. Note that reduced space sampling provides simpler receiver architectures, but the flexibility of processing the signal will be reduced; therefore, there is a performance vs complexity trade-off. In a TH-IR-UWB system, sub-Nyquist sampling can be done in several rates:

- *Chip Spaced Sampling*—this allows analysis of the samples of the correlator outputs at chip rate. Note that the chip rate is considerably lower than the Nyquist rate (can be more than an order of magnitude lower depending on the pulse shape). Chip-spaced sampling can be very useful for implementing practical interference cancellation schemes, and for combining the samples corresponding to symbols optimally.
- *Frame Spaced Sampling*—frame space sampling is lower than chip-spaced sampling by the number of chips per frame. The statistics of the received signal at each frame could be useful for interference cancellation, optimal pulse combining, etc. For example, in employing some MAI cancellation techniques, sampling at frame rate would be useful for chip discarding [1–4]. Also, minimum-mean-squared-error (MMSE) combining of the pulse within a symbol can exploit the statistics available in each frame. MMSE combining has the ability to reduce various interference effects as will be discussed later.
- *Symbol Spaced Sampling*—provides a low sampling rate with reasonably good performance receivers. Some interference cancellation techniques can still be employed at the symbol-spaced sample outputs. However, information like which pulse is more interfered with than others cannot be obtained from these samples. Also, the statistics across different pulses will be lost.

7.3.2 UWB Channel and Channel Parameters Estimation

In UWB, the transmission bandwidth is extremely large, which leads to multiple resolvable paths. For a given transmitted power, the power is distributed over extremely large bandwidths, and hence causes very low interference to narrowband users.

In time domain, the high resolvability due to wide bandwidth can affect the receiver performance. Since the total power is distributed over many multipath components, the power in each of these individual paths will be very low [5].

Note also that, when a wave reflects off an object or penetrates through a material in the process of multipath propagation, the effects are frequency-sensitive and therefore the transmitted waveform is filtered in some way. Hence, the resulting single multipath component may actually be represented by several or many terms in the model. This suggests that a channel model for UWB signals that is more closely related to physical propagation paths and the pulse shape associated with a propagation path depends on that path.

If the receiver knows the effect of the channel perfectly on all these paths, a capacity close to AWGN channel can be obtained [6, 7]. However, in reality, for conventional coherent receivers, the effect of the channel on the transmitted signal must be estimated to recover the transmitted information. As long as the receiver is able to estimate what the channel did to the transmitted signal, it can accurately recover the information sent. The channel estimation includes estimation of the multipath delays, multipath coefficients, and the estimation of the received pulse due to the effect of the channel. Since the channel is random and time-varying, the estimation process needs to be continuous to be able to adapt the changes in the channel.

In coherent receivers, the error in channel parameter estimation can degrade the performance significantly. As a result, if the transceivers are not designed properly, increasing the bandwidth indefinitely might reduce capacity [6]. Therefore, there are several main issues that need to be taken into account regarding the estimation of multipath delays, coefficients, and received pulse in UWB systems. The first is the large number of multipath components, which implies a huge amount of parameter estimation. The second issue is the very low power, and hence low SNR, in each of these multipath components, making the accurate estimation very difficult. The third issue is the time-variation of the channel, which requires continuous update of the estimate. Finally, fast and accurate channel parameter estimation requires the transmission of training bits along with the data bits. The number and frequency of these training bits depend on several things, including the SNR, number of resolvable paths, and desired performance. Increasing the training bits reduces the data rates and capacity while improving the receiver performance.

Synchronization Synchronization can be roughly described as the process of providing the same time reference for the receiver as is used for the transmitter. In other words, the synchronization operation is a search operation, where the receiver searches for the correct timing of the transmitted signal and locks onto it. In UWB, synchronization is more difficult compared with other narrowband systems. Fast and accurate acquisition with low overhead is desired. Without a correct timing synchronization, demodulation and data detection are not possible.

Synchronization in general can be grouped as course and fine synchronization. Course synchronization involves detection of the existence of the signal and aligning the receiver with the correct transmitted pulse and symbol sequence. The fine synchronization is locking the receiver to the correct pulse timing. In rake receivers, the receiver needs to lock into the appropriate path positions. Estimation of the proper finger positions is part of the synchronization process. Note that often timing estimation of the strongest path is not enough for dispersive NLOS channels as the energy from the other paths is not exploited. Therefore, estimation of multiple finger locations increases the synchronization complexity. In less coherent schemes (TR, differential detector, and energy detector), the receiver is not required to lock into the individual multipath components. Instead, the receiver locks into where the clusters of the multipath components are gathered (the start and end point for the integration region). More important than this, these receivers, as will be discussed later, are less sensitive to timing jitters. Note that, due to the very narrow pulses, a slight error in the estimation of delays will lead to significant performance degradation for rake and correlator receivers. On the other hand, in less coherent schemes, a slight shift in the integration region will not result in a significant performance degradation, making the system more robust to synchronization errors and allowing the use of less complex synchronization algorithms.

In addition to time synchronization, clock synchronization is also important. If the receiver and transmitter clocks have the same frequency, the position of the pulse and template do not change with respect to each other across the different pulses. Therefore, clock synchronization is an integral part of the receiver. The clock difference at transmitter and receiver needs to be estimated and compensated continuously.

Estimation of Multipath Coefficients Similar to estimation of the multipath delays, the estimation of multipath coefficients is very difficult from multipath components that have very low SNR. Also, estimation of several hundred multipath coefficients is not feasible for all-rake-type receivers. Instead, the coefficients over selected taps need to be estimated. The channel estimation can be avoided for some modulation options (like PPM, OOK) by implementing noncoherent reception.

Received Pulse Shape Estimation In receivers that employ a local template for the correlations (i.e., rake and correlator receivers), knowledge about the received pulse is important to be able to correlate the received signal with a local template. The local template (which is known from previous work) is the estimated pulse shape [8]. Note that the received pulse shape is not the same as the transmitted pulse shape [9–11]. The channel and antennas can distort the pulse. As mentioned before, the distortion due to the channel can be different for different multipath delays. This might create many problems in employing optimal coherent reception. Estimation of the pulse shape at each multipath delay in addition to the above estimates is extremely difficult. One common approach is to assume the distortion to be the same over all multipath delays and estimate a single received pulse shape. However, the pulse shape needs to be estimated. Other simplifying assumptions involve assuming the received pulse to be the same as the transmitted pulse, or just using the derivative of the transmitted pulse [12]. However, these assumptions degrade the performance of the coherent receivers if the received pulse does not match the template used in the receiver [8, 13]. The noncoherent receivers (TR, differential detector, or energy detector) do not require a local template, and avoid the need to estimate the pulse shape.

7.3.3 Interference in UWB

In addition to the radio channel, the interference in UWB systems affects the receiver performance and the receiver choice significantly. Interference can be due to various sources including MAI, NBI, ISI, and IFI. Note that, compared with the AWGN, these interferences are colored and the receivers can take advantage of the correlation for improving the receiver design. Coherent detection allows cancellation of several sources of interference. However, as mentioned before, many interference cancellation routines require additional *a priori* information about interference statistics, like operation frequency, power, time/frequency/space correlation, and code of the interfering signal. Some of the major interference sources are as follows.

IFI This is the interference that arises when the minimum pulse-to-pulse duration is shorter than the maximum excess delay of the channel. If frame-spaced samples (or better) are available, some form of interference cancellation can be employed. IFI cancellation increases the complexity but the performance and data rates can be improved significantly.

ISI This is a problem for high-data-rate systems. Equalizers can be employed to handle ISI. Symbol-spaced samples will be adequate to employ equalization. There are several possible equalization techniques. Simple and efficient equalization is a key issue. The techniques that are used for narrowband systems can be employed here as well.

NBI The very low transmission power and large bandwidth enable UWB systems to co-exist with other narrowband systems without interfering with them. However, the effect of the narrowband signals on the UWB signal can be significant and may jam the UWB receiver completely. Even though the narrowband signals interfere with only a small fraction of the UWB signal spectrum, due to the large relative power of the narrowband signals with respect to the UWB signal, the performance and capacity of UWB systems can be affected considerably. Therefore, the UWB receivers might need to employ NBI suppression techniques to improve the performance, capacity, and range of communication.

The current trend in NBI is to avoid the transmission (using multicarrier and multiband approaches) of the UWB signal over the part of the frequency where NBI is strong. However, this restricts the transmission of UWB waveforms. The NBI can also be suppressed at the receiver, which relaxes the transmission format. However, this increases the receiver complexity. Analog bandpass filtering before the reception of the signal [14], notch filtering and peak clipping [15], and MMSE combining [16–20] are some of the approaches that are considered for cancelling NBI. Further research is needed in NBI cancellation. NBI is discussed in more detail in Chapter 11.

MAI The coexistence of a large number of UWB transmitters in a dense environment is very important. The transmitted signals of each user share the same spectrum, and simultaneous transmissions by multiple users are popularly achieved by TH or DS spreading codes. Ideally, it is desired to have orthogonal codes for each user. However, in practice the received signal from different users is not orthogonal because of multipath, asynchronous transmission. Also, designing perfect codes with zero auto- and cross-correlation properties for all shifts is not possible. As a result, MAI in UWB communication systems is a major problem.

Multiuser interference cancellation receivers have been studied extensively for narrowband TDMA-based systems [21–23] as well as CDMA-type wider band wireless communication systems [24]. The effectiveness of interference cancellation receivers relies on the ability to separate the desired signal from the interferer(s). In UWB, multiuser interference cancellation has not been studied extensively. Approaches that use MMSE combining of rake fingers [25] and the techniques that are used for CDMA are recently being considered for UWB [26, 27].

7.3.4 Other Receiver-Related Issues

Even though some important issues that impact the receiver design are given above, there are many other factors that affect the receiver design and choice. For example, the modulation that is used at the transmitter impacts the receiver design. If the transceiver complexity and cost are the primary concerns, a scheme that enables noncoherent demodulation (OOK, positive PAM, PPM, and M-ary PPM) is preferable. On the other hand, some other modulations like BPSK, M-ary PAM, and QAM require coherent demodulation and have the potential to provide better performance.

Similarly, the specific application in using UWB also impacts the receiver design. Some applications require high data rates and capacity, motivating the use of high performance receivers. Some other applications require low data rates, low cost, and low powers where simple noncoherent receivers can be employed. In some applications, the system might be noise-limited compared with others where the system performance is affected by strong interference sources. Ideally, a receiver that works for all kinds of conditions, for a variety of data rates, for a variety of coverage scenarios, with very low cost, low power and small size, is desired. Accommodating all this in a single structure might not be possible. Therefore, adaptive transceiver design, which has already gained a significant amount of interest for the new generation of wireless standards, also needs to be considered for the design of efficient UWB systems and transceivers.

7.4 TH-IR-UWB RECEIVER OPTIONS

A generic UWB receiver structure is shown in Figure 7.2. Depending on the implementation, the received signal can be sampled in various places. Two possible sampling points, before the multiplier and after the integrator, are shown in the

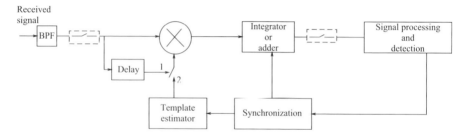

Figure 7.2 A generic UWB receiver structure.

figure. As mentioned before, sampling after the integrator allows a low sampling rate with analog front end circuitry. Sampling before the multiplier allows all digital, flexible, and software-defined radio-based receiver designs.

The optimal matched filtering and simplified versions implemented as rake and simple single correlator receivers correlate the received signal with a local template. The local template can be a pulse template (pulse-matched filter), a frame template (frame matched filter), or a template that includes multiple pulses (symbol matched filter) to include the relative delays between pulses based on the TH code. In either case, the template needs to be estimated locally based on the received signal by transmitting some training sequences or blindly. This type of receiver, which correlates the received signal with a template, is often referred as "locally generated reference" systems [28] or simply as "correlation" receivers [29]. Alternatively, the received signal can be correlated by itself, which leads to "autocorrelation" receiver structures [30]. There are several possible autocorrelation receiver structures. If the delay element in the figure is zero, this will end up being an energy detector, where the multiplier can be used as a square device. In differential detectors, the delay is the time difference between consecutive symbols. In transmit reference schemes, the delay is the difference between the reference and data-bearing symbols.

Note that in rake reception often each possible finger can be implemented in parallel branches and these fingers are combined after the integrator. In this case, Figure 7.2 represents one of these parallel branches. Alternatively, the local template might include all the finger elements. Note also that Figure 7.2 shows a generic receiver. Several other alternative implementations, which cannot be represented in this structure, are also possible.

After the bandpass filter (BPF) the signal and noise components of the received signal given in Equation (7.2) will be affected. In the signal component, the filter effect can be folded into the channel. Therefore, in the received signal model given in Equation (7.2), we assume that the channel includes the received and transmitted filter effects. In the noise component at the output of the filter, the noise statistics might change, as the noise spectrum after the filter will include the filter response. Therefore, even if the noise can be assumed to be white before the filter, the noise after the filter will be colored due to the filter response. However, for similar reasons to those mentioned before (to simplify the receiver design and analysis), the noise after the filter is also commonly assumed to be white and

Gaussian distributed. In this chapter, we will use the same signal model as given in Equation (7.2) after the receiver filter, as the noise in that model is general enough that the filter effect can also be folded into the noise as needed.

7.4.1 Optimal Matched Filter

In matched filtering, the basic principle is to estimate the noiseless replica of the received waveform and to use this estimate as a filter to maximize the SNR at the output of the matched filter. In AWGN channels (no other user's interference and self-interference), the matched filtering provides an optimal receiver in the sense that it minimizes the probability of error.

For matched filtering, the local template used in Figure 7.2, based on the received signal model given in Equation (7.2), can be written as:

$$\sum_{l=1}^{L} \hat{\gamma}_l(t) \hat{\omega}_l^{rx}(t - jT_f - c_j T_h - \hat{\tau}_l). \tag{7.3}$$

Note that the local template estimate requires the estimate of the channel coefficients $\hat{\gamma}_l$ and corresponding delays $\hat{\tau}_l$ as well as the received pulse shape estimate $\hat{\omega}_l^{rx}$ for all L resolvable multipath components. Imperfect estimation of these channel coefficients will affect the performance greatly. The time-hopping code sequence is assumed to be known at the receiver.

The matched filter can be implemented in a rake receiver (multiple parallel correlators and optimal combining) structure, as shown in Figure 7.3, where the local templates are the estimated received pulse shapes in each parallel branch. The template used in each branch just matches to the received pulse shape with the estimated delays and TH codes. The estimated tap coefficients are then used for combining different branches optimally by maximal ratio combining (MRC).

As mentioned before, various implementations are possible depending on the desired sampling rate at the output of the correlator. For example, the local template can be a train of (N_s) pulses with the delays between them according to the used TH code. Each correlator output will be the symbol-spaced samples. Note that the integration time in the correlators is one symbol period. The starting point of the integrators in each branch will be determined by the tap delay estimate. If the local correlator is a single pulse (instead of a pulse train), frame-spaced samples can be obtained. In this case, the integrator needs to be triggered depending on the tap-delays and relative delays due to the time hopping codes used for each pulse. The advantage of the frame-spaced sampling is that the samples obtained for each frame can be combined optimally to improve the detection of the symbols.

Approximate Solutions Derived from Matched Filtering Even though the matched filtering receiver described in previous section performs well, it comes at the expense of computational and hardware complexity, especially for UWB

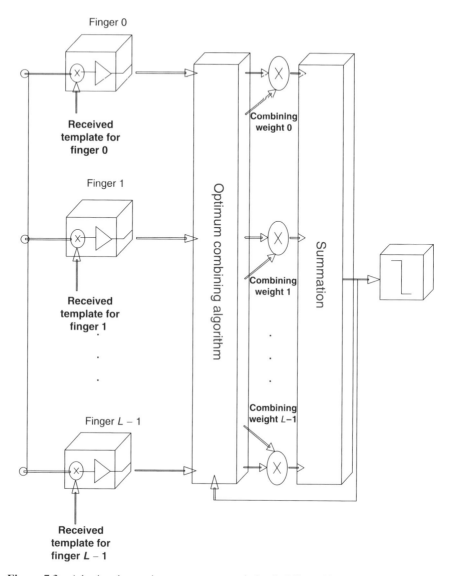

Figure 7.3 A basic rake receiver structure: correlation is followed by weighted combining. In the correlation, each parallel branch is tuned to a different multipath location. The local templates in each branch require the knowledge of the received pulse shape. The outputs of the correlator are sampled in frame or symbol-spaced depending on whether the template is a single pulse or a train of pulses.

signal reception. Depending on the available side-information, there are various suboptimal ways of implementing rake reception.

The received pulse shape, which is one of the most challenging things to estimate, can be assumed to be the same for all the fingers so that only a single pulse can be estimated and used in each branch. Even though this reduces the complexity, a common received pulse shape estimation for all the fingers is still not an easy task. Alternatively, a fixed pulse shape can be used in the receiver, which avoids the need for pulse shape estimation. For the fixed received pulse template, the transmitted pulse can be used at the receiver. Several earlier publications also considered the use of a fixed received pulse that takes the antenna effects into account (like the use of the first derivative of the transmitted pulse). However, various antennas might have a different effect on the transmitted pulse shape.

Another simplification over the optimal matched filtering can be made by avoiding the need for channel tap coefficients estimation. However, this requires the use of noncoherent modulation options like PPM and OOK. Also, optimal combining is not possible if individual tap coefficients and tap powers are not available.

In order to fully exploit all the available multipath components, the rake receiver implementation given in the previous section includes a finger associated for all possible paths (all-rake). As discussed before, in UWB, the number of possible resolvable paths can run into hundreds. Much lower complexity rake reception can be obtained by limiting the number of fingers and only considering the energies in the stronger paths [6, 31]. Such receivers are referred as selective-rake (S-rake). In the limiting case, if only a single finger that is locked to the strongest path is considered, a single-correlator receiver is obtained. However, a single correlator receiver can be operating a 10–15 dB signal energy disadvantage relative to the all-rake. This performance gap will close as the number of fingers that include the significant paths increases [6, 31], giving a performance vs complexity trade-off.

Extension of Optimal Receiver to Colored Noise and Interference Scenarios So far, the noise term at the receiver is assumed to be AWGN; self-interference like ISI and IFI, and other user interferences like multiple access interference and narrowband interference are ignored. When the noise is not white, as in the case of narrowband interference or multiple access interference, the matched filtering will not provide the optimal solution. The correlation of the interference across different fingers, or across different UWB pulses within a symbol, can be exploited for better receivers. An interference rejection combination that is robust against MAI and NBI provides good performance at the expense of additional computational complexity. Also, this requires additional information about the interference statistics, which need to be estimated from the received signal. For example, the autocorrelation of the interference across the correlator outputs needs to be estimated. Interference combining reduces to MRC when the noise across the fingers (correlator outputs) is uncorrelated, as in the case of AWGN.

Since DS-CDMA and TH-IR both employ user-specific multiple access sequences to share the spectrum, almost all of the interference cancellation methods used in DS-CDMA (such as those in [32]) can be considered for TH-IR. However, it is worth mentioning that interference cancellation is *practically* a very challenging task for UWB. Therefore, the unique signaling scheme of TH-IR systems needs to be exploited for efficient implementation of low-complexity interference cancellation methods. For example, in MAI cancellation, the sparse signaling scheme of UWB can be exploited. If there is interference from a certain user, it only affects certain pulses, but not all the pulses of the desired user. This fact implies that it may be sufficient to consider only certain users and certain pulses when designing the interference canceler, which considerably decreases the complexity of the receiver.

In multiuser detection, if the receiver knows the TH sequences and amplitudes of all the users, and if the system is synchronous (or the receiver can estimate the delays of all the users and incorporate this into the algorithm), it is possible to identify which pulses are corrupted by which users. As only certain users significantly interfere with the desired user, the complexity of the multiuser detection algorithm which considers only these users will be very low. The ML detector, decorrelator, and MMSE receivers can therefore be much more practical compared with their direct implementations that consider all the users [33]. Similarly, subtractive (successive and parallel) interference cancellation approaches [32] benefit from the signaling scheme of TH-IR systems. One other approach for multiuser interference suppression in TH-IR is MMSE *combining* of the N_s pulses that represent a bit, and/or MMSE combining of rake fingers [34]. However, this requires an interference correlation matrix across the correlator outputs and/or across the rake fingers. If the *a priori* information of code correlations, received amplitudes, etc., are already available at the receiver, sampling the correlator output at the symbol rate may be enough for these discussed algorithms. Alternatively, frame-spaced matched filter outputs can be processed to obtain signal correlations. However, as mentioned earlier, estimation of correlation information is extremely difficult in UWB systems compared with cellular CDMA.

In terms of the receiver complexity, probably the most practical multiuser interference cancellation approach for TH-IR systems is pulse discarding receivers. The basic idea behind these type of receivers is simply to discard the pulses which are corrupted. These receivers work well if the number of pulses per symbol, N_s, is large, and the number of corrupted pulses is relatively low. A blinking receiver [33] assumes knowledge of the TH sequences of all the users to determine if a pulse is corrupted or not, and then simply discards the pulse if it is corrupted. Chip discriminators [1] set an optimal threshold (which is chosen so as to minimize the BER) on the frame-spaced matched filter outputs, and if they are larger than the threshold, they discard these corrupted pulses prior to symbol decision. However, setting an appropriate threshold requires accurate estimation of the channel, which is also affected by multiuser interference and near/far effects, and often requires transmission of training symbols.

Similar to multiuser interference, NBI can also be canceled by employing several approaches at the receiver. The NBI has been discussed in detail in another chapter in this book. Therefore, readers are referred to that chapter.

The ISI and IFI problems have been avoided in much previous research by assuming the minimum pulse-to-pulse duration to be larger than the maximum excess delay of the channel. However, as mentioned before, this will not be the case in high-data-rate applications and for heavily dispersive channels, hence, these interferences will be unavoidable in certain cases. In [35], the received signal passed through a prefilter (time-reversal filter) before processing to reduce the effective channel delay spread. By employing such a filter, the majority of the energy can be collected in a smaller number of multipath components. However, the prefilter requires the estimate of the channel impulse response, which is also not an easy task. The MMSE-type receivers can also be used to cancel the self interference problem. Additionally, equalization can be used to cancel self-interference. The design of simple and high-performance equalizers for UWB needs to be studied further. Also, adaptive transceivers that change the pulse-to-pulse duration depending on the delay spread information can be employed. This reduces the IFI, while maximizing the data rate. Of course this is a relatively easy task in a single user scenario. Employing such an adaptive strategy jointly for a multiuser scenario is a challenging task that needs to be studied further.

7.4.2 TR-Based Scheme

The interest in the TR scheme in UWB transceiver design has grown recently. There is significant amount of effort going into improving the performance of TR schemes while keeping the complexity at the reasonable level. The basic principle in TR-based schemes is to transmit a reference (unmodulated) pulse along with the data (modulated) pulses. The reference pulses and the data pulses are transmitted with a delay between. When the delay is less than the coherence time of the channel, the reference and data pulses can be assumed to be affected similarly due to the channel. Therefore, instead of using a local template as described in the previous section, the TR scheme uses the reference pulses as the template for correlating the data pulses, and for the demodulation of the transmitted information (Figure 7.4). In this way, an explicit channel parameters estimation is not needed at the receiver. As a result, TR communication possesses some advantages when transmitting through an unknown channel, which severely distorts the transmitted waveforms.

The TR-based scheme has the ability to capture the energy from all multipath components of the received signal with a simple receiver structure. However, this gain comes at the expense of a higher noise power, due to the structure of the

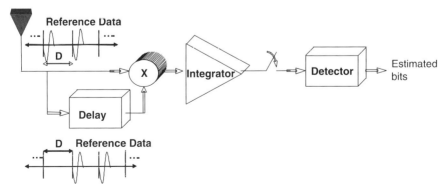

Figure 7.4 A simple TR receiver structure.

172 ULTRA WIDEBAND RECEIVER ARCHITECTURES

TR receiver, but the trade-off is an advantageous one in an environment with sufficiently high and dense multipaths. In sparse channels and in channels that have few multipaths, the conventional TR-based schemes are not desirable compared with the high-performance rake and correlator receivers.

Even though the TR scheme is often considered as a noncoherent scheme that avoids the estimation of pulse shape and other channel parameters, one can also interpret it as a coherent scheme where the reference pulses are used to estimate the channel parameters. Essentially, the channel is estimated using these reference (or pilot) pulses. The estimate over multiple reference symbols can be combined to improve the quality of the template. There are various way of implementing the template pulses [28, 36–38]. Templates and data pulses can be transmitted back-to-back, or the template pulses can be transmitted as a group followed by a group of data pulses, as shown in Figure 7.5(a).

One of the most striking features of the TR scheme is that the timing requirement is less stringent compared with a local correlator receiver. Note that the local template type of receivers try to match the timing of the received pulse exactly with the template. However, in the TR scheme, there is no need for a fine timing estimate of all the multipath components. This has both advantages and disadvantages. The advantage is less simple timing and more immunity to timing errors. The disadvantage is that, without the fine timing, both the noise and signal over a window

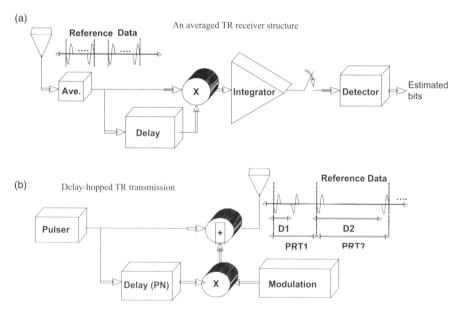

Figure 7.5 Some alternative TR schemes: (a) the reference and data pulses are send as groups; the reference pulses are averaged to obtain a clean template; (b) the transmitter of a delay-hopped TR schemes. The delays between reference and data pulses are not fixed; instead pseudorandom delays are used, which allow a multiplexing capability along with the random another pseudorandom duration between each group of pulses within a bit.

(whether there is a multipath component or not) are absorbed. This is one of the reasons why TR schemes do not provide good performance, as more noise than an ideal matched filtering is received. In essence, the TR scheme assumes that everything is useful over the integration window. In reality, some samples contain energy, and some contain noise. If one integrates all of them, than the integrator is not collecting the energies optimally. One way to solve this problem is to control the integrator. If the locations of multipath components are known, then the energies only from these received samples where the multipath components are located need to be collected. Then, this will end up being a rake reception. As a result, one can make a TR scheme work as well as a rake receiver if additional parameters about the channel are known. A simple step towards this is adaptation of the integration interval depending on the maximum excess delay of the channel [36]. Previous work has shown that adaptation of the integration interval improves the BER performance of the receiver. A further step for performance improvement is weighting the output of the multiplier with some side-information regarding the expected channel delay profile.

Note that, even though the TR scheme does not require the exact timing of the received pulses, it requires an accurate delay line. A mismatch between the delays that are used at the transmitter and receiver will affect the performance significantly. Matching the receiver delay to the transmitter delay can be challenging for practical implementations. Figure 7.6 shows the effect of the delay lines mismatch at the transmitter and receiver. As can be seen, as the mismatch increases, the signal component of the normalized integrator output decreases, which reduces the SNR and increases the BER. Notice that the mismatch effect is related to the received pulse shape correlation.

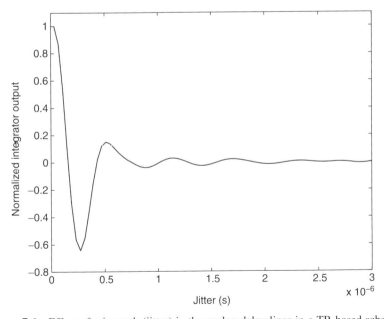

Figure 7.6 Effect of mismatch (jitter) in the analog delay lines in a TR-based scheme.

Another important issue in TR scheme is the use of the noisy templates for the correlation [29, 37, 39]. This problem can be avoided by averaging the reference pulses to obtain a clean template. However, averaging using analog circuitry is not a straightforward task. In particular, for TH-IR, the averaging circuit might require multiple delay lines, or programmable delay lines. In addition, during the averaging over multiple reference pulses, the channel variation should be taken into account unless the channel is time-invariant over several reference pulses. Reducing the noise in the reference signal and using a clean template is one area that has been studied significantly. In addition to the reference pulses, the data pulses are also exploited in clean template estimation by incorporating the reliability of the detected symbols associated with the data pulses. In addition, optimizing the reference pulses in terms of their power with respect to the data pulses, optimal number and placement of reference pulses among the data pulses, etc., is also studied to improve the spectral and power efficiency [37].

The samples at the integrator output in a TR scheme can be represented as

$$y = \int_{\tau}^{\tau+T_o} [\omega(t) + n_1(t)][\beta\omega(t) + n_2(t)] \, dt$$

$$= \underbrace{\beta \int \omega^2(t) \, dt}_{\text{signal term}} + \underbrace{\beta \int \omega(t)n_1(t) \, dt + \int \omega(t)n_2(t) \, dt + \int n_1(t)n_2(t) \, dt}_{\text{noise term}} \quad (7.4)$$

where τ is the integration start point, T_o is the integration interval, and $\omega(t)$ is the effective channel that contains all the multipath components. Notice that the noise part of the received samples consist of three terms due to the noise in the template and the noise in the signal. If the template is not averaged and assuming AWGN, the noise components in template [$n_1(t)$] and signal [$n_2(t)$] will have the same variance. When the template is averaged over several pulses, the noise variance in the template will be lower depending on the averaging size. The power of the signal component that is captured depends on the integration window. Ideally (when the template is perfect), the integration interval should be large enough to capture all the energy in all the multipath components. However, in practice, due to the noise, there is an optimal integration interval that optimizes the performance of the receiver. Assuming white Gaussian noise for the noise components and representing the signal term as E_s, the total noise variance and SNR at the output of the integrator can be given as

$$\sigma_{\text{total}}^2 = E_s \left(\frac{N_o}{2} + \frac{N_o}{2N_{\text{av}}} \right) + \frac{N_o^2}{2N_{\text{av}}} T_o B_w$$

$$\text{SNR} = \frac{E_s^2}{\sigma_{\text{total}}^2} \quad (7.5)$$

where $N_o/2$ is the double-sided power spectral density of the AWGN in the received signal, N_{av} is the averaging size (if averaging is employed) for the template, and B_w is the single-sided bandwidth of the receiver. Note that the channel is assumed to be constant over the averaging window size N_{av}. The averaging window size should be chosen carefully to make sure that the channel is constant over the averaging window. Notice that, when N_{av} goes to infinity and for a constant channel assumption, the performance of the TR scheme approaches perfect matched filtering performance. Notice also that, as N_{av} increases for a given $N_o/2$ and channel power delay profile, the integration interval can be increased to improve the SNR.

In addition to single user performance, multiple accessing capability of TR schemes and techniques to improve multiuser performance is also important. If not designed properly, TR schemes are expected to be more susceptible to MAI and to other interference sources. First of all, the TR schemes experience more interference due to the integration of the multiplier output over a window that both desired pulses and possible interfering sources exist. This requires careful design of the multiple access codes so that the number of hits by other sources will be reduced. Also, many of the interference cancellation techniques that are suitable for rake-type receivers cannot be implemented in TR schemes. This requires the development of new interference cancellation approaches that are suitable for TR schemes, as well as new multiuser code designs. Note that the delays between the reference and data pulse can be pseudo-random to introduce a multiple accessing capability. Figure 7.5(b) shows a delay-hopped TR scheme [40].

There have been many recent efforts in further improving the performance of the TR schemes by employing additional *a priori* information regarding the channel and noise characteristics [36, 41–43]. In essence, the effort is going in the direction of pushing the TR schemes to be more coherent to close the gap with respect to the matched filtering and rake reception. For example, the TR receiver can be further improved by weighting the integrator input (or the received template) with a function related to the channel's power delay profile, which can be interpreted as partial channel-state information. However, estimation of this partial channel state information might not be easy. Also, if one can compact the energy of the received signal (i.e., the delay spread of the effective channel) over a shorter time interval, it would greatly enhance the performance of the TR-reference schemes, as the same amount of transmitted energy can be received over a shorter interval with shorter integration interval requirement at the receiver. Such a strategy will also reduce IFI and hence increase the date rate of the communication.

7.4.3 Differential Detector

The differential detector is similar to the TR scheme. Instead of sending reference pulses for correlating the received signal, the data pulses corresponding to the previous symbol is used as the template [30, 39, 44]. This requires differentially encoding the transmitted bits before modulation, so that the information is in the difference of the two consecutive symbols. Avoiding the use of reference pulses in differential detectors increases spectral efficiency and hence doubles the data rates compared with the TR scheme.

Similar concerns as mentioned in TR scheme are also valid for differential detector receivers. The previous symbols that are used as a reference are noisy, degrading the performance of the receiver. The delay used for differential detector is a symbol length delay which is less robust to timing errors. The accuracy of the delays used at the transmitter and receiver affect the performance, as in the case of TR scheme. A slight difference in the delays will degrade the performance significantly. Note also that the channel needs to be time-invariant over the two-symbol period. For high mobility and larger symbol duration, this might be an issue that needs to be taken into account.

Even though it has been mentioned that the differential detector provides a 3 dB advantage compared with the TR scheme, this might not be true in a practical system design. First of all, the differential detector does not provide the flexibility of TR scheme in terms of adjusting the power, position, and number of the reference signal. Averaging the reference information in a differential detector is not possible. Also, as mentioned in the previous section, the TR scheme can also use the data pulses for improving the template waveform, which indirectly allows improvement of the spectral efficiency of TR schemes. As a result, even though in its basic form, the differential detector might have the advantage over the TR scheme, the TR scheme seems to have more potential for future enhancement and provides a better path towards more coherent receivers as the technology evolves.

7.4.4 Energy Detector

An energy detector (see Figure 7.7) is a simple suboptimal noncoherent UWB-IR receiver scheme, which can be implemented with modulations like OOK [45] or PPM [46]. When OOK is employed, a threshold must be set and used, where the bit decision is 0 if the received energy is less than the threshold, and 1 if it is larger than the threshold. The optimum threshold is the intersection of probability density functions corresponding to energies for 0 and 1. On the other hand, if the PPM is used, the threshold problem disappears, but that may sacrifice the data rate, as the pulses that represent different bits should be adequately separated (more than maximum excess delay of the channel).

Similar to the TR scheme, the UWB energy detector receiver approach generally requires only coarse synchronization, which makes the system robust against clock jitter and triggering inaccuracy [47, 48]. It is also not sensitive to distortion and phase nonlinearity of devices like antennas, amplifiers or filters [47].

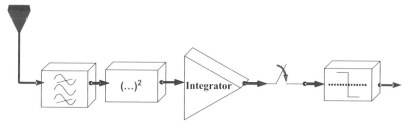

Figure 7.7 Block diagram for a simple energy detector receiver.

In spite of its simplicity, conventional energy detectors as implemented in fiber-optic channels perform poorly [49]. This is one of the reasons why energy detectors have not been heavily considered for UWB applications. However, the performance can be improved with a careful receiver design. One of the improvements would be in averaging the received signal before the square device to reduce the noise effect (i.e., improve the signal-to-noise-ratio). The averaging algorithm can be implemented using a single delay line with variable pulse-to-pulse duration or using multiple delay lines using fixed pulse-to-pulse durations. Similarly, averaging can be done using digital samples if the sampling is employed before the square device. However, this will increase the sampling rate and ADC requirements. Averaging improves the performance significantly, as it reduces the noise effect.

Another improvement is in finding an accurate and simple threshold estimation technique for OOK modulation. The choice of the threshold plays an extremely important role in the performance of the receiver. The conventional receivers assume a fixed threshold for decision-making. However, the radio channel and noise statistics vary significantly, suggesting the use of an adaptive threshold that depends on the variation of the signal power and noise power.

Similar to TR and differential detector, the integration interval can also be adapted depending on the maximum excess delay of the channel [50]. Choosing an integration interval that sacrifices the insignificant multipath components in order to decrease the collected noise energy will improve the performance [36, 46]. Therefore, an optimal integration interval that minimizes the BER exists, and depends on the channel statistics and the noise variance. Figure 7.8 shows the

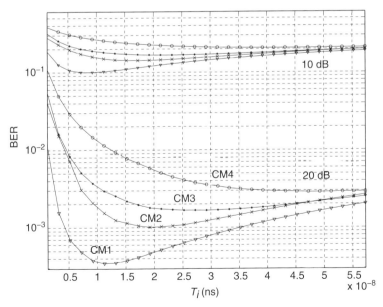

Figure 7.8 The effect of integration interval on BER performance of the energy detector receivers.

effect of the integration interval on the BER performance. As can be seen, depending on the SNR and the multipath delay profile, the optimal integration interval changes. An adaptive receiver can measure the channel, signal, and noise statistics and change the integration interval to optimize the performance.

The output of integrate and dump circuitry for each symbol in the energy detector can be written as

$$y_j = \int_{\tau}^{\tau+T_o} [\beta_j s(t) + n(t)]^2 \, dt,$$

$$y_j = \underbrace{\beta_j \int_{\tau}^{\tau+T_o} s^2(t) \, dt}_{\text{Signal term}} + \underbrace{2\beta_j \int_{\tau}^{\tau+T_o} s(t) n(t) \, dt + \int_{\tau}^{\tau+T_o} n^2(t) \, dt}_{\text{Noise term}} \quad (7.6)$$

where $s(t)$ is the noiseless received signal after passing through multipath channel, β_j is the OOK-modulated symbols (assuming OOK is used at the transmitter), T_o is the integration interval, and τ is the starting point of the integration. Note that the first part in the noise term is a zero mean Gaussian distributed random variable, and the second part of the noise has chi-square distribution with a variance of $k(N_o^2/2)$, where k is the degree of freedom, which can be given as $2T_o B_w + 1$. If the degree of freedom is large, the chi-square distribution can be approximated by Gaussian distribution [49].

7.5 CONCLUSION

In this chapter, several receiver options for TH-IR based UWB systems have been discussed. It is argued that each receiver option has several trade-offs in terms of performance, cost, hardware, and computational complexity, throughput, multiple access capability, etc. Depending on the application and the transceiver requirements, different receivers might be preferable.

Currently, the implementation of fully digital and fully coherent reception is not feasible for UWB radio. On the other hand, the performance gap between the low complex receivers and high-performance receivers is very wide. The research opportunities for the transceiver design in UWB radio are wide open. The research is relatively immature compared with other narrowband radio counterparts. There has been some recent efforts to reduce the complexity of the coherent matched filtering-based receivers. These efforts include reducing the number of fingers in a rake reception while trying to keep the performance close as to the all-rake receiver, avoiding the intensive estimation of the time varying channel parameters, developing simple and computationally efficient channel parameter estimation techniques. In parallel, there are also recent efforts to improve the performance of noncoherent receivers to close the performance gap with respect to the fully coherent receivers. Mainly, the approaches for improving the performance of the noncoherent receivers are based on estimating some additional parameters regarding the channel parameters, and making these receivers more coherent.

REFERENCES

1. W. M. Lovelace and J. K. Townsend, "Chip discrimination for large near-far power ratios in UWB networks," in *Proc. IEEE Military Communications Conf.*, vol. 2, Boston, MA, October 2003, pp. 868–873.
2. W. M. Lovelace and J. K. Townsend, "Adaptive rate control with chip discrimination in UWB networks," in *Proc. IEEE Ultrawideband Systems Technology (UWBST)*, Reston, VA, November 2003, pp. 195–199.
3. Z. Tian, H. Ge, and L. L. Scharf, "Low complexity multiuser detection and reduced-rank Wiener filters for ultra-wideband multiple access," in *Proc. IEEE Int. Conf. on Acoustics, Speech, and Signal Processing (ICASSP)*, vol. 3, Baltimore, MA, March 2005, pp. 621–624.
4. I. Guvenc and H. Arslan, "UWB channel estimation with various sampling rate options," in *Proc. IEEE Sarnoff Symp.*, Princeton, NJ, April 2005, pp. 229–232.
5. D. Cassioli, M. Z. Win, and A. F. Molisch, "Effects of spreading bandwidth on the performance of UWB RAKE receivers," in *Proc. IEEE Int. Conf. Communications (ICC)*, vol. 5, May 2003, pp. 3545–3549.
6. M. Z. Win, G. Chrisikos, and N. R. Sollenberger, "Performance of RAKE reception in dense multipath channels: implications of spreading bandwidth and selection diversity order," *IEEE Journal of Selected Areas in Communications*, vol. 18, no. 8, pp. 1516–1525, August 2000.
7. M. S. W. Chen and R. W. Brodersen, "The impact of a wideband channel on UWB system design," in *Proc. IEEE Military Communication Conf. (MILCOM)*, vol. 1, Monterey, CA, October 2004, pp. 163–168.
8. R. D. Wilson and R. A. Scholtz, "Template estimation in ultra-wideband radio," in *Proc. IEEE Asilomar Conf. Signals, Systems, Computers*, vol. 2, Pacific Grove, CA, November 2003, pp. 1244–1248.
9. R. C. Qiu, "A generalized time domain multipath channel and its application in ultra-wide-band (UWB) wireless optimal receiver design: system performance analysis," in *Proc. Wireless Communications Networking Conf. (WCNC)*, vol. 2, Atlanta, GA, April 2004, pp. 901–907.
10. T. Wang, Z. N. Chen, and K. Chen, "Effect of selecting antenna and template on ber performance in pulsed UWB wireless communication systems," in *Proc. IEEE Int. Workshop on Antenna Technology (IWAT)*, Marina Mandarin, Singapore, March 2005, pp. 446–449.
11. Z. N. Chen, X. H. Wu, H. F. Li, N. Yang, and M. Y. W. Chia, "Considerations for source pulses and antennas in UWB radio systems," *IEEE Transactions on Antennas and Propagation*, vol. 52, no. 7, pp. 1739–1748, July 2004.
12. M. Hamalainen, V. Hovinen, R. Tesi, J. H. J. Iinatti, and M. Latva-aho, "On the UWB system coexistence with GSM900, UMTS/WCDMA, and GPS," *IEEE Journal on Selected Areas in Communications*, vol. 20, no. 9, pp. 1712–1721, December 2002.

13. K. Taniguchi and R. Kohno, "Design and analysis of template waveform for receiving UWB signals," in *Proc. IEEE Conf. on Ultrawideband Systems Technology (UWBST)*, Kyoto, May 2004, pp. 125–129.
14. T. Ikegami and K. Ohno, "Interference mitigation study for UWB impulse radio," in *Proc. IEEE Personal, Indoor, Mobile Radio Communications (PIMRC)*, vol. 1, September 2003, pp. 583–587.
15. R. D. Wilson, R. D. Weaver, M. H. Chung, and R. A. Scholtz, "Ultra wideband interference effects on an amateur radio receiver," in *Proc. IEEE Conf. UWB Systems Technology (UWBST)*, vol. 3, Baltimore, MD, May 2002, pp. 315–319.
16. I. Bergel, E. Fishler, and H. Messer, "Narrowband interference suppression in time-hopping impulse radio systems," in *Proc. IEEE Conf. Ultrawideband Systems Technology (UWBST)*, Baltimore, MD, May 2002, pp. 303–307.
17. H. Sheng, A. M. Haimovich, A. F. Molisch, and J. Zhang, "Optimum combining for time hopping impulse radio UWB rake receivers," in *Proc. IEEE Ultrawideband Systems Technology (UWBST)*, Reston, VA, November 2003, pp. 224–228.
18. G. Durisi, J. Romme, and S. Benedetto, "Performance of TH and DS UWB multiaccess systems in presence of multipath channel and narrowband interference," in *Proc. Int. Workshop Ultrawideband Systems*, Oulu, June 2003.
19. D. Cassioli, M. Z. Win, F. Vatalaro, and A. F. Molisch, "Performance of low-complexity RAKE reception in a realistic UWB channel," in *Proc. IEEE Int. Conf. Communications (ICC)*, vol. 2, New York, April 2002, pp. 763–767.
20. L. Rusch and H. V. Poor, "Multiuser detection techniques for narrowband interference suppression in spread spectrum communications," *IEEE Transactions on Communications*, vol. 43, no. 2–4, pp. 1725–1737, 1995.
21. S. W. Wales, "Technique for co-channel interference suppression in TDMA mobile radio systems," *IEE Proceedings, Communications*, vol. 142, no. 2, pp. 106–114, April 1995.
22. H. Arslan and K. Molnar, "Co-channel interference suppression with successive cancellation in narrowband systems," *IEEE Communications Letters*, vol. 5, pp. 37–39, February 2001.
23. H. Arslan, S. Gupta, G. Bottomley, and S. Chennakeshu, "New approaches to adjacent channel interference suppression in FDMA/TDMA mobile radio systems," *IEEE Transactions on Vehicular Technology*, vol. 49, no. 4, pp. 1126–1139, July 2000.
24. S. Verdu, *Multiuser Detection*, 1st edn. Cambridge: Cambridge University Press, 1998.
25. Q. Li and L. A. Rusch, "Multiuser receivers for DS-CDMA UWB," in *Proc. IEEE Ultrawideband Systems and Technology (UWBST)*, Baltimore, MD, May 2002, pp. 163–167.
26. Y. C. Yoon and R. Kohno, "Optimum multi-user detection in ultra-wideband (UWB) multiple-access communication systems," in *Proc. IEEE Int. Conf. Communications*, vol. 2, May 2002, pp. 812–816.
27. A. Muqaibel, B. Woerner, and S. Riad, "Application of multi-user detection techniques to impulse radio time hopping multiple access systems," in *Proc. IEEE Ultrawideband Systems and Technology (UWBST)*, Baltimore, MD, May 2002, pp. 169–173.
28. T. Quek and M. Win, "Ultrawide bandwidth transmitted-reference signaling," in *Proc. IEEE Int. Conf. on Communications*, Paris, June 2004, pp. 3409–3413.
29. J. D. Choi and W. E. Stark, "Performance of ultra-wideband communications with sub-optimal receivers in multipath channels," *IEEE Journal of Selected Areas in Communications*, vol. 20, no. 9, pp. 1754–1766, December 2002.

30. M. Pausini and G. Janssen, "Analysis and comparison of autocorrelation receivers for IR-UWB signals based on differential detection," in *Proc. IEEE Int. Conf. on Acoustics, Speech, and Signal Processing*, Netherlands, May 2004, pp. 1520–6149.
31. D. Cassioli, M. Z. Win, F. Vatalaro, and A. F. Molisch, "Performance of low-complexity RAKE reception in a realistic uwb channel," in *Proc. IEEE Int. Conf. Communications (ICC)*, vol. 2, New York, May 2002, pp. 763–767.
32. S. Verdu, *Multiuser Detection*, 1st edn. Cambridge: Cambridge University Press, 1998.
33. E. Fishler and H. V. Poor, "Low complexity multi-user detectors for time hopping impulse radio systems," *IEEE Transactions on Signal Processing*, vol. 52, no. 9, pp. 2561–2571, September 2004.
34. S. Gezici, H. Kobayashi, H. V. Poor, and A. F. Molisch, "Optimal and suboptimal linear receivers for time-hopping impulse radio systems," in *Proc. IEEE Ultrawideband Systems and Technology (UWBST)*, Kyoto, May 2004, pp. 11–15.
35. T. Strohmer, M. Emami, J. Hansen, G. Papanicolaou, and P. J. Arogyaswami, "Application of time-reversal with MMSE equalizer to UWB communications," in *Proc. IEEE Global Communications Conf. (GLOBECOM)*, Dallas, TX, 2004.
36. S. Franz and U. Mitra, "Integration interval optimization and performance analysis for UWB transmitted reference systems," in *Proc. IEEE Ultrawideband Systems Technology (UWBST)*, Kyoto, May 2004, pp. 26–30.
37. L. Yang and G. Giannakis, "Optimal pilot waveform assisted modulation for ultrawideband communications," *Proc. IEEE Transactions on Wireless Communications*, vol. 3, no. 4, pp. 1236–1249, July 2004.
38. M. H. Chung and R. A. Scholtz, "Comparison of transmitted- and stored-reference systems for ultra-wideband communications," in *Proc. IEEE Military Communications Conf. (MILCOM)*, vol. 1, Monterey, CA, October 2004, pp. 521–527.
39. Y.-L. Chao and R. A. Scholtz, "Optimal and suboptimal receivers for ultra-wideband transmitted reference systems," in *Proc. IEEE Global Telecommunications Conf. (Globecom)*, December 2003, pp. 759–763.
40. R. Hoctor and H. Tomlinson, "Delay-hopped transmitted-reference communications," in *Proc. IEEE Conf. Ultra Wideband Systems and Technologies*, May 2002, pp. 265–269.
41. F. Tufvesson and A. F. Molisch, "Ultra-wideband communication using hybrid matched filter correlation receivers," in *Proc. IEEE Vehicular Technology Conf. (VTC 2004 Spring)*, Milan, May 2004.
42. G. Leus and A.-J. van der Veen, "Noise suppression in UWB transmitted reference systems," in *Proc. Fifth IEEE Workshop on Signal Processing Advances in Wireless Communications*, Lisbon, July 2004.
43. F. Dowla, F. Nekoogar, and A. Spiridon, "Interference mitigation in transmitted-reference ultra-wideband (UWB) receivers," in *Proc. IEEE Antennas and Propagation Society Symp.*, Monterey, CA, June 2004, pp. 1307–1310.
44. M. Ho, V. S. Somayazulu, J. R. Foerster, and S. Roy, "A differential detector for an ultra-wideband communications system," in *Proc. IEEE Vehicular Technology Conf. (VTC)*, vol. 4, Alabama, Spring 2002, pp. 1896–1900.
45. L.-M. A. S. Paquelet and B. Uguen, "impulse radio asynchronous transceiver for high data rates," in *Proc. Int. Workshop on UltraWideband Systems joint Conf. on UltraWideband Systems and Technologies*, Japan, May 2004.

46. M. Weisenhorn and W. Hirt, "Robust noncoherent receiver exploiting UWB channel properties," in *Proc. IEEE Ultrawideband Systems Technology (UWBST)*, Kyoto, May 2004, pp. 156–160.
47. S. Paquelet, L. M. Aubert, and B. Uguen, "An impulse radio asynchronous transceiver for high data rates," in *Proc. IEEE Ultrawideband Systems Technology (UWBST)*, Kyoto, May 2004, pp. 1–5.
48. A. Rabbachin and I. Oppermann, "Synchronization analysis for UWB systems with a low-complexity energy collection receiver," in *Proc. IEEE Ultrawideband Systems Technology (UWBST)*, Kyoto, May 2004, pp. 288–292.
49. P. A. Humblet and M. Azizoglu, "On the bit error rate of lightwave systems with optical amplifiers," *Journal of Lightwave Technology*, vol. 9, no. 11, pp. 1576–1582, November 1991.
50. H. Akahori, Y. Shimazaki, and A. Kasamatsu, "Examination of the automatic integration time length selection system using PPM in UWB," in *Proc. IEEE Ultrawideband Systems Technology (UWBST)*, Kyoto, May 2004, pp. 268–272.

CHAPTER 8

Ultra Wideband Channel Modeling and Its Impact on System Design

CHIA-CHIN CHONG

8.1 INTRODUCTION

Accurate knowledge of the wireless propagation channel is of great importance when designing radio systems. A realistic radio channel model that provides insight into the radio wave propagation mechanisms is essential for the design and successful deployment of wireless systems. If an accurate channel model is available, it is possible to design receiver (RX) algorithms that achieve good performance by exploiting the properties of the channel. Unfortunately, the mechanisms that govern radio propagation in a wireless communication channel are complex and diverse. Therefore, a better understanding of the propagation mechanisms and effects is key for the development of a realistic channel model. Consequently, channel modeling has been a subject of intense research for a long time [1–9]. Early channel modeling work aimed to develop models which could provide an accurate estimate of the mean received power (or path loss) and study the behavior of the received signal envelope. This led to the conventional statistical models of the fading signal envelope [2–4]. Since these models were typically developed for narrowband systems, the temporal domain such as delay spread was largely neglected. As the need for higher data rates increased, larger bandwidths were necessary. In order to accurately model wideband systems, narrowband channel models were enhanced to include the prediction of the temporal domain properties such as TOA and delay spread. Such wideband models are important when analyzing digital modulation over wireless communication links and for cell planning in digital mobile radio [7–9].

The type of channel model that is desired depends critically on the carrier frequency, bandwidth, the type of environment, and the system under consideration. For example, different types of channel models are needed for indoor and outdoor

---

Ultra Wideband Wireless Communication. Edited by Arslan, Chen, and Di Benedetto
Copyright © 2006 John Wiley & Sons, Inc.

environments, and for narrowband, wideband and UWB systems. The UWB wireless communication system has been the subject of extensive research in recent years [10–13]. In particular, UWB technology has emerged as one of the most promising candidates for many WPAN applications such as security systems, sensor networks, and wireless home networking. Several new standards are currently under development that consider UWB technology such as IEEE 802.15.3a and IEEE 802.15.4a. The former standard is capable of providing high data rate for short-range applications (i.e., up to 110 Mbps at 10 m and 480 Mbps at shorter distances), while the latter standard aims to provide low data rate for longer range applications (i.e., approximately 1 Mbps at up to 30 m) with high-precision ranging and location capability [14, 15]. The emergence of UWB technology for both short-range high-rate and long-range low-rate communication systems implies that understanding the characteristics of the UWB propagation channel is essential. Accurate channel models are vital for performance evaluation and the design of UWB communications systems. Given the very wideband nature of UWB signals (i.e., up to tens of GHz of frequency bandwidth), the conventional channel models developed for narrowband transmissions are inadequate for UWB transmission. This implies that more UWB channel measurements are required in order to gain more profound knowledge of the UWB channel behaviors.

This chapter gives an overview of the UWB propagation channel modeling work and its impact on the UWB communication system design. Section 8.2 establishes the fundamental concepts and background for modeling the UWB multipath propagation channel; Section 8.3 discusses the two commonly used channel sounding techniques; Section 8.4 describes the UWB statistical-based channel modeling work; Section 8.5 details the impact of UWB channel on the system design; finally, in Section 8.6 appropriate conclusions are drawn.

8.2 PRINCIPLES AND BACKGROUND OF UWB MULTIPATH PROPAGATION CHANNEL MODELING

8.2.1 Basic Multipath Propagation Mechanisms

In a wireless system, the transmitted signal interacts with the physical environment in a complex manner. The signal arriving at the RX is in general a summation of both direct LOS and several multipath components (MPCs). Multipath occurs due to the three basic multipath propagation mechanisms, namely, *reflection, diffraction, and scattering of the transmitted signal*. All three of these phenomena cause radio signal distortions and give rise to signal fades, as well as additional signal propagation losses in a wireless communication system [16]. The relative importance of these propagation mechanisms depends on the particular environment. For example, if there is a direct LOS between terminals, then reflection dominates the propagation, whilst if the mobile is in a heavily cluttered area with no LOS path, scattering and diffraction usually play a major role. Combinations of these mechanisms will be seen to account for all of the observed effects in the UWB propagation

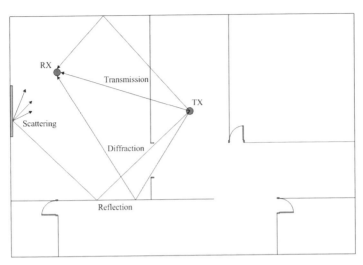

Figure 8.1 Illustration of the basic propagation mechanisms that cause multipath effects.

channel, as will be discussed in this chapter. Figure 8.1 shows the basic propagation mechanisms that cause multipath effects.

8.2.2 Classification of UWB Channel Models

The requirement to model many different types of wireless propagation channels has resulted in a large number of different modeling approaches reported in the literature [9]. One reason for the abundance of modeling approaches is the complex phenomena encountered by a transmitted signal. The transmitted signal will usually arrive at the RX via several paths, that is, multipaths, where the signal encounters various propagation mechanisms such as reflection, scattering, and/or diffraction. Therefore, many different types of simplifications and approximations are necessary in order to obtain a simple yet accurate and reliable model of the wireless communications channel. Signal propagation theory is well established for both narrowband and wideband systems [17, 18]. As for the UWB channel, many different types of channel models have been proposed recently. In general, the UWB propagation channel models can be classified as *deterministic* and *statistical*.

Deterministic Models Deterministic models apply an electromagnetic simulation tool such as ray tracing techniques to obtain nearly exact propagation characteristics for a specified geometry. These models try to describe the physics of the propagation mechanisms such as path loss, reflection, diffraction, and scattering. Ray tracing models are based on exact computations making use of a detailed database of the geometry of the specific physical environment under study. Ultimately, the accuracy of ray tracing models relies on the accuracy and detail of the site-specific representation of the propagation medium [19]. For example, indoor channel modeling usually relies on the availability of a three-dimensional (3-D) database. In addition to the geometry, the electromagnetic parameters of the materials also need to be included in the database. The major advantage of these models is that they offer great accuracy with site-specific results. Any site can be modeled if its physical characteristics are available, and any parameter can be calculated by adjusting these models. However, these models have several disadvantages. Firstly, the topographical and building data is always tied to a particular site and thus a huge amount of such data is required in order to obtain a comprehensive set of different propagation environments. Secondly, they are usually computationally intensive, especially when the environment is complex. Thus, detailed physical characteristics of the simulated environment must be known beforehand which is often time-consuming and impractical.

Numerous UWB channel models have been developed under this category. For example, in [20] the ray tracing approach is combined with the uniform theory of diffraction (UTD) to model the received signal as a superposition of rays. This model has the advantage of being versatile and of general validity. However, it has high complexity in the channel modeling which leads to a high computational load. Another UWB channel model based on UTD technique is reported in [21, 22]. This model consists of three basic ray mechanisms of geometrical optics and time-domain UTD (i.e., directed ray, multireflected rays from lossy surfaces and diffracted ray from lossy edge). This approach can determine both the signal attenuation and the waveform distortion in terms of pulse shape and pulse duration. A different approach has been pursued in [23], where the frequency and distance characteristics of a two-ray propagation model have been modeled using a deterministic approach. Examples of other UWB deterministic channel models reported in the literature are in [24–29].

8.2.2.2 Statistical Models Statistical models are normally less complex than the deterministic models, and can provide sufficiently accurate channel information. Statistical models attempt to generate synthetic channel responses that are representative of real propagation channels. Typically, such models can be tuned to imitate various propagation environments by setting appropriate values for the channel model parameters. Note that fixed parameter settings do not produce identical outputs on each simulation run but stochastic processes are used to create variability within a fixed environment type. For example, a particular set of parameters might generate a representative set of propagation scenarios found in indoor environments. Statistical models may be formed based on the basic principals of wave propagation for random communications channels and by assuming a statistical distribution of the channel parameters, and computing the required statistical moments from the data collected from the real-time measurements. This category of models has the ability to provide accurate statistical information, without the complexity of detailed deterministic approaches.

Several statistical-based UWB channels have been proposed recently. For example, Ghassemzadeh et al. [30] proposes a model based on the frequency-domain channel measurements results in 23 residential environments (based on extensive propagation studies in 23 homes). Measurements were conducted in the frequency range 4.375–5.625 GHz. A path loss model and a second-order autoregressive model are proposed for frequency response generation of the UWB indoor channel. Probability distributions of the model parameters for different locations and time-domain results such as root mean square (RMS) delay spread and percentage of captured power are reported in [30]. Foerster and Li [31] proposed a statistical-based multipath channel model based upon measurements collected in a condominium in the frequency range 2–8 GHz. Three channel models were considered, namely the Rayleigh tap delay line model, the Δ-K model and the Saleh-Valenzuela (S-V) model. The comparisons showed that the S-V model gives the best fit to the measured channel characteristics such as the mean excess delay, mean RMS delay spread and the mean number of significant paths within 10 dB of the peak multipath arrival. In addition, the log-normal distribution was proposed to model the amplitude distribution, which was shown to give a better fit than the Rayleigh distribution. Chong et al. [32–35] proposed a statistical-based UWB channel model which incorporates the clustering of MPC phenomenon observed in the measurement data collected in the frequency band 3–10 GHz in various types of high-rise apartments in Korea. Both the large-scale and small-scale channel statistics were characterized. The large-scale parameters included the distance and frequency dependency of path loss as well as the shadowing fading statistics. The small-scale parameters included the temporal domain parameters, the distribution of clusters and MPCs, clusters, and MPC arrivals statistics. A new distribution, namely, mixtures of two Poisson processes, was proposed to model the ray arrival times. Furthermore, the small-scale fading amplitude statistics as well as the temporal correlation between adjacent path amplitude were also investigated in the model. Other statistical-based channel models reported in the literature include those in [36–46].

8.3 CHANNEL SOUNDING TECHNIQUES

Channel measurements are the most direct method of studying radio wave propagation and to verify propagation theory. This can then provide invaluable input to the development of realistic channel models. Generally, there are two possible techniques to perform the UWB channel measurements that is, the *time-domain measurement technique* and the *frequency-domain measurement technique*.

8.3.1 Time-Domain Technique

This technique is based on the impulse transmission, that is, employing a narrow pulse to probe the channel. A narrow single impulse or a train of impulses is sent through the channel and the channel impulse response (CIR), $h(t)$, is measured using a digital sampling oscilloscope. The measurement resolution is equal to the width of transmitted pulse. The pulse repetition period should be chosen in such a way as to allow observation of the time-varying response of individual propagation paths and at the same time to ensure that all MPCs are received between successive pulses. The advantage of this technique is that the environment does not need to be

188 ULTRA WIDEBAND CHANNEL MODELING AND ITS IMPACT ON SYSTEM DESIGN

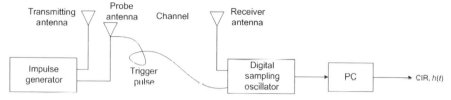

Figure 8.2 Time-domain measurement technique based on pulse transmission.

static within the recordings, but as long as the coherence time assumptions is still valid. The corresponding train of impulses can also be generated using a conventional direct sequence spread spectrum-based measurement system with a correlation receiver. However, the main drawback in this technique is that it needs very high chip rates to achieve bandwidths over several GHz. Figure 8.2 illustrates the time-domain measurement technique. Several UWB channel measurements have been performed using this technique [43, 44, 45, 47].

8.3.2 Frequency-Domain Technique

This technique is based on the frequency sweeping in which the radio channel is sounded by sweeping a sinusoidal carrier over the frequency band of interest. A vector network analyzer (VNA) is then used to control a synthesized frequency sweeper. The bandwidth centered around the frequency of interest is scanned by the synthesizer through discrete frequencies. For each frequency step, a known sinusoidal signal is transmitted and the channel transfer function (CTF), $H(f)$, can be measured by the VNA. The long sweeping time requires the channel coherence time to be relatively large, and in practice measurements can be carried out only in static environments. The CIR can be obtained by taking the inverse Fourier transform. Figure 8.3 illustrates the frequency-domain measurement technique. Several

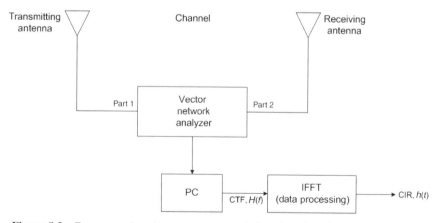

Figure 8.3 Frequency-domain measurement technique based on frequency sweeping.

UWB channel measurements have been performed based on this technique [30, 32–35, 39, 46, 48].

8.4 UWB STATISTICAL-BASED CHANNEL MODELING

8.4.1 Modeling Philosophy and Mathematical Framework

Due to the broad variation of real environments, in addition to the time-varying nature of the multipath channels, it is difficult to model the wireless channels deterministically. Therefore, the channel properties are best captured using statistical approaches. Generally, a statistical-based channel model requires the following four important steps:

1. Setting up a generic mathematical framework for the channel model.
2. Identifying the channel parameters that have to be determined for a full description of the channel model.
3. Performing the measurement campaigns in the relevant environments.
4. Post-processing the measurement data in order to extract the required channel parameters statistically.

The aim here is to introduce the mathematical framework for a generic UWB channel model. The model is characterized by the CIR, $h(t)$, which consists of a sum of contribution from the MPCs given as follows:

$$h(t) = \sum_{k=0}^{K} a_k \delta(t - \tau_k) \tag{8.1}$$

where $\delta(\cdot)$ is the Dirac delta function, K is the number of MPCs, and a_k and τ_k are the multipath gain coefficient and delay of the kth MPC, respectively.

Based upon the apparent existence of clusters in most of the UWB measurement campaigns reported in the literature [7, 8, 31–35, 43], a UWB channel model which accounts for the clustering of MPCs is proposed by modifying Equation (8.1) based on the conventional S-V channel model [6]. Now, the clustering CIR can be expressed as follows [34, 35]:

$$h(t) = \sum_{l=0}^{L} \sum_{k=0}^{K_l} a_{k,l} \delta(t - T_l - \tau_{k,l}) \tag{8.2}$$

where L is the number of clusters and K_l is the number of MPCs within the lth cluster, $a_{k,l}$ is the multipath gain coefficient of the kth component in the lth cluster, T_l is the delay of the lth cluster which is defined as the TOA of the first arriving MPC within the lth cluster and $\tau_{k,l}$ is the delay of the kth MPC relative to the lth cluster arrival time, T_l. The clustering channel model relies on two classes of parameters, namely, *inter-cluster* and *intra-cluster parameters*, which characterize the cluster and MPC, respectively. From Equation (8.2), $\{L, T_l\}$ and $\{K_l, \tau_{k,l}, a_{k,l}\}$ are classified as the inter-cluster and intra-cluster parameters, respectively.

8.4.2 Large-Scale Channel Characterization

In general, wireless multipath propagation can be characterized by two major effects according to the variability of power level, namely, *large-scale fading* and *small-scale fading*. *Large-scale fading* can be categorized by the *path loss* (PL) and the *shadowing*. PL is an important characteristic for link budget calculation and system design which is defined as the ratio between the received and transmitted power. In the conventional narrowband channel, the Friis transmission formula is used to model the received signal power, P_r in free space given by [16]

$$P_r(d) = \frac{P_t G_t G_r \lambda^2}{(4\pi d)^2} \qquad (8.3)$$

where P_t is the transmitted power, G_t and G_r is the transmitter (TX) and RX antenna gain, respectively, λ is the wavelength and d is the TX–RX separation distance. The Friis transmission formula shows that the received signal power falls off with the square of the TX–RX separation distance. From Equation (8.3), the term PL is given by $(\lambda/4\pi d)^2$, which predicts that signal power will decrease with the square of increasing frequency. These show the presence of both distance and frequency dependency in PL. Note that the latter has a minimal effect on the narrowband systems and thus is negligible. However, for UWB systems, due to the large communication bandwidth (>500 MHz), the frequency dependency in PL can be significant. For such a large bandwidth system, the channel would distort the spectrum of the signal which directly related to the signal distortions, that is, pulse shape distortion. Therefore, the conventional PL model used in narrowband analysis needs to be investigated more closely in order to justify its application to UWB systems. PL modeling can be simplified by assuming that the frequency dependence and the distance dependence can be treated independently of each other [32, 33, 35, 40, 41]:

$$PL(f, d) = PL(f) \cdot PL(d). \qquad (8.4)$$

Distance Dependence Path Loss The distance dependency of PL, $PL(d)$, describes the attenuation of the median power as a function of distance traveled. From the frequency-domain measurement, $PL(d)$ can be obtained directly from the calibrated measured CTFs by frequency averaging over the total measurement bandwidth for each data set at each RX location. The PL in dB as a function distance

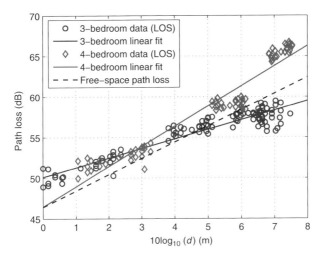

Figure 8.4 Path loss as a function of distance for three- and four-bedroom apartments under an LOS scenario.

can be expressed by [16]

$$PL(d) = PL_0 + 10 \cdot n \cdot \log_{10}\left(\frac{d}{d_0}\right) + S; \quad d \geq d_0 \qquad (8.5)$$

where $PL(d)$ represents the received power at a distance d, computed relative to a reference distance d_0, for example, $d_0 = 1m$ and PL_0 is the free-space PL in the far-field of the antennas at a reference distance $d_0 \cdot PL_0$ is the interception point and is usually calculated based on the mid-band frequency. n is the PL exponent and S is the shadowing fading parameter. The PL exponent is obtained by performing least squares linear regression on the logarithmic scatter plot of averaged received powers vs distance to Equation (8.5). Figure 8.4 shows an example of a scatters plot of PL as a function of distance.

Frequency Dependence Path Loss Most of the measurement results in the literature have reported that the narrowband model can be used to approximate the PL for UWB systems, that is, the PL is independent of the frequency [30, 31, 37]. However, several more recently published works reported that the PL of UWB channel models vary as a function of frequency [32, 33, 35, 38–41]. This clearly implies that more work should be done in order to further characterize the frequency dependency of the PL. The measurement results in [38] showed that the frequency dependency of the PL can be modeled by

$$PL(f) \propto \exp(-\delta_1 \cdot f) \qquad (8.6)$$

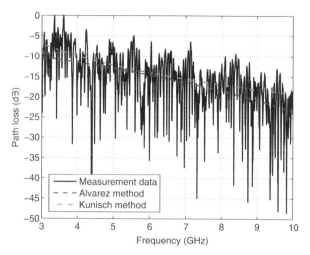

Figure 8.5 Path loss as a function of frequency for the four-bedroom apartment under an NLOS scenario.

with δ_1 varying between 1 and 1.4, while measurement results in [39] modeled the frequency dependence of the PL by a simple power law as follows:

$$\sqrt{PL(f)} \propto f^{-\delta_2} \tag{8.7}$$

with δ_2 lying between 0.8 and 1.4. The *frequency decaying factor*, δ, is the average slope of the CTF. In [32, 33, 35], the values of both δ_1 and δ_2 are calculated using the least-square curve fitting of Equations (8.6) and (8.7), respectively, on the measured data. It shows that both Equations (8.6) and (8.7) can model the frequency dependency PL well, with Equation (8.7) currently being adopted in the IEEE 802.15.4a standard channel model final report [40, 41]. The values of δ_1 and δ_2 are found to lie the ranges 0.08–0.14 and 0.98–1.54, respectively. Figure 8.5 shows an example of the plot of frequency dependency of the PL.

Shadowing The dynamic evolution of propagation paths wherein new paths arise and old paths disappear can occur due to variations in the surrounding environments. Therefore, PL observed at any given point will deviate from its average value [16]. This phenomenon is called *shadowing* and is defined as the slow variation of the local mean signal strength. The term "shadowing" is descriptive since typically these changes are due to the appearance or disappearance of shadowing objects on signal paths. It has been reported by many measurements in the literature to follow a log-normal distribution [30, 32, 33, 35, 37, 49]. Referring to Equation (8.6), shadowing fading parameter is given by the term S that varies randomly from one location to another location within any home. It is a zero-mean Gaussian distributed random variables in dB with standard deviation σ_S which is also in dB. Different values of σ_S have been reported in the literature, ranging from 0.83 dB to 6 dB [40, 41], strongly depending on the measurement environments and scenarios.

8.4.3 Small-Scale Channel Characterization

Number of Clusters and MPCs The number of clusters, L, is modeled by a Poisson distribution as proposed in [40, 41]. From the collection of measurement results reported in the literature, the average number of clusters, $\bar{L}$, was approximately 3–14 and 1–11 under the LOS and non-LOS scenarios, respectively [40, 41]. The clustering phenomenon is a result of the superstructure (e.g., furniture, walls, doors) of the environment. Therefore, the presence of furniture can increase the number of clusters in the environment under consideration. On the other hand, the number of MPCs per cluster, K_l, was found can be well modeled by an exponential p.d.f., $f(K_l)$, in [34, 35]. In general, K_l increases from LOS to NLOS scenarios. The number of MPCs per cluster (thus, the number of clusters) is dependent on several factors such as the resolution of the parameter estimation technique, the TX–RX separation and location, the physical layout of the environment, as well as the dynamic range of the measurement system. More clusters and MPCs were observed in a heavily cluttered environment (e.g., a fully furnished room) because many MPCs will undergo more complex propagation mechanisms such as multiple-order reflections, scattering, and diffraction. The number of resolvable MPCs is an important parameter since it has a direct relationship to the complexity of the channel simulator and thus, the whole communication systems.

Clusters and MPC Arrival Times According to the conventional S-V channel model [6] and the statistical-based UWB channel model proposed in [34–36], the distributions of the cluster arrival times, T_l, and the ray arrival times, $\tau_{k,l}$, are given by two Poisson processes. Thus, the cluster inter-arrival times and ray intra-arrival times are described by two independent exponential PDFs as follows:

$$p(T_l|T_{l-1}) = \Lambda \exp[-\Lambda(T_l - T_{l-1})], \quad l > 0 \tag{8.8}$$

$$p(\tau_{k,l}|\tau_{(k-1),l}) = \lambda \exp[-\lambda(\tau_{k,l} - \tau_{(k-1),l})], \quad k > 0 \tag{8.9}$$

where Λ is the mean cluster arrival rate and λ is the mean ray arrival rate. Typically, each cluster consists of many rays where $\lambda \gg \Lambda$. However, measurement results reported in [34, 35, 42] show that the single Poisson process given in Equation (8.9) is insufficient to model the ray arrival times in the indoor residential, indoor office, and outdoor office environments. Thus, mixtures of two Poisson processes is proposed in [34, 35] in order to model the ray arrival times

$$p(\tau_{k,l}|\tau_{(k-1),l}) = \beta\lambda_1 \exp[-\lambda_1(\tau_{k,l} - \tau_{(k-1),l})] \\ + (1-\beta)\lambda_2 \exp[-\lambda_2(\tau_{k,l} - \tau_{(k-1),l})], \quad k > 0 \tag{8.10}$$

where β is the mixture probability, while λ_1 and λ_2 are the ray arrival rates. This new distribution gives an excellent match to the ray arrival times for the above-mentioned three environments. The interested reader is referred to [34, 35] for more detailed information.

In order to estimate the Poisson cluster arrival rate, Λ, the first arrival in each cluster was considered to be the beginning of the cluster, regardless of whether or not it had the largest amplitude. The arrival time of each cluster was subtracted from its successor, so that the conditional probability distribution given in Equation (8.8) could be estimated. Estimates for Λ were done by fitting the cluster inter-arrival times, ΔT, to an exponential distribution. The fitting was done using a linear least mean squares criterion of the complementary cumulative distribution function (CCDF) of ΔT, $S(\Delta T)$ in the natural logarithmic scale. The CCDF of the exponential distribution in the natural logarithmic domain can be expressed by

$$\ln[S(\Delta T)] = -\Lambda \cdot \Delta T. \tag{8.11}$$

Note that Equation (8.11) gives a linear relationship in Λ which can be estimated by linear least mean squares fit through the origin based on $\ln[S(\Delta T)]$. Similarly, the single Poisson process ray arrival rate, λ, and the mixtures of two Poisson processes ray arrival rates, λ_1 and λ_2, were estimated based on the average separation time between ray arrivals within a cluster. Using a similar approach, by replacing ΔT and Λ with $\Delta \tau$ and λ (or λ_1 and λ_2), respectively, in Equation (8.11), where $\Delta \tau$ is the ray intra-arrival time, λ or λ_1 and λ_2 can be estimated by linear least mean squares fit through the origin based on $\ln[S(\Delta \tau)]$, where $S(\Delta \tau)$ is the CCDF of $\Delta \tau$. Figure 8.6 shows the fitting of a single Poisson process (red solid line) and

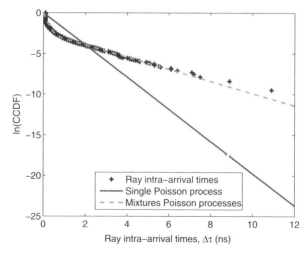

Figure 8.6 Logarithmic CCDF of the ray intra-arrival times for the three-bedroom apartment under an LOS scenario.

mixtures of two Poisson processes (green dashed line) to the ray intra-arrival times for measurement results adopted from [34, 35]. This graph clearly reveals that the mixtures of two Poisson processes model the ray intra-arrival times better. This result has been further validated in [42] for measurement in indoor and outdoor office environments.

Clusters and MPC Power Decaying Phenomenon The average power of both the clusters and the rays within the clusters are assumed to decay exponentially, such that the average power of a MPC at a given delay, $T_k + \tau_{k,l}$, is given by

$$\overline{a_{k,l}^2} = \overline{a_{0,0}^2} \cdot e^{-T_l/\Gamma} \cdot e^{-\tau_{k,l}/\gamma} \tag{8.12}$$

where $\overline{a_{0,0}^2}$ is the expected value of the power of the first arriving MPC, Γ, is the decay exponent of the clusters and γ is the decay exponent of the rays within a cluster. Typically, $\Gamma > \gamma$, which indicates that the expected power of the rays in a cluster decays faster than the expected power of the first ray of the next cluster. Figure 8.7 illustrates the typical power delay profile (PDP) of the S-V channel model with the double exponential power decay phenomenon.

The cluster and ray decay time constants, Γ and γ, can be estimated from the measurement data by superimposing clusters and rays with normalized amplitudes

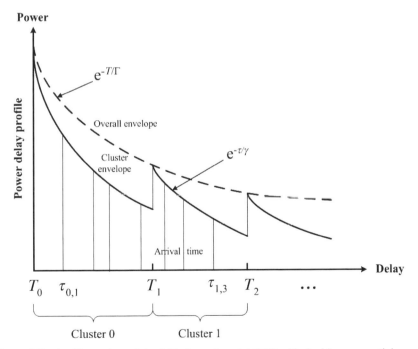

Figure 8.7 An illustration of the S-V channel model PDP with double exponential power decay phenomenon [6].

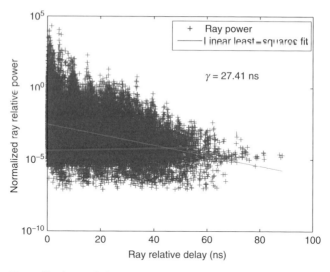

Figure 8.8 Normalized ray relative power vs relative delay for the four-bedroom apartment under an NLOS scenario.

and time delays and selecting their mean decay rates. For example, in order to estimate Γ, the first cluster arrival in each data set was normalized to an amplitude of one and a time delay of zero and all other cluster arrivals in the same data set are expressed relative to this time. The estimate for Γ was found by linear least-squares curve fitting the data in the logarithmic domain such that the mean squared error was minimized. Similarly, in order to estimate γ, the first arrival in each cluster was set to a time of zero and amplitude of one, and all other ray arrivals within the clusters were then adjusted accordingly and superimposed. Following this model, an MMSE linear fit in the logarithmic domain was determined from the ray relative powers and ray relative delays. This attempts to estimate the value of γ by the slope of the MMSE linear fit. Figure 8.8 shows an example of the normalized ray relative powers within a cluster falling off as a function of the relative delay.

Small-Scale Amplitude Fading Statistics Amplitude fading in a multipath propagation channel may follow different distributions depending on the measurement environments and scenarios. In the conventional narrowband channel models, two distributions that are widely used to describe the amplitude fading statistics are the Rayleigh and Rice distributions. Rayleigh distribution usually models the extreme multipath situations when there is no direct LOS between the TX and RX, while under the LOS condition, the Rice distribution is usually a better candidate. The Rayleigh fading model is the result of coherent interaction of sinusoidal signals due to the assumption that several paths arrive at certain delay that is irresolvable by the measurement system bandwidth, whereas in the UWB propagation, the wide frequency bandwidth corresponds to a high temporal resolution capability, which enables a single path arriving at a certain delay bin to be resolved. Thus, the amplitude distribution in each delay bin differs from the Rayleigh fading distribution.

In order to evaluate the small-scale amplitude fading statistics, the relative amplitudes over small-scale areas are calculated. Empirical data from bins at specific excess delays were matched to some typical theoretical distributions for amplitude fading statistics such as log-normal, Nakagami, Rayleigh, Rice, and Weibull distributions. The measurement results reported in [31, 50] show that the small-scale amplitude fading can be modeled by a log-normal distribution, while measurement results reported in [43, 51] show that Rayleigh distribution gives a better fit. In [39], the Rice distribution is reported to give the best fit for the amplitude fading statistics. Other measurement campaigns such as [44, 52] have found that the Nakagami distribution can be a good fit for small-scale amplitude fading statistics. The advantage of Nakagami distribution is that, for special cases such as $m = 1$ and for very large values of m, it can be generalized to become the Rayleigh and log-normal distributions, respectively. Furthermore, a conversion from the Nakagami distribution to the Rice distribution is also possible and can be approximately expressed as follows [40, 41]:

$$m = \frac{(K_r + 1)^2}{(2K_r + 1)} \qquad (8.13)$$

and

$$K_r = \frac{\sqrt{m^2 - m}}{m - \sqrt{m^2 - m}} \qquad (8.14)$$

where K_r and m are the Rice factor and Nakagami-m factor, respectively.

Hypothesis tests such as Kolmogorov–Smirnov (K–S) and chi-square (χ^2) tests are typically used to elaborate the goodness-of-fit of these distributions. Detailed results that compare the passing rates of the K–S and χ^2 tests for the above distributions were reported in [34, 35] for the indoor residential environment. It was found that the small-scale amplitude statistics can be well modeled by the log-normal, Nakagami or Weibull distributions in which the parameters of these distributions, that is, the standard deviation of the log-normal distribution, m-parameter of the Nakagami distribution and b-shape parameter of the Weibull distribution were found to be log-normal distributed random variables, respectively. Variations of these parameters with increasing excess delay were reported to be negligible [34, 35]. Measurement results reported in [53] also show that Weibull distribution gives a good fit for the amplitude fading statistics.

8.4.4 Temporal Dispersion and Correlation Properties

The channel temporal dispersion is assessed in terms of the *RMS delay spread* and *mean excess delay*, while the temporal correlation properties are determined by the *temporal correlation coefficient*.

Temporal Dispersion Two figures of merit often used to characterize the temporal dispersion of the multipath channel are the *RMS delay spread* and the *mean excess delay*. These two parameters are important for evaluating the performance of digital systems. Delay spreads restrict the transmitted data rates and could limit the capacity of the system when multi-user systems are considered. The *mean excess delay*, τ_m, is defined as the first moment of the PDP and is defined as [16]

$$\tau_m = \frac{\sum_k a_k^2 \tau_k}{\sum_k a_k^2} = \frac{\sum_k P(\tau_k) \tau_k}{\sum_k P(\tau_k)} \tag{8.15}$$

where a_k, τ_k and $P(\tau_k)$ are the gain coefficient, delay and PDP of the kth MPC, respectively. The *RMS delay spread*, τ_{rms}, is the square root of the second central moment of the PDP and is defined as [16]

$$\tau_{rms} = \sqrt{\tau_m^2 - (\tau_m)^2} \tag{8.16}$$

where

$$\tau_m^2 = \frac{\sum_k a_k^2 \tau_k^2}{\sum_k a_k^2} = \frac{\sum_k P(\tau_k) \tau_k^2}{\sum_k P(\tau_k)}. \tag{8.17}$$

The measurement results reported in [54] show that there is no correlation between delay spread and TX–RX separation distance, while the measurement results reported in [33, 47, 50] show that correlation exists between these two domains where higher values of temporal dispersion was found when the TX–RX separation increased or in a more cluttered environment, particularly when the LOS path was obstructed. The RMS delay spread has been shown to decrease considerably when using directional antennas in comparison to using omnidirectional antennas [37, 55, 56]. This implies that the directional antennas can be used to mitigate the effects of multipath in the UWB channel as compared to omni-directional antennas. Table 8.1 summarizes the average value of the delay spread for various environments such as indoor residential, indoor office, indoor industrial, outdoor, and agricultural environments under both LOS and NLOS scenarios.

Temporal Correlation The temporal correlation between adjacent path amplitudes in a PDP can be characterized by the *temporal correlation coefficient*, $\rho_{a_{k,k+1}}$, defined as follows:

$$\rho_{a_{k,k+1}} = \frac{E\{(a_k - \overline{a_k})(a_{k+1} - \overline{a_{k+1}})\}}{\sqrt{E\{(a_k - \overline{a_k})^2 (a_{k+1} - \overline{a_{k+1}})^2\}}} \tag{8.18}$$

TABLE 8.1 Average Values of the Mean Excess Delay and RMS Delay Spread

Source	Environment	LOS τ_m (ns)	LOS τ_{rms} (ns)	NLOS τ_m (ns)	NLOS τ_{rms} (ns)
[30–35, 48, 57–59]	Indoor residential	2.15–7.52	3.55–14.00	6.93–36.09	7.35–38.61
[37, 42, 48, 55, 56, 59–63]	Indoor office	0.44–8.80	0.53–15.60	2.09–32.20	2.30–23.60
[64, 65]	Indoor industrial	—	30	—	40
[42, 66]	Outdoor	24.10	55.10	83.50	94.00
[29]	Agriculture	18.20	21	—	—

where $E\{\cdot\}$ denotes the expectation, a_k and a_{k+1} are the amplitude of the kth and $(k+1)$th bin, respectively, while $\overline{a_k}$ and $\overline{a_{k+1}}$ are their respective mean values. Analysis of the measurement results in [34, 35, 50] shows that $\rho_{a_{k,k+1}}$ is relatively small and can be assumed to be negligible for practical purposes. This is reasonable as, in an indoor environment, the large numbers of scattering objects present are uniformly distributed in the surrounding environment and different MPCs due to different scatterers would result a low degree of correlation.

8.5 IMPACT OF UWB CHANNEL ON SYSTEM DESIGN

The UWB propagation channel can have a significant impact on system design and performance evaluation. For example, the ultra-short pulses (due to the ultra wide bandwidth) traveling along different paths arrive at the RX almost without overlapping. The analysis of UWB measurement results revealed that the available signal power arriving at the RX varies insignificantly if the position is changed over a limited range, that is, within the small-scale region. Thus, the available signal power at a certain position only depends on the large-scale properties of the environment. This particular characteristic implies that only a very small fading margin is required in the UWB impulse radio system design. A low degree of temporal correlation between adjacent path amplitudes in a PDP has been reported in the literature [34, 35, 50]. Therefore, interference caused by pulse transmitted from the same origin is negligible if a certain guard interval of about the pulse duration is maintained. However, this property can only be exploited with perfect knowledge of the channel associated with the desired transmitted signal. Hence, a reliable channel estimation algorithm is required.

In the UWB propagation channel, the high temporal resolution capability enables every single path arriving at the RX at a certain delay bin to be resolved. This implies that multipath arrivals will undergo less fading as compared with the conventional narrowband system. Some of the UWB propagation environments have relatively long delay spreads such as outdoor and harsh industrial environments. For such environments, the average total energy is usually distributed in this delay region between a large number of MPCs. Thus, unique systems need to be designed so that the designated RX has the energy capture capability which can then take the advantage of that energy. For example, a RAKE RX can be deployed [67]. However, since the complexity of the RAKE RX increases with the number of RAKE fingers, they need to be designed according to the knowledge of the channel conditions.

8.6 CONCLUSION

In this chapter, an overview of the UWB propagation channel modeling work and the effect of UWB channels on system design was presented. In the first part, the basic multipath propagation mechanisms and characteristics were studied before the development and classification of various channel models were discussed. This was followed by a brief review of some common channel-sounding techniques. Then, the UWB statistical-based channel modeling work was described in detail, which included model mathematical framework, large-scale and small-scale channel characteristics, as well as the temporal dispersion and correlation properties. Finally, the impact of the UWB channel models on communication system design was studied. As a whole, this chapter acts a stepping stone for the remainder of the book.

REFERENCES

1. P. A. Bello, "Characterization of randomly time-variant linear channels," *IEEE Trans. Commun. Syst.*, vol. 11, no. 4, pp. 360–393, December 1963.
2. R. H. Clarke, "A statistical theory of mobile radio reception," *Bell Syst. Tech. J.*, vol. 47, pp. 957–1000, July 1968.
3. W. Jakes, *Microwave Mobile Communications*, Wiley Interscience, New York, 1974.
4. T. Aulin, "A modified model for the fading signal at a mobile radio channel," *IEEE Trans. Vehicular Technol.*, vol. 28, no. 3, pp. 182–203, August 1979.
5. G. L. Turin, F. D. Clapp, T. L. Johnston, S. B. Fine, and D. Lavry, "A statistical model of urban multipath propagation," *IEEE Trans. Vehicular Technol.*, vol. 28, no. 3, pp. 182–203, August 1979.
6. A. A. M. Saleh and R. A. Valenzuela, "A statistical model for indoor multipath propagation," *IEEE J. Select. Areas Commun.*, vol. 5, no. 2, pp. 128–137, February 1987.
7. C.-C. Chong, C.-M. Tan, D. I. Laurenson, S. McLaughlin, M. A. Beach, and A. R. Nix, "A new wideband spatio-temporal channel model for 5-GHz band WLAN systems," *IEEE J. Select. Areas Commun.*, vol. 21, no. 2, pp. 139–150, February 2003.
8. C.-C. Chong, C.-M. Tan, D. I. Laurenson, S. McLaughlin, M. A. Beach, and A. R. Nix, "A novel wideband dynamic directional indoor channel model based on a Markov process," *IEEE Trans. Wireless Commun.*, vol. 4, no. 4, pp. 1539–1552, July 2005.

9. B. H. Fleury and P. E. Leuthold, "Radiowave propagation in mobile communications: an overview of European research," *IEEE Commun. Mag.*, vol. 34, no. 2, pp. 70–81, February, 1996.
10. M. Z. Win and R. A. Scholtz, "Impulse radio: How it works," *IEEE Commun. Lett.*, vol. 2, no. 2, pp. 36–38, February 1998.
11. M. Z. Win and R. A. Scholtz, "On the robustness of ultra-wide bandwidth signals in dense multipath environments," *IEEE Commun. Lett.*, vol. 2, no. 2, pp. 51–53, February 1998.
12. M. Z. Win and R. A. Scholtz, "On the energy capture of ultra-wide bandwidth signals in dense multipath environments," *IEEE Commun. Lett.*, vol. 2, no. 9, pp. 245–247, September 1998.
13. M. Z. Win and R. A. Scholtz, "Ultra-wide bandwidth time-hopping spread-spectrum impulse radio for wireless multiple-access communications," *IEEE Trans. Commun.*, vol. 48, no. 4, pp. 679–691, April 2000.
14. IEEE 802.15.3a Task Group, "WPAN high rate alternative PHY," www.ieee802.org/15/pub/TG3a.html.
15. IEEE 802.15.4a Task Group, "WPAN low rate alternative PHY," www.ieee802.org/15/pub/TG4a.html.
16. T. S. Rappaport, *Wireless Communications: Principles and Practice*, 2nd edn, Prentice Hall, Upper Saddle River, NJ, 2002.
17. H. Bertoni, W. Honcharenko, L. R. Maciel, and H. Xia, "UHF propagation prediction for wireless personal communications," *Proc. IEEE*, vol. 82, no. 9, pp. 1333–1359, September 1994.
18. R. Vaughan and J. B. Andersen, *Channels, Propagation and Antennas for Mobile Communications*, IEE Electromagnetic Wave Series, IEEE, New York, 2003.
19. T. S. Rappaport and S. Sandhu, "Radio-wave propagation for emerging wireless personal-communication systems," *IEEE Ant. Propagat. Mag.*, vol. 36, no. 5, pp. 14–24, October 1994.
20. B. Uguen, E. Plouhinec, Y. Lostanlen, and G. Chassay, "A deterministic ultra wideband channel model," in *Proc. IEEE Conf. UWB Systems and Technologies (UWBST02)*, Baltimore, MD, May 2002, pp. 1–6.
21. R. Yao, W. Zhu, and Z. Chen, "An efficient time-domain ray model for UWB indoor multipath propagation channel," in *Proc. IEEE Vehicular Technologies Conf. (VTC 2003-Fall)*, Orlando, FL, September 2003, pp. 1293–1297.
22. R. Yao, G. Gao, Z. Chen, and W. Zhu, "UWB multipath channel model based on time-domain UTD technique," in *Proc. IEEE Global Telecommunications Conf. (GLOBECOM03)*, San Francisco, CA, December 2003, pp. 1205–1210.
23. A. Armogida, B. Allen, M. Ghavami, M. Porretta, G. Manara, and H. Aghvami, "Path loss modelling in short range UWB transmissions," in *Proc. Int. Workshop on Ultra Wideband Systems (IWUWBS03)*, Oulu, June 2003.
24. R. C. Qiu, "A study of the ultra-wideband wireless propagation channel and optimum UWB receiver design," *IEEE J. Select. Areas Commun.*, vol. 20, no. 9, pp. 1628–1637, December 2002.
25. Y. Zhao, H. Hao, A. Alomainy, and C. Parini, "UWB on-body radio channel modeling using ray theory and subband FDTD method," *IEEE Trans. Microwave Theory and Techniques*, vol. 54, no. 4, pp. 1827–1835, April 2006.
26. M. Terre, A. Hong, G. Guibe, and F. Legrand, "Major characteristics of UWB indoor transmission for simulation," in *Proc. IEEE Vehicular Technologies Conf. (VTC 2003-Spring)*, Jeju, May 2003, pp. 19–23.

27. G. Schiavone, P. Wahid, R. Palaniappan, J. Tracy, E. van Doorn, and B. Lonske, "Outdoor propagation analysis of ultra wide band signals," in *Proc. IEEE Antenna Propagation. Symp. (APS03)*, Columbus, OH, June 2003, pp. 999–1002.

28. H. Sugahara, Y. Watanabe, T. Ono, K. Okanoue, and S. Yarnazaki, "Development and experimental evaluations of "RS-2000"—a propagation simulator for UWB systems," in *Proc. IEEE Conf. UWB Systems and Technologies (UWBST04)*, Kyoto, May 2004, pp. 76–80.

29. S. Emami, C. A. Corral, and G. Rasor, "An ultra-wideband channel model and coverage for farm/open-area applications," IEEE 802.15-04-0325-00-004a, July 2004.

30. S. S. Ghassemzadeh, R. Jana, C. Rice, W. Turin, and V. Tarokh, "Measurement and modeling of an ultra-wide bandwidth indoor channel," *IEEE Trans. Commun.*, vol. 52, no. 10, pp. 1786–1796, October 2004.

31. J. R. Foerster and Q. Li, "UWB channel modeling contribution from Intel," IEEE P802.15-02/279r0-SG3a, June 2002.

32. C.-C. Chong, Y. Kim, and S.-S. Lee, "Statistical characterization of the UWB propagation channel in various types of high-rise apartments," in *Proc. IEEE Wireless Communications Networking Conf. (WCNC05)*, New Orleans, LA, March 2005, pp. 944–949.

33. C.-C. Chong, Y. Kim, S. K. Yong, and S.-S. Lee, "Statistical characterization of the UWB propagation channel in indoor residential environment," *Wireless Commun. Mobile Comput.*, (Special Issue on UWB Communications), August 2005, vol. 5, no. 5, pp. 503–512.

34. C.-C. Chong and S. K. Yong, "A generic statistical based UWB channel model for high-rise apartments," *IEEE Trans. Ant. Propagat.*, vol. 53, no. 8, pp. 2389–2399, August 2005.

35. C.-C. Chong, Y. Kim, and S.-S. Lee, "UWB channel model for indoor residential environment," IEEE 802.15-04-0452-01-004a, September 2004.

36. J. R. Foerster, "Channel modeling sub-committee report (final)," IEEE P802.15-02/490r1-SG3a, February 2003.

37. B. M. Donlan, S. Venkatesh, V. Bharadwaj, R. M. Buehrer, and J.-A Tsai, "The ultra-wideband indoor channel," in *Proc. IEEE Vehicular Technologies Conf. (VTC 2004-Spring)*, Milan, May 2004, pp. 208–212.

38. A. Alvarez, G. Valera, M. Lobeira, R. Torres and J. L. Garcia, "New channel impulse response model for UWB indoor system simulations," in *Proc. IEEE Vehicular Technologies Conf. (VTC 2003-Spring)*, Jeju, May 2003, pp. 1–5.

39. J. Kunisch and J. Pamp, "Measurement results and modeling aspects for the UWB radio channel," in *Proc. IEEE Conf. UWB Systems and Technologies (UWBST02)*, Baltimore, MD, May 2002, pp. 19–23.

40. A. F. Molisch, K. Balakrishnan, D. Cassioli, C.-C. Chong, S. Emami, A. Fort, J. Karedal, J. Kunisch, H. Schantz, and K. Siwiak, "A comprehensive model for ultrawideband propagation channels," in *Proc. IEEE Global Telecommunications Conf. (GLOBECOM05)*, St. Louis, MO, December 2005, pp. 3648–3653.

41. A. F. Molisch, B. Kannan, C.-C. Chong, S. Emami, A. Fort, J. Karedal, J. Kunisch, H. Schantz, U. Schuster and K. Siwiak, "IEEE 802.15.4a channel model—final report," IEEE 802.15-04-0662-00-004a, November 2004.

42. B. Kannan, C. W. Kim, X. Sun, L. C. Chiam, F. Chin, Y. H. Chew, C. C. Chai, T. T. Tjhung, X. Peng, M. Ong, and K. Sivanand, "Characterization of UWB channels: small-scale parameters for indoor and outdoor office environment," IEEE 802.15-04-0385-00-04a, July 2004.

43. R. J.-M. Cramer, R. A. Scholtz, and M. Z. Win, "Evaluation of an ultrawide-band propagation channel," *IEEE Trans. Ant. Propagat.*, vol. 50, no. 5, pp. 561–570, May 2002.
44. D. Cassioli, M. Z. Win, and A. F. Molisch, "The ultra-wide bandwidth indoor channel: from statistical model to simulations," *IEEE J. Select. Areas Commun.*, vol. 20, no. 6, pp. 1247–1257, August 2002.
45. M. Z. Win and R. A. Scholtz, "Characterization of ultra-wide bandwidth wireless indoor communications channel: a communication theoretic view," *IEEE J. Select. Areas Commun.*, vol. 20, no. 9, pp. 1613–1627, December 2002.
46. V. Hovinen, M. Hamalainen, and T. Patsi, "Ultra wideband indoor radio channel models: preliminary results," in *Proc. IEEE Conf. UWB Systems and Technologies (UWBST02)*, Baltimore, MD, May 2002, pp. 75–80.
47. S. Yano, "Investigating the ultra-wideband indoor wireless channel," in *Proc. IEEE Veh. Technol. Conf. (VTC 2002-Spring)*, Birmingham, AL, May 2002, pp. 1200–1204.
48. J. Keignart and N. Daniele, "Subnanosecond UWB channel sounding in frequency and temporal domain," in *Proc. IEEE Conf. UWB Systems and Technologies (UWBST02)*, Baltimore, MD, May 2002, pp. 25–30.
49. S. S. Ghassemzadeh, L. J. Greenstein, A. Kavcic, T. Sveinsson, and V. Tarokh, "UWB indoor path loss model for residential and commercial buildings," in *Proc. IEEE Vehicular Technologies Conf. (VTC 2003-Fall)*, Orlando, FL, September 2003, pp. 629–633.
50. Q. Li and W. S. Wong, "Measurement and analysis of the indoor UWB channel," in *Proc. IEEE Vehicular Technologies Conf. (VTC 2003-Fall)*, Orlando, FL, September 2003, pp. 1–5.
51. P. Pagani, P. Pajusco, and S. Voinot, "A study of the ultra-wide band indoor channel: propagation experiment and measurement results," in *Proc. Int. Workshop on Ultra-Wideband Systems (IWUWBS03)*, Oulu, June 2003.
52. F. Zhu, Z. Wu, and C. R. Nassar, "Generalized fading channel model with application to UWB," in *Proc. IEEE Conf. UWB Systems and Technologies (UWBST02)*, Baltimore, MD, May 2002, pp. 13–18.
53. P. Pagani and P. Pajusco, "Experimental assessment of the UWB channel variability in a dynamic indoor environment," in *Proc. IEEE Intl. Symp. Personal, Indoor and Mobile Radio Commun. (PIMRC04)*, Barcelona, September 2004, pp. 2973–2977.
54. A. H. Muqaibel, A. Safaai-Jazi, A. M. Attiya, A. Bayram, and S. M. Raid, "Path-loss and time dispersion parameters for indoor UWB propagation," *IEEE Trans. Wireless Commun.*, vol. 5, no. 3, pp. 550–559, March 2006.
55. D. R. McKinstry and R. M. Buehrer, "UWB small scale channel modeling and system performance," in *Proc. IEEE Vehicular Technologies Conf. (VTC 2003-Fall)*, Orlando, FL, September 2003, pp. 6–10.
56. J. A. Dabin, A. M. Haimovich, and H. Grebel, "A statistical ultra-wideband indoor channel model and the efffects of antenna directivity on path loss and multipath propagation," *IEEE J. Select. Areas Commun.*, vol. 24, no. 4, pp. 852–758, April 2006.
57. S. S. Ghassemzadeh, L. J. Greenstein, A. Kavcic, T. Sveinsson, and V. Tarokh, "UWB indoor delay profile model for residential and commercial environments," in *Proc. IEEE Vehicular Technologies Conf. (VTC 2003-Fall)*, Orlando, FL, September 2003, pp. 3120–3125.
58. M. Pendergrass and W. C. Beeler, "Empirically based statistical UWB channel model," IEEE P802.15-02/240SG3a, July 2002.

59. J. Keignart, J.-B. Pierrot, N. Daniele, A. Alvarez, M. Lobeira, J. L. Garcia, G. Valera, and R. P. Torres, "Radio channel sounding results and model," Deliverable D31, IST-2001-32710-U.C.A.N., November 2002.
60. J. Keignart and N. Daniele, "Channel sounding and modeling for indoor UWB communications," *International Workshop on Ultra Wide Band Systems 2003 (IWUWBS03)*, Oulu, June 2003.
61. R. M. Buehrer, W. A. Davis, A. Safaai-Jazi, and D. Sweeney, "Characterization of the ultra-wideband channel," in *Proc. IEEE Conf. UWB Systems and Technologies (UWBST03)*, Reston, VA, November 2003, pp. 26–31.
62. Virginia Tech, "UWB channel measurements and modeling for DARPA NETEX," www.darpa.mil/ato/solicit/NETEX/documents.htm.
63. T. Jämsä, V. Hovinen, and L. Hentilä, "Comparison of wideband and ultra-wideband channel measurements," COST 273 TD(04)080, Gothenburg, June 2004.
64. J. Karedal, S. Wyne, P. Almers, F. Tufvesson, and A. F. Molisch, "Statistical analysis of the UWB channel in an industrial environment," in *Proc. IEEE Vehicular Technologies Conf. (VTC 2004-Fall)*, Los Angeles, CA, September 2004, pp. 81–85.
65. J. Karedal, S. Wyne, P. Almers, F. Tufvesson, and A. F. Molisch, "UWB channel measurements in an industrial environment," in *Proc. IEEE Global Telecommunications Conf. (GLOBECOM04)*, Dallas, TX, November 2004, pp. 3511–3516.
66. B. Kannan, C. W. Kim, X. Sun, L. C. Chiam, F. Chin, Y. H. Chew, C. C. Chai, T. T. Tjhung, X. Peng, M. Ong, and K. Sivanand, "UWB channel characterization in outdoor environments," IEEE 802.15-04-0440-00-04a, September 2004.
67. M. Z. Win, G. Chrisikos, and N. R. Sollenberger, "Performance of Rake reception in dense multipath channels: Implications of spreading bandwidth and selection diversity order," *IEEE J. Select. Areas Commun.*, vol. 18, no. 8, August 2000, pp. 1516–1525.

CHAPTER 9

MIMO and UWB

THOMAS KAISER

9.1 INTRODUCTION

UWB and MIMO systems are key technologies for future high-data-rate wireless communications. MIMO systems can be divided into three major categories:

- Spatial multiplexing;
- Spatial diversity;
- Beamforming.

While *beamforming* requires correlated wavefronts to reduce interference and enhance coverage, *space-time coding* or *spatial multiplexing* schemes demand distinctive scattering environments to improve link quality or increase data rates, respectively. In turn, UWB communication systems can be suitably classified by the type of modulation:

- Impulse radio with time domain equalization (e.g., rake receiver);
- Single carrier transmission with frequency domain equalization (SC-FDE);
- Multicarrier transmission supported by appropriate coding.

While IR can be seen as the classical UWB approach, the multicarrier schemes offer—despite their potential high number of subcarriers—several advantages for applications due to their flexible configurability. In contrast, single carrier schemes with frequency domain equalization have not attracted much attention yet in UWB communications. Note that UWB wireless systems mainly target *extremely high data rates* for indoor communications, but *positioning* with even subcentimeter resolution and *sensing* are considered as having great potential for future applications as well.

Ultra Wideband Wireless Communication. Edited by Arslan, Chen, and Di Benedetto
Copyright © 2006 John Wiley & Sons, Inc.

At first glance, the aims of MIMO and UWB do partially overlap, mainly with focus on high data rate, but are also partially unrelated. Hence, the aim of this contribution is to first shed light on the marriage of MIMO with UWB. The acronym used in the following is MIMO&UWB, under which all combinations of the above categories are summarized, for example, beamforming with impulse radio or multiple carrier systems with spatial multiplexing.

This chapter is organized as follows. First, the potential benefits of MIMO&UWB in terms of range extension, data-rate improvement, interference rejection, and potential technological simplifications are briefly introduced. Second, a literature review on ultra wideband multiantenna techniques, subdivided in spatial multiplexing, spatial diversity, beamforming, and related topics is given to illustrate the state of the art. Then, spatial ultra wideband channel measurements and modeling are highlighted to provide a solid basis for algorithmic design of MIMO&UWB transceivers. Finally, our own recent results on MIMO&UWB are presented. This contribution closes by an outlook summarizing the major challenges in this interdisciplinary research field.

9.2 POTENTIAL BENEFITS OF MIMO AND UWB

At any large conference on communications, special sessions on ultra-wideband systems and multiantenna techniques are rather common, indicating that both approaches belong to the few key emerging technologies in wireless communications. Against this background the question arises as to whether their combination is principally feasible and, if so, what are the possibilities. Some potential benefits of MIMO&UWB can be easily foreseen:

- *Greater Coverage*—because of the low transmit power spectral density, the coverage of UWB communication systems is rather limited. Multiantenna systems (MAS) provide an *array gain* of 3 dB (or even up to 6 dB peak gain for a large *bandwidth-antenna spacing*, see below) per doubling of the number of antennas, therefore increase the coverage. In principle, a potential *diversity gain* can alternatively boost the SNR and therefore indirectly enhance the coverage. Both properties have to be further examined, however. First, because the diversity gain tends to degrade with increasing bandwidth B simply due to diminished small-scale fading, and second, since the 6 dB peak gain has to be proven still in practice. Lastly, multiple antennas at the transmitter, for example, for beamforming, may potentially violate the EIRP (effective isotropic radiated power) requirements dictated by UWB regulations.
- *Highest Data Rate*—it has been recently shown [1] that the MIMO&UWB ergodic channel capacity increases approximately linearly with the minimum number of transmit and receive antennas under typical UWB Nakagami fading channels. Since the same property for narrow or wideband MIMO Rayleigh fading channels has motivated the enormous research activities on

MIMO communication systems, MIMO will also become of relevance for ultra wideband channels as soon as spectral efficiency matters. Moreover, typical UWB impulses of short duration admit a high time resolution and render up to several tens or even hundreds of resolvable paths in typical indoor environments. As a result, such rich scattering makes the adoption of spatial multiplexing schemes for UWB systems rather promising, since signals launched from different spatial positions can probably be distinguished by their different spatial signature. Lastly, since the demand on data rate has increased exponentially with time for more than 20 years [2], MIMO&UWB systems will probably become relevant for practice earlier than in the far future. Note that MIMO&UWB promises a data rate close to the highest possible data rate for wireless indoor communications, since most of the frequencies penetrating through walls (say below 10 GHz) are occupied by UWB, while MIMO further pushes spectral efficiency to its limits.

- *Interference Rejection*—in addition to the deployment of beamforming for coverage extension, it can also significantly reduce interferers by *nulling* of the most dominant and undesired interfering wavefronts impinging on the receiver array. In case of UWB, such nulling must be generally[1] valid for a large frequency range, which requires adequate filters on behalf of single true time delays or even simplistic phase shifters in each antenna branch. These filters might be of analog type, because digital beamforming is prohibited for today's applications due to necessary sampling frequencies in the double-digit giga-Hertz domain. Moreover, although the channel coherence time in indoor environments is large, the analog filter coefficients should be tracked continuously, which further complicates a successful realization. However, for research purposes, a full digital recording is even possible today with state of the art digital storage oscilloscopes. Such a sophisticated device allows not only for analyzing all analog signals inherent in the transmission link, but even more importantly, it enables arbitrary signal processing applied to the stored data. A stepwise optimization of a MIMO&UWB system becomes feasible. Last but not least, the dense multipath environment may severely limit the usefulness of UWB beamforming, but the so-called *double-dB gain* and the *clustering phenomenon* in UWB indoor propagation may favor it. Here again, more research is required in order to gain deeper insight and reliable general statements.

- *Technological Aspects*—the technical feasibility of UWB is by definition a question of bandwidth. For example, the design of compact antennas and amplifiers for high bandwidths and, at the same time, large gain and low power consumption, represents a current research topic and a technological challenge. Moreover, for multi-carrier schemes the FFT consumes more digital hardware resources with increasing bandwidth because of subcarrier multiplicity. Taking this into account, MIMO&UWB may even offer a technological alternative to

---

[1] Except for narrowband interferers like WLAN access points.

an UWB single antenna system with doubled bandwidth requirements. This technical aspect is of low relevance for narrowband systems with its comparably low bandwidth requirements (in the order of a very few tens of MHz), but might be worthwhile for MIMO&UWB; for instance, due to the bisection of bandwidth while keeping the data rate the same, the amplifier gain is approximately doubled[2] and the halved antenna bandwidth further relieves the technological burdens. Moreover, even signal processing hardware may become less costly; for example, suppose P subcarriers, then two transmit/receive antennas, admit only two $P/2$ point FFTs instead of one P point FFT for a SISO (single input single output) link. Hence, in contrast to narrowband systems, where MIMO systems seem to be successful only if spectral efficiency reaches its limits, here MIMO may even facilitate the technical design of UWB systems to some extent.

Finally, we have to define what is meant by "MIMO&UWB". Current WLANs operate at a bandwidth of 20 MHz, optionally at 40 MHz under the latest standard (www.heise/newsticker/meldung/50045), and aiming at a 100 MHz bandwidth. Per FCC definition, UWB communication systems start at a minimal bandwidth of 500 MHz and end at 7.5 GHz. Hence, we define MIMO&UWB as *any multiantenna technique requiring a minimum bandwidth of 500 MHz*.

In conclusion, the vital question of MIMO&UWB is how conventional multiantenna-based techniques, that is, spatial multiplexing, spatial diversity, and beamforming, depend on bandwidth by taking into account not only the dependence of the spatial channel on bandwidth, but also related impacts of analog and digital transceiver hardware.

9.3 LITERATURE REVIEW OF UWB MULTIANTENNA TECHNIQUES

The aim of this section is to give a short overview about the state of the art in MIMO&UWB by briefly summarizing the major results of several references. Because MIMO&UWB can be understood as the bandwidth dependance of conventional MIMO, this section is again subdivided according to the three major MIMO categories.

9.3.1 Spatial Multiplexing

There are only a few references on spatial multiplexing for UWB systems. Recently, we demonstrated [1] that by using N transmit and N receive antennas the MIMO&UWB ergodic channel capacity increases linearly with N. This confirms our conjecture that MIMO and UWB are in this way complementary to each other. Keeping in mind that frequencies above 10 GHz are severely attenuated in

---

[2]Keeping costs fixed, the gain–bandwidth product of an amplifier is approximately constant.

indoor environments because of walls etc., this result leads us to another conclusion: The FCC-regulated frequency range from 3.1 GHz to 10.6 GHz already occupies 70% of the whole available bandwidth; hence, the marriage of MIMO with UWB approaches *the upper physical limit in terms of maximum indoor data rate*. If we assume the validity of *Edholm's law of bandwidth* [2] for the next decade, data rates of several tens of Gbps can theoretically be expected from MIMO&UWB. A possible application is, therefore, ultrafast wireless data exchange of huge amounts of data (e.g., video movies) between portable devices in an extremely short time. First investigations of the VBLAST (vertical Bell Laboratory layered space–time) algorithm in conjunction with UWB systems are given in Kumar and Buehrer [3].

9.3.2 Spatial Diversity

Analog UWB space–time coding (STC) schemes provide a diversity gain by exploitation of spatial diversity and simultaneously balance fluctuations in sampling points to a certain extent [4, 5]. A possible unfavorable electromagnetic coupling between antenna elements has been proven to be small, even for marginal antenna separations [6]. Simulation results given in Sibille [7] confirm that the *diversity order* of UWB-MAS equals the product $N : M$ (N, M are the number of transmit and receive antennas, respectively) under the assumption of an unrealistic single-path channel model. In the case of multipath propagation, the diversity order constitutes $N\ M\ L$, where L is the number of resolvable multipaths. Therefore, an increment of the number of antennas promises a limited diversity gain due to the already existing distinct UWB multipath diversity [8]. UWB systems with a receive antenna array [SIMO&UWB (single input multiple output)] are discussed in Tan et al. [9], where multiuser interferences and multipaths are taken into consideration. Even though a power increment (array gain) by multiple antennas is noticed, a general investigation on bandwidth dependence substantiated by quantitative results is missing so far. In general, it is evident that a boost of bandwidth is tendentially accompanied by a diminished small-scale fading. Hence, a *threshold region* will probably exist. Exceeding this region makes SIMO/MISO&UWB (multiple input single output) less promising because only the array gain persists.

9.3.3 Beamforming

UWB systems for indoor communications are characterized by a dense multipath environment. At a first glance, the huge number of multipaths seems to limit the usefulness of conventional beamforming algorithms. Even if the mainlobe perfectly steers towards the direction of arrival (DoA) of the desired signal, all the other numerous paths are interfering, so that the array gain might be too small for successful signal separation. However, there are at least two reasons to shed more light on this topic:

- First, several test series [10, 11] have documented that in typical UWB indoor environments the multipaths occur in a *clustered way*, meaning that the transmitted signal impinges from a few relevant main directions. Based on this observation, some type of *cluster beamforming* becomes attractive, where the mainlobe width fits the cluster width so as to collect enough relevant energy, while the undesired clusters are suppressed with the array gain. In this way, not only the signal-to-interference-plus-noise ratio (SINR) will be increased (the multiple paths are considered as interferers here), but also the channel delay spread will be shortened, so that higher data rates become feasible to some extent. Moreover, in a multiuser scenario true interferers might be suppressed. Only two references have been published on UWB beamforming [12, 13], but there are numerous contributions covering wideband DoA estimation [14–17] and wideband beamforming [18–22].

- Second, UWB beamforming shows one special property—the so-called *double-dB gain* [23], which will be explained later in more detail. For the moment notice only that, due to an extremely short pulse duration w.r.t. the travel time accross the array, the conventional array gain might be doubled on a decibel scale, meaning squaring on a linear scale. Hence, interferers can be superiorly cancelled, making UWB beamforming worthwhile even in a dense multipath environment.

To avoid reinventing the wheel, several ideas can be applied from acoustic indoor transmission, where the environment is also rich in terms of scattering, the relative bandwidth is larger than 100%, and beamforming is deployed by microphone arrays for SNR improvements. This analogy can be substantiated by the following table.

First, it is obvious that both delay spreads divided by the corresponding wave velocity are of the same magnitude. Hence, both channels are characterized by dense multipath propagation, if the bandwidth is fully exploited. Moreover, indoor environments are non-stationary, that is, time-variant, and the wavefronts are of spherical shape in general, but can be approximated as planar if the receive array size is much smaller than the distance to the transmitter. Also, the typical SNR (including spreading gain) is a few decibels to insure satisfying speech recovery or moderate bit error rates. Besides these similarities, there are some differences. While for acoustical processing the complete signal shape has to be determined in general, in UWB processing data detection is sufficient. In addition, the type of distortion is different: broadband interferers and correlated noise are dominant in acoustic transmissions, whereas narrowband and UWB interference, and uncorrelated receiver noise characterize electromagnetic transmission. Last but not least, a voice activity detector is needed for speech reconstruction, because signal duration is not known, while a synchronization circuit is sufficient for UWB data transfer. In short, several ideas from microphone array processing can be applied to UWB beamforming.

9.3.4 Related Topics

In addition to the benefits of MIMO&UWB for communications, ultra wideband multiantenna techniques can be also applied for ranging [24] and for sensing.

TABLE 9.1 Speech Processing vs UWB

	Speech Processing	UWB
Bandwidth	~10 kHz (180%)	~10 GHz (110%)
Delay spread	~100 ms	~100 ns
Environment	Indoor, nonstationary	Indoor, nonstationary
Challenges	Highly reverberant	Multipath propagation
Wavefront	Spherical or planar?	Spherical or planar?
Typical SNR	6–12 dB (moderately positive)	6–12 dB
Type	Waveform estimation	Detection (matched filter)
Interferer	Broadband	Broadband and narrowband
Noise	Correlated (e.g., ventilator)	Uncorrelated (thermal)
Other	Signal duration not known	Signals duration known

Ultra short pulses allow spatial resolution even down to subcentimeter range and are therefore another current UWB research topic (see e.g., [25]). By means of multi-antenna techniques, additional spatial parameters can be extracted (e.g., direction of arrival or direction of departure), leading to enhanced precision in ranging. Initial applications cover the detection of breast cancer [26] or mine localization [27], as well as the detection of fires by active UWB radiation [28].

9.4 SPATIAL CHANNEL MEASUREMENTS AND MODELING

Without doubt, *spatial channel measurements* and *spatial channel modeling* are of particular importance for a thorough investigation of MIMO&UWB systems. For that reason we briefly revisit these two topics.

9.4.1 Spatial Channel Measurements

Beside basic theoretical considerations, *channel measurements* are of particular importance to meet the claim of a comprehensive investigation of MIMO&UWB, especially as a function of bandwidth. The *channel models* introduced in the following sections only apply to certain bandwidths, and therefore just admit random checks of ulterior gained insights w.r.t. bandwidth. In the following, two databases are illustrated, where the second one is provided for unrestricted use (www.impulse.usc.edu):

- *IMST Database*—in 2001 the IMST GmbH (Kamp-Lintfort, Germany) accomplished spatial UWB channel measurements in the frequency range between 1 GHz and 11 GHz within the scope of the European whyless.com project. The transmitter was positioned on a 1.5 m × 30 cm rectangle with a grid width of 1 cm. This resulted in 4681 different transmitter positions for a single scenario. The receiver was fixed on a single spot. Therefore the measurements basically reflect a MISO&UWB system, which—thanks to the antenna and channel reciprocity—could be also interpreted as a SIMO&UWB system. Altogether four different scenarios were recorded:

- Pure LOS.
- Mixed LOS/NLOS, that is, parts of the transmitter positions have inter visibility with the receiver, others do not;
- Pure NLOS, transmitter and receiver in the same room;
- Pure NLOS, transmitter and receiver in different rooms.

Furthermore, it is worth mentioning that these channel measurements are also suitable for analyzing MIMO scenarios under LOS conditions to some extent, since numerous MIMO impulse responses between grid points on the above-quoted rectangle could be recovered by convolution via the receiver. Since all measurements are carried out separately by systematic adjustment of single transmit antennas, the recorded data does not contain any coupling among antennas. As mentioned earlier the antenna coupling in ultra wideband transmission can be neglected even for distances no larger than a few centimeters.

- *Intel Database*—the Intel Corporation (San Jose, California) carried out 730,000 MIMO&UWB indoor channel measurements in the frequency range between 2 GHz and 8 GHz. The magnitude and phase of the complex baseband channel impulse response are available for numerous different LOS and NLOS scenarios. A huge number of different measurements features reliable resulted, for example, for channel modeling or algorithm testing. The measuring setup contained an antenna installed under the ceiling (to simulate a wireless access point, AP) and a mobile antenna to simulate a free to move radio unit (MRU). The AP and the MRU were positioned on a 1.83 m × 91 cm and a 46 cm × 46 cm large square, respectively and their distance was varied between 3 m and 25 m.

A first analysis of the IMST measurement data was conducted in our department concerning the correlation for different environments and bandwidths (in analogy to the simulation results given in Weisenhorn and Hirt [8]). The normalized correlation is

$$\varphi_{h_1,h_2}(\Delta_d, \tau) = \frac{|\int h_1^*(t) h_2(t+\tau)\,dt|}{\sqrt{\int |h_1(t)|^2\,dt}\sqrt{\int |h_2(t)|^2\,dt}}, \qquad (9.1)$$

where $h_1(t)$, $h_2(t)$ are two impulse responses measured between two transmitters spaced by Δ_d and one receiver. Figure 9.1 shows the results for an LOS and an NLOS environment with bandwidths of 500 MHz and 7.5 GHz. Obviously, the dominant path in the LOS environment yields an extensive correlation for both bandwidths. Hence, under these conditions beamforming seems to be promising for SNR improvement. In pure NLOS environments, such a clear characteristic cannot be observed. Note also, that the normalized correlation for 500 MHz decreases at a much slower rate than that for 7.5 GHz. Such spatially correlated wavefronts limit

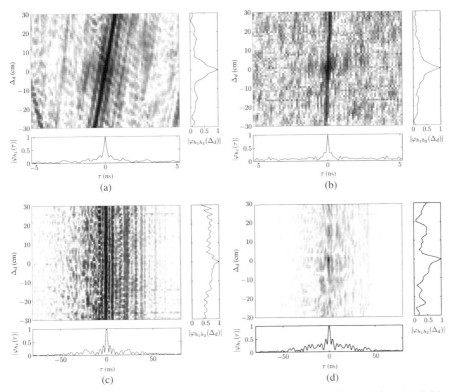

Figure 9.1 Spatial Correlation for different environments and bandwidths. (a) LOS, 7.5 GHz; (b) NLOS, 7.5 GHz; (c) LOS, 500 MHz; (d) NLOS, 500 MHz.

the data rate improvement of MIMO algorithms under the theoretical maximum in case of no correlation. For example, numerous narrowband measurements [29, 30] have shown that only approximately 70% of channel capacity can be achieved because of such spatially correlated wavefronts. Hence, MIMO&UWB systems seem to converge closer to the theoretical maximum, simply because of larger bandwidth, but further investigations are without doubt necessary for statements of more general value.

9.4.2 Spatial Channel Modeling

Channel models reflect major properties of recorded measurement data in order to admit coarse forecasts of algorithms performance. As mentioned earlier, the UWB channel is characterized by a dense multipath propagation and the clustering phenomenon. The IEEE standard model 802.15.3a [31] reflects both properties adequately. It is based on the Saleh–Valenzuela model being widely used for indoor channel modeling. The channel impulse response is composed of L clusters and

K rays in one cluster as follows:

$$h(t) = \sum_{l=1}^{L} \sum_{k=1}^{K} \alpha_{k,l} \delta(t - T_l - \tau_{k,l}), \tag{9.2}$$

where $\alpha_{k,l}$ denotes the amplitude of the kth ray of cluster l, T_l is the arrival time of the lth cluster and $\tau_{k,l}$ is the delay of ray k of cluster l relative to the cluster start T_l. Although this standard model mirrors important characteristics of the UWB channel, for example, delay spread and power delay profile, the spatial component is still entirely missing. A straightforward generalization [11] motivates the spatial channel impulse response:

$$h(t,\theta) = \sum_{l=1}^{L} \sum_{k=1}^{K} \alpha_{k,l} \delta(t - T_l - \tau_{k,l}, \theta - \Theta_l - \vartheta_{k,l}), \tag{9.3}$$

which assigns a DoA $\Theta_l + \vartheta_{k,l}$ to every ray. The *average angle* Θ_l of the lth cluster is assumed to be uniformly distributed, whereas the *individual angles* $\vartheta_{k,l}$ are modeled as Laplace-distributed random variables with constant variances being independent on k and l. Note that this coarse spatial model reflects neither the antenna spacings nor their individual positions with respect to the transmitter. Another model [32], relying on a quite different philosophy, avoids these disadvantages. Here, the spatial channel impulse response is composed of a statistical component for reflecting the dense multipaths and a simple quasideterministic method to model distinct individual echoes. For that reason, so-called *virtual transmitters (sources)* are placed in a *generic room* and are obtained by mirroring the actual transmitter positions, r_t, at the walls of this room. These mirror positions allow the generation of relevant echoes; its selection is still the object of current research. The distance $d(k, r_r)$ between the kth virtual source and the receiver, r_r, leads to the delay

$$\tau(k, r_r) = \frac{d(k, r_r)}{c}$$

and the corresponding magnitude

$$G(k, r_r) = \left(\frac{c}{4\pi d_0 F}\right)^2 \left[\frac{d(k, r_r)}{d_0}\right]^{-\alpha}$$

of the respective kth path. Note that α is the path loss, F, is the center frequency and d_0 is a reference distance which allows further model flexibility. In order to incorporate the clustering phenomenon, any specific echo is accompanied by a cluster of multipaths, similar to model (9.3). Observe, that the resulting impulse response depends on the transmitter and receiver position, r_t and r_r. Hence, the model is well suited for MIMO&UWB [8]; however, the explicit dependency on bandwidth is missing, or, more precisely, the upper and lower cut-off frequencies are not yet incorporated. Although this deficiency can be mitigated by proper shaping of the individual signals traveling along the paths, the above model neither includes dispersion caused through the walls, nor the bandwidth dependence in the magnitude $G(k, r_r)$.

9.5 SPATIAL MULTIPLEXING

As outlined in the introduction, we now proceed to the discussion of preliminary results on MIMO&UWB. Starting with spatial multiplexing and related channel capacity issues, we continue with spatial diversity aspects and conclude with UWB beamforming and direction of arrival estimation.

It has been recently demonstrated that the ergodic channel capacity increases linearly with N, even for ultra wideband transmission, using N transmit and N receive antennas. This result corresponds to the narrowband case with Rayleigh or Ricean fading, while for an UWB channel Nakagami fading seems to be more appropriate, since the number of scattering components within one resolvable path is of the average order of only two to three. Hence, the central limit theorem causing the Rayleigh or Ricean distribution for the narrowband case is not longer applicable. An alternative to the Nakagami probability density function (p.d.f.) is the log–normal p.d.f. because both pass a Kolmogorov–Smirnov test with a significance level of 1% [31]. However, similar results of general significance for the log–normal distribution turn out to be more difficult to obtain. The far-reaching statement given above can be proven in *closed form* for the rather unrealistic assumption of a frequency flat channel, where the ergodic channel capacity C_e is expressed as

$$\lim_{N \to \infty} \frac{C_e}{NB} = \log(\mathrm{SNR}_1) - 1 + \frac{\sqrt{1 + 4\mathrm{SNR}_1} - 1}{2\mathrm{SNR}_1} + 2\tanh^{-1}\frac{1}{\sqrt{1 + 4\mathrm{SNR}_1}} \quad (9.4)$$

with SNR_1 as a modified SNR including the fading coefficients power. Observe that C_e increases linearly with N for a huge number of antennas. This result can be generalized for a two-path channel also in closed form. For arbitrary frequency selective channels only a numerical solution for calculating the outage probability P_{out} is derived. The behavior is illustrated in the following two figures, which show on the one hand the bit rate scaled by the bandwidth vs the number of antennas for selected outage probabilities and on the other hand the outage probability for a selected number of antennas. Apparently, even for $N \times N$ MIMO&UWB transmission and frequency selective fading the bit rate increases linearly with slope one against the antenna number N. This result also applies to a small number of antennas and therefore extends Equation (9.4) in a straightforward manner. Consequently, in terms of *spatial multiplexing*, MIMO&UWB systems promise the same improvement of spectral efficiency as conventional narrowband MIMO systems.

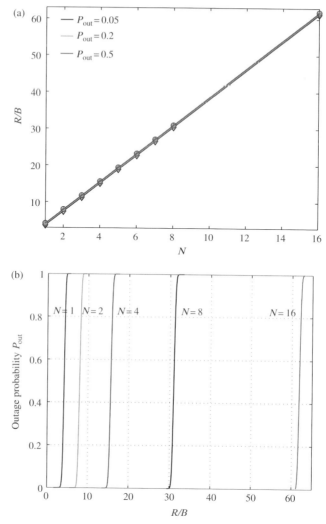

Figure 9.2 Ergodic channel capacity of UWB communication systems under Nakagami frequency selective fading. (a) Scaled bit rate vs number of transmit and receive antennas. (b) Outage probability vs scaled bit rate.

9.6 SPATIAL DIVERSITY

As mentioned earlier, the usefulness of spatial diversity decreases with increasing bandwidth, since signal cancelation caused by fading becomes more unlikely with larger bandwidth. Therefore, the central question is to find the *threshold bandwidth* to which spatial diversity is still worth exploiting. In order to avoid any misleading array gain and to investigate the diversity gain only, Alamouti-STC with two

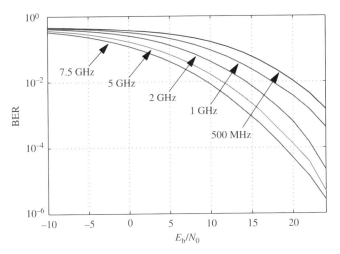

Figure 9.3 Alamouti-STC for different bandwidths.

transmit and a single receive antenna has been investigated, so that even channel state information (CSI) at the transmitter is useless (for that reason the array gain equals zero). The frequency selective UWB channel is divided into numerous frequency flat subchannels with an individual bandwidth of 6.25 MHz via OFDM. Figure 9.3 shows the bit error ratio obtained by real channel measurements and permits a preliminary conclusion about spatial diversity gain. Note that the transmit power doubles when the bandwidth is doubled and therefore a gain of 3 dB can be noticed, but not more. Even for a low UWB bandwidth of 500 MHz the diversity gain seems to be negligible. Again, a closer investigation is recommended since these results are not representative. For that reason, suppose that for each subcarrier the magnitude of the fading coefficient a is described by the Nakagami distribution

$$p_{|a|}(x) = \begin{cases} 2\left(\frac{\kappa}{2\Omega}\right)^{\kappa/2} \frac{1}{\Gamma\left(\frac{\kappa}{2}\right)} x^{\kappa-1} e^{-\kappa x^2/2\Omega} & \text{when } x \geq 0, \\ 0 & \text{when } x < 0 \end{cases} \quad (9.5)$$

where $\kappa \geq 1$, and the sign of a is equally distributed on $+1$ and -1. Note that the parameter Ω is related to the strength of the fading a as

$$\Omega = \mathcal{E}(a^2),$$

and the parameter κ is related with the variance of a in a complex way, which can be written as

$$\text{var}(a) = \left[1 - \left(\frac{\Gamma(2\kappa + \frac{1}{2})}{\Gamma(2\kappa)}\right)^2 \frac{1}{2\kappa}\right] \Omega. \quad (9.6)$$

It can be shown that for *fixed* Ω, the variance of a decreases monotonically with κ, achieving its maximum value when $\kappa = 1$. On the other hand, we have argued that, with the increase in the system bandwidth, the channel shows less fading. In the extreme case, when the system bandwidth approaches infinity, there will be no fading so that κ should approach infinity. Thus, we will have var$(a) \to 0$, as can be seen from Equation (9.6). Therefore the parameter κ should be related to the system bandwidth somehow, but an explicit relationship is not yet known. Hence, we preliminarily assume a simple linear dependence described by

$$\kappa = 1 + \beta(B - B_{\min}), \tag{9.7}$$

where β is a suitable scaling parameter and $B_{\min}$ is the minimum bandwidth, below which the channel model passes over from UWB to narrowband or wideband systems. Hence, $B_{\min} = 100\,\text{MHz}$ seems to be reasonable. Since measurements have shown that κ takes typical values from 1 to 9 for a system bandwidth of $B = 500\,\text{MHz}$, it follows that $\beta = 1/100\,\text{MHz}$ seems to be an appropriate choice. Contrary to the previous example on transmit diversity, we are now able to investigate also the case of receive diversity. Maximum ratio combining (MRC) is applied in the following yielding an array gain of 3 dB per each antenna doubling and undetermined diversity gain. The following figures show the BER as a function of SNR for one transmit, two receive antennas, bandwidths of 500 MHz, 7.5 GHz, and $\Omega = 1, 10$. Note that here κ is given by Equation (9.7).

From Figures 9.4 and 9.5, it can be observed that, for $\Omega = 1$ and for wideband wireless communications with 20 MHz bandwidth, the diversity gain obtained by two receive antennas with the MRC algorithm only about 1 dB for low SNR

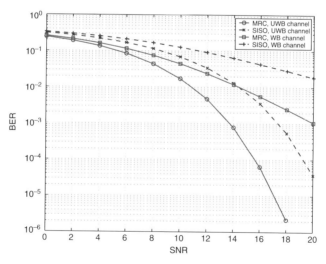

Figure 9.4 UWB SIMO wireless systems compared with wideband (WB) SIMO wireless systems for $B = 500\,\text{MHz}$ and $\Omega = 1$.

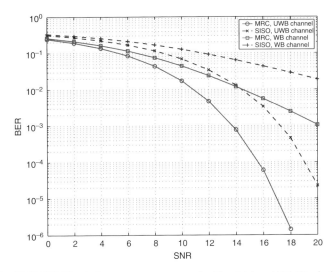

Figure 9.5 UWB SIMO wireless systems compared with wideband SIMO wireless systems for $B = 7.5$ GHz and $\Omega = 1$.

(<10 dB), but becomes remarkable for higher SNR. In contrast, for UWB wireless communications, only a marginal diversity gain can be observed for both bandwidths. However, by significantly increasing the fading by expanding Ω to 10, significant diversity gains can be achieved also for UWB wireless systems (Figures 9.6 and 9.7) almost irrespective of the bandwidth.

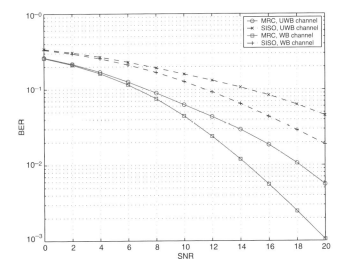

Figure 9.6 UWB SIMO wireless systems compared with WB SIMO wireless systems for $B = 500$ MHz and $\Omega = 10$.

220 MIMO AND UWB

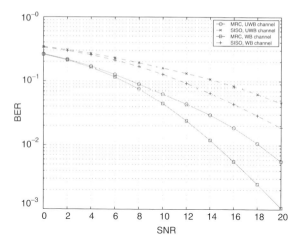

Figure 9.7 UWB SIMO wireless systems compared with WB SIMO wireless systems for $B = 7.5$ GHz and $\Omega = 10$.

The above observations may raise some doubt about the usefulness of equipping a UWB wireless system with multiple receive antennas to combat fading, but further work is still needed here to validate this conjecture in detail. For example, the above model (9.7) for κ is more or less heuristic; hence, more attention should focus on model validation for κ or even Ω as a function of bandwidth. Then, based on such models, the diversity gain cannot only be investigated by measurements, but also by analytical methods. All these steps are currently under progress in our UWB research group.

9.7 BEAMFORMING

Whereas the properties of spatial multiplexing in MIMO&UWB are quite obvious and for spatial diversity the threshold bandwidth is the major open issue, the benefits of UWB beamforming are harder to examine. While the beampattern of an UWB impulse beamformer can be analytically investigated and shows some attractive properties [33, 34]. The identification of all directions of arrivals seems to be challenging in a dense multipath environment. However, due to the clustering phenomenon—well-known from cellular wireless communication systems—the most relevant bearings can probably be determined in the case of UWB transmission even in an NLOS indoor environment [35]. In order to substantiate this, Figures 9.8 and 9.9 show the time-dependent beampattern (BP)

$$\mathrm{BP}(t_0, \Delta t, \theta) := \left(\int_{t_0}^{t_0 + \Delta t} |y(t, \theta)|^2 \, dt \right)^{\frac{1}{2}}, \qquad (9.8)$$

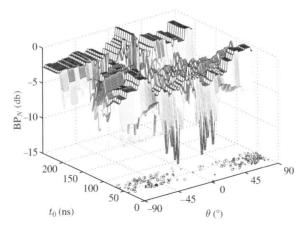

Figure 9.8 Normalized beampattern BP(t_0, Δt, θ)/ $\max_\theta$ [BP(t_0, Δt, θ)] for an LOS environment. $\Delta t = 50$ ns and $t_0 = 0$ ns (5 ns) 240 ns.

normalized to its maximum magnitude w.r.t. θ. Here, $y(t)$ is the output of a conventional *delay and sum beamformer*, Δt the observation interval, and t_0 represents the time dependence.

The plus in the ground plane indicates a positive path amplitude of the spatial impulse response, whereas the circle stands for a negative path amplitude. First, observe that the LOS can be easily detected, and second, even in an NLOS environment, the relevant cluster bearings can be identified. These observations confirm our previous statement (for succeeding discussions see Ries et al. [34]).

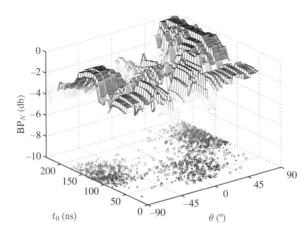

Figure 9.9 Normalized beampattern BP(t_0, Δt, θ)/ $\max_\theta$ [BP(t_0, Δt, θ)] for an NLOS environment. $\Delta t = 50$ ns and $t_0 = 0$ ns (5 ns) 220 ns.

Since the estimation of cluster bearings seems not to be unrealistic, UWB beamforming is also worth investigating further. In fact, because the UWB impulse duration is shorter then the average travel time across the array, UWB beamforming shows several interesting phenomena. These are briefly summarized as follows:

- *Mainlobe Width*—it can be proven that the UWB mainlobe width greatly depends on the ratio of center wavelength over array size, which corresponds to conventional narrow- and broadband beamforming.
- *Grating Lobes*—grating lobes disappear for UWB signals because of broad averaging over frequency. Taking into account also the previous statement, the UWB mainlobe width can be made almost arbitrarily narrow simply by increasing the spacing between two antennas. This is opposite to narrow- and broadband beamforming, where more antennas are needed to sharpen the mainlobe without undesired ambiguities caused by grating lobes.
- *Sidelobe Level*—the UWB sidelobe level is fixed for a rather wide range of bearings. This occurs because, for bearings outside the mainlobe direction, the travel time across the array may become larger than the short impulse duration, so that an overlap at the beamformers output is prevented. Hence, for UWB beamforming *unequal prefiltering* among the antenna branches is not really meaningful. This is in contrast to narrow- and broadband beamforming, where unequal prefilters are used to enhance the main- to sidelobe ratio.
- *Peak Gain*—the UWB peak gain is doubled compared with conventional narrow- and wideband beamforming. This feature is of particular relevance for applications because it principally allows for halving the number of antennas. It can be exemplified as follows. Suppose a delay-and-sum beamformer with two antennas, large antenna spacing and zero delays. Under this set-up a pulse-shaped wavefront arriving from the broadside is added constructively by the beamformer, leading to an output signal with doubled magnitude and therefore four times the peak (or instantaneous) power of the signal itself. In contrast, for another pulse-shaped wavefront arriving from end-fire, the two pulses traveling along the two antenna branches do not overlap at the beamformer output (because of the large antenna spacing), so that the peak power equals the single signal power only. Hence, even with only two antennas, the peak power *ratio* is four, which means a *doubled-dB gain*.
- *Interference Rejection*—opposite to sidelobe level suppression, rejection of UWB interferers requires filters on behalf of delays in each antenna branch in order to enable nulling over a wide frequency range for a few selected directions. However, since UWB is characterized by a dense multipath environment, distinct interference rejection is hard to achieve.

In conclusion, UWB beamforming shows several interesting properties, simply because the travel time across the array is larger than the pulse duration. Related proofs, more simulations, and further discussions are given in Ries and Kaiser [34].

9.8 CONCLUSION AND OUTLOOK

The aim of this contribution is to highlight major features of MIMO&UWB. Some preliminary conclusions are:

- A *spatial channel model* adequately reflecting the dependence of bandwidth is still missing; this is a precondition for a thorough analytical and numerical investigation of MIMO&UWB.
- The linear dependence of the ergodic channel capacity on the number of transmit/receive antennas principally holds true also for MIMO&UWB.
- MIMO&UWB seems to be a promising approach for achieving highest data rates and extending coverage, but noticeable improvements of link quality by exploiting spatial diversity should not be expected.
- MIMO&UWB represents the upper physical bound for data rates in indoor environments.

However, in order to substantiate these conclusions and to extend them further, numerous issues have to be addressed in more detail, for example:

- Adequate transceiver architectures for spatial multiplexing and space time coding;
- Advanced cluster beamforming;
- Interference cancellation by UWB beamforming with low number of antennas;
- Transmit beamforming, digital beamforming;
- Adjustable precise *analog* true time delays;
- Pulse shaping, modulation, and synchronization for MIMO&UWB systems;
- Ranging with MIMO&UWB systems;
- Technical aspects, impairments and their cancellation of MIMO&UWB systems;
- Multiuser MIMO&UWB systems.

In conclusion, MIMO&UWB constitutes an interdisciplinary field of challenging long term strategic research with rich potential for future applications.

REFERENCES

1. Z. Feng and T. Kaiser, On channel capacity of multi-antenna UWB indoor wireless systems, *IEEE International Symposium on Spread Spectrum Techniques and Applications*, 30 August to 2 September, 2004, Sydney.
2. S. Cherry, Edholm's law of bandwidth, *IEEE Spectrum*, vol. 41, no. 7, 2004 July, pp. 58–60.
3. N. A. Kumar and R. M. Buehrer, Application of layered space–time processing to ultra-wideband communication, *IEEE Midwest Symposium on Circuits and Systems*, Tulsa, OK, 4–7 August, 2002, pp. 597–600.

4. L. Yang and G. B. Giannakis, Space–time coding for impulse radio, *IEEE Conf. on Ultra Wideband Systems and Technologies*, 2002.
5. L. Yang and G. B. Giannakis, Analog space–time coding for multi-antenna ultra-wideband transmissions, *IEEE Transactions on Communications*, vol. 24, no. 3, pp. 507–517, March 2004.
6. A. Sibille and S. Bories, Spatial diversity for UWB communications, *5th European Personal Mobile Communications Conf.*, Glasgow, April 2003.
7. A. Sibille, MIMO diversity for ultra wide band communications, Technical Report COST 273 TD (03) 071, Barcelona, 2003.
8. M. Weisenhorn and W. Hirt, Performance of binary antipodal signaling over the indoor UWB MIMO channel, *IEEE Int. Conf. on Communications*, vol. 26, no. 1, pp. 2872–2878, May 2003.
9. S. Tan, B. Kannan, and A. Nallanathan, Performance of UWB multiple access impulse radio systems in multipath environment with antenna array, *GLOBECOM*, December 2003.
10. J. M. Cramer, R. A. Scholtz, and M. Z. Win, On the analysis of UWB communication channels, *IEEE Military Communications Conference*, No. 1, October 1999, pp. 1191–1195.
11. Q. Spencer, B. Jeffs, M. Jensen, and A. Swindlehurst, Modeling the statistical time and angle of arrival characteristics of an indoor multipath channel, *IEEE Journal on Selected Areas in Communications*, vol. 18, no. 3, pp. 347–360, March 2000.
12. M. Hussain, Principles of space–time array processing for ultrawide-band impulse radar and radio communications, *IEEE Transactions on Vehicular Technology*, vol. 51, no. 3, pp. 393–403, May 2002.
13. T. Sato, T.F. de Abreu, and R. Kohno, Beamforming array antenna for ultra wideband communications, *IWUWB2003*, Oulu, June 2003.
14. H. Wang and M. Kaveh, Coherent signal-subspace processing for the detection and estimation of angles of arrival of multiple wide-band sources, *IEEE Transactions on Acoustic, Speech, and Signal Processing*, vol. ASSP-33, no. 4, pp. 823–831, August 1985.
15. M. Ghavami and R. Kohno, Direction finding of broadband signals for frequency-selective fading channels, *IEEE VTC*, 1999.
16. M. Ghavami and R. Kohno, Bearing estimation with uniform resolution in broadband environments, IEEE IZS, 2000b.
17. L. C. Godara, Smart antennas, CRC Press, Boca Raton, FL, 2004.
18. M. Ghavami and R. Kohno, Recursive fan filters for a broad-band partially adaptive antenna, *IEEE Transactions on Communications*, vol. 48, no. 2, February 2000a.
19. M. Ghavami, An adaptive wideband array using a single real multiplier for each antenna element, *IEEE PIMRC*, 2002a.
20. M. Ghavami, Wideband smart antenna theory using rectangular array structures, *IEEE Transactions on Signal Processing*, vol. 50, no. 9, September 2002b.
21. T. Do-Hong and P. Russer, Frequency-invariant beam-pattern and spatial interpolation for wideband beamforming in smart antenna systems, *Proceedings of 2nd VDE World Microtechnologies Congress (MICRO.tec 2003)*, October 2003a, pp. 626–631.
22. T. Do-Hong and P. Russer, A new design method for digital beamforming using spatial interpolation, *IEEE Antennas and Wireless Propagation Letters*, vol. 2, no. 1, pp. 177–181, 2006.

23. S. Ries and T. Kaiser, Ultra wideband impulse beamforming: it's a different world, Special issue on "*Signal Processing in UWB Communications*", invited paper, Elsevier Science, 2005 (in press).
24. M. G. di Benedetto, W. Hirt, T. Kaiser, A. Molisch, I. Oppermann, and D. Porcino, *UWB Communication Systems—a Comprehensive Overview*, EURASIP Book Series, Hindawi Publisher, April 2005.
25. R. Zetik, J. Sachs, and P. Peyerl, UWB radar: distance and positioning measurements, *Int. Conf. on Electromagnetics in Advanced Applications*, Torino, September 2003.
26. E. J. Bond, X. Li, S. C. Hagness, and B. D. Van Veen, Microwave imaging via space–time beamforming for early detection of breast cancer, *IEEE Transactions on Antennas and Propagation*, vol. 51, no. 8, pp. 1690–1705, August 2003.
27. J. Sachs, P. Pegerl, P. Rauschenbach, F. Tkac, M. Kmec, and S. Crabbe, Integrated digital UWB-radar, *AMEREM-2002*, Annapolis, MD, p. 8, June 2002.
28. I. Willms, J. Sachs, and Th. Kaiser, Active UWB fire detection, 13. *Internationale Konferenz ueber Automatische Brandentdeckung, AUBE '04*, Duisburg, 14–16 September, 2004.
29. A. van Zelst, R. van Nee, and G. Awater, Turbo-BLAST and its performance, *Proc. IEEE VTC*, vol. 2, pp. 1282–1286, Rhodes, May 2001.
30. A. Molisch, M. Win, and J. Winters, Capacity of MIMO systems with antenna selection, *Proc. IEEE ICC*, vol. 2, pp. 570–574, Helsinki, June 2001.
31. A. Molisch, J. Foerster, and M. Pendergrass, Channel models for ultrawideband personal area networks, *IEEE Wireless Communications*, December 2003, pp. 14–21.
32. J. Kunisch and J. Pamp, An ultra-wideband space-variant multipath indoor radio channel model, *UWBST*, Reston, VA, November 2003.
33. S. Ries and T. Kaiser, Towards beamforming for UWB signals, *EUSIPCO*, Vienna, 7–10 September 2004.
34. S. Ries, C. Senger, and T. Kaiser, UWB beamforming and DoA-estimation, in *UWB Communication Systems—a Comprehensive Overview*, EURASIP Book Series, Hindawi Publisher, 2005.
35. C. Buchholz and T. Kaiser, Is DoA estimation feasible for dense multipath UWB indoor channels?, *16th International Conference on Wireless Communications*, 12–14, July 2004, Calgary, Canada.

CHAPTER 10

Multiple-Access Interference Mitigation in Ultra Wideband Systems

SINAN GEZICI, HISASHI KOBAYASHI, and H. VINCENT POOR

10.1 INTRODUCTION

Ultra wideband systems offer many advantages for communications, such as high data rate transmission, robustness against small-scale fading, and low probability of interception. Moreover, their large spreading factor results in large multiuser capacity; that is, it is theoretically possible to accommodate a large number of users in a UWB multiple-access environment [1–6].

In this chapter, we consider MAI mitigation for multiaccess impulse radio UWB (IR-UWB) systems, and present signal processing techniques for combatting the effects of interfering users on the detection of information symbols. We consider a scenario, in which all the users transmit their signals at the same time by relying on their multi-access codes, which can be considered as a CDMA scheme [7, 8]. However, in some UWB systems, such as the IEEE 802.15.3a-based high-data-rate UWB physical layer for PANs [9–11], the transmissions from different users are time division multiplexed so that no two users in a given piconet transmit at the same time. However, even with such time-division multiplexing, there is still MAI from neighboring piconets, thus MAI mitigation is still an issue.

The organization of this chapter is as follows: after introducing the signal model in Section 10.2, we analyze MAI mitigation techniques for IR-UWB systems under two categories. The first includes the multiuser detection (MUD) algorithms at the receiver side (Section 10.3), where we consider ML detection, linear receivers, iterative algorithms, and other common MAI mitigating receivers. Because of the similarity between random CDMA (RCDMA) and IR-UWB systems [8, 12], the well-known MAI mitigation algorithms for CDMA systems can be employed for IR-UWB systems with no change. However, the special signaling structure of IR-UWB systems facilitates the design of special MAI mitigation algorithms that cannot be used for ordinary CDMA systems. Since multiuser receivers for CDMA

Ultra Wideband Wireless Communication. Edited by Arslan, Chen, and Di Benedetto
Copyright © 2006 John Wiley & Sons, Inc.

systems have been thoroughly investigated, we mainly consider here the MAI mitigation algorithms specifically designed for IR-UWB systems, and summarize the well-known CDMA approaches briefly. For the second category of MAI mitigation techniques, in Section 10.4, we investigate different multiaccess code design algorithms at the transmitter side; namely, TH and polarity code designs. The chapter concludes in Section 10.5 with a discussion of the MAI mitigation algorithms.

10.2 SIGNAL MODEL

10.2.1 Transmitted Signal

We consider a TH IR-UWB system with K users, in which the transmitted signal from user k is represented by

$$s_{\text{tx}}^{(k)}(t) = \sqrt{\frac{E_k}{N_{\text{f}}}} \sum_{j=-\infty}^{\infty} d_j^{(k)} b_{\lfloor j/N_{\text{f}} \rfloor}^{(k)} w_{\text{tx}}(t - jT_{\text{f}} - c_j^{(k)} T_{\text{c}} - a_{\lfloor j/N_{\text{f}} \rfloor}^{(k)} \Delta), \quad (10.1)$$

where $w_{\text{tx}}(t)$ is the transmitted UWB pulse, E_k is the bit energy of user k, T_{f} is the "frame" time, and N_{f} is the number of pulses representing one information symbol. For binary PAM, $b_{\lfloor j/N_{\text{f}} \rfloor}^{(k)} \in \{+1, -1\}$ and $a_{\lfloor j/N_{\text{f}} \rfloor}^{(k)} = 0$, $\forall j, k$, and for M-ary PPM, $b_{\lfloor j/N_{\text{f}} \rfloor}^{(k)} = 1$ and $a_{\lfloor j/N_{\text{f}} \rfloor}^{(k)} \in \{0, 1, \ldots, M-1\}$ with Δ denoting the modulation index [6, 13].

In order to allow the channel to be shared by many users without causing catastrophic collisions, a TH sequence $\{c_j^{(k)}\}$ is assigned to each user, where $c_j^{(k)} \in \{0, 1, \ldots, N_{\text{c}} - 1\}$ with N_{c} being the number of chips in a frame, that is, $N_{\text{c}} = T_{\text{f}}/T_{\text{c}}$.

The polarity codes, $d_j^{(k)}$'s, are binary (± 1) random variables that help reduce the spectral lines in the power spectral density of the transmitted signal [14] and mitigate the effects of MAI [7]. The receiver for user k is assumed to know its polarity code.

Considering binary PAM[1] and defining a sequence $\{s_j^{(k)}\}$ as

$$s_j^{(k)} = \begin{cases} d_{\lfloor j/N_{\text{c}} \rfloor}^{(k)}, & j - N_{\text{c}} \lfloor j/N_{\text{c}} \rfloor = c_{\lfloor j/N_{\text{c}} \rfloor}^{(k)} \\ 0, & \text{otherwise} \end{cases} \quad (10.2)$$

we can express Equation (10.1) as

$$s_{\text{tx}}^{(k)}(t) = \sqrt{\frac{E_k}{N_{\text{f}}}} \sum_{j=-\infty}^{\infty} s_j^{(k)} b_{\lfloor j/(N_{\text{f}} N_{\text{c}}) \rfloor}^{(k)} w_{\text{tx}}(t - jT_{\text{c}}), \quad (10.3)$$

which indicates that an IR-UWB system with polarity codes can be regarded as an RCDMA system [15] with a "generalized spreading" sequence $\{s_j^{(k)}\}$ [7, 8]. Note that the main difference of the signal model in Equation (10.1) from the RCDMA model [16–18] is the use of $\{-1, 0, +1\}$ as the spreading sequence, instead of

---
[1]Extension to the PPM case is also possible [8].

Figure 10.1 A TH IR-UWB signal with pulse-based polarity randomization where $N_f = 6$, $N_c = 4$ and the TH sequence is $\{2, 1, 2, 3, 1, 0\}$. Assuming that $+1$ is currently being transmitted, the polarity codes for the pulses are $\{+1, +1, -1, +1, -1, +1\}$.

$\{-1, +1\}$. Also the elements of the spreading sequence are usually modeled as independent and identically distributed (i.i.d.) for RCDMA systems, whereas they are dependent for IR-UWB systems [12]. The signal model given by Equation (10.3) can represent an RCDMA system with a processing gain of N_f, by considering the special case when $T_f = T_c$.

An example IR-UWB signal is shown in Figure 10.1, where six pulses are transmitted for each information symbol ($N_f = 6$) with the TH sequence $\{2, 1, 2, 3, 1, 0\}$.

10.2.2 Received Signal

Consider the discrete presentation of the channel, $\alpha^{(k)} = [\alpha_1^{(k)} \cdots \alpha_L^{(k)}]$ for user k, where L is assumed to be the number of multipath components for each user, and T_c is the multipath resolution. Then, the received signal can be expressed as

$$r(t) = \sum_{k=1}^{K} \sqrt{\frac{E_k}{N_f}} \sum_{j=-\infty}^{\infty} \sum_{l=1}^{L} \alpha_l^{(k)} s_j^{(k)} b_{\lfloor j/(N_f N_c) \rfloor}^{(k)} w_{rx}[t - jT_c - (l-1)T_c] + \sigma_n n(t), \quad (10.4)$$

where $w_{rx}(t)$ is the received unit-energy UWB pulse, which is usually modeled as the derivative of $w_{tx}(t)$ due to the effects of the antenna, and $n(t)$ is zero mean white Gaussian noise with unit spectral density.

Consider a filter matched to the UWB pulse $w_{rx}(t)$, as shown in Figure 10.2. When the output of this filter is sampled at the instant when the lth path of the jth frame arrives for the ith information bit, and then despread[2] by the polarity code of the user of interest, say user 1, the discrete signal is expressed, for a synchronous

Figure 10.2 Matched-filtering, sampling and despreading of the received signal.

---

[2] In the context of IR-UWB systems, spreading by random polarity codes is not intended for expanding the bandwidth of the signal. It mainly helps reduce the effect of MAI [7] and eliminates the spectral lines [14].

system, as[3]

$$r_{l,j} = \mathbf{s}_{l,j}^T \mathbf{A} \mathbf{b}_i + \mathbf{n}_{l,j}, \quad (10.5)$$

where $\mathbf{\Lambda} = \text{diag}\{\sqrt{E_1/N_f}, \ldots, \sqrt{E_K/N_f}\}$, $\mathbf{b}_i = [b_i^{(1)} \cdots b_i^{(K)}]^T$ and $n_{l,j} \sim \mathcal{N}(0, \sigma_n^2)$. The signature vector $\mathbf{s}_{l,j}$ is a $K \times 1$ vector, which can be expressed as a sum of the desired signal part (SP), IFI and MAI terms:

$$\mathbf{s}_{l,j} = \mathbf{s}_{l,j}^{(SP)} + \mathbf{s}_{l,j}^{(IFI)} + \mathbf{s}_{l,j}^{(MAI)}, \quad (10.6)$$

where the kth elements can be expressed as

$$\left[\mathbf{s}_{l,j}^{(SP)}\right]_k = \begin{cases} \alpha_l^{(1)}, & k = 1 \\ 0, & k = 2, \ldots, K \end{cases} \quad (10.7)$$

$$\left[\mathbf{s}_{l,j}^{(IFI)}\right]_k = \begin{cases} d_j^{(1)} \sum_{(n,m) \in \mathcal{A}_{l,j}} d_m^{(1)} \alpha_n^{(1)}, & k = 1 \\ 0, & k = 2, \ldots, K \end{cases} \quad (10.8)$$

$$\left[\mathbf{s}_{l,j}^{(MAI)}\right]_k = \begin{cases} 0, & k = 1 \\ d_j^{(1)} \sum_{(n,m) \in \mathcal{B}_{l,j}^{(k)}} d_m^{(k)} \alpha_n^{(k)}, & k = 2, \ldots, K \end{cases} \quad (10.9)$$

with

$$\mathcal{A}_{l,j} = \{(n,m) : n \in \{1, \ldots, L\}, m \in \mathcal{F}_i, m \neq j, mN_c + c_m^{(1)} + n$$
$$= jN_c + c_j^{(1)} + l\} \quad (10.10)$$

and

$$\mathcal{B}_{l,j}^{(k)} = \{(n,m) : n \in \{1, \ldots, L\}, m \in \mathcal{F}_i, mN_c + c_m^{(k)} + n = jN_c + c_j^{(1)} + l\}, \quad (10.11)$$

where $\mathcal{F}_i = \{iN_f, \ldots, (i+1)N_f - 1\}$.

Note that $\mathcal{A}_{l,j}$ is the set of frame and multipath indices of pulses from user 1 that originate from a frame different from the jth one and collide with the lth path of the jth pulse of user 1. Similarly, $\mathcal{B}_{l,j}^{(k)}$ is the set of frame and path indices of pulses from user k that collide with the lth path of the jth pulse of user 1. Also note that for simplicity of the analysis, we assume a guard interval between information symbols that is equal to the length of the channel impulse response [23], which avoids ISI. Therefore, for bit i, we only consider the interference from the pulses in the frames of the current symbol i, namely, from the pulses in frames $iN_f, \ldots, (i+1)N_f - 1$.

---
[3]Note that the dependence of $r_{l,j}$ on the index of the information bit, i, is not shown explicitly.

Special Case: AWGN Channel For the simple AWGN channel case, $\alpha_1^{(k)} = 1$ and $\alpha_l^{(k)} = 0$ for $l > 1$ and $\forall k$. Therefore, there is no IFI and the received samples from an information bit, say the 0th bit, of the desired user (user 1) are denoted as

$$\mathbf{r} = [r_{1,0}\, r_{1,1} \cdots r_{1,N_f-1}]^T, \qquad (10.12)$$

where $r_{1,j}$ is as given in Equation (10.5), with the kth element of $\mathbf{s}_{1,j}$ given by

$$[\mathbf{s}_{1,j}]_k = \begin{cases} 1, & k = 1 \\ d_j^{(1)} d_j^{(k)} \mathbf{I}_{\{c_j^{(k)} = c_j^{(1)}\}}, & k = 2, \ldots, K \end{cases} \qquad (10.13)$$

Here $\mathbf{I}_{\mathcal{D}}$ denotes an indicator function that is equal to one inside $\mathcal{D}$, and zero elsewhere.

The received signal in Equation (10.12) can also be expressed as

$$\mathbf{r} = \mathbf{SAb} + \mathbf{n}, \qquad (10.14)$$

where $\mathbf{b} = [b_0^{(1)} \cdots b_0^{(K)}]^T$, $\mathbf{n}$ is a $K \times 1$ vector of i.i.d. Gaussian noise components, $\mathbf{n} \sim \mathcal{N}(\mathbf{0}, \sigma_n^2 \mathbf{I})$, and $\mathbf{S}$ is the $N_f \times K$ signature matrix, the jth row of which is $\mathbf{s}_{1,j}^T$, given by Equation (10.13).

Due to the low duty-cycle nature of the IR-UWB signals, some of the users may not be interfering with the desired user. If the kth user does not collide with the pulses of user 1, the kth column of $\mathbf{S}$ becomes an all-zero vector. Therefore, we can ignore that user in the signal model and define a simpler model. If K_1 is the number of users colliding with the pulses of user 1, then we obtain [12]

$$\mathbf{r} = \mathbf{S}_1 \mathbf{A}_1 \mathbf{b}_1 + \mathbf{n}, \qquad (10.15)$$

where $\mathbf{b}_1$ is a $(K_1 + 1) \times 1$ vector containing the information symbols from the first user and the users colliding with that user; $\mathbf{A}_1$ is a diagonal matrix with the first element being the amplitude of the signal from user 1 and the remaining elements being the amplitudes of the users' signals colliding with user 1; and the $N_f \times (K_1 + 1)$ signature matrix $\mathbf{S}_1$ is obtained from $\mathbf{S}$ in Equation (10.14) by removing the columns corresponding to elements that do not collide with the first user.

10.3 MULTIPLE-ACCESS INTERFERENCE MITIGATION AT THE RECEIVER SIDE

As we have seen in Section 10.2.1, an IR-UWB system can be modeled as an RCDMA system with generalized signature sequences. Therefore, the classical MUD techniques developed for RCDMA systems can be applied directly to IR-UWB systems [12–20]. However, the complexity of these techniques is often quite high, and the signaling structure of IR-UWB systems is suitable for less complex MUD algorithms specifically designed to exploit that structure.

Moreover, chip-rate sampling is not very suitable for UWB systems due to the need for sampling rates as high as tens of GHz. Therefore, algorithms based on frame-rate or symbol-rate sampling are more suitable in order to have low-power receiver architectures.

In this section, we consider different MUD techniques for IR-UWB systems. Since basically any MUD technique for an RCDMA system can be applied to IR-UWB systems, we focus mainly on multi-user receivers specifically designed for IR-UWB systems. The conventional MUD techniques for RCDMA systems can be found in many references, such as [19, 21].

10.3.1 Maximum-Likelihood Sequence Detection

The optimal receiver that minimizes the probability of error chooses the information symbols that maximize the *a posteriori* log-likelihood. However, the complexity of this optimal scheme grows exponentially with the number of users K, namely $\mathcal{O}(2^K)$ [19, 20]. However, if we consider the samples at instants only when the pulses from the desired user, user 1, arrives, we can design a suboptimal but simpler receiver given by

$$\hat{b}^{(1)} = \arg\max_{b^{(1)} \in \pm 1} \sum_{\tilde{\mathbf{b}} \in \{-1,1\}^{K_1}} \|\mathbf{r} - \mathbf{S}_1[b^{(1)}\tilde{\mathbf{b}}]^{\mathrm{T}}\|^2, \qquad (10.16)$$

where $\mathbf{r}$ and $\mathbf{S}_1$ are as in Equation (10.15), and K_1 is the number of users colliding with the first user.

This receiver is called the "quasi-ML" receiver [12]. Note that the complexity of this receiver is $\mathcal{O}(2^{K_1})$. Also it is the optimal receiver given the received samples at the instants when the pulses from user 1 arrive. The reduction in complexity is the result of using one sample in each frame and ignoring the users that do not collide with the desired user at the sampling instants in the frames. This also results in a performance loss compared with the optimal ML receiver based on chip-rate samples (total of $N_f N_c$ samples), which form a sufficient statistic.

10.3.2 Linear Receivers

The main difficulty with the ML-based multiuser receiver is its complexity. One way to have a low complexity receiver is to consider linear receiver structures. When we observe Equation (10.14), we see that the received samples are related to the symbols by a linear model, but the fitting of this model is difficult due to the integer constraint on the information symbols. The main idea behind the linear receivers is to ignore this integer constraint and to consider the unconstrained linear model. In this way, low complexity receiver designs become possible, with some loss in performance.

10.3 MULTIPLE-ACCESS INTERFERENCE MITIGATION AT THE RECEIVER SIDE

Pulse Discarding Receivers A basic class of linear multiuser detectors specifically suited to IR-UWB systems includes blinking receivers (BRs) [12] and chip discriminators [22]. Due to the signaling structure of an IR-UWB signal, usually not all N_f pulses for a symbol from a given user are corrupted by pulses of interfering users. Therefore, by discarding the pulses that are (significantly) corrupted by the interferers, and only considering the remaining pulses for symbol detection, MAI can be rejected to a certain degree. Therefore, those type of receivers are also called "pulse-discarding" receivers.

The BR ignores all the pulses corrupted by any of the interference and makes use of only the uncorrupted pulses. If we consider the received samples over the AWGN channel given by Equation (10.12), then the bit estimate is given by [12]

$$\hat{b} = \text{sgn}\{\mathbf{w}_{BR}^T \mathbf{r}\}, \tag{10.17}$$

where $\text{sgn}(x) \triangleq x/|x|$ is the sign function, and $\mathbf{w}_{BR} = [w_1 \cdots w_{N_f}]^T$ is a weighting vector given by

$$w_j = \begin{cases} 1, & \text{if } [\mathbf{s}_{1,j}]_2 = \cdots = [\mathbf{s}_{1,j}]_K = 0 \\ 0, & \text{otherwise} \end{cases} \tag{10.18}$$

Although the BR is a very simple multiuser detector, it still requires knowledge of TH sequences to determine whether there occur collisions between the pulses. Also, there should be a sufficient number of uncorrupted pulses in order to make a reliable decision. Moreover, in some cases, discarding pulses that are colliding with weak interferers may cause the BR to perform worse than a conventional matched filter (MF) receiver, which is characterized by $\mathbf{w}_{MF} = \mathbf{1}$, with $\mathbf{1}$ denoting a vector of unit elements [24].

Unlike the BR, the chip discriminator [22] uses only the samples that are smaller than a certain threshold. In this way, the problem of the BR that slightly corrupted pulses are discarded is resolved. However, in order to be able to set an optimal threshold value to minimize the bit error probability (BEP), accurate estimation of the channel is required.

Quasi-Decorrelator By the similarity between IR-UWB and RCDMA systems, the decorrelating receiver [19] for RCDMA systems can be implemented for IR-UWB systems as well. However, it requires the inversion of a $K \times K$ matrix every time the spreading sequences change. Therefore, it might be very complex for IR-UWB systems when there is a large number of users, but when we consider the simplified signal structure in Equation (10.15) by considering the specific signaling structure of IR-UWB systems, we can obtain the following quasi-decorrelator receiver [12]:

$$\hat{b} = \text{sgn}\left\{ \left[(\mathbf{S}_1^T \mathbf{S}_1)^{-1} (\mathbf{S}_1)^T \mathbf{r} \right]_1 \right\}, \tag{10.19}$$

which requires the inversion of a $(K_1 + 1) \times (K_1 + 1)$ matrix, where K_1 is the number of users colliding with the user of interest, the first user.

It is shown in [12] that, under appropriate conditions, the quasi-decorrelating receiver can provide considerable complexity reduction. However, its performance is practically equal to that of the BR, and degrades significantly when the number of users is large. It also performs worse than the MF receiver when the interference is not very strong.

Quasi-MMSE Similar to the quasi-decorrelator, the quasi-MMSE receiver is defined by [12]

$$\hat{b} = \text{sgn}\left(\left\{[\mathbf{S}_1^T\mathbf{S}_1 + \sigma_n^2(\mathbf{A}_1)^{-2}]^{-1}(\mathbf{S}_1)^T\mathbf{r}\right\}_1\right). \tag{10.20}$$

As in the decorrelating receiver, the quasi-MMSE receiver requires a matrix inversion of size $(K_1 + 1)$. However, it also considers the thermal noise component and therefore solves the problem of the quasi-decorrelator receiver in a weak interference scenario.

When the SNR is high, the quasi-MMSE receiver and the BR perform similarly. This is expected because, at high SNR, the MMSE receiver has the same performance as the decorrelating receiver, which performs practically the same as the BR [12]. On the other hand, at low SNR, the quasi-MMSE receiver performs almost the same as the MF receiver.

Optimal and Suboptimal Pulse Combining Schemes The linear receivers we have considered in the previous sections use the low duty cycle property of IR-UWB signals to reduce complexity. We have considered the AWGN channel model in order to have succinct explanations. The same ideas can be extended to multipath channels as well. However, especially in indoor environments, the multipath channel can have hundreds of multipath components due to the high resolution of UWB signals. In such cases, the previous algorithms cannot have the desired complexity reduction since more collisions will occur through multipath components. Also in order to have sufficient energy collection more than one multipath component needs to be combined at the receiver, and the need for rake receivers arises. Therefore, in a multipath environment, the multipath diversity from different signal components needs to be considered in addition to the repetition diversity due to the transmission of the same information bit over N_f UWB pulses.

In this section, we consider a linear receiver with M parallel correlators (i.e., rake with M fingers), where frame-rate sampling is employed at each branch, as shown in Figure 10.3. The frame-rate sampling can be used if we consider a frame-long template signal at each branch, with the pulse position determined by the TH sequence for that frame and the delay of the multipath component for that branch (finger).

In a typical UWB channel, there is a large number of multipath components. Therefore, it is not practically possible to combine all the multipath components of the incoming signal; that is, to design an all-rake receiver [26]. Hence, a subset

10.3 MULTIPLE-ACCESS INTERFERENCE MITIGATION AT THE RECEIVER SIDE

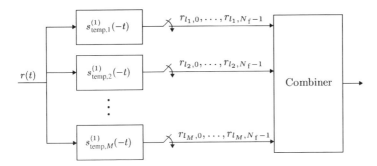

Figure 10.3 Rake receiver with M branches where frame-rate sampling is employed at each branch.

of the multipath components is employed by the rake receiver, which is then called a selective-rake receiver [26]. Let the set of multipath components that are used as the rake fingers be denoted by $\mathcal{L} = \{l_1, \ldots, l_M\}$ with $M \leq L$. Then, at the mth branch of the rake, the received samples from the 0th information bit are denoted as $r_{l_m,0}, \ldots, r_{l_m,N_f-1}$, where $r_{l,j}$ is as given by Equation (10.5).

In order to mitigate the MAI as much as possible, two main issues need to be considered for a selective-rake receiver:

- Optimal selection of the multipath components to employ at the receiver (the optimal finger selection problem);
- Optimal combining of signals at different rake fingers.

The optimal finger selection problem for linear MMSE selective-rake receivers is considered in [25]. The optimal scheme which maximizes the overall SINR of the systems is shown to be an NP-hard problem, but two suboptimal algorithms with polynomial complexity are proposed based on Taylor series approximation and integer relaxation techniques.

Note that the optimal finger selection problem depends on the combining scheme at the rake receiver. In order to understand the tradeoffs among different combining schemes, we assume that the set of optimal fingers for the rake receiver is known beforehand, and consider the pulse combining problem in the remainder of this section.

Linear MMSE Combining When we consider the optimal combining of received samples from the M branches of the rake receiver in Figure 10.3, we can first use a linear receiver for user 1 that combines all the samples from the received signal optimally, according to the MMSE criterion [27].

Let $\mathbf{r}$ be an $N \times 1$ vector denoting the distinct samples $r_{l,j}$ for $(l,j) \in \mathcal{L} \times \mathcal{F}_i$:

$$\mathbf{r} = \left[r_{l_1, j_1^{(1)}} \cdots r_{l_1, j_{m_1}^{(1)}} \cdots r_{l_M, j_1^{(M)}} \cdots r_{l_M, j_{m_M}^{(M)}} \right]^T, \quad (10.21)$$

where $\sum_{i=1}^{M} m_i = N$ denotes the total number of distinct samples, with $N \leq MN_f$, and $\mathcal{F}_i$ is the set of frames corresponding to the ith information bit $\mathcal{F}_i = \{iN_f, \ldots, (i+1)N_f - 1\}$. Note that, due to the multipath channel, some sampling instants can overlap, and the number of distinct samples N can be smaller than MN_f. Also note that, for an AWGN channel, $\mathbf{r}$ in Equation (10.21) reduces to $\mathbf{r}$ in Equation (10.12).

From Equation (10.5), $\mathbf{r}$ can be expressed as

$$\mathbf{r} = \mathbf{SAb}_i + \mathbf{n}, \qquad (10.22)$$

where $\mathbf{A}$ and $\mathbf{b}_i$ are as in Equation (10.5) and $\mathbf{n} \approx \mathcal{N}(\mathbf{0}, \sigma_n^2 \mathbf{I})$. $\mathbf{S}$ is a signature matrix, which has $\mathbf{s}_{l,j}^T$ [see Equations (10.6)–(10.9)] for $(l,j) \in \mathcal{C}$ as its rows, where $\mathcal{C} = \{(l_1, j_1^{(1)}), \ldots, (l_1, j_{m_1}^{(1)}), \ldots, (l_M, j_1^{(M)}), \ldots, (l_M, j_{m_M}^{(M)})\}$.

From Equations (10.6)–(10.9), $\mathbf{S}$ can be expressed as $\mathbf{S} = \mathbf{S}^{(SP)} + \mathbf{S}^{(IFI)} + \mathbf{S}^{(MAI)}$. Then, after some manipulation, $\mathbf{r}$ becomes

$$\mathbf{r} = b_i^{(1)} \sqrt{\frac{E_1}{N_f}} (\boldsymbol{\alpha} + \tilde{\mathbf{e}}) + \mathbf{S}^{(MAI)} \mathbf{Ab}_i + \mathbf{n}, \qquad (10.23)$$

where $\boldsymbol{\alpha} = [\alpha_{l_1}^{(1)} \mathbf{1}_{m_1}^T \cdots \alpha_{l_M}^{(1)} \mathbf{1}_{m_M}^T]^T$, with $\mathbf{1}_m$ denoting an $m \times 1$ vector of all ones, and $\tilde{\mathbf{e}}$ is an $N \times 1$ vector whose elements are $e_{l,j} = d_j^{(1)} \sum_{(n,m) \in \mathcal{A}_{l,j}} d_m^{(1)} \alpha_n^{(1)}$ for $(l,j) \in \mathcal{C}$.

A linear receiver combines the elements of $\mathbf{r}$ and obtains a decision variable as follows:

$$y_1 = \boldsymbol{\theta}^T \mathbf{r}, \qquad (10.24)$$

where $\boldsymbol{\theta}$ is the weighting vector.

The MMSE weights that maximize the SINR of the received signal in Equation (10.23) can be obtained as [19]

$$\boldsymbol{\theta}_{MMSE} = \mathbf{R}_{\mathbf{w}_1}^{-1} (\boldsymbol{\alpha} + \tilde{\mathbf{e}}) \qquad (10.25)$$

where $\mathbf{w}_1 = \mathbf{S}^{(MAI)} \mathbf{Ab}_i + \mathbf{n}$ and $\mathbf{R}_{\mathbf{w}_1} = E\{\mathbf{w}_1 \mathbf{w}_1^T\}$. Assuming equiprobable information symbols, the correlation matrix can be expressed as

$$\mathbf{R}_{\mathbf{w}_1} = \mathbf{S}^{(MAI)} \mathbf{A}^2 (\mathbf{S}^{(MAI)})^T + \sigma_n^2 \mathbf{I}. \qquad (10.26)$$

Then, the linear MMSE receiver becomes

$$\hat{b}_i^{(1)} = \text{sgn}\{\mathbf{r}^T \mathbf{R}_{\mathbf{w}_1}^{-1} (\boldsymbol{\alpha} + \tilde{\mathbf{e}})\}. \qquad (10.27)$$

Note that this receiver requires the inversion of an $N \times N$ matrix. It combines the samples from different frame and multipath components in an optimal manner among the class of linear receivers. In other words, it exploits both the repetition and the multipath diversity. The main problem about this receiver is the complexity

10.3 MULTIPLE-ACCESS INTERFERENCE MITIGATION AT THE RECEIVER SIDE

of inverting the $N \times N$ correlation matrix. If we can sacrifice some optimality from the repetition diversity or the multipath diversity domains, we can reduce the computational complexity. Two such receivers are considered in the following sections.

Optimal Frame Combining Consider suboptimal combining of different multipath components in each frame for complexity reduction. In this case, the multipath components in each frame are added according to the maximal ratio combining (MRC) criterion. Then, those combined components in the frames are combined according to the MMSE criterion, as shown in Figure 10.4. That is, the decision variable is given by

$$y_2 = \sum_{j=iN_f}^{(i+1)N_f - 1} \gamma_j \sum_{l \in \mathcal{L}} \alpha_l^{(1)} r_{l,j}, \qquad (10.28)$$

where $\gamma_{iN_f}, \ldots, \gamma_{(i+1)N_f-1}$ are the weighting factors for the ith bit.

From Equation (10.5), y_2 can be expressed as

$$y_2 = \boldsymbol{\gamma}_i^T \left(\sum_{l \in \mathcal{L}} \alpha_l^{(1)} \hat{\mathbf{S}}_l \mathbf{A} \mathbf{b}_i + \sum_{l \in \mathcal{L}} \alpha_l^{(1)} \hat{\mathbf{n}}_l \right), \qquad (10.29)$$

where $\boldsymbol{\gamma}_i = [\gamma_{iN_f} \cdots \gamma_{(i+1)N_f-1}]^T$ is the vector of weighting coefficients, $\hat{\mathbf{n}}_l = [n_{l,iN_f} \cdots n_{l,(i+1)N_f-1}]^T$ is the noise vector, which is distributed as $\mathcal{N}(\mathbf{0}, \sigma_n^2 \mathbf{I})$, and $\hat{\mathbf{S}}_l$

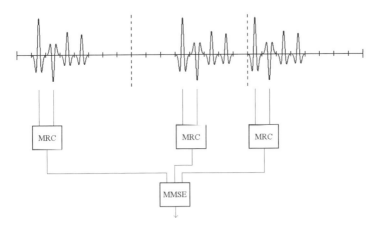

Figure 10.4 The optimal frame combining (OFC) scheme, in which the multipath components in each frame are added according to the MRC criterion, and the combined components in the frames are combined according to the MMSE criterion. In the figure, two of the multipath components are assumed to be employed by the rake receiver, and three frames are transmitted per symbol. That is, $M = 2$, $\mathcal{L} = \{1, 2\}$ and $N_f = 3$.

is an $N_f \times K$ matrix, whose jth row is $\mathbf{s}_{l,iN_f+j-1}^T$. Using Equations (10.6)–(10.9), $\hat{\mathbf{S}}$ can be expressed as $\hat{\mathbf{S}} = \hat{\mathbf{S}}^{(SP)} + \hat{\mathbf{S}}^{(IFI)} + \hat{\mathbf{S}}^{(MAI)}$. Then, we get

$$y_2 = \boldsymbol{\gamma}_i^T \left\{ b_i^{(1)} \sqrt{\frac{E_1}{N_f}} \left[\sum_{l \in \mathcal{L}} (\alpha_l^{(1)})^2 \mathbf{1}_{N_f} + \sum_{l \in \mathcal{L}} \alpha_l^{(1)} \hat{\mathbf{e}}_l \right] + \mathbf{w}_2 \right\}, \qquad (10.30)$$

where $\hat{\mathbf{e}}_l$ is an $N_f \times 1$ vector whose jth element is $e_{l,iN_f+j-1} = d_{iN_f+j-1}^{(1)} \sum_{(n,m) \in \mathcal{A}_{l,iN_f+j-1}} d_m^{(1)} \alpha_n^{(1)}$ and $\mathbf{w}_2 = \sum_{l \in \mathcal{L}} \alpha_l^{(1)} \hat{\mathbf{S}}_l^{(MAI)} \mathbf{A} \mathbf{b}_i + \sum_{l \in \mathcal{L}} \alpha_l^{(1)} \hat{\mathbf{n}}_l$.

From Equation (10.30), the MMSE weights can be obtained as

$$\boldsymbol{\gamma}_{MMSE} = \mathbf{R}_{\mathbf{w}_2}^{-1} \left[\sum_{l \in \mathcal{L}} (\alpha_l^{(1)})^2 \mathbf{1}_{N_f} + \sum_{l \in \mathcal{L}} \alpha_l^{(1)} \hat{\mathbf{e}}_l \right], \qquad (10.31)$$

where

$$\mathbf{R}_{\mathbf{w}_2} = \sum_{l \in \mathcal{L}} \alpha_l^{(1)} \hat{\mathbf{S}}_l^{(MAI)} \mathbf{A}^2 \sum_{l \in \mathcal{L}} \alpha_l^{(1)} (\hat{\mathbf{S}}_l^{(MAI)})^T \\ + \sum_{l_1 \in \mathcal{L}} \sum_{l_2 \in \mathcal{L}} \alpha_{l_1}^{(1)} \alpha_{l_2}^{(1)} E\{\hat{\mathbf{n}}_{l_1} \hat{\mathbf{n}}_{l_2}^T\}. \qquad (10.32)$$

It is straightforward to show that $E\{\hat{\mathbf{n}}_{l_1} \hat{\mathbf{n}}_{l_2}^T\} = \sigma_n^2 \mathbf{I}$ for $l_1 = l_2$. When $l_1 \neq l_2$, the element at row j_1 and column j_2, $[E\{\hat{\mathbf{n}}_{l_1} \hat{\mathbf{n}}_{l_2}^T\}]_{j_1 j_2}$, is equal to σ_n^2 if $j_1 N_c + c_{j_1}^{(1)} + l_1 = j_2 N_c + c_{j_2}^{(1)} + l_2$ and zero otherwise [$j_1 = iN_f, \ldots, (i+1)N_f - 1$ and $j_2 = iN_f, \ldots, (i+1)N_f - 1$].

We note from Equations (10.30) and (10.31) that the OFC receiver, $\hat{b}_i^{(1)} = \text{sign}\{y_2\}$, requires the inversion of an $N_f \times N_f$ matrix. The reduction in complexity compared with the optimal linear MMSE receiver of the previous section is due to the suboptimal combination of the multipath components.

The SINR of the system can be expressed as

$$\text{SINR}_{OFC} = \frac{E_1}{N_f} \mathbf{x}_2^T \mathbf{R}_{\mathbf{w}_2}^{-1} \mathbf{x}_2. \qquad (10.33)$$

where $\mathbf{x}_2 = \sum_{l \in \mathcal{L}} (\alpha_l^{(1)})^2 \mathbf{1}_{N_f} + \sum_{l \in \mathcal{L}} \alpha_l^{(1)} \hat{\mathbf{e}}_l$.

Optimal Multipath Combining (OMC) Now consider a receiver that combines different multipath components optimally, according to the MMSE criterion, while employing equal gain combining (EGC) for contributions from different frames, as shown in Figure 10.5. In this case, the decision variable is given by

$$y_3 = \sum_{l \in \mathcal{L}} \beta_l \sum_{j=iN_f}^{(i+1)N_f - 1} r_{l,j}, \qquad (10.34)$$

where $\boldsymbol{\beta} = [\beta_{l_1} \cdots \beta_{l_M}]^T$ is the weighting vector.

10.3 MULTIPLE-ACCESS INTERFERENCE MITIGATION AT THE RECEIVER SIDE

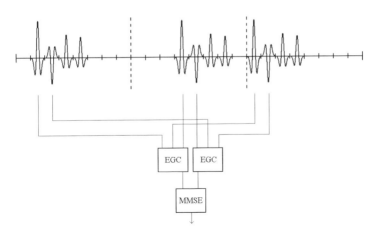

Figure 10.5 The optimal multipath combining scheme, where different multipath components are combined optimally according to the MMSE criterion, and EGC is employed for contributions from different frames. In the figure, two of the multipath components are assumed to be employed by the rake receiver, and three frames are transmitted per symbol. That is, $M = 2$, $\mathcal{L} = \{1, 2\}$ and $N_f = 3$.

Using Equation (10.5), y_3 can be expressed as

$$y_3 = \boldsymbol{\beta}^T \left(\sum_{j=iN_f}^{(i+1)N_f - 1} [\tilde{\mathbf{S}}_j \mathbf{A} \mathbf{b}_i + \tilde{\mathbf{n}}_j] \right), \tag{10.35}$$

where $\tilde{\mathbf{n}}_j = [n_{l_1,j} \cdots n_{l_M,j}]^T$ is the noise vector, which is distributed as $\mathcal{N}(\mathbf{0}, \sigma_n^2 \mathbf{I})$, and $\tilde{\mathbf{S}}_j$ is an $M \times K$ signature matrix, whose mth row is $\mathbf{s}_{l_m,j}^T$. Using Equations (10.6)–(10.9), $\tilde{\mathbf{S}}$ can be expressed as $\tilde{\mathbf{S}} = \tilde{\mathbf{S}}^{(SP)} + \tilde{\mathbf{S}}^{(IFI)} + \tilde{\mathbf{S}}_t^{(MAI)}$. Then, we get

$$y_3 = \boldsymbol{\beta}^T \left[b_i^{(1)} \sqrt{\frac{E_1}{N_f}} \left(N_f \tilde{\boldsymbol{\alpha}} + \sum_{j=iN_f}^{(i+1)N_f - 1} \tilde{\mathbf{e}}_j \right) + \mathbf{w}_3 \right], \tag{10.36}$$

where $\tilde{\boldsymbol{\alpha}} = [\alpha_{l_1}^{(1)} \cdots \alpha_{l_M}^{(1)}]^T$, $\tilde{\mathbf{e}}_j$ is an $M \times 1$ vector whose mth element is $e_{l_m,j} = d_j^{(1)} \sum_{(n,m) \in \mathcal{A}_{l_m,j}} d_m^{(1)} \alpha_n^{(1)}$, and $\mathbf{w}_3 = \sum_{j=iN_f}^{(i+1)N_f - 1} \tilde{\mathbf{S}}_j^{(MAI)} \mathbf{A} \mathbf{b}_i + \sum_{j=iN_f}^{(i+1)N_f - 1} \tilde{\mathbf{n}}_j$.

From Equation (10.36), the MMSE weights are chosen as

$$\boldsymbol{\beta}_{MMSE} = \mathbf{R}_{\mathbf{w}_3}^{-1} \left(N_f \tilde{\boldsymbol{\alpha}} + \sum_{j=iN_f}^{(i+1)N_f - 1} \tilde{\mathbf{e}}_j \right), \tag{10.37}$$

where

$$\mathbf{R}_{w_3} = \sum_{j=iN_f}^{(i+1)N_f-1} \tilde{\mathbf{S}}_j^{(MAI)} \mathbf{A}^2 \sum_{j=iN_f}^{(i+1)N_f-1} (\tilde{\mathbf{S}}_j^{(MAI)})^T \quad (10.38)$$

$$+ \sum_{j_1=iN_f}^{(i+1)N_f-1} \sum_{j_2=iN_f}^{(i+1)N_f-1} E\{\tilde{\mathbf{n}}_{j_1} \tilde{\mathbf{n}}_{j_2}^T\}. \quad (10.39)$$

It can be observed that $E\{\tilde{\mathbf{n}}_{j_1} \tilde{\mathbf{n}}_{j_2}^T\} = \sigma_n^2 \mathbf{I}$ for $j_1 = j_2$. When $j_1 \neq j_2$, the element at row l_1 and column l_2, $[E\{\hat{\mathbf{n}}_{j_1} \hat{\mathbf{n}}_{j_2}^T\}]_{l_1 l_2}$, is equal to σ_n^2 if $l_1 N_c + c_{l_1}^{(1)} + j_1 = l_2 N_c + c_{l_2}^{(1)} + j_2$ and zero otherwise ($l_1 \in \mathcal{L}$ and $l_2 \in \mathcal{L}$).

We note from Equations (10.36) and (10.37) that the OMC receiver, $\hat{b}_i^{(1)} = \text{sign}\{y_3\}$, needs to invert the $M \times M$ matrix $\mathbf{R}_{w_3}$. The reduction in the complexity compared to the optimal linear receiver is the result of suboptimal combination of the contributions from different frames.

The SINR of the system can be expressed as

$$\text{SINR}_{OMC} = \frac{E_1}{N_f} \mathbf{x}_3^T \mathbf{R}_{w_3}^{-1} \mathbf{x}_3. \quad (10.40)$$

where $x_3 = N_f \tilde{\alpha} + \sum_{j=iN_f}^{(i+1)N_f-1} \tilde{\mathbf{e}}_j$.

From the previous equations, it can be shown [27] that for a single user system where the pulses in a frame never collide with any pulse in another frame (i.e., no IFI and MAI), the expressions for SINR reduce to

$$\text{SINR}_{OFC} = \text{SINR}_{OMC} = \frac{E_1}{\sigma_n^2} \sum_{l \in \mathcal{L}} \alpha_l^2. \quad (10.41)$$

Other Suboptimal Combining Schemes The OFC and OMC are just two possible techniques, among many others, for reducing the complexity of the optimal MMSE receiver. For example, instead of combining the samples by a two-stage approach, a multistage MMSE combining approach can be adopted as follows: first, N distinct outputs from the rake fingers are divided into a number of groups, and the elements in each group are combined by the MMSE approach. Then, the new combined samples are grouped again and the same procedure is repeated until the final output is obtained. In this way, low-complexity suboptimal combining schemes can be developed. The number of stages can be determined considering the tradeoff between the computation complexity and the performance.

10.3.3 Iterative (Turbo) Algorithms

Iterative MUD algorithms exchange soft information (i.e., posterior probabilities) between MUD and channel decoding in order to provide low-complexity and near-optimal demodulation in coded multiple-access channels [28]. This turbo principle of iteration among the two decision algorithms, that is, soft MUD and soft channel decoding, can also be applied to IR-UWB systems, when channel coding is employed (e.g., IR-UWB systems with convolutional codes [29–31]). However, for uncoded IR-UWB systems, there is no obvious structure for the application of a turbo-like algorithm. However, by observing the inherent signal structure that employs N_f pulses per information symbol, we can still adopt turbo-like receivers for IR-UWB systems [12, 32].

In this section, we first describe an iterative interference cancellation and decoding scheme for coded IR-UWB systems [30], and then consider the novel pulse-symbol iterative detector for uncoded systems of [12].

Iterative Interference Cancellation and Decoding for Coded IR-UWB

A low complexity iterative receiver for a convolutionally coded IR-UWB system is considered in [30], where soft information is exchanged between soft interference canceller-likelihood calculators (SICLCs) and soft-input soft-output (SISO) channel decoders.

Considering an AWGN channel,[4] the proposed receiver mainly consists of pulse correlators, SICLCs and SISO channel decoders. Also there are interleavers and deinterleavers between the SICLCs and the SISO channel decoders [30].

The pulse correlator for user k correlates the received signal $r(t)$ with the received pulse shape $w_{rx}(t)$, and sends the correlation outputs to the SICLC block. In the SICLC step, the soft information about the interfering signals provided by the SISO channel decoders is used to perform parallel soft interference cancellation. In other words, for user k, the total interference from all other users is calculated based on the soft information from the SISO channel decoders, and is subtracted from the correlation output corresponding to user k. Then, from the resulting output for user k, the log-likelihood ratio (LLR) for bit k is calculated by a single-user likelihood calculator [30]. This value is the soft (extrinsic) information that is delivered to the kth SISO channel decoder, which uses it as the *a priori* information and calculates an update of LLRs for the coded bits based on the code constraint. Then, these updated LLRs are sent to the SICLC block for the next iteration. After a number of iterations, the bit decisions are made based on the LLRs calculated by the SISO channel decoders.

Although this receiver is probably the first iterative receiver structure proposed for coded IR-UWB systems, it does not exploit the unique characteristics of IR-UWB signals, such as the low duty cycle and pulse repetition properties. In other words, the receiver design is suitable for any coded CDMA system including the coded IR-UWB system. Further complexity reductions and/or performance improvements can be obtained by using the special signalling structure of IR-UWB systems, as is considered in the next section.

---

[4]The same approach can easily be extended to multipath channels [31]. The AWGN case is considered here for the sake of exposition.

Pulse-Symbol Iterative Detector When we consider turbo receiver design for an uncoded IR-UWB system, there is not an obvious two-stage structure in which soft information can be exchanged iteratively between MUD and channel decoding. However, by observing the repetition of N_f pulses per information symbol in IR-UWB systems, we can design an iterative receiver, in which the first stage assumes that different pulses from a given user carry independent and identically distributed (i.i.d.) information symbols, while the second stage exploits the information that N_f pulses from the same user correspond to the same information symbol [12]. In other words, the first step ignores some *a priori* information about the information symbols in order to have a two-stage iterative structure.

Pulse Detector This is the first stage of the turbo multiuser detector, which computes the posterior LLR of the information symbol corresponding to the jth pulse of the kth user, $\tilde{b}_j^{(k)}$, $\forall j, k$ by using the received signal, the information about the transmitted bits from the other users, and the *a priori* information about $\tilde{b}_j^{(k)}$ provided by the symbol detector. It can be shown that the *a posteriori* LLR of $\tilde{b}_j^{(k)}$ at the nth iteration is given by [12]

$$L_1^n\left(\tilde{b}_j^{(k)}\right) \triangleq \log \frac{\Pr\{\tilde{b}_j^{(k)} = 1 | r_j^{(k)}\}}{\Pr\{\tilde{b}_j^{(k)} = -1 | r_j^{(k)}\}} \tag{10.42}$$

$$= \log \frac{f\left(r_j^{(k)} | \tilde{b}_j^{(k)} = 1\right)}{f\left(r_j^{(k)} | \tilde{b}_j^{(k)} = -1\right)} + \log \frac{\Pr\{\tilde{b}_j^{(k)} = 1\}}{\Pr\{\tilde{b}_j^{(k)} = -1\}} \tag{10.43}$$

$$= \lambda_1^n\left(\tilde{b}_j^{(k)}\right) + \lambda_2^{n-1}\left(\tilde{b}_j^{(k)}\right), \tag{10.44}$$

where $r_j^{(k)}$ is the discrete received signal from the jth frame of the kth user,[5] $\lambda_2^{n-1}(\tilde{b}_j^{(k)})$ represents the *a priori* LLR of $\tilde{b}_j^{(k)}$, which is obtained from the $(n-1)$th iteration of the symbol detector, and $\lambda_1^n(\tilde{b}_j^{(k)})$ is the soft information provided by the pulse detector about the transmitted symbol.

Symbol Detector The second stage of the receiver exploits the information about the information symbols from the same user. In other words, considering the 0th symbol without loss of generality, the symbol detector uses the fact that $\tilde{b}_0^{(k)} = \cdots = \tilde{b}_{N_f-1}^{(k)} = b_0^{(k)}$, for $k = 1, \ldots, K$, where $b_i^{(k)}$ is as in Equation (10.1). Using this information, and the soft information from the pulse detector, the

---

[5] For transmission over multipath channels, $r_j^{(k)}$ can be chosen as the received signal from the strongest multipath component of the kth user's channel [32].

symbol detector calculates the *a posteriori* LLR of $\tilde{b}_j^{(k)}$ as follows [12]:

$$L_2^n\left(\tilde{b}_j^{(k)}\right) \triangleq \log \frac{\Pr\left\{\tilde{b}_j^{(k)} = 1 | \lambda_1^n\left(\tilde{b}_j^{(k)}\right), j = 0, \ldots, N_f - 1\right\}}{\Pr\left\{\tilde{b}_j^{(k)} = -1 | \lambda_1^n\left(\tilde{b}_j^{(k)}\right), j = 0, \ldots, N_f - 1\right\}} \quad (10.45)$$

$$= \lambda_2^n\left(\tilde{b}_j^{(k)}\right) + \lambda_1^n\left(\tilde{b}_j^{(k)}\right), \quad (10.46)$$

which can be seen as the summation of the prior information from the pulse detector, $\lambda_1^n(\tilde{b}_j^{(k)})$, and the soft information about $\tilde{b}_j^{(k)}$, $\lambda_2^n(\tilde{b}_j^{(k)})$, which is obtained from the information about the received samples other than $r_j^{(k)}$ [32]. In the next iteration, the soft information $\lambda_2^n(\tilde{b}_j^{(k)})$ is sent to the pulse detector, which is used as *a priori* information about the *j*th pulse of user *k*.

Other Iterative Detectors Similar to the algorithm in Section 1.3.3.1, when any kind of forward error correction (FEC) coding is applied to an IR-UWB system, iterative (turbo) multiuser detectors can be designed. A coded IR-UWB system can be considered to employ serial concatenated coding where the inner code is the spreading code inherently present in the system and the outer code is the FEC code [33]. For such serially concatenated coding the iterative decoding between the inner and outer decoder results in additional coding gain by means of a SISO decoding algorithm [34, 35]. Therefore, by applying various types of FEC coding, iterative receiver structures can be designed for IR-UWB systems [34, 36].

10.3.4 Other Receiver Structures

In addition to the previous receiver structures we have discussed, other MAI mitigating receiver designs are also possible for IR-UWB systems. One way to mitigate MAI is to consider frequency-domain processing of the received signal. For example, in [39], the FT of the incoming signal is calculated, which transforms the delay estimation problem (assuming a PPM scheme) in the time domain to a phase detection problem in the frequency domain. Therefore, a linear system model can be obtained, and typical linear receivers, such as the MMSE and decorrelating receivers, can be employed.

A large class of MAI mitigation schemes includes subtractive interference cancellation techniques, which are not mentioned in a separate section since there has not been any special application of them to IR-UWB systems; that is, the same techniques for CDMA systems can be directly adopted [19, 21, 37]. The main idea behind the subtractive interference cancelation techniques is to estimate the MAI and to subtract it from the received signal. There are different implementations of this idea depending on the performance and complexity issues. In successive interference cancellation (SIC), the interference due to each user is estimated and subtracted from the received signal sequentially. In its simplest form, the SIC receiver first estimates the strongest signal by a conventional single-user receiver, and subtracts it from the received signal and continues the same procedure with the second strongest signal, etc., until all the users' signals are detected. The main disadvantage of the SIC approach is the delay associated with demodulating the signals sequentially. In contrast to the SIC method, a parallel interference cancellation (PIC) scheme detects all the signals in parallel and subtracts the interference *estimate* for each user (sum of all the signal estimates except the desired user's) from the received signal. This procedure can be repeated a number of times in order to obtain better performance, by using the results of the previous step to regenerate the interference. Apart from the SIC and PIC approaches, we can consider the multistage detection principle, which is in fact a symmetrized version of the SIC technique [19]. Finally, the decision feedback approach combines several important characteristics of the SIC and multistage receivers. It basically uses the final decisions as the feedback, and uses both the linear and nonlinear techniques for MAI mitigation [19].

For all the previous receiver structures we have considered, the receiver is assumed to know the channel. In practice, one way to obtain this information is to perform channel estimation, before the multiuser detection stage, by employing training symbols (data-aided channel estimation) [38]. Alternatively, channel estimation and data detection can be performed jointly without requiring any training data, which is called a *blind* (non-data aided) algorithm. In view of the similarity between the IR-UWB and CDMA systems, application of blind MUD algorithms to IR-UWB systems is possible, as considered in [40–42].

Finally, subspace approaches can be useful for UWB systems to provide low-complexity designs, when the rank of the covariance matrix of the discrete received signal is large. For example, the implementation of the optimal MMSE receiver in Section 10.3.2 can require the inversion of a very large matrix when many samples are to be combined optimally. Therefore, the subspace schemes determine a low-rank subspace spanned by the columns of the covariance matrix, where the design of a simpler MMSE scheme becomes possible. One way to implement this rank-reduction is principal component analysis [43, 44], where the eigen-decomposition of the covariance matrix is employed to determine a signal subspace spanned by the eigenvectors associated with the largest eigenvalues and a noise subspace spanned by the eigenvectors associated with the remaining eigenvalues. Then, the received signal is projected onto this signal subspace. The application of this subspace approach to IR-UWB systems is investigated in [45]. Another technique for rank-reduction is the multistage Wiener filter (MSWF) approach [46], which does not need any eigen-decomposition, and usually outperforms other rank-reduction approaches [47]. This reduced-rank approach based on the MSWF can be adopted for IR-UWB systems without any significant modification [48].

10.4 MULTIPLE-ACCESS INTERFERENCE MITIGATION AT THE TRANSMITTER SIDE

The MAI mitigation techniques investigated in Section 10.3 do not require any change in the generic transmitted signal structure described in Section 10.2.1. In other words, they aim to mitigate MAI by signal processing techniques at the receiver side. In this section, we consider MAI mitigation techniques at the transmitter side, which mainly aim to design multiple-access signatures so as to mitigate interference between different users. Although these techniques are called MAI mitigation techniques at the transmitter side, they usually require appropriate receiver structures in order to exploit the specific code designs used at the transmitter side.

10.4.1 Time-Hopping Sequence Design for MAI Mitigation

Consider the signal model in Equation (10.1). If we consider a synchronous IR-UWB system over a flat fading channel, it is possible to design N_c orthogonal TH codes, and hence, to have interference-free communications with N_c users; that is, the MAI can be mitigated completely in this case. The orthogonal TH sequences mean that $c_j^{(k_1)} \neq c_j^{(k_2)}$ for $k_1 \neq k_2$. Among possible orthogonal code constructions for the synchronous flat fading case are linear, quadratic, cubic or hyperbolic congruence codes (LCC, QCC, CCC, or HCC). For example, a variant of linear congruence codes can be expressed as [49]

$$c_j^{(k)} = (k+j-1) \bmod(N_c), \qquad (10.47)$$

for $j \in \{0, 1, \ldots, N_f - 1\}$ and $k \in \{1, \ldots, N_c\}$, where mod denotes the modulo operator. With this code construction, it is possible to have N_c users with orthogonal signals.

Since UWB channels have usually many multipath components, the previous TH design approach for flat fading channels needs to be generalized. Assuming the knowledge of the maximum excess delay of the channel and the number of users K, the following approach is proposed in [50]

$$c_j^{(k)} = \left((k-1)D + j + \left\lfloor \frac{k-1}{N_f} \right\rfloor \right) \bmod(N_c), \qquad (10.48)$$

where $D = \lceil \tau_d/T_c + 1 \rceil$, with τ_d being the maximum excess delay, and $\lfloor \cdot \rfloor$ and $\lceil \cdot \rceil$ denote the integer floor and integer ceiling operations, respectively. For the proposed code, the number of pulses per symbol is selected as $N_f = N_c/D$ so that the multipath components do not destroy the orthogonal construction, and it is possible to have MAI-free communications for $K \leq N_f$. For $K > N_f$, the codes are shifted by $\lfloor (k-1)/N_f \rfloor$ to construct a new group of orthogonal codes (see [50] for details).

The suitability of IR-UWB signaling for adaptive systems can result in situations, in which different users in the network assign different numbers of pulses per information symbol in order to satisfy certain quality of service (QoS) requirements [51]. In such cases, in order to facilitate the design of orthogonal TH sequences, one can consider a more general IR-UWB signaling structure, where the constraint of inserting pulses into certain frame intervals is removed [50]. Let $N_f^{(k)}$ denote the number of pulses per information symbol of the kth user. Then, define a common symbol duration in terms of the chip duration as $N'_c = \sum_{k=1}^{K} N_f^{(k)}$ and let $\mathcal{S} = \{1, \ldots, N'_c\}$. The code construction algorithm is described in Table 10.1, where $\mathbf{c}^{(k)} = \text{rand}(\mathcal{S}, N_f^{(k)})$ chooses $N_f^{(k)}$ random elements from the set $\mathcal{S}$ and inserts them into the vector $\mathbf{c}^{(k)}$, and $\mathcal{S} - \mathbf{c}^{(k)}$ excludes the elements of $\mathbf{c}^{(k)}$ from the set $\mathcal{S}$.

For the asynchronous case, the users' symbol transition instants are not aligned and it is not possible to design orthogonal TH sequences. In such cases, the aim is to have TH sequences with good auto-correlation and cross-correlation properties. Owing to the similarity between the design of time-hopping and frequency-hopping codes, LCC, QCC, CCC, and HCC can be considered for IR-UWB systems [52]. The analysis in [50] shows that QCC have reasonably good cross-correlation *and* auto-correlation characteristics compared with the other options.

10.4.2 Pseudochaotic Time Hopping

In order to generate TH sequences with a random distribution of inter-pulse intervals, which results in a smooth power spectral density of the transmitted signal, pseudochaotic TH (PCTH) is proposed for IR-UWB systems in [54]. In this scheme, a pseudochaotic encoder driven by i.i.d. binary information symbols generates the amount of shift (time hopping) to be applied to each UWB pulse. Since the TH sequence depends on the information symbols, the resulting code is aperiodic.

In a typical PCTH system, the i.i.d. information bits are stored in an M-bit shift register, and the state of the system is represented as

$$x = 0.b_1 b_2 \cdots b_M = \sum_{i=1}^{M} 2^{-i} b_i, \qquad (10.49)$$

where $b_i \in \{0,1\}$, and $x \in I = [0,1]$. Dividing the interval I into $I_0 = [0, 0.5]$ and $I_1 = [0.5, 1]$, the binary information bits are assigned to different intervals, which means that if a pulse is in the first half of a symbol interval, information 0 is being transmitted and if it is in the second half, a 1 is being transmitted. Dividing the symbol interval into $N_f = 2^M$ slots, the pulse can reside in any N_f positions in the symbol interval. For each new information bit, the binary bits in the representation of state x in Equation (10.49) are shifted leftwards by discarding the old

TABLE 10.1 The Th Code Construction Algorithm for Synchronous IR-UWB Users with Different Numbers of Pulses per Symbol [50]

Code Construction
for $k = 1: K$
$\quad \mathbf{c}^{(k)} = \text{rand}(\mathcal{S}, N_f^{(k)})$
$\quad \mathcal{S} = \mathcal{S} - \mathbf{c}^{(k)}$
end

10.4 MULTIPLE-ACCESS INTERFERENCE MITIGATION

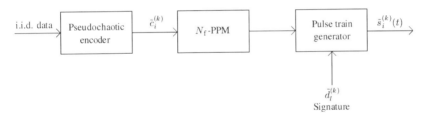

Figure 10.6 A simplified block diagram of the transmitter for user k in a PCTH system.

most significant bit (MSB), b_1, and assigning the new bit as the least significant bit (LSB), b_M.

In order to provide MAI mitigation in PCTH systems, the following signaling scheme is proposed in [55], in which the transmitted signal from user k for the ith information symbol is given by

$$\tilde{s}_i^{(k)}(t) = \sum_{l=0}^{N_c-1} \tilde{d}_l^{(k)} w_{tx}(t - lT_c - \tilde{c}_i^{(k)} T_f), \qquad t \in [0, T_s], \qquad (10.50)$$

where T_s is the symbol interval, which is divided into N_f frames with T_f being the frame interval that consists of N_c chips ($T_f = N_c T_c$), $\tilde{d}_l^{(k)} \in \{0,1\}$ is the signature of the kth user, and $\tilde{c}_i^{(k)} \in \{0,1,\ldots,N_f-1\}$ is the output of the pseudochaotic encoder that is determined by the incoming sequence of information bits, as shown in Figure 10.6. In other words, depending on $\tilde{c}_i^{(k)}$, each user transmits its pulses in one frame, which is different from the conventional IR-UWB scheme, where each user transmits one pulse per frame. If two users transmit their pulses in different frames, there occurs no interference. If they send their pulses in the same frame, the pulses can overlap, but the effects of this overlap can be reduced by careful design of the users' signature sequences $\tilde{d}_l^{(k)}$, for $l \in \{0,1,\ldots,N_c-1\}$, and $k = 1,\ldots,K$.

One of the main disadvantages of the PCTH scheme is that the self-interference from the pulses of a given user can be significant in multipath channels, since all the pulses are transmitted in the same frame interval. Also, the synchronization can be difficult since PCTH results in aperiodic TH sequences due to the dependence of the pulse positions on the incoming information symbols.

10.4.3 Multistage Block-Spreading UWB Access

As considered in Section 10.4.1, it is possible to provide interference-free communications for up to N_c users over flat fading channels by designing orthogonal TH sequences. However, the total processing gain of an IR-UWB system is given by $N_f N_c$, assuming UWB pulses with duration T_c, which indicates that a much larger multiuser capacity is available [53]. The multistage block-spreading (MSBS) approach in [8] uses this large user capacity of IR-UWB systems by means of polarity codes [$d_j^{(k)}$ in Equation (10.1)] in addition to the TH codes.

Assuming that the total number of users satisfies $K \leq N_\mathrm{f} N_\mathrm{c}$, a TH sequence is assigned to a group of $\lfloor K/N_\mathrm{c} \rfloor$ (or $\lceil K/N_\mathrm{c} \rceil$) users. Then, the polarity codes (forming a "multiuser address") are employed to distinguish among the users in the same group. Moreover, the users in different groups are separated by their TH codes. Therefore, the same polarity codes can be assigned to the users in different groups. By this joint use of the TH and polarity sequences, $N_\mathrm{f} N_\mathrm{c}$ orthogonal user signals can be constructed [8].

In an MSBS IR-UWB system, the transmitter first spreads a block of symbols, and then chip-interleaving is performed. In this way, the mutual orthogonality between different users can be preserved even in multipath environments. The receiver first despreads the received signal by a linear filtering stage, which essentially reduces the multiple-access channel into a set of single-user ISI channels. Then, an equalizer is sufficient for a given user before the symbol detection without any need for additional multi-user signal processing.

Compared to the previous MAI mitigation techniques at the transmitter side, the MSBS approach has the advantage of supporting many more active users, since it effectively uses both the polarity and the TH codes for multiuser separation.

10.5 CONCLUDING REMARKS

In this chapter, we considered MAI mitigation techniques for IR-UWB systems. Due to the similarities between the RCDMA and IR systems, many multiuser detectors for CDMA systems can be directly employed in IR-UWB systems. However, the special signalling structure of IR-UWB systems facilitates the design of special multiuser detection algorithms that cannot be used for standard CDMA systems. Since the multiuser receivers for CDMA systems have been investigated thoroughly over the past two decades, we focused mainly on the special multiuser detection algorithms for IR-UWB systems in this study.

We first considered MAI mitigation techniques at the receiver side, which depend on the signal processing algorithms at the receiver only, without requiring any modification at the transmitter side. One of the main practical constraints in the receiver design is the sampling rate. In CDMA systems, chip rate sampling is employed and the MUD algorithms are run based on chip-rate samples. However, for UWB systems, chip-rate sampling can turn out to be on the order of billions of samples per second, which results in very high power consumption. Therefore, receivers based on frame-rate or symbol-rate samples are more realistic for UWB systems. The first receiver we considered is the "optimal" ML receiver in Section 10.3.1, which is based on the frame-rate samples from the user of interest. Because of the computational complexity of this receiver, we also presented a class of linear receivers, such as the pulse-discarding receiver, quasi-decorrelator, and quasi-MMSE receiver, which all employ frame-rate sampling. Moreover, in order to collect sufficient energy in the presence of multipath propagation, we considered optimal and suboptimal linear combining schemes. In suboptimal schemes, a multistage combining approach is employed to reduce computational complexity. After the linear receivers, we investigated the iterative (turbo) receivers, which can attain the single user performance with only a few iterations [12], but which are computationally more complex than the linear receivers [32]. Finally, in Section 10.3.4, we summarized some other multiuser detection algorithms, which are not specific to UWB systems but can be applied to them by considering the similarity between CDMA and IR-UWB systems. Specifically, we considered the frequency domain receivers, subtractive interference cancellation algorithms, blind, and subspace approaches.

In addition to the MAI mitigation techniques at the receiver side, we considered MAI mitigation techniques at the transmitter side, which mainly aim to design appropriate multiple-access sequences. We first considered the TH sequence design problem and presented TH codes for synchronous and asynchronous systems. Then, a multiple-access scheme using PCTH codes was considered. However, this scheme could not provide enough MAI mitigation in the presence of multipath propagation. Finally, the MSBS approach, which exploits both the TH and the polarity codes to increase the multiuser capacity, was described. This approach has the advantage of supporting many more active users than the previous ones.

REFERENCES

1. M. Z. Win and R. A. Scholtz, "Impulse radio: how it works," *IEEE Communications Letters*, vol. 2, no. 2, pp. 36–38, February 1998.
2. M. Z. Win and R. A. Scholtz, "On the energy capture of ultra-wide bandwidth signals in dense multipath environments," *IEEE Communications Letters*, vol. 2, pp. 245–247, September 1998.
3. M. L. Welborn, "System considerations for ultra-wideband wireless networks," *Proc. IEEE Radio and Wireless Conference*, pp. 5–8, Boston, MA, August 2001.
4. R. A. Scholtz, "Multiple access with time-hopping impulse modulation," *Proc. IEEE Military Communications Conference, 1993 (MILCOM'93)*, vol. 2, pp. 447–450, Bedford, MA, October 1993.
5. US Federal Communications Commission, "First Report and Order 02–48," Washington, DC, 2002.
6. M. Z. Win and R. A. Scholtz, "Ultra-wide bandwidth time-hopping spread-spectrum impulse radio for wireless multiple-access communications," *IEEE Transactions on Communications*, vol. 48, no. 4, pp. 679–691, April 2000.
7. E. Fishler and H. V. Poor, "On the tradeoff between two types of processing gain," *IEEE Transactions on Communications*, vol. 53, no. 10, pp. 1744–1753, October 2005.
8. L. Yang and G. B. Giannakis, "Multi-stage block-spreading for impulse radio multiple access through ISI channels," *IEEE Journal on Selected Areas in Communications*, vol. 20, no. 9, pp. 1767–1777, December 2002.

9. A. F. Molisch, Y. P. Nakache, P. Orlik, J. Zhang, Y. Wu, S. Gezici, S. Y. Kung, H. Kobayashi, H. V. Poor, Y. G. Li, H. Sheng, and A. Haimovich, "An efficient low-cost time-hopping impulse radio for high data rate transmission," *EURASIP Journal on Applied Signal Processing (Special Issue on UWB—State of the Art)*, vol. 2005, no. 3, pp. 397–412, March 2005.
10. J. Balakrishnan, A. Batra, and A. Dabak, "A multi-band OFDM system for UWB communication," *Proc. IEEE Conference on Ultra Wideband Systems and Technologies (UWBST 2003)*, pp. 354–358, Reston, VA, November 2003.
11. P. Runkle, J. McCorkle, T. Miller, and M. Welborn, "DS-CDMA: the modulation technology of choice for UWB communications," *Proc. IEEE Conference on Ultra Wideband Systems and Technologies (UWBST 2003)*, pp. 364–368, Reston, VA, November 2003.
12. E. Fishler and H. V. Poor, "Low-complexity multiuser detectors for time-hopping impulse-radio systems," *IEEE Transactions on Signal Processing*, vol. 52, no. 9, pp. 2561–2571, September 2004.
13. C. J. Le Martret and G. B. Giannakis, "All-digital impulse radio for wireless cellular systems," *IEEE Transactions on Communications*, vol. 50, no. 9, pp. 1440–1450, September 2002.
14. Y.-P. Nakache and A. F. Molisch, "Spectral shape of UWB signals—influence of modulation format, multiple access scheme and pulse shape," *Proc. IEEE 57th Vehicular Technology Conference (VTC 2003-Spring)*, vol. 4, pp. 2510–2514, Jeju, April 2003.
15. U. Madhow and M. L. Honig, "On the average near-far resistance for MMSE detection for direct sequence CDMA signals with random spreading," *IEEE Transactions on Information Theory*, vol. 45, pp. 2039–2045, September 1999.
16. J. S. Lehnert and M. B. Pursley, "Error probabilities for binary direct-sequence spread spectrum communications with random signature sequences," *IEEE Transactions on Communications*, vol. COM-35, pp. 87–98, January 1987.
17. E. Geraniotis and B. Ghaffari, "Performance of binary and quaternary direct-sequence spread-spectrum multiple-access systems with random signature sequences," *IEEE Transactions on Communications*, vol. 39, no. 5, pp. 713–724, May 1991.
18. G. Zang and C. Ling, "Performance evaluation for band-limited DS-CDMA systems based on simplified improved Gaussian approximation," *IEEE Transactions on Communications*, vol. 51, no. 7, pp. 1204–1213, July 2003.
19. S. Verdú. *Multiuser Detection*, Cambridge University Press, Cambridge, 1998.
20. Y. C. Yoon and R. Kohno, "Optimum multi-user detection in ultra-wideband (UWB) multiple-access communication systems," *Proc. IEEE International Conference on Communications*, pp. 812–816, New York, April 2002.
21. S. Moshavi, "Multi-user detection for DS-CDMA communications," *IEEE Communications Magazine*, vol. 34, no. 10, pp. 124–136, October 1996.
22. W. M. Lovelace and J. K. Townsend, "Chip discrimination for large near-far power ratios in UWB networks," *Proc. IEEE Military Communications Conference (MILCOM 2003)*, vol. 2, pp. 868–873, Boston, MA, 13–16, October 2003.
23. J. Evans and D. N. C. Tse, "Large system performance of linear multiuser receivers in multipath fading channels," *IEEE Transactions on Information Theory*, vol. IT-46, pp. 2059–2078, September 2000.

24. S. Gezici, H. Kobayashi, and H. V. Poor, "A comparative study of pulse combining schemes for impulse radio UWB systems," *Proc. IEEE Sarnoff Symposium 2004*, pp. 7–10, Princeton, NJ, 26–27, April 2004.
25. S. Gezici, M. Chiang, H. V. Poor, and H. Kobayashi, "Optimal and suboptimal finger selection algorithms for MMSE Rake receivers in impulse radio ultra-wideband systems," *Proc. IEEE Wireless Communications and Networking Conference (WCNC 2005)*, vol. 2, pp. 861–866, New Orleans, LA, 13–17, March 2005.
26. D. Cassioli, M. Z. Win, F. Vatalaro, and A. F. Molisch, "Performance of low-complexity RAKE reception in a realistic UWB channel," *Proc. IEEE International Conference on Communications (ICC 2002)*, vol. 2, pp. 763–767, New York, 28 April to 2 May 2002.
27. S. Gezici, H. Kobayashi, H. V. Poor, and A. F. Molisch, "Optimal and suboptimal linear receivers for time-hopping impulse radio systems," *Proc. IEEE Conference on Ultra Wideband Systems and Technologies (UWBST 2004)*, pp. 11–15, Kyoto, 18–21 May 2004.
28. H. V. Poor, "Iterative multiuser detection," *IEEE Signal Processing Magazine*, vol. 21, no. 1, pp. 81–88, January 2004.
29. A. R. Forouzan, M. Nasiri-Kenari, and J. A. Salehi, "Performance analysis of time-hopping spread-spectrum multiple-access systems: uncoded and coded schemes," *IEEE Transactions on Wireless Communications*, vol. 1, no. 4, pp. 671–681, October 2002.
30. A. Bayesteh and M. Nasiri-Kenari, "Iterative interference cancellation and decoding for a coded UWB-TH-CDMA system in AWGN channel," *Proc. 7th IEEE International Symposium on Spread Spectrum Techniques and Applications*, vol. 1, pp. 263–267, Prague, September 2002.
31. A. Bayesteh and M. Nasiri-Kenari, "Iterative interference cancellation and decoding for a coded UWB-TH-CDMA system in multipath channels using MMSE filters," *Proc. 14th IEEE International Symposium on Personal, Indoor and Mobile Radio Communications (PIMRC 2003)*, vol. 2, pp. 1555–1559, 7–10 September 2003.
32. E. Fishler and H. V. Poor, "Iterative ("turbo") multiuser detectors for impulse radio systems," preprint.
33. J. Hagenauer, "Forward error correcting for CDMA systems," *Proc. IEEE 4th International Symposium on Spread Spectrum Techniques and Applications*, vol. 2, pp. 566–569, Mainz, 22–25 September 1996.
34. K. Takizawa and R. Kohno, "Combined iterative demapping and decoding for coded UWB-IR systems," *Proc. IEEE Conference on UltraWideband Systems and Technologies (UWBST 2003)*, pp. 423–427, Reston, VA, 16–19 November 2003.
35. S. Benedetto, D. Divsalar, G. Montorsi, and F. Pollara, "Serial concatenation of interleaved codes: performance analysis, design, and iterative decoding," *IEEE Transactions on Information Theory*, vol. 44, no. 3, pp. 909–926, May 1998.
36. N. Yamamoto and T. Ohtsuki, "Adaptive internally turbo-coded ultra wideband-impulse radio (AITC-UWB-IR) system," *Proc. IEEE International Conference on Communications (ICC 2003)*, vol. 5, pp. 3535–3539, Anchorage, AK, 11–15 May 2003.
37. A. Muqaibel, B. Woerner, and S. Riad, "Application of multiuser detection techniques to impulse radio time hopping multiple access systems," *Proc. IEEE Conference on UltraWideband Systems and Technologies (UWBST 2002)*, pp. 169–173, Baltimore, MD, 21–23, May 2002.

38. V. Lottici, A. D'Andrea, and U. Mengali, "Channel estimation for ultra-wideband communications," *IEEE Journal on Selected Areas in Communications*, vol. 20, no. 9, pp. 1638–1645, December 2002.
39. Z. Xu, J. Tang, and P. Liu, "Frequency-domain estimation of multiple access ultra wideband signals," *Proc. IEEE Workshop on Statistical Signal Processing*, pp. 74–77, St Louis, MO, 28 September–01 October, 2003.
40. Z. Xu, P. Liu, and J. Tang, "Blind multiuser detection for impulse radio UWB systems," *Proc. IEEE Topical Conference on Wireless Communication Technology*, pp. 453–454, Honolulu, HI, 15–17 October 2003.
41. P. Liu, Z. Xu, and J. Tang, "Minimum variance multiuser detection for impulse radio UWB systems," *Proc. IEEE Conference on Ultra Wideband Systems and Technologies*, pp. 111–115, Reston, VA, 16–19 November 2003.
42. P. Liu and Z. Xu, "Performance of POR multiuser detection for UWB communications," *Proc. IEEE International Conference on Acoustics, Speech, and Signal Processing (ICASSP 2005)*, Philadelphia, PA, 19–23 March 2005.
43. H. Hotelling, "Analysis of a complex of statistical variables into principal component," *Journal of Education Psychology*, vol. 24, pp. 417–441 and 498–520, 1933.
44. C. Eckart and G. Young, "The approximation of one matrix by another of lower rank," *Psychometrica*, vol. 1, pp. 211–218, 1936.
45. P. Liu, Z. Xu, and J. Tang, "Subspace multiuser receivers for UWB communication systems," *Proc. IEEE Conference on Ultra Wideband Systems and Technologies*, pp. 116–120, Reston, VA, 16–19 November 2003.
46. J. S. Goldstein, I. S. Reed, and L. L. Scharf, "A multistage representation of the Wiener filter based on orthogonal projections," *IEEE Transactions on Information Theory*, vol. 44, no. 7, pp. 2943–2959, November 1998.
47. M. L. Honig and W. Xiao, "Performance of reduced-rank linear interference suppression," *IEEE Transactions on Information Theory*, vol. 47, no. 5, pp. 1928–1946, July 2001.
48. W. Sau-Hsuan, U. Mitra, and C.-C. J. Kuo, "Multistage MMSE receivers for ultra-wide bandwidth impulse radio communications," *Proc. IEEE Conference on Ultra Wideband Systems and Technologies (UWBST 2004)*, pp. 16–20, Kyoto, 18–21 May 2004.
49. M. S. Iacobucci and M. G. D. Benedetto, "Multiple access design for impulse radio communication systems," *Proc. IEEE International Conference on Communications (ICC 2002)*, vol. 2, pp. 817–820, New York, 28 April to 2 May 2002.
50. I. Guvenc and H. Arslan, "Design and performance analysis of TH sequences for UWB-IR systems," *Proc. IEEE Wireless Communications and Networking Conference (WCNC 2004)*, vol. 2, pp. 914–919, Atlanta, GA, 21–25 March 2004.
51. I. Guvenc, H. Arslan, S. Gezici, and H. Kobayashi, "Adaptation of multiple access parameters in time hopping UWB cluster based wireless sensor networks," *Proc. IEEE 1st International Conference on Mobile Ad-hoc and Sensor Systems (MASS 2004)*, pp. 235–244, Fort Lauderdale, FL, 25–27 October 2004.
52. O. Moreno and S. V. Maric, "A new family of frequency-hop codes," *IEEE Transactions on Communications*, vol. 48, no. 8, pp. 1241–1244, August 2000.
53. L. Yang and G. B. Giannakis, "Ultra-wideband communications: an idea whose time has come," *IEEE Signal Processing Magazine*, vol. 21, no. 6, pp. 26–54, November 2004.

54. G. M. Maggio, N. Rulkov, and L. Reggiani, "Pseudo-chaotic time hopping for UWB impulse radio," *IEEE Transactions on Circuits and Systems I: Fundamental Theory and Applications*, vol. 48, no. 12, pp. 1424–1435, December 2001.
55. G. M. Maggio, D. Laney, F. Lehmann, and L. Larson, "A multi-access scheme for UWB radio using pseudo-chaotic time hopping," *Proc. IEEE Conference on Ultra Wideband Systems and Technologies (UWBST 2002)*, pp. 225–229, Baltimore, MD, 21–23 May 2002.

CHAPTER 11

Narrowband Interference Issues in Ultra Wideband Systems

HÜSEYIN ARSLAN and MUSTAFA E. ŞAHIN

11.1 INTRODUCTION

Ultra wideband is becoming an attractive solution for wireless communications, particularly for short- and medium-range applications. UWB systems operate over extremely wide frequency bands (wider than 500 MHz), where various narrowband technologies also operate with much higher power levels (illustrated in Figure 11.1). The unlicensed usage of a very wide spectrum that overlaps with the spectra of narrowband technologies brings about some concerns. Therefore, significant research has been carried out lately to quantify the effect of UWB signals on narrowband systems [1].

The transmitted power of UWB devices is controlled by the regulatory agencies (such as the FCC in the United States), so that narrowband systems are affected by UWB signals only at a negligible level. This way, UWB systems are able to co-exist with narrowband technologies. However, looking at the fact from the other side, the influence of narrowband signals on the UWB system can still be significant, and in the extreme case, these signals may jam the UWB receiver completely. Even though narrowband signals interfere with only a small fraction of the UWB spectrum, due to their relatively high power with respect to the UWB signal, the performance and capacity of UWB systems can be affected considerably [2]. Recent studies show that the BER performance of the UWB receivers is greatly degraded due to the impact of narrowband interference (NBI) [3–8]. The high processing gain of the UWB signal can cope with the narrowband interferers to some extent. However, in many cases, even the large processing gain alone is not sufficient to suppress the effect of the high power interferers. Therefore, either the UWB system design needs to consider avoiding the transmission of the UWB signal over the frequencies of strong narrowband interferers, or the UWB receivers need to employ NBI

Ultra Wideband Wireless Communication. Edited by Arslan, Chen, and Di Benedetto
Copyright © 2006 John Wiley & Sons, Inc.

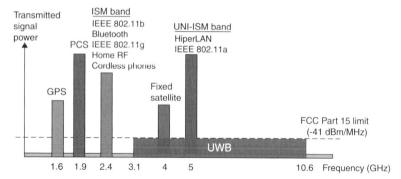

Figure 11.1 Spectrum crossover of the narrowband interferers in UWB systems.

suppression techniques to improve the performance, the capacity, and the range of the UWB communications.

NBI is not a new problem. It has been studied extensively for wideband systems like direct sequence spread spectrum–code division multiple accessing (DSSS-CDMA) based wireless communications [9], and for the operation of broadband OFDM systems in unlicensed frequency bands [10]. In DSSS-CDMA systems, NBI is partially handled with the processing gain as well as by employing interference cancelation techniques. Approaches including notch filtering [11], linear and nonlinear predictive techniques [12–17], adaptive methods [18–21], MMSE detectors [22, 23], and transform domain techniques [24–30] have been investigated extensively for interference suppression. Similarly, in OFDM systems, interference cancelation as well as interference avoidance techniques have been studied [10, 31–34]. Compared with these wideband systems, NBI suppression in UWB is a more challenging problem because of the restricted power transmission and the higher number of narrowband interferers due to the extremely wide band occupied. More significantly, in carrier modulated wideband systems, before demodulating the received signal both the desired wideband and the narrowband interfering signals are down-converted to the baseband, and the baseband signal is sampled at least with the Nyquist rate. Sampling at the Nyquist rate allows numerous efficient narrowband interference cancelation algorithms based on advanced digital signal processing techniques to be employed. However, in UWB, the desired signal is already in the baseband, while the narrowband interferer is in radio frequency (RF). Sampling the received signal at the Nyquist rate before the pulse correlator requires an extremely high sampling frequency, which is not possible with the existing technology. In addition to the high sampling rate, the ADC must support a very large dynamic range to resolve the signal from the strong narrowband interferers. Currently, such ADCs are far from being practical. An alternative is to apply analog (notch) filtering before the pulse correlation is considered. However, this method requires a number of narrowband analog filter banks, since the frequency and power of the narrowband interferers can vary. Therefore, employing analog filtering adds complexity, cost, and size to the UWB receivers. Also, adaptive implementation of the analog filters is not straightforward. As a result, many of the NBI suppression techniques applied to other wideband systems are either not applicable for UWB, or the complexities of these methods are too great for the UWB receiver requirements.

Given the low complexity requirements in both hardware and computation, and considering the other limitations such as low-power and low-cost transceiver design in many UWB applications, the NBI problem needs to be handled more carefully, and effective techniques that are able to cope with NBI need to be developed. One approach to deal with NBI is to avoid the transmission of the UWB signal over the frequencies of possible strong narrowband interferers. Attempts toward this goal include approaches like multiband-UWB (both using impulse radio based and OFDM based techniques) [35, 36]. Another approach to handling NBI is to design interference canceling receivers. However, as mentioned previously, the interference cancelation approach in UWB has more limitations compared with the conventional NBI cancelation approaches employed for other wideband systems. Very recently, some NBI cancelation techniques, most of which are based on the previous methods implemented in CDMA systems, have been considered for UWB. Analog bandpass filtering has been applied before the correlation receiver in [37]. As discussed above, fixed analog filtering is not an efficient solution, unless the interferer is fixed (i.e., the frequency, bandwidth, power, and channel of the interferer are constant), and always exists. In [26], notch filtering (or peak clipping) is applied by carrying out a high-speed sampling before the correlation. The frequency domain signal is obtained from these digital samples through front-end FFT. Then, the narrowband interferers in the frequency domain, which are the collection of large peaks in the frequency, are clipped or notch-filtering is applied on these locations. However, as discussed above, high sampling rate before correlation makes the practical and cost-effective implementation of this technique difficult. Modifying and estimating the optimal receiver template for the correlation of the received signal is another solution that is proposed for partial suppression of NBI [37, 38]. By far the most popularly considered approach is the use of a rake (multiple correlators) receiver along with MMSE combining [39–42]. MMSE combining is known to perform well when the noise is not white (i.e., noise on different rake fingers is correlated). The performance of MMSE depends on the number of fingers. Note also that rake receivers are much more complex compared with the correlation receivers, and their complexity increases with the number of rake fingers.

In this chapter, NBI in UWB systems will be studied. First, the effect of NBI on the performance of UWB transmission will be discussed. Appropriate models for NBI sources will be investigated. Then, techniques for avoiding NBI in UWB system design will be reviewed. Approaches including multiband/multicarrier transmission and pulse shaping for avoiding NBI will be discussed briefly. In Section 11.4, NBI handling approaches based on interference cancelation will also be investigated for relaxing the system and transmission requirements. Finally, Section 11.5 will conclude the chapter with a discussion of some future research areas.

11.2 EFFECT OF NBI IN UWB SYSTEMS

According to the modern definition, UWB transmission is not limited to the impulse radio. Any technology that has a bandwidth greater than 500 MHz or a fractional bandwidth greater than 0.2 can be considered a UWB system. Therefore, depending on the access technology, the signal and interference models might vary. In general, the UWB signal bandwidth is extremely large and the transmitted signal power is very low. In contrast, the narrowband signal occupies a much smaller bandwidth, where its power spectrum is very high. Another distinction is that the narrowband signal is modulated with a carrier, and it is a continuous time signal, whereas the UWB signal can be a baseband signal composed of discrete short-time pulses as well as a carrier modulated signal.

Impulse radio-based UWB transmission has some similarities to the widely used spread spectrum (SS) systems. In the SS systems such as direct sequencing, the bandwidth occupied is larger than the bandwidth required for transmitting the data bits for a single user. Each user is assigned a PN sequence of N chips (where the chip durations are much shorter than the actual symbol duration) to transmit a symbol. Without the spreading, the same transmission bandwidth can be used to transmit N information symbols, but the spreading operation allows simultaneous transmission of information from multiple users on the same bandwidth without interfering with each other, leading to the CDMA type of multiplexing. Therefore, even though the peak data rate is reduced for a single user, the capacity of the system is preserved to a great extent by allowing multiple users in the system. In addition to these, spreading provides immunity to interference sources like NBI, reduces the power of the transmitted signal (so that it causes less interference to other systems sharing the same band), allows path diversity in the presence of multipath signals that are longer than a chip period, and last but not least, provides covert communications.

The NBI jamming resistance of DSSS systems has been studied extensively [6]. The jamming resistance in these systems is provided by the processing gain, which is obtained by spreading. The larger the spreading ratio (the ratio of bandwidths of the spread signal and the original information), the higher the processing gain, and hence, the better jamming resistance is obtained. At the receiver, the transmitted spread spectrum signal and the narrowband interferer go through a despreading operation, where the receiver takes the wideband spread spectrum signal and collapses it back to the original data bandwidth, while spreading the interferer to a wide spectrum. As a result, within the data bandwidth, the effect of interferer is mitigated, a fact which is referred as the jamming resistance or the natural interference immunity of the spread spectrum signals.

Similar to the DSSS systems, impulse radio based UWB also has inherent immunity to NBI. Time-hopping UWB systems can be considered as an example. The processing gain of TH-UWB signal is mainly obtained by transmitting very narrow pulses with a very low duty cycle. Figure 11.2(a) demonstrates a simple scenario that shows the UWB pulses and the continuous time narrowband interferer.

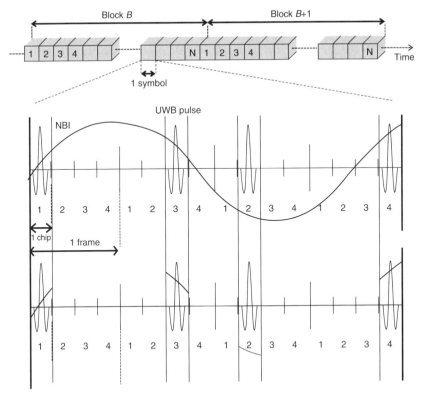

Figure 11.2 (a) TH-UWB pulses along with a narrowband interferer. (b) Reduced interference power by means of time gating.

During the reception of TH-UWB signals, using a matched filter that basically operates as a time gate (i.e., lets the UWB signal along with interference pass over the duration of the expected pulses, and blocks the rest of the received signal), the power of the interfering signal is reduced significantly [Figure 11.2(b)]. As a result, jamming resistance against NBI is obtained. Note that there will be still partial interference at the output of the matched filter depending on the processing gain, the power of the interferer, and other factors.

It is necessary to investigate the models of the UWB signal and narrowband interferers for a thorough understanding of NBI effects on UWB systems. Considering a binary pulse position modulated (BPPM) time-hopping UWB signal, the transmitted waveform can be modeled as [43]

$$s_{\text{tr}}(t) = \sum_{i=-\infty}^{+\infty} \omega_{\text{tr}}(t - iT_{\text{f}} - c_i T_{\text{c}} - \delta d) \quad (11.1)$$

where ω_{tr} denotes the UWB pulse, T_f is the pulse repetition duration, c_i is the time-hopping code in the ith frame, T_c is the chip time, δ is the pulse position offset regarding BPPM, and d represents the data, which is a binary number.

Depending on its type, the narrowband interference can be modeled in various ways. For example, it can be considered to consist of a single tone interferer, which can be modeled as

$$i(t) = \gamma\sqrt{2P_i} \cos(2\pi f_c t + \phi_i), \tag{11.2}$$

where γ is the channel gain, P_i is the average power, f_c is the frequency of the sinusoid, and ϕ_i is the phase.

NBI can also be thought as the effect of a band limited interferer; then the corresponding model is a zero-mean Gaussian random process and its power spectral density is as follows:

$$S_i(f) = \begin{cases} P_{int}, & f_c - \dfrac{B}{2} \leq |f| \leq f_c + \dfrac{B}{2} \\ 0, & \text{otherwise} \end{cases} \tag{11.3}$$

where B and f_c are the bandwidth and the center frequency of the interferer, respectively, and P_{int} is the power spectral density.

Since the narrowband signal has a bandwidth much smaller than the coherence bandwidth of the channel, the time domain samples of the NBI are highly correlated with each other. Therefore, for the investigation of the narrowband interferers, the correlation functions are of primary interest, rather than the time- or frequency-domain representations. The correlation functions corresponding to the single tone and band limited cases can be written as

$$R_i(\tau) = P_i \, |\gamma|^2 \cos(2\pi f_c \tau), \tag{11.4}$$

$$R_i(\tau) = 2P_{int} \, B \cos(2\pi f_c \tau) \, \text{sinc}(B\tau), \tag{11.5}$$

respectively. The resulting correlation matrices for the kth and lth interference samples are [44]

$$[\mathbf{R_i}]_{k,l} = 4 \, N_s P_i \, |\gamma|^2 \, |W_r(f_c)|^2 \, [\sin(\pi f_c \delta)]^2 \, \cos[2\pi f_c(\tau_k - \tau_l)] \tag{11.6}$$

for the single tone interferer, and

$$\begin{aligned}[\mathbf{R_i}]_{k,l} = {} & 2 \, N_s \, P_{int} \, B \, |W_r(f_c)|^2 \\ & \times \{2\cos[2\pi f_c(\tau_k - \tau_l)] \, \text{sinc}[B(\tau_k - \tau_l)] \\ & - \cos[2\pi f_c(\tau_k - \tau_l - \delta)] \, \text{sinc}[B(\tau_k - \tau_l - \delta)] \\ & - \cos[2\pi f_c(\tau_k - \tau_l + \delta)] \, \text{sinc}[B(\tau_k - \tau_l + \delta)]\}\end{aligned} \tag{11.7}$$

for the case of band limited interference, where $|W_r(f_c)|^2$ is the power spectral density of the received signal at the frequency f_c.

Another strong candidate for UWB communications beside the impulse radio is the multicarrier approach, which can be implemented using OFDM. OFDM has become a very popular technology for wireless communications due to its special features such as robustness against multipath interference, ability to allow frequency diversity with the use of efficient FEC coding, capability of capturing the multipath energy efficiently, and ability to provide high bandwidth efficiency through the use of sub-band adaptive modulation and coding techniques. A strong motivation for employing OFDM in UWB applications is its resistance to NBI, and its ability to turn the transmission *on* and *off* on separate carriers depending on the level of interference. The NBI models that can be considered for OFDM include one or more tone interferers, as well as a zero-mean Gaussian random process that occupies certain carriers along with white noise as

$$S_n(k) = \begin{cases} \dfrac{N_i + N_w}{2}, & \text{if } k_1 < k < k_2 \\ \dfrac{N_w}{2}, & \text{otherwise} \end{cases} \quad (11.8)$$

where k is the carrier index, K is the total number of carriers, and $N_i/2$ and $N_w/2$ are the spectral densities of the narrowband interferer and white noise, respectively.

11.3 AVOIDING NBI

NBI can be avoided at the receiver by properly designing the transmitted UWB waveform. If the statistics regarding the NBI are known, the transmitter can adjust the transmission parameters appropriately. NBI avoidance can be achieved in various ways, and it depends on the type of access technology.

11.3.1 Multicarrier Approach

Multicarrier approach can be one way of avoiding NBI. OFDM, which was mentioned in the previous section, is a well-known example of multicarrier techniques. In OFDM-based UWB, NBI can be avoided easily by an adaptive OFDM system design. As the simple interference scenario illustrated in Figure 11.3 shows, NBI will corrupt only some carriers in OFDM spectrum. Therefore, only the information that is transmitted over these frequencies will be affected by the interference. If the interfered carriers can be identified, transmission over these carriers can be avoided. In addition, by sufficient FEC and frequency interleaving, jamming resistance against NBI can be obtained easily. Avoiding or adapting the transmission over the strongly interfered carriers can provide more spectrum and power efficiency, as they increase the immunity against NBI, and hence relax the FEC coding power requirement.

262 NARROWBAND INTERFERENCE ISSUES IN ULTRA WIDEBAND SYSTEMS

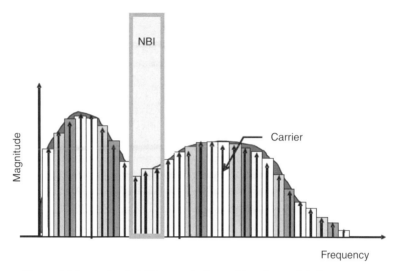

Figure 11.3 A simple NBI scenario for multicarrier modulation systems.

At the OFDM receiver, the signal is received along with noise and interference. After synchronization and removal of the cyclic prefix, FFT is applied to convert the time-domain received samples to the frequency-domain signal. The received signal at the kth sub-carrier of the nth OFDM symbol can then be written as

$$Y_{n,k} = S_{n,k} H_{n,k} + \underbrace{I_{n,k} + W_{n,k}}_{\text{NBI+AWGN}}, \tag{11.9}$$

where $S_{n,k}$ is the transmitted symbol which is obtained from a finite set (e.g., QPSK or QAM), $H_{n,k}$ is the value of the channel frequency response, $I_{n,k}$ is the NBI, and $W_{n,k}$ denotes the uncorrelated Gaussian noise samples. The impairments due to imperfect synchronization, transceiver nonlinearities, etc. can be folded into the noise term $W_{n,k}$.

In OFDM, in order to identify the interfered carriers, the transmitter requires a feedback from the receiver. The receiver should have the ability to identify these interfered carriers. Once the receiver estimates these carriers, the relevant information will be sent back to the transmitter. The transmitter will then adjust the transmission accordingly. Note that, in such a scenario, the interference statistics need to be constant for a certain period of time. If the interference statistics change very fast, by the time the transmitter receives feedback, and adjusts the transmission parameters, the receiver might observe different interference characteristics.

The feedback information can be manifold, including the interfered carrier index, in some cases the amount of interference on these carriers, the center frequency of NBI, and the bandwidth of NBI. The identification of the interfered carriers can be accomplished by different means. One simple technique is to look at the average signal power in each carrier, and compare it with a threshold. If the average received signal power of a subcarrier is greater than the threshold, that channel can be regarded as severely interfered with NBI. Instead of making a hard-decision on whether a carrier is interfered or not, soft estimation of NBI power can also be done [45].

11.3.2 Multiband Schemes

Similar to the multicarrier approach, multiband schemes are also considered for avoiding NBI. Rather than employing a UWB radio that uses the entire 7.5 GHz band to transmit information, by exploiting the flexibility of the FCC definition of the minimum bandwidth of 500 MHz, the spectrum can be divided into smaller sub-bands. The combination of these sub-bands can be used freely for optimizing the system performance. By partitioning the spectrum into smaller chunks (which are still larger than 500 MHz), a better co-existence with other current and future wireless technologies can be achieved. This approach will also enable worldwide inter-operability of the UWB devices, as the spectral allocation for UWB could possibly be different in various parts of the world. In multiband systems, information on each of the sub-bands can be transmitted using either single-carrier (pulse-based) or multicarrier (OFDM) techniques. Figure 11.4 shows some representative multi-band schemes. The pulse-based approach [as shown in Figure 11.4(a)] uses dual-band with bandwidths in each band exceeding 1 GHz [46]. The lower band occupies the spectrum from 3.1 GHz to 4.85 GHz, and the upper band occupies the spectrum from 6.2 GHz to 9.7 GHz. The spectrum in between upper and lower bands is not used for UWB transmission, since potential interference sources like IEEE 802.11a operate in this unlicensed band. The OFDM-based multiband approach

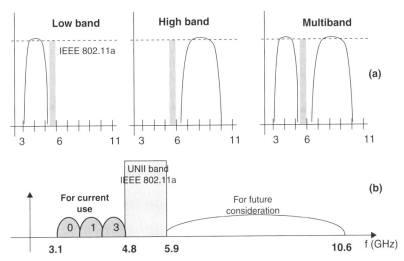

Figure 11.4 Some proposed multiband approaches for WPAN: (a) the Xtreme Spectrum-Motorola proposal of a dual-band approach [46]. (b) Multiband OFDM [47].

[shown in Figure 11.4(b)] uses 528 MHz channels in each band, where the three lower band channels are for initial deployments and mandatory, and the upper bands are optional and for future use [47]. As the radio frequency technology improves, the upper bands are expected to be included into the system gradually.

11.3.3 Pulse Shaping

Another technique for avoiding narrowband interference is pulse shaping. As can be seen in Equations (11.6) and (11.7), the effect of interference is directly related to the spectral characteristics of the receiver template pulse waveform. That means, if the transmission at the frequencies where NBI is present can be avoided, the influence of interference on the received signal can be mitigated significantly. Therefore, designing the transmitted pulse shape properly, such that the transmission at some specific frequencies is omitted, NBI avoidance can be realized. An excellent example for the implementation of this approach is the Gaussian doublet [48]. A Gaussian doublet, representing one bit, consists of a pair of narrow Gaussian pulses with opposite polarities. Considering the time delay T_d between the pulses, the doublet can be represented as

$$s_d(t) = \frac{1}{\sqrt{2}}[s(t) - s(t - T_d)]. \qquad (11.10)$$

The corresponding spectral amplitude of the doublet is then

$$|S_d(f)|^2 = 2|S(f)|^2 \sin^2(\pi f T_d), \qquad (11.11)$$

where $|S(f)|^2$ is the power spectrum of a single pulse. Notice that, due to the sinusoidal term in Equation (11.11), the power spectrum will have nulls at $f = n/T_d$, where n can be any integer (shown in Figure 11.5). The basic idea for avoiding NBI is adjusting the location of these nulls in such a way that they overlap with the peaks created by narrowband interferers. By modifying the time delay T_d, a null can be obtained at the specific frequency where NBI exists, and this way the strong effect of the interferer can be avoided. If T_d is adjusted to 2 ns, for example, the interferences located at the integer multiples of 500 MHz can be suppressed.

The purpose of avoiding NBI through abstaining transmission at frequencies of interference can also be carried out by making use of notch filters in the transmitter. To accomplish this, the parameters of the filters have to be adjusted such that the notches they create overlap with the frequencies of strong NBI. When notch filters are employed in the transmitter, the transmitted pulse is shaped in such a way (Figure 11.6) that the correlation of NBI with the pulse template in the receiver is minimized.

Pulse shaping techniques are not limited to the Gaussian doublet and notch filtering. Another feasible method is the adjustment of the PPM modulation parameter δ. Revisiting the correlation matrix for the single tone interferer given in Equation (11.6), it is seen that $[\mathbf{R}_i]_{k,l} = 0$ for $\delta = n/f_c$, where $n = 1, 2, \ldots, M$, M

11.3 AVOIDING NBI

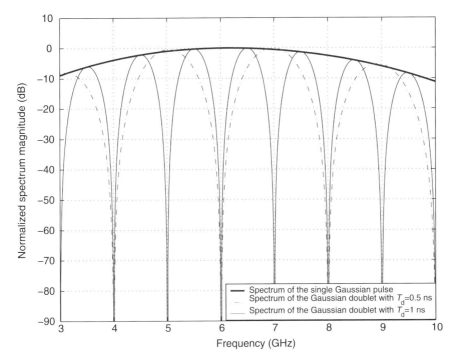

Figure 11.5 Normalized spectra for the single Gaussian pulse and two different Gaussian doublets.

being the number of possible pulse positions. Therefore, an effective interference avoidance can be attained by setting δ to n/f_c. Similarly, considering the correlation matrix corresponding to the band limited interference (11.7), it is seen that $\cos[2\pi f_c(\tau_k - \tau_l \pm \delta)] = \cos[(2\pi f_c(\tau_k - \tau_l)]$, when $\delta = n/f_c$. Also, in the light of the knowledge that the bandwidth of the interference (B) is much smaller than its center frequency (f_c), the assumption $\text{sinc}[B(\tau_k - \tau_l \pm \delta)] \simeq \text{sinc}[B(\tau_k - \tau_l)]$ can be made for $\delta = n/f_c$. These two facts lead to the conclusion that $[\mathbf{R_i}]_{k,l}$ in Equation (11.7) becomes zero for the band limited interference case, too, when δ is set to n/f_c.

Although the adjustment of the PPM modulation parameter δ is a straightforward way of avoiding NBI, it has an important drawback. The correlation output is also dependent on δ, and for a certain value of it a maximum signal correlation can be obtained. However, this value of δ does not necessarily have to be equal to $1/f_c$. For the AWGN case (without considering the NBI), the bit-error-rate function from which the optimum δ can be determined is [49]

$$Q\left(\sqrt{\frac{N_s A E_p}{N_0}} R_{\text{opt}}\right), \tag{11.12}$$

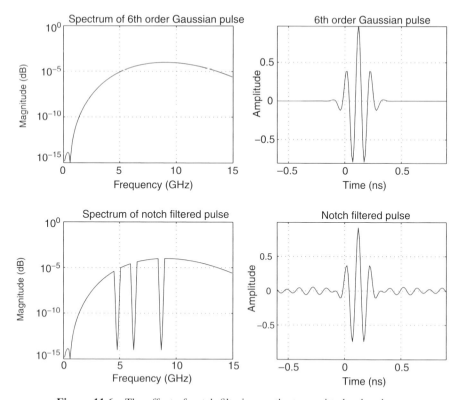

Figure 11.6 The effect of notch filtering on the transmitted pulse shape.

where $R_{opt} = R(0) - R(\delta_{opt})$, N_s is the number of pulses per symbol, A is the pulse amplitude, $N_0/2$ is the double-sided power spectral density of AWGN, and $R(\Delta t)$ is the autocorrelation function of the received pulse. Therefore, there is an obvious trade-off between maximizing R_{opt} and avoiding NBI, when determining the δ parameter. Depending on the level of NBI and AWGN, this parameter can be adjusted to provide an optimal performance.

11.3.4 Other NBI Avoidance Methods

For the time-hopping UWB systems, it is possible to avoid NBI by placing notches in the spectrum by adjusting the time-hopping code [50]. In [51], a PAM UWB signal is considered. Each symbol has a duration of T_s and is composed of N_s pulses, giving rise to N_s frames, which last for $T_f = T_s/N_s$ and are divided into chips with a duration of T_c. The pseudorandom TH code determines the position of the pulse inside the frame by selecting the chip where to place the pulse. In short, a PAM UWB signal over a symbol duration can be written as

$$u(t) = A \sum_{n=0}^{N_s-1} \omega(t - c_n T_c - n T_f - T_s), \qquad (11.13)$$

where $A \varepsilon \{-1, 1\}$ denotes the amplitude of the pulse, and c_n is the TH code. In [50], the spectrum shape for the multisymbol case is given by

$$P_u(f) = |W(f)|^2 \sum_{k=0}^{N_b-1} |T_k(f)|^2, \qquad (11.14)$$

where $W(f)$ is the Fourier transform of the transmitted pulse, N_b is the total number of different TH codes used, k is the symbol index, and

$$T_k(f) = \sum_{n=0}^{N_s-1} \exp[-j2\pi f(c_{n,k}T_c + nT_f + kT_s)]. \qquad (11.15)$$

From Equation (11.15), it is seen that changing the time-hopping code causes the spectrum of the transmitted signal to vary. This means that, by employing various methods, the TH code can be adjusted in such a way that spectral notches are created at frequencies of strong NBI, allowing the system to avoid interference.

In addition to the methods mentioned, physical solutions can also be considered for avoiding NBI. In [52], an NBI avoidance technique depending on antenna design is proposed. The main idea is to create frequency notches by intentionally adding a narrowband resonant structure to the antenna, and thus making it insensitive to some particular frequencies. This technique is more economical than the explicit notch filtering method since it does not require additional notch filters. In [52], a frequency-notched UWB antenna suitable for avoiding NBI is explained in detail. This special-purpose antenna is obtained by employing planar elliptical dipole antennas and incorporating a half-wave resonant structure, which is obtained by implementing triangular and elliptical notches. It is necessary to note that the performance of the antenna is reduced with increasing number of notches. This leads to the idea that the frequency notched antenna may not be successful enough in avoiding numerous simultaneously existing narrowband interferers.

11.4 CANCELING NBI

Although most of the avoidance methods mentioned seem to have a high feasibility, they may not be implemented under all circumstances. The main limitation on these methods is their dependency on the exact knowledge about narrowband interferers. Without having the accurate information about the center frequency of the interference, suppressing NBI is not possible by means of any of the avoidance techniques explained. Even if the complete knowledge about the NBI is available, if there is an abundant number of interferers, methods like employing notch filters or changing the parameters of the transmitted pulse may lose their practicality. If it is not possible to avoid NBI at the transmission stage for any reason, one should make effort at the receiver side to extract and eliminate it from the received signal.

Throughout the previous section, methods of avoiding NBI have been discussed and limitations on their realization have been mentioned. In practice, UWB systems that employ only avoidance techniques are not totally successful in eliminating NBI.

268 NARROWBAND INTERFERENCE ISSUES IN ULTRA WIDEBAND SYSTEMS

In this section, an overview of different types of NBI cancelation methods will be provided.

11.4.1 MMSE Combining

One of the popular receivers considered for UWB is the rake receiver. Rake receivers are designed to collect the energy of strong multipath components, and with this purpose they employ *fingers*. In each rake finger, there is a correlation receiver synchronized with one of the multipath components. The correlation receiver is followed by a linear combiner whose weight is determined depending on the combination algorithm used. The output of the receiver for the *i*th pulse can be denoted as [39]

$$y_i = \sum_{k=0}^{M-1} d_i c_k \psi \beta_k + c_k n_k, \qquad (11.16)$$

where M is the number of rake fingers, d_i is the data bit transmitted on the *i*th pulse, c_k, β_k, and n_k are the weight used by the combiner, the channel gain, and the noise for the *k*th multipath component, respectively, and

$$\psi = \int_{t=-\infty}^{\infty} \omega_{\text{rx}}(t) v(t) \, dt, \qquad (11.17)$$

where $\omega_{\text{rx}}(t)$ denotes the received waveform, and $v(t)$ is the correlating function.

In the traditional rake receiver, which employs MRC, the weight of the combiner is the conjugate of the gain of the particular multipath component ($c = \beta^*$). Such a selection maximizes the SNR in the absence of NBI. However, when NBI exists, since interference samples are correlated, MRC is no longer the optimum method. MMSE combining, which is an alternative approach, depends on varying these weights in such a way that the mean square error between the required and actual outputs is minimized. In the existence of interference, the SNR is maximized when MMSE weight vector is used [53]:

$$\mathbf{c} = \alpha \mathbf{R_n}^{-1} \beta, \qquad (11.18)$$

where $\mathbf{c} = [c_1 c_2 \cdots c_M]^T$, α is the scaling constant, $\mathbf{R_n}^{-1}$ is the inverse of the correlation matrix of noise plus interference, and $\beta = [\beta_1 \beta_2 \cdots \beta_M]^T$ is the channel gain vector

The NBI cancelation methods other than MMSE combining can be grouped in three categories as frequency domain, time–frequency domain, and time domain approaches.

11.4.2 Frequency Domain Techniques

Cancellation techniques in the frequency domain can be exemplified by notch filtering in the receiver side. Having an estimation about the frequencies of powerful narrowband interferers, notch filters can be used to suppress NBI. The pleasant fact about this method is that it can be utilized in almost all kind of receivers, so that the UWB system is not forced to employ a correlation-based receiver. The main weakness of frequency domain methods, on the other hand, is that they are useful only when the received signal, which is a superposition of the UWB signal and NBI from various sources, exhibits stationary behavior. If the received signal has a time-varying nature, methods that analyze the frequency content taking the temporal changes into account are required. These methods are called the time–frequency approaches.

11.4.3 Time–Frequency Domain Techniques

The most commonly employed time-frequency domain method for interference suppression is the wavelet transform. Similar to the well-known Fourier transform, the wavelet transform also employs basis functions, and expresses any time domain signal as a combination of them. However, these basis functions, which are called wavelets, are different from the complex exponentials used by the Fourier transform in the sense that they are not time unlimited. Hence, the wavelet transform is able to represent the time local characteristics of signals, and is not limited to stationary signals like the Fourier transform. A wavelet is defined as

$$\psi_{ab}(t) = \frac{1}{|\sqrt{a}|} \psi\left(\frac{t-b}{a}\right), \quad (11.19)$$

where a and b are the scaling and shifting parameters, respectively. If these parameters are set as $a = 1$ and $b = 0$, the mother wavelet is obtained. By dilating and shifting the mother wavelet, a family of daughter wavelets are formed. The continuous wavelet transform can be expressed as

$$W(a, b) = \int_{-\infty}^{+\infty} f(t)\psi_{ab}(t)\, dt. \quad (11.20)$$

The version of the wavelet transform that is appropriate for computer implementation is the discrete wavelet transform (DWT), which is defined as [25]

$$d_{m,n} = \frac{1}{\sqrt{a_0^m}} \int f(t)\psi\left(\frac{t}{a_0^m} - nb_0\right) dt, \quad (11.21)$$

where m and n are integers.

Computers realize the DWT not by using wavelets, but by employing filters. An effective algorithm for performing DWT based on using filters was proposed by Mallat [54]. The Mallat algorithm results in a detailed analysis, in which the lowest frequency component is expressed with the smallest number of samples, whereas the largest number of samples expresses the highest frequency component.

One possible way of suppressing the narrowband interference using the wavelet transform is to have the transmitter part of the UWB system estimate the electromagnetic spectrum, and set a proper threshold for interference detection [55]. The interference level at each frequency component is then determined with the wavelet transform, and compared with this threshold in order to distinguish between the interfered and not interfered frequency components. According to the results of this comparison step, the transmitter does not transmit at frequencies where strong NBI exists. Obviously, this method is quite similar to the multicarrier approach in NBI avoidance techniques.

Methods employing the wavelet transform in the receiver side of the system also exist [56, 57]. In these methods, wavelet transform is applied to the received signal, and frequency components with a considerably high energy are considered to be affected by narrowband interference. These components are then suppressed using conventional methods like notch filtering.

Although the discrete wavelet transform is a very useful tool for eliminating NBI, the inability of current ADCs to sample the UWB signals at the Nyquist rate sets a practical limit on the feasibility of this method. Therefore, the effectiveness of DWT at the frame-rate and symbol-rate sampling has to be investigated thoroughly to be able to decide about the usefulness of this approach with the existing technology.

11.4.4 Time Domain Techniques

The third group of NBI cancelation methods is the time domain approaches, which can also be called predictive methods. Predictive methods are based on the assumption that the predictability of narrowband signals is much higher than the predictability of wideband signals, because wideband signals have a nearly flat spectrum [9]. Hence, in a UWB system, a prediction of the received signal is expected to primarily reflect the narrowband interference rather than the UWB signal. This fact leads to the consequence that NBI can be canceled by subtracting the predicted signal from the received signal.

Predictive methods can be classified as linear and nonlinear techniques. Linear techniques employ transversal filters in order to obtain an estimate of the received signal depending on the previous samples and model assumptions [16]. If one-sided taps are used, the filter employed is a linear prediction filter, whereas it is a linear interpolation filter if the taps are double-sided. It is worth noting that interpolation filters proved more effective in canceling NBI.

Common examples for linear predictive methods are the Kalman–Bucy prediction, which is based on the Kalman–Bucy filter with infinite impulse response (IIR), and least-mean-squares (LMS) algorithm based on a finite impulse response (FIR) structure.

Nonlinear methods are found to provide a better solution than linear ones for DS systems because they are able to make use of the highly non-Gaussian structure of the DS signals [9]. However, for UWB systems, this is not the case because such a non-Gaussianity does not exist in UWB signals.

Adaptive prediction filters are considered as a powerful tool against NBI. When an interferer is detected in the system, the adaptation algorithm creates a notch to suppress the interference caused by this source. However, if the interferer vanishes suddenly, since there is no mechanism to respond immediately to remove the notch created, the receiver continues to suppress the portion of the wanted signal around the notch. If narrowband interferers enter and exit the system in a random manner, this shortcoming reduces the performance of the adaptive system dramatically. A more useful algorithm is proposed in [16], where a hidden-Markov model (HMM) is employed to keep track of the interferers entering and exiting the system. In this algorithm, the frequency locations where an interferer is present are detected by an HMM filter, and a suppression filter is put there. When the system detects that the interferer has vanished, the filter is removed automatically.

11.5 CONCLUSION AND FUTURE RESEARCH

In this chapter, an overview of narrowband interference in UWB systems is given. The significance of the NBI problem for UWB systems has been discussed, different models for NBI have been analyzed, and the effects of NBI on UWB communications have been addressed. Methods of dealing with NBI have been examined under two separate categories as NBI avoidance and NBI cancelation algorithms. NBI avoidance methods including multicarrier approaches and multiband schemes, as well as alternative solutions based on pulse shaping, time-hopping code adjustment, and antenna design have been investigated. Among the cancelation techniques, details of MMSE combining algorithm are presented. Frequency domain techniques such as notch filtering, time–frequency methods like wavelet transform and time domain approaches, particularly linear techniques, have been discussed in separate sections.

As of now, none of the avoidance or cancelation methods has proved to be the optimum solution to the NBI problem. It seems that the most inexpensive and successful way of suppressing NBI can be achieved by employing an adaptive method combining the avoidance and cancelation approaches. The UWB communications can be initially started by applying the proper avoidance methods in the transmitter side; then in the light of the feedback provided by the receiver, the effectiveness of interference excision can be determined, and if it is found that the interferers cannot be suppressed satisfactorily, NBI cancelation methods can be run in the receiver side of the system. Considering that the computational burden related to the cancelation methods is generally much higher than avoidance methods, such an adaptive approach can be very useful in terms of wise usage of resources.

REFERENCES

1. R. Johnk, D. Novotny, C. Grosvenor, N. Canales, and J. Veneman, "Time-domain measurements of radiated and conducted UWB emissions," *IEEE Aerospace Electron. Syst. Mag.*, vol. 19, 2004, August 2004, pp. 18–22.

2. J. Foerster, "Ultra-wideband technology enabling low-power, high-rate connectivity (invited paper)," in *Proc. IEEE Workshop Wireless Commun. Networking*, Pasadena, CA, September 2002.
3. J. R. Foerster, "The performance of a direct-sequence spread ultra-wideband system in the presence of multipath, narrowband interference, and multiuser interference," in *Proc. IEEE Vehicular Technology Conf.*, vol. 4, Birmingham, AL, May 2002, pp. 1931–1935.
4. J. Choi and W. Stark, "Performance of autocorrelation receivers for ultra-wideband communications with PPM in multipath channels," in *Proc. IEEE Ultrawideband Systems and Technology (UWBST)*, Baltimore, MD, May 2002, pp. 213–217.
5. W. Tao, W. Yong, and C. Kangsheng, "Analyzing the interference power of narrowband jamming signal on UWB system," in *Proc. IEEE Personal, Indoor, Mobile Radio Commun. (PIMRC)*, Singapore, September 2003, pp. 612–615.
6. L. Zhao and A. Haimovich, "Performance of ultra-wideband communications in the presence of interference," *IEEE J. Select. Areas Commun.*, vol. 20, pp. 1684–1691, December 2002.
7. G. Durisi and S. Benedetto, "Performance evaluation of TH-PPM UWB systems in the presence of multiuser interference," *IEEE Commun. Lett.*, vol. 7, no. 5, pp. 224–226, May 2003.
8. R. Tesi, M. Hamelainen, J. Iinatti, and V. Hovinen, "On the influence of pulsed jamming and coloured noise in uwb transmission," in *Proc. Finnish Wireless Commun. Workshop (FWCW)*, Espoo, May 2002.
9. X. Wang and H. V. Poor, *Wireless Communication Systems: Advanced Techniques for Signal Reception*, 1st edn, Prentice Hall, Englewood Cliffs, NJ, 2004.
10. D. Zhang, P. Fan, and Z. Cao, "Interference cancellation for OFDM systems in presence of overlapped narrow band transmission system," *IEEE Trans. Consum. Electron.*, vol. 50, pp. 108–114, February 2004.
11. J. Choi and N. Cho, "Narrow-band interference suppression in direct sequence spread spectrum systems using a lattice IIR notch filter," in *Proc. IEEE Int. Conf. Acoustics, Speech, Signal Processing (ICASSP)*, vol. 3, Munich, April 1997, pp. 1881–1884.
12. L. Rusch and H. Poor, "Narrowband interference suppression in CDMA spread spectrum communications," *IEEE Personal Commun. Mag.*, vol. 42, pp. 1969–1979, April 1994.
13. L. Rusch and H. Poor, "Multiuser detection techniques for narrowband interference suppression in spread spectrum communications," *IEEE Trans. Commun.*, vol. 42, pp. 1727–1737, April 1995.
14. J. Proakis, "Interference suppression in spread spectrum systems," in *Proc. IEEE Int. Symp. on Spread Spectrum Techniques and Applications*, vol. 1, September 1996, pp. 259–266.
15. L. Milstein, "Interference rejection techniques in spread spectrum communications," in *Proc. IEEE*, vol. 76, June 1988, pp. 657–671.
16. C. Carlemalm, H. V. Poor, and A. Logothetis, "Suppression of multiple narrowband interferers in a spread-spectrum communication system," *IEEE J. Select. Areas Commun.*, vol. 18, no. 8, pp. 1365–1374, August 2000.
17. P. Azmi and M. Nasiri-Kenari, "Narrow-band interference suppression in CDMA spread-spectrum communication systems based on sub-optimum unitary transforms," *IEICE Trans. Commun.*, vol. E85-B, no.1, pp. 239–246, January 2002.

18. T. J. Lim and L. K. Rasmussen, "Adaptive cancelation of narrowband signals in overlaid CDMA systems," in *Proc. IEEE Int. Workshop Intelligent Signal Processing and Communication Systems*, Singapore, November 1996, pp. 1648–1652.
19. H. Fathallah and L. Rusch, "Enhanced blind adaptive narrowband interference suppression in dsss," in *Proc. IEEE Global Telecommun. Conf. (GLOBECOM)*, vol. 1, London, November 1996, pp. 545–549.
20. W.-S. Hou, L.-M. Chen, and B.-S. Chen, "Adaptive narrowband interference rejection in DS-CDMA systems: a scheme of parallel interference cancellers," *IEEE J. Select. Areas Commun.*, vol. 20, pp. 1103–1114, June 2001.
21. P.-R. Chang, "Narrowband interference suppression in spread spectrum CDMA communications using pipelined recurrent neural networks," in *Proc. IEEE Int. Conf. Universal Personal Commun. (ICUPC)*, vol. 2, October 1998, pp. 1299–1303.
22. H. V. Poor and X. Wang, "Code-aided interference suppression in DS/CDMA spread spectrum communications," *IEEE Trans. Commun.*, vol. 45, no. 9, pp. 1101–1111, September 1997.
23. S. Buzzi, M. Lops, and A. Tulino, "Time-varying mmse interference suppression in asynchronous DS/CDMA systems over multipath fading channels," in *Proc. IEEE Int. Symp. on Personal, Indoor and Mobile Radio Communications*, September 1998, pp. 518–522.
24. M. Medley, "Narrow-band interference excision in spread spectrum systems using lapped transforms," *IEEE Trans. Commun.*, vol. 45, pp. 1444–1455, November 1997.
25. A. Akansu, M. Tazebay, M. Medley, and P. Das, "Wavelet and subband transforms: fundamentals and communication applications," *IEEE Commun. Mag.*, vol. 35, pp. 104–115, December 1997.
26. R. D. Weaver, "Frequency domain processing of ultra-wideband signals," in *Proc. IEEE Asilomar Conf. Signals, Systems Computers*, Pacific Grove, CA, November 2003, pp. 1221–1224.
27. B. Krongold, M. Kramer, K. Ramchandran, and D. Jones, "Spread spectrum interference suppression using adaptive time-frequency tilings," in *Proc. IEEE Int. Conf. Acoustics, Speech, Signal Processing (ICASSP)*, vol. 3, Munich, April 1997, pp. 1881–1884.
28. Y. Zhang and J. Dill, "An anti-jamming algorithm using wavelet packet modulated spread spectrum," in *Proc. IEEE Military Commun. Conf.*, vol. 2, November 1999, pp. 846–850.
29. E. Pardo, J. Perez, and M. Rodriguez, "Interference excision in DSSS based on undecimated wavelet packet transform," *IEEE Electron. Lett.*, vol. 39, no. 21, pp. 1543–1544, October 2003.
30. T. Kasparis, "Frequency independent sinusoidal suppression using median filters," in *Proc. IEEE Int. Conf. Acoustics, Speech, Signal Processing (ICASSP)*, vol. 3, Toronto, April 1991, pp. 612–615.
31. R. Lowdermilk and F. Harris, "Interference mitigation in orthogonal frequency division multiplexing (OFDM)," in *Proc. IEEE Int. Conf. Universal Personal Commun. (ICUPC)*, vol. 2, Cambridge, MA, September 1996, pp. 623–627.
32. R. Nilsson, F. Sjoberg, and J. LeBlanc, "A rank-reduced lmmse canceller for narrowband interference suppression in OFDM-based systems," *IEEE Trans. Commun.*, vol. 51, no. 12, pp. 2126–2140, December 2003.
33. S. Vogeler, L. Broetje, K.-D. Kammeyer, R. Rueckriem, and S. Fechtel, "Blind bluetooth interference detection and suppression for OFDM transmission in the ISM band,"

in *Proc. IEEE Asilomar Conf. on Signals, Systems, Computers*, vol. 1, Pacific Grove, CA, November 2003, pp. 703–707.

34. M. Ghosh and V. Gadam, "Bluetooth interference cancellation for 802.11 g WLAN receivers," in *Proc. IEEE Int. Conf. Communications (ICC)*, vol. 2, Anchorage, AK, May 2003, pp. 1169–1173.

35. S. Roy, J. Foerster, V. Somayazulu, and D. Leeper, "Ultrawideband radio design: the promise of high-speed, short-range wireless connectivity," *IEEE Proceedings*, vol. 92, no. 2, pp. 295–311, February 2004.

36. A. Batra, J. Balakrishnan, G. Aiello, J. Foerster, and A. Dabak, "Design of a multiband OFDM system for realistic UWB channel environments," *IEEE Trans. Microwave Theory and Techniques*, vol. 52, no. 9, pp. 2123–2138, September 2004.

37. T. Ikegami and K. Ohno, "Interference mitigation study for uwb impulse radio," in *Proc. IEEE Personal, Indoor, Mobile Radio Communications (PIMRC)*, vol. 1, September 2003, pp. 583–587.

38. R. Wilson and R. Scholtz, "Template estimation in ultra-wideband radio," in *Proc. IEEE Asilomar Conf. on Signals, Systems, Computers*, vol. 2, Pacific Grove, CA, November 2003, pp. 1244–1248.

39. I. Bergel, E. Fishler, and H. Messer, "Narrowband interference suppression in impulse radio systems," in *IEEE Conf. on UWB Systems Technology*, Baltimore, MD, May 2002, pp. 303–307.

40. H. Sheng, A. Haimovich, A. Molisch, and J. Zhang, "Optimum combining for time hopping impulse radio UWB Rake receivers," in *Proc. IEEE Ultrawideband Systems Technology (UWBST)*, Reston, VA, November 2003.

41. N. Boubaker and K. Letaief, "A low complexity MMSE-RAKE receiver in a realistic UWB channel and in the presence of NBI," in *Proc. IEEE Wireless Communications Networking Conf. (WCNC)*, vol. 1, New Orleans, LA, March 2003, pp. 233–237.

42. D. Cassioli, M. Z. Win, F. Vatalaro, and A. F. Molisch, "Performance of low-complexity RAKE reception in a realistic UWB channel," in *Proc. IEEE Int. Conf. Communications (ICC)*, vol. 2, New York, April 2002, pp. 763–767.

43. M. Z. Win and R. A. Scholtz, "Impulse radio: How it works," *IEEE Commun. Lett.*, vol. 2, no. 2, pp. 36–38, February 1998.

44. X. Chu and R. Murch, "The effect of NBI on UWB time-hopping systems," *IEEE Trans. Wireless Commun.*, vol. 3, no. 5, pp. 1431–1436, September 2004.

45. T. Yücek and H. Arslan, "Noise plus interference power estimation in adaptive OFDM systems," in *Proc. IEEE Vehicular Technology Conf.*, vol. 2, Stockholm, May 2005, pp. 1278–1282.

46. R. Fisher, R. Kohno, H. Ogawa, H. Zhang, and K. Takizawa, "IEEE P802.15 working group for wireless personal area networks (WPANs), DS UWB proposal update," May 2004. Available at: www.uwbforum.org/documents/15-04-0140-04-003a-merger2-proposal-ds-uwb-presentation.ppt

47. "Ultrawideband: High-speed, short-range technology with far-reaching effects, multiband OFDM alliance," September 2004. Available at: www.multibandofdm.org/papers/MBOA-UWB-White-Paper.pdf

48. A. Taha and K. Chugg, "A theoretical study on the effects of interference on UWB multiple access impulse radio," in *Proc. IEEE Asilomar Conf. on Signals, Systems, Computers*, vol. 1, Pacific Grove, CA, November 2002, pp. 728–732.

49. I. Guvenc and H. Arslan, "Performance evaluation of UWB systems in the presence of timing jitter," in *Proc. IEEE Ultra Wideband Systems Technology Conf.*, Reston, VA, November 2003, pp. 136–141.
50. L. Piazzo and J. Romme, "Spectrum control by means of the TH code in UWB systems," in *Vehicular Technology Conf.*, vol. 3, April 2003, pp. 1649–1653.
51. L. Piazzo and J. Romme, "On the power spectral density of time-hopping impulse radio," in *IEEE Conf. Ultrawideband Systems Technology (UWBST)*, May 2002, pp. 241–244.
52. H. Schantz, G. Wolenec, and E. Myszka, "Frequency notched UWB antennas," in *IEEE Conf. Ultrawideband Systems Technology (UWBST)*, vol. 3, November 2003, pp. 214–218.
53. S. Verdu, *Multiuser Detection*, 1st edn, Cambridge University Press, Cambridge, 1998.
54. S. Mallat, "A theory for multiresolution signal decomposition: the wavelet representation," *IEEE Pattern Anal. Machine Intell.*, vol. 11, no. 7, pp. 674–693, 1989.
55. R. Klein, M. Temple, R. Raines, and R. Claypoole, "Interference avoidance communications using wavelet domain transformation techniques," *Electron. Lett.*, vol. 37, no. 15, pp. 987–989, July 2001.
56. M. Medley, G. Saulnier, and P. Das, "Radiometric detection of direct-sequence spread spectrum signals with interference excision using the wavelet transform," in *IEEE Int. Conf. on Communication (ICC 94)*, vol. 3, May 1994, pp. 1648–1652.
57. J. Patti, S. Roberts, and M. Amin, "Adaptive and block excisions in spread spectrum communication systems using the wavelet transform," in *Asilomar Conf. on Signals, Systems, Computers*, vol. 1, November 1994, pp. 293–297.

CHAPTER 12

Orthogonal Frequency Division Multiplexing for Ultra Wideband Communications

EBRAHIM SABERINIA and AHMED H. TEWFIK

12.1 INTRODUCTION

Orthogonal frequency-division multiplexing is an effective multicarrier modulation technique that has been adopted in several current communication systems like IEEE 802.11a and 802.11g wireless local area networks [1, 2]. It is a major candidate for future wireless cellular systems like 4G. OFDM offers several desirable features that make it attractive for high-bit-rate communications over wireless multipath fading channels. These features include: ISI-free high-data-rate transmission using a cyclic prefix, simple (one-tap) channel equalization by converting a frequency-selective fading channel to several parallel flat fading channels, simple timing and synchronization by changing a serial transmission to several parallel transmissions and expanding the time of transmission of a single symbol, and all-digital transceiver implementation in base-band using FFT and inverse FFT (IFFT) algorithms [3]. OFDM is also well known for its robustness to the multipath fading.

Research on adapting OFDM modulation to UWB communications started at the beginning of 2002 in both academia [4–6] and industry [7, 8]. The results of both studies were presented as two separate proposals to the IEEE 802.15.3a Wireless Personal Area Network Standardization Committee [8, 9]. Later on, these two proposals and several other non-OFDM proposals were merged together and an OFDM-based UWB system named multiband OFDM (MB-OFDM) was introduced [10]. The MB-OFDM system is currently one of the leading proposals for the IEEE 802.15.3a standard and is supported by more than 100 large companies and universities. In this chapter, we will discuss the MB-OFDM system in detail. Note that the standard is not finalized and therefore the MB-OFDM proposal is still evolving.

Ultra Wideband Wireless Communication. Edited by Arslan, Chen, and Di Benedetto
Copyright © 2006 John Wiley & Sons, Inc.

Our discussion of MB-OFDM will be based on the sixth version of the proposal. The reader can look for the most recent updates on the IEEE 802.15.3a web site.

In the second part of the chapter we will discuss an enhancement to the MB-OFDM system that leads to considerable savings in terms of complexity and power consumption. The complexity and power consumption of the transceivers is a very important issue in the applications considered for these high bit rate wireless personal area networks [11]. The new scheme, MB-pulsed-OFDM, is more suitable for high-speed wireless personal area networking applications such as those envisioned by the IEEE 802.15.3a standard. Instead of using pulses with duty cycle one, as in normal OFDM systems, the MB-pulsed-OFDM system uses pulses with duty cycle less than one [12, 13]. We show that, by using the additional spreading gain achieved by pulsation, we can reduce the number of carriers and coding overhead in the MB-OFDM system. Hence, we can design transceivers with much lower complexity and power consumption than the baseline MB-OFDM system and with comparable or better performance in realistic indoor multipath environments.

12.2 MULTIBAND OFDM SYSTEM

12.2.1 Band Planning

In the MB-OFDM system, the whole available UWB spectrum is divided into several sub-bands with smaller bandwidth. This simplifies the design of the analog RF front end and analog-to-digital and digital-to-analog converters (ADCs and DACs). It also decreases overall power consumption. The bandwidth of each sub-band is larger than 500 MHz in compliance with the FCC rules for UWB transmission. Table 12.1 shows the band planning for the current MB-OFDM system. The bandwidth of each sub-band is equal to 528 MHz. The sub-bands are assigned into five different groups. Groups 1–4 have three sub-bands each and group 5 has two sub-bands.

Devices operating in band group 1 (the three lowest frequency bands) are denoted mode 1 devices, and it is mandatory for all devices to support mode 1 operation. Support for the other band groups is optional at this time and will be added in the future. The sub-bands of band group 1 are shown in Figure 12.1.

12.2.2 Sub-Band Hopping

Every device uses only one sub-band group to transmit and receive data. In any time slot, an OFDM symbol is transmitted in one of the sub-bands. The system switches to another sub-band within the group to transmit next symbol. Fast switching between bands is achieved by using a single oscillator and a frequency divider network. The transmitted signal can be presented as follows:

$$x(t) = \text{Re}\left\{\sum_{r=0}^{R-1} s_r(t - rT_{\text{OFDM}}) \exp(j2\pi f_r t)\right\} \quad (12.1)$$

12.2 MULTIBAND OFDM SYSTEM

TABLE 12.1 MB-OFDM Band Planning: UWB Spectrum is Divided into 14 Sub-Bands

Band Group	Sub-Band	Lower Frequency (MHz)	Center Frequency (MHz)	Upper Frequency (MHz)
1	1	3168	3432	3696
	2	3696	3960	4224
	3	4224	4488	4752
2	4	4752	5016	5280
	5	5280	5544	5808
	6	5808	6072	6336
3	7	6336	6600	6864
	8	6864	7128	7392
	9	7392	7656	7920
4	10	7920	8184	8448
	11	8448	8712	8976
	12	8976	9240	9504
5	13	9504	9768	10032
	14	10032	10296	10560

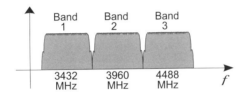

Figure 12.1 Sub-bands assigned for operation of a mode 1 device.

where R is the number of transmitted OFDM symbols and $s_r(t)$ is the baseband signal representing the rth OFDM symbol occupying a symbol interval of length T_{OFDM}. Let f_r be the carrier frequency corresponding to the sub-band in which the rth OFDM symbol is transmitted. The carrier frequency f_r hops between bands according to a time–frequency code (TFC) of length 6. Therefore, f_r is periodic with period 6. The system specifies four 3-band TFCs and two 2-band TFCs. Table 12.2 shows the TFCs for three lower sub-bands (group 1). By using these sequences, four simultaneously operating piconets can coexist with minimal

TABLE 12.2 Time Frequency Codes and Associated Preamble Patterns

TFC1	1	2	3	1	2	3
TFC2	1	3	2	1	3	2
TFC3	1	1	2	2	3	3
TFC4	1	1	3	3	2	2

interference in any 3-band group while two simultaneously operating piconets can coexist in the 2-band group.

12.2.3 OFDM Modulation

As mentioned earlier, one OFDM symbol is transmitted in each time slot within a sub-band. Quadrature phase shift keying (QPSK) modulation is used for OFDM. Therefore, the base-band OFDM signal transmitted in the rth time slot is given by:

$$s_r(t) = p(t) \sum_{k=0}^{N-1} b_k^r e^{j2\pi k f_0 t} \qquad (12.2)$$

where N is number of subcarriers and $p(t)$ is a lowpass pulse shaping signal with duration T_p. The QPSK symbol that is transmitted in the rth time slot and over the kth subcarrier is denoted by b_k^r. The subcarrier spacing is denoted by f_0 and is equal to $1/T_p$.

In the current MB-OFDM scheme, the number of subcarriers is $N = 128$. Of these 128 subcarriers, 100 subcarriers are used to transmit data and 12 subcarriers are used as pilots. Ten subcarriers at the edge of the spectrum are defined as guard carriers and are used to shape the transmitted signal spectrum. The remaining six subcarriers are not used by the system and carry a null (zero) signal. The OFDM modulation is done using an IFFT in the base-band. The subcarrier spacing is equal to $f_0 = 4.125$ MHz ($T_p = 242.42$ ns). In the initial proposal, a cyclic prefix (CP) of length 37 was used. The current scheme instead adds 37 zeroes to the output of the IFFT to generate an output with 165 symbols generating OFDM symbol interval of $T_{OFDM} = 312.5$ ns. This zero padding (ZP) scheme enhances the signal spectrum as compared with the scheme that uses a CP.

12.2.4 Frequency Repetition Spreading

At lower supported bit rates, only 50 QAM symbols are transmitted in one OFDM symbol. While these symbols modulate 50 subcarriers, their conjugates modulate another 50 subcarriers providing a conjugate symmetric input to the IFFT module. Therefore, every QAM symbol is carried over two different subcarriers providing a frequency spreading factor of 2. Furthermore, it ensures that the output of the IFFT is a real-valued signal and there is no need for the Q branch in the RF section of the transmitter.

12.2.5 Time Repetition Spreading

At some of lower bit rates, a time-domain spreading operation is also considered in order to provide diversity and enhance system performance in multipath fading and the presence of interference from simultaneously operating piconets. The time-domain spreading is performed by transmitting the same information over two

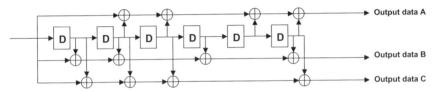

Figure 12.2 Rate $R = 1/3$; convolutional encoder used in MB-OFDM scheme with constraint length $K = 7$.

OFDM symbols. The repeated OFDM symbol is coded to ensure a flat power spectral density (without impulses because of repetition) for the transmitted signal.

12.2.6 Coding

The MB-OFDM system uses convolutional error correction codes with rates of 1/3, 1/2 and 5/8 in different supported bit rates in order to mitigate distortion caused by the channel. All of these codes are generated from a systematic convolutional encoder with rate 1/3 shown in Figure 12.2. Other coding rates are derived from the rate $R = 1/3$ convolutional code by employing "puncturing." Puncturing is a procedure for omitting some of the encoded bits at the transmitter (thus reducing the number of transmitted bits and increasing the coding rate). Decoding is performed with the Viterbi algorithm at the receiver. For the punctured codes dummy "zeros" are inserted in place of the omitted bits before decoding.

12.2.7 Supported Bit Rates

The MB-OFDM system supports bit rates of 53.3, 80, 106.7, 160, 200, 320, and 400 Mbps by changing the coding, frequency spreading, and time spreading rates as shown in Table 12.3. Other optional bit rates, for example 39.4 Mbps and

TABLE 12.3 Different Bit Rates in MB-OFDM System and Corresponding Modulation Scheme, Coding Rate, Frequency Spreading and Time Spreading to Achieve Those Bit Rates

Data Rate (Mbps)	Modulation	Coding Rate (R)	Conjugate Symmetric Input to IFFT	Time Spreading Factor (TSF)	Overall Spreading Gain	Coded Bits Per OFDM Symbol (N_{CBPS})
53.3	QPSK	1/3	Yes	2	4	100
80	QPSK	1/2	Yes	2	4	100
106.7	QPSK	1/3	No	2	2	200
160	QPSK	1/2	No	2	2	200
200	QPSK	5/8	No	2	2	200
320	DCM	1/2	No	1 (No spreading)	1	200
400	DCM	5/8	No	1 (No spreading)	1	200

480 Mbps, are considered as well. For bit rates of 320 Mbps and 400 Mbps, dual-carrier modulation (DCM) is used instead of QPSK. In DCM a block of 200 coded bits first generate 200 bipolar bits by mapping "0"s into "−1"s. Then, 100 complex-valued symbols are generated as follows:

$$\begin{bmatrix} y_n \\ y_{n+50} \end{bmatrix} = \frac{1}{\sqrt{10}} \begin{bmatrix} 2 & 1 \\ 1 & -2 \end{bmatrix} \begin{bmatrix} x_{k(n)} + jx_{k(n)+50} \\ x_{k(n)+1} + jx_{k(n)+51} \end{bmatrix} \quad n = 0, 1, \ldots, 49 \quad (12.3)$$

where x_m is the mth bipolar bit, y_m is the mth output symbol and $k(n)$ is an index mapping defined as follows:

$$k(n) = \begin{cases} 2n & n = 0, 1, \ldots, 24 \\ 2n + 50 & n = 25, 26, \ldots, 49. \end{cases} \quad (12.4)$$

The outputs $[y(0), y(1), \ldots, y(99)]$ are used as symbols to be transmitted over the subcarriers of the OFDM symbol. When combined with soft-input Viterbi decoding at the receiver, DCM increases frequency diversity and overall performance.

12.2.8 MB-OFDM Transceiver

The transmitter and receiver structures of the MB-OFDM system are presented in Figure 12.3. Except for the hopping carrier frequency, the structure is similar to that of standard OFDM transceivers presented in textbooks like [14]. At the transmitter, after channel coding and interleaving, the coded bits are mapped into QPSK constellations and OFDM modulation is performed using an IFFT. After ZP, the signal is converted to analog and modulated by the carrier signal. The frequency of the carrier signal hops between OFDM blocks from one sub-band to

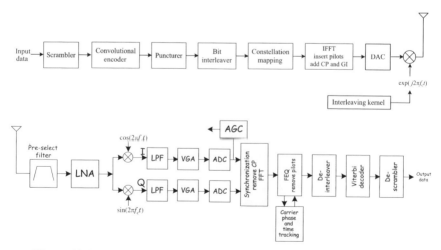

Figure 12.3 Transmitter and receiver structures of multiband OFDM system.

another according to the TFC code. At the receiver, after down conversion to base band and low pass filtering by the bandwidth of a sub-band, the signal is sampled at the Nyquist rate. The OFDM demodulation is performed using an FFT. The effect of fading channel is removed by a one-tap equalizer for each sub-carrier in the frequency domain and a hard decision is implemented to detect the transmitted QPSK symbols. After de-mapping into a binary stream and de-interleaving, the signal is sent to the decoder.

12.2.9 Improvement to MB-OFDM

Since publication of the MB-OFDM proposal, several ideas have been proposed to enhance this system. One area that has received much attention is enhancing the frequency spreading techniques used in MB-OFDM. Note that MB-OFDM is a UWB system and the transmitted signal bandwidth is much higher than the data rate. Hence, different frequency spreading techniques can be used to fill the spectrum. The processing gain that is achieved from this spreading is used to mitigate the multipath fading and other interferences. As mentioned earlier, the current MB-OFDM scheme uses a frequency spreading of factor of 2 by sending the conjugate of the symbol in different subcarriers at low data rates. It also uses time repetition of OFDM symbols at some data rates and provides another spreading factor of 2. Further, it relies on band hopping between symbols and strong forward error correction codes with interleaving between subcarriers and sub-bands to combat fading and exploit frequency diversity. While different traditional schemes can be considered to replace the current spreading schemes, the main difference between them lies in the complexity of implementation. The main reason that the authors of the MB-OFDM scheme have chosen simple repetition in frequency and time over other better performing spreading techniques is complexity. Unlike other UWB transmission schemes that rely on explicit frequency spreading [15], the MB-OFDM approach does not require the use of a rake receiver to capture partial or full multipath diversity. The implementation of a rake receiver in UWB systems is complex because of the stringent timing requirements imposed by the very short pulses that are typically used in UWB and the potentially relatively large number of paths that need to be captured to guarantee acceptable performance.

One of the ideas that have been proposed to enhance the frequency spreading characteristics of the MB-OFDM system is the pulsed-OFDM scheme [12, 13]. This scheme uses the same spreading technique as the original UWB systems (using low duty cycle pulses) and marries it to OFDM modulation. Pulsed-OFDM effectively spreads the frequency content of the baseband OFDM signal over a much wider band. This spreading leads to diversity which can be exploited to reduce the number of carriers used in the MB-OFDM system and decrease its coding overhead. Like the MB-OFDM scheme, this system does not require a rake receiver to exploit diversity. In fact, the complexity of the new scheme is much lower than that of the original MB-OFDM system. We will discuss pulsed-OFDM system in more detail in the next section.

12.3 MULTIBAND PULSED-OFDM UWB SYSTEM

The multiband pulsed-OFDM system preserves the band planning of the original MB-OFDM scheme. It only replaces every OFDM symbol with a pulsed-OFDM symbol. The pulsed-OFDM symbol is generated by replacing the pulse-shaping lowpass signal with a regular train of pulses with low duty cycle. Specifically, the pulsed-OFDM symbol to be transmitted in the rth time slot can be represented with the same formula as nonpulsed OFDM signal in Equation (12.2):

$$s_r(t) = p(t) \sum_{k=0}^{N-1} b_k^r e^{j2\pi k f_0 t} \tag{12.5}$$

while in this case, $p(t)$ is a train of pulses with duty cycle less than one, that is,

$$p(t) = \sum_{n=0}^{N-1} s(t - nT). \tag{12.6}$$

In the above equation, N is number of subcarriers, $s(t)$ is a monopulse with duration T_s and $T = T_p/N$ is pulse separation time larger than T_s.

12.3.1 Pulsed-OFDM Transmitter

The pulsed-OFDM can be simply generated by replacing the DAC in an OFDM transmitter with a pulse train generator. The generator produces amplitude modulated pulses with duty cycle less than one. If the inverse of the duty cycle is integer, the same signal can also be generated by up-sampling the digital baseband OFDM modulated signal before sending it to a conventional DAC. The up-sampling is done by inserting $K-1$ zeroes between samples of the signal. The resulting pulsed-OFDM signal is then a pulse train with duty cycle equal to $1/K$. Since this latter point of view is mathematically more useful, we shall retain it in the remainder of this chapter. We also refer to parameter K as the processing gain of pulsed-OFDM system. Both transmitter structures are shown in Figure 12.4.

12.3.2 Pulsed-OFDM Signal Spectrum

The spectrum of the pulsed-OFDM signal is easily derived from the impulse response $s(t)$ of the DAC, or equivalently from the pulse train $p(t)$. Specifically, following [14, p. 208], we have:

$$S_{\text{POFDM}}(f) = \sum_{k=0}^{N-1} \left| P\left(f - \frac{k}{NT}\right) \right|^2, \tag{12.7}$$

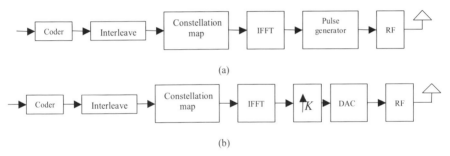

Figure 12.4 Pulsed-OFDM transmitter structure: (a) implementation using low duty cycle pulse generator; (b) implementation using up-sampling in digital domain.

where $P(f)$ is the Fourier transform of $p(t)$ which, by taking Fourier transform from both sides of Equation (12.6), can be shown to be given by

$$|P(f)|^2 = |S(f)|^2 \frac{\sin^2(\pi N f T)}{\sin^2(\pi f T)}. \qquad (12.8)$$

Figure 12.5 shows the spectrum of a pulse train with number of pulses equal to $N = 4$ and duty cycle of $(1/5)$ and the spectrum of the corresponding pulsed-OFDM signal having $N = 4$ subcarriers. Equations (12.7) and (12.8) show that the bandwidth of a pulsed-OFDM signal with symbol rate $1/T$ and processing gain of K is approximately equal to $(K+1)/T$. Note that Equation (12.7) shows the spectrum of the pulsed-OFDM signal in a single sub-band.

Spectrum of the pulsed-OFDM can be also calculated noticing the fact that the pulsed-OFDM signal generated by up-sampling a normal OFDM signal. As is well known, the upsampling process spreads the frequency of the signal over a band K times larger than the original by repeating the original signal spectrum in frequency domain [16]. Then we can compute the spectrum of the pulsed-OFDM signal as:

$$S_{\text{POFDM}}(f) = \sum_{m=0}^{M-1} S_{\text{OFDM}}\left(f - \frac{m}{T_s}\right), \qquad (12.9)$$

where $S_{\text{OFDM}}(f)$ is the spectrum of the original OFDM signal. This discussion indicates that the pulsed-OFDM leads to a simple frequency repetition scheme. The main advantages of the proposed scheme compared with traditional frequency repetition schemes such as [17] are that it is easy to implement as it requires up-sampling or the use of DACs with low duty cycle, it automatically guarantees that the minimum distance between any two subcarriers carrying the same symbols is maximized and it provides an advantageous trade-off between complexity and performance. In particular, as we will show below, it leads to considerable complexity and power consumption savings while achieving an adequate

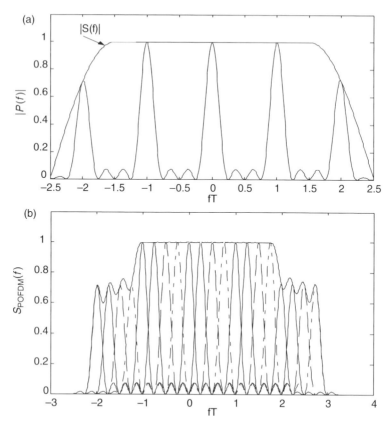

Figure 12.5 (a) Spectrum of a pulse train with $N = 4$ pulses per train and a duty cycle of $1/5$; and (b) spectrum of the corresponding pulse-OFDM signal.

performance in IEEE 802.15.3a environments that is higher than that of many of the more complex schemes. In general, however, one expects the more complex and more sophisticated frequency encoding schemes to exhibit better performance.

12.3.3 Digital Equivalent Model and Diversity of Pulsed-OFDM

It is advantageous to use a digital equivalent model to derive some of the properties of pulsed-OFDM modulation. Assuming that the received signal is sampled at the same rate as that of the transmitter DAC, the entire transmission system after constellation mapping can be represented by the digital equivalent model of Figure 12.6(a). Here $H_d(z)$ is the digital equivalent channel. The impulse response of this channel $h_d(n)$ is a sampled version of the equivalent analog channel that consists of the physical channel, transmitter filter, and receiver filter. We assume that $H_d(z)$ is an FIR filter with $L+1$ taps. The main difference between pulsed-OFDM and normal OFDM is the up-sampling operation after the IFFT. We will show

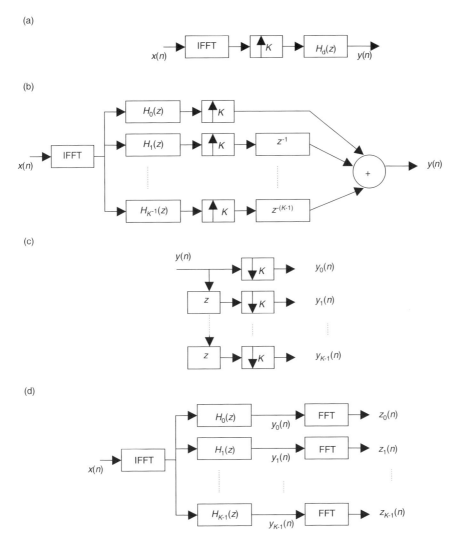

Figure 12.6 Digital equivalent models for pulsed-OFDM transmitter and channel: (a) basic model; (b) another model by replacing channel polyphase decomposition in the basic model; (c) a circuit to separate diversity branches at the model (b); and (d) final model presenting K diversity branches captured by the pulsed-OFDM system.

that the up-sampling operation provides K branches of diversity that can be separated at the receiver. To establish this fact, we use the polyphase decomposition of the digital equivalent channel $H_d(z)$:

$$H_d(z) = \sum_{k=0}^{K-1} z^{-k} H_k(z^K). \qquad (12.10)$$

In Equation (12.10) $H_k(z)$, $k = 0, \ldots, K − 1$, are the polyphase factors of $H_d(z)$. Substituting this polyphase decomposition in Figure 12.6(a) and using conventional multirate signal processing principles, we obtain the digital equivalent model of Figure 12.6(b) [16]. This model shows that the pulsed-OFDM received signal consists of a parallel to serial conversion of K parallel normal OFDM received signals. Each of these normal OFDM received signals is the output of a *different* channel driven by the same normal OFDM input signal. Hence, a pulsed-OFDM system with processing gain K provides K branches of diversity at the receiver. These branches can be separated by a simple serial to parallel conversion structure, as shown in Figure 12.6(c). Combining the structures in Figure 12.6(b) and (c), we derive our final digital equivalent model of a pulsed-OFDM system as depicted in Figure 12.6(d). We will propose a receiver structure for the pulsed-OFDM based on this model.

12.3.4 Pulsed-OFDM Receiver

The diversity branches provided by the pulsation can be combined at the receiver to combat fading in multipath channels. Since each branch is equivalent to a normal OFDM received signal, we can demodulate each branch using normal OFDM demodulation by applying the FFT to the received signal in that branch. As with other OFDM systems, a CP of length larger than the maximum subchannel length is added after the IFFT at the transmitter and discarded from the received signals before the FFT in each branch. The cyclic prefix eliminates ISI and ICI in all branches. In particular, the output of each sub-carrier in each branch is equal to the symbol transmitted in that subcarrier modulated by a flat fading channel gain equal to the Fourier transform of the subchannel impulse response evaluated at the subcarrier center frequency. Alternatively, zero-padding can be used.

Let $\mathbf{x} = [x_0, \ldots, x_{N-1}]^T$ be an N-element block of the input data stream $x(n)$. After removing the CP and applying an FFT, the output of each branch is given by [18, p. 180]:

$$\mathbf{z}_i = \mathbf{D}_i \mathbf{x} + \mathbf{w}_i \quad i = 0, \ldots, K − 1$$
$$\mathbf{D}_i = \mathrm{diag}\{H_i(0), H_i(f_0), \ldots, H_i[(N − 1)f_0]\} \quad (12.11)$$

where $H_i(f) = \sum_{n=0}^{L'} h_i(n) e^{-j2\pi f n}$ is the Fourier transform of the ith virtual digital subchannel impulse response and f_0 is the subcarrier center frequency separation. The vector $\mathbf{w}_i$ consists of samples of the filtered AWGN.

Equation (12.11) shows that, for each transmitted symbol, we receive K symbols affected by different flat fading amplitudes. We can combine these branches before making a decision about that symbol. Any kind of diversity combining method can be used. The optimum way of combining diversity branches is Maximal Ratio Combining (MRC) [18, Chapter 6]. In this approach, for each subcarrier n, the demodulated outputs corresponding to the different diversity branches are combined as:

$$\hat{b}(n) = \frac{\sum_{i=0}^{k-1} \hat{d}_i^*(n) z_i(n)}{\sum_{i=0}^{k-1} \left|\hat{d}_i^*(n)\right|^2} \quad (12.12)$$

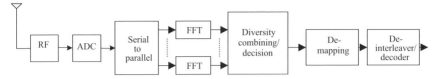

Figure 12.7 Receiver structure for pulsed-OFDM system.

before symbol detection. In the above expression, $d_i(n)$ is the nth diagonal entry of matrix $\mathbf{D}_i$ and $z_i(n)$ is the nth entry of vector z_i. These parameters can be estimated from the training data, as is normally done while adaptively constructing the one-tap equalizers in OFDM receivers. In addition, MRC requires a sampling rate equal to the bandwidth of the sub-bands and the evaluation of KN-point FFTs. Other diversity combining methods like Equal Gain Combining (EGC) or Selection Combining (SC) can be also used to reduce the complexity of the receiver. Figure 12.7 shows the block diagram of a typical pulsed-OFDM receiver.

12.3.5 Selecting the Up-Sampling Factor

The analysis presented above leads to the important questions of how large the up-sampling factor K can be and how we select a suitable value for the up-sampling factor K in a given environment. To answer these questions, first observe that, to get maximal benefit from the K diversity branches with minimal complexity, we would want to make sure that they are uncorrelated. This condition holds as long as the bandwidth of each subchannel is larger than the coherence bandwidth of the channel. In other words, the up-sampling factor K needs to be smaller than or equal to an upper limit $K_{\max}$ given by:

$$K_{\max} = \left\lfloor \frac{w}{B_c} \right\rfloor = \lfloor w T_{\text{spread}} \rfloor \qquad (12.13)$$

where w is the total channel bandwidth, B_c is the coherence bandwidth of the channel and T_{spread} is its maximum delay spread. Here, $\lfloor x \rfloor$ denotes the largest integer that is smaller than x. In a given channel setting we may then look for the optimum K in the range $K = 1, \ldots, K_{\max}$.

The selection of the up-sampling factor K in a given scenario can be done once a suitable design criterion is chosen. In [18], we address this issue using the concept of outage capacity [19, 20] of the pulsed-OFDM system in fading channels. The advantage of this approach is that it leads to results that can be applied regardless of the choice of coding, interleaving, and modulation schemes. In particular, we provide an algorithm to choose the optimal up-sampling rate for a given set of requirements and channel conditions. For example, the results of [18] show that it is best not to use up-sampling (i.e., select $K = 1$) in the IEEE 802.15.3a environment for transmissions at 480 Mbps at 1 m, while a value of $K = 4$ is more suitable for transmission at 106 Mbps at 10 m. The latter value will be retained for discussion in the section where we compare MB-OFDM and pulsed-OFDM systems in terms of performance, complexity, and power consumption.

12.4 COMPARING MB-OFDM AND MB-PULSED-OFDM SYSTEMS

In order to compare pulsed and non-pulsed OFDM modulation in a realistic situation, we use the IEEE 802.15.3a standard physical layer as a framework. In this section we compare an MB-pulsed-OFDM system designed for the 106.7 Mbps transmission rate requirements with the MB-OFDM system at the same setting.

12.4.1 System Parameters

By using the results of [18], we select a processing gain of $K = 4$ with $N = 32$ subcarriers for a pulsed-OFDM system to run at 106.7 Mbps data rate. Twenty-five subcarriers are used to send 25 QPSK symbol generated from 50 coded bits. The remaining subcarriers are used as pilots and guard subcarriers. To compensate the lower coded bit rate in the pulsed-OFDM system, we use a convolutional error correcting code of rate equal to $2/3$ (instead of rate $1/3$). So both systems have equal bit rates.

12.4.2 Complexity Comparision

Both the pulsed-OFDM and non-pulsed-OFDM systems have the same RF front end. Thus, in the analog domain, possibly except for the DAC and ADC parts depending on implementation, the two systems have similar complexity and power consumption. However, in the digital domain, we can achieve considerable saving in terms of power consumption and complexity using pulsed-OFDM system.

The reduction in complexity came from the inherent advantage of the pulsed-OFDM scheme using reduced number of sub-carriers. In particular, only a 32-point IFFT is required at the transmitter side. This is much simpler than the 128-point IFFT used in the baseline system. It is known that the complexity of an FFT or IFFT processor is proportional to $N\log N$, where N is the size of the data block. At the receiver side, the pulsed-OFDM requires four 32-point FFTs while the nonpulsed-OFDM system only requires a single 128-point FFT. Again, the complexity of the pulsed OFDM receiver is slightly lower than that of the MB-OFDM receiver.

12.4.3 Power Consumption Comparison

In addition to its lower complexity, the pulsed-OFDM system has a big advantage over nonpulsed OFDM in terms of power consumption. The power consumption of a VLSI chip is determined by its clock rate, the supply voltage and the capacitance of the circuit. It can be roughly computed as:

$$P = C_{\text{total}} V_0^2 f, \tag{12.14}$$

where C_{total} denotes the total capacitance of the circuit, V_0 is the supply voltage, and f is the clock frequency. According to the analysis of the previous sections, the baseband section of the pulsed-OFDM runs at a lower clock frequency than that of MB-OFDM. This fact is illustrated in Table 12.4, where the clock rates of different parts of transmitter and receiver are listed for both pulsed and nonpulsed systems. Actually, due to the simplicity of the circuits in pulsed-OFDM, such as FFT processors and Viterbi decoders, the C_{total} in pulsed-OFDM is smaller than the one in nonpulsed-OFDM, leading to further reduction in power consumption.

12.4.4 Chip Area Comparison

Complexity and power consumption are not the only criteria for a good design. The amount of chip area consumed by different functions is also of interest. A straightforward implementation of four parallel FFT structures will occupy significantly larger area than a single 128-point FFT processor. Fortunately, it is possible to reduce four parallel FFT structures down to one without introducing significant extra complexities. In Figure 12.8(a), a widely used hardware implementation structure of an FFT processor is illustrated [21]. This structure is called a radix-4 multipath delay commutator (R4MDC) structure. In the figure, the block marked with C4 represents a four-input–four-output commutator, the block marked with BF4 is a four-point butterfly structure and the symbol $\otimes$ represents a multiplier. The blocks with numbers represent the delay elements in the path. One important weakness of this hardware structure is its low hardware utilization. All of the hardware elements in the FFT processor, such as commutators, butterfly structures, and multipliers, are only utilized 25% of the time. This means that, most of the time, the hardware elements are not used. In pulsed-OFDM, the low hardware efficiency of the structure makes it possible to combine multiple parallel FFT processors into the same structure by utilizing time multiplexing. As stated in [21], four parallel inputs, such as in the receiver of the pulsed-OFDM, can be combined using multiplexers and buffers into the same hardware structure. Figure 8(b) and (c) shows the structure of a modified commutator with four parallel inputs. By utilizing this modified FFT structure the hardware efficiency can be improved to 100%. Hence, compared with nonpulsed-OFDM, which uses a single 128-point FFT structure, the four-parallel 32-point FFT processor not only has lower computational and hardware complexity but also has higher hardware efficiency.

Another challenge in design of a transceiver for the MB-pulsed-OFDM system in the same chip area of the MB-OFDM arises when designing Viterbi decoder for the rate-2/3 convolutional code. Direct implementation of the Viterbi decoder for the rate-2/3 systematic convolutional code has much higher complexity than that of a rate-1/3 convolution code used in MB-OFDM system. This is due to the fact

TABLE 12.4 Clock Rate in Different Parts of Transceiver for Nonpulsed and Pulsed OFDM

	Input Data (MHz)	After Coding with Puncturing (MHz)	After Puncturing (Mbps)	After Constellation Mapping (Mbps)	Output of IFFT (Mbps)	Input to FFT (Mbps)	Input to Decoder (Mbps)
Nonpulsed	110	330	320	160	320	320	320
Pulsed	110	220	160	80	80	320	160

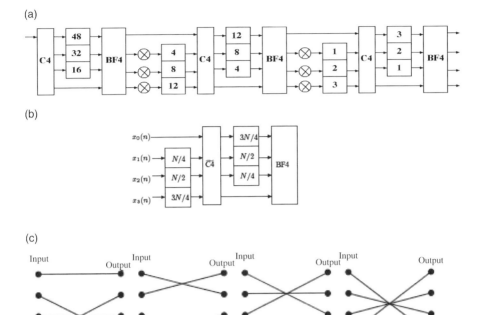

Figure 12.8 (a) A 64-point R4MPC FFT implementation structure [21]; (b) The new commutator for the structure in part (a) with four parallel inputs; and (c) new connections in the commutator.

that each point in the trellis representation of a rate-1/3 convolutional code has only two inputs, while a point in the trellis representation of a rate-2/3 convolutional code has four inputs. The increase in the number of inputs will significantly increase the complexity of the add–compare–select (ACS) unit. To address this issue, we use punctured codes. It can be shown that the performance difference between the optimal rate-2/3 code and the punctured code is less than 0.5 dB at a BER $= 10^{-5}$. In the simulation results shown in next sub-section we have used the low complexity punctured code.

12.4.5 Performance Comparison

In order to compare the performance of the pulsed and nonpulsed systems, we ran a complete simulation of both systems over the channel models CM3 and CM4 of the IEEE 802.15.3a. Figure 12.9 shows the results of the simulation of both the nonpulsed and pulsed-OFDM systems when operating at 106.7 Mbps over a CM-4 channel. This channel model has a delay spread of 250 ns. In this figure the BER is plotted vs distance for both systems. The BER must be less than 10^{-5} in order

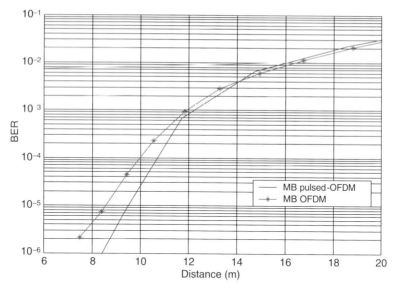

Figure 12.9 Bit error rate versus distance for pulsed and nonpulsed-OFDM systems in channel CM4.

to achieve a packet error rate less than 8% as required by the call for proposal (CFP) document. Figure 12.9 shows that the pulsed-OFDM system can operate at a range of 9.7 m on CM4 type channels while the range of normal OFDM is only 8.8 m. Figure 12.10 shows the results of the simulation of both the nonpulsed and

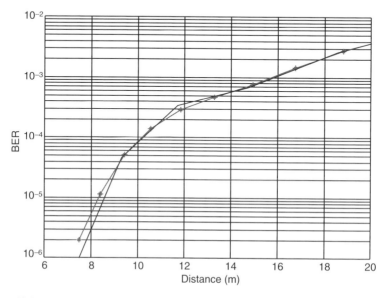

Figure 12.10 Bit error rate versus distance for pulsed and nonpulsed-OFDM systems in channel CM3.

pulsed-OFDM systems when operating at over CM3 channel. This channel model has a delay spread of 150 ns. Figure 12.10 shows that, even in this more favorable channel, and despite its lower complexity and power consumption, the pulsed-OFDM system has a slightly longer range than the nonpulsed system at a BER of 10^{-5}. Hence, we conclude that pulsed-OFDM outperforms nonpulsed-OFDM in dense multipath channel under identical bit rate and bandwidth conditions.

12.5 CONCLUSION

MB-OFDM system uses advantages of OFDM modulation to implement a high-bit-rate UWB system for indoor dense multipath wireless channels. The system has the ability to exploit frequency diversity and capture multipath energy without implementing a high-complexity rake receiver. Simple frequency and time repetition schemes along with heavy channel coding and interleaving is used to fill available UWB bandwidth and use the result processing gain to mitigate multipath and multiuser interference. Even better spreading schemes can be used to improve the performance or reduce the complexity. Pulsed-OFDM is a simple frequency spreading technique that can be used to reduce the complexity of the MB-OFDM system by lowering coding overhead and reducing the number of subcarriers while maintaining processing gain and overall performance.

REFERENCES

1. J. A. C. Bingham, "Multicarrier modulation for data transmission: an idea whose time has come," *IEEE Commun. Mag.*, pp. 5–14, May 1990.
2. IEEE wireless local area network standardaization work group official web site; http://grouper.ieee.org/groups/802/11/
3. L. J. Cimini Jr and N. R. Sollenberger, "OFDM with diversity and coding for high-bit-rate mobile data applications," *Proc. 3rd International Workshop on Mobile Multimedia Communications*, September 1996.
4. E. Saberinia and A. H. Tewfik, "Synchronous UWB-OFDM," *IEEE Symp. on Advances in Wireless Communications 2002 (ISWC'02)*, pp. 41–42, Vancouver, September 2002.
5. A. H. Tewfik and E. Saberinia, "High bit rate ultra-wideband OFDM," *Proc. IEEE Global Telecommunications Conference 2002 (GLOBECOME'02)*, Taipei, November 2002.
6. E. Saberinia and A. H. Tewfik, "Single and multi-carrier UWB communications," *Proc. IEEE Seventh International Symposium on Signal Processing and its Applications 2003 (ISSPA'03)*, Paris, July 2003.
7. J. Balakrishnan, A. Batra and A. Dabak, "A multi-band OFDM system for UWB communication," *IEEE Conference on Ultra Wideband Systems and Technologies 2003*, pp. 354–358, November 2003.
8. Anuj Batra et al., "TI physical layer proposal: time–frequency Interleaved OFDM", IEEE 802.15 work group official web site; http://grouper.ieee.org/groups/802/15/pub/2003/Mar03/, Dallas, TX, March 2003.

9. A. H. Tewfik and E. Saberinia, "University of Minnesota proposal: Multi-carrier UWB", IEEE 802.15 work group official web site; http://grouper.ieee.org/groups/802/15/pub/2003/Mar03/, Dallas, TX, March 2003.
10. Anuj Batra et al., "Multi-band OFDM: merged proposal #1," Merged proposal for the IEEE 802.15.3a standard, IEEE 802.15 work group official web site; http://grouper.ieee.org/groups/802/15/pub/2003/Jul03/, San Francisco, CA, July 2003.
11. "Summary of the eight application presentations from the Study Group IEEE 802.15.3a call for applications," IEEE 802.15 work group official web site; http://grouper.ieee.org/groups/802/15/pub/2003/Jan03/
12. E. Saberinia and A. H. Tewfik, "Pulsed and non-pulsed ultra wideband wireless personal area networks," *Proc. of the IEEE Conf. on Ultra Wideband Systems and Technologies (UWBST 2003)*, Reston, VA, November 2003.
13. E. Saberinia, J. Tang, A. H. Tewfik, and K. Parhi, "Pulsed OFDM modulation for ultra wideband communications" *IEEE Int. Symp. on Circuits and Systems 2004 (ISCAS'04)*, Vancouver, May 2004.
14. G. L. Stuber, *Principales of Mobaile Communications*, 2nd edn, Kluwer Academic, Dordrecht, February 2001.
15. Matt Welborn, "XtremeSpectum proposal for IEEE 802.15.3a," IEEE 802.15 work group official web site; http://grouper.ieee.org/groups/802/15/pub/2003/Jul03/, San Francisco, CA, July 2003.
16. P. P. Vaidyanathanm, *Multirate Systems and Filter Banks*, 1st edn, Prentice Hall, Englewood Cliffs, NJ, September 1992.
17. D. Gerakoulis and P. Salmi, "An interference suppressing OFDM system for ultra wide-bandwidth radio channels," *IEEE Conf. on Ultra Wideband Systems and Technologies (UWBST'02)*, May 2002.
18. E. Saberinia and A. H. Tewfik, "Outage capacity of pulsed OFDM ultra wideband communications," *Joint IEEE Conf. on Ultra Wideband Systems and Technologies and Int. Workshop on Ultra Wideband Systems* (UWBST & IWUWBS 2004), Tokyo, May 2004.
19. E. Bieglieri, J. Proakis, and S. Shamai, "Fading channels: information-theoretic and communications aspects," *IEEE Trans. Inform. Theory*, vol. 44, no. 6, pp. 2619–2692, October 1998.
20. L. H. Ozarow, S. Shamai, and A. D. Wayner, "Information-theoretic considerations for cellular mobile radio," *IEEE Trans. Vehich. Tech.*, vol. 43, no. 2, pp. 359–378, May 1994.
21. L. R. Rabiner and B. Gold, *Theory and Application of Digital Signal Processing*, Prentice-Hall, Englewood Cliffs, 1975.

CHAPTER 13

UWB Networks and Applications

KRISHNA M. SIVALINGAM and ANIRUDDHA RANGNEKAR

13.1 INTRODUCTION

Recent approval by the FCC has led to considerable interest in developing UWB communications on an unlicensed basis in the 3.1–10.6 GHz band [1–7]. UWB technology is defined as any transmission scheme whose instantaneous bandwidth is greater than 20% of a center frequency or where the available bandwidth is greater than 500 MHz. The FCC has currently set an emissions mask that will limit the radiated emissions for UWB signals. This mask will enable the simultaneous operation of UWB devices with existing narrowband systems, thereby increasing the efficiency of spectrum reuse.

In UWB, the data is transmitted over a wide range of frequency bands, resulting in high data rates. Since the signal energy is spread very thinly over the entire bandwidth, the energy density is very low. This helps in reducing the probability of detection and interception. Another important property of UWB signal is the high immunity to multipath fading [8]. Multipath fading is a phenomenon observed in continuous wave signals. It occurs due to the reflection of the signals off objects resulting in destructive cancellation and constructive addition. Since UWB is not a continuous wave technology, it is not affected since the reflections can be resolved in time. As a matter of fact, the narrow pulses used in UWB transmission allow multipath resolution and hence can be used to effectively detect the transmitted symbol in a multipath environment.

There are two main differences between UWB and other narrowband and wideband systems. First, the bandwidth of UWB systems in 20% of a center frequency. This is much greater than the bandwidth of any currently used technology. Second, regular narrowband systems use radio frequency (RF) carriers to move the frequency of the signal from baseband to the carrier frequency. UWB radio, on

---

Ultra Wideband Wireless Communication. Edited by Arslan, Chen, and Di Benedetto
Copyright © 2006 John Wiley & Sons, Inc.

the other hand, is implemented in a carrier-less fashion. It involves the transmission of very short (subnanosecond) pulses that are emitted in periodic sequences. This sharp rise and fall time of the pulse results in a waveform that occupies a much larger bandwidth.

To summarize, some of UWB's potential advantages include: (i) low-power operation since transceiver circuitry power requirements are low [4]; (ii) UWB transmissions are below the noise level thereby providing low probability of detection (LPD); (iii) low probability of jamming (LPJ) capabilities due to the low energy per frequency band and the use of precisely timed patterns; (iv) ability to penetrate walls and vegetation due to the lower frequencies used; (v) higher immunity to multipath fading effects due to increased diversity; and (vi) availability of precise location information, since UWB uses precise pico-second pulses for transmission and tight synchronization between the communicating nodes, which enables centimeter-accurate location determination.

However, UWB has a few disadvantages such as long signal acquisition times (up to a few milliseconds [9, 10]), and FCC regulatory issues. There are also several technical challenges at the physical layer to be resolved such as: antenna design, propagation and channel modeling, devices and circuits design, and waveform design.

UWB-based networking is currently being predominantly considered for WPANs, which are defined as networks formed by low-power wireless devices with relatively short transmission distances. The technology for WPANs is in its infancy and is undergoing rapid development as part of several standards projects including IEEE 802.15.3a and the Multiband OFDM Alliance (MBOA, [11, 12]). In addition; the application of UWB for wireless sensor networks is also being explored.

In this chapter, we present a survey of UWB based networks and some of their applications. The chapter is organized as follows. Background material on UWB technology is presented in Section 13.2. In Section 13.3, research on medium access control protocols for UWB networks is presented. In Section 13.4, some of the applications of UWB networks are presented; and Section 13.5 presents a summary and discussion.

13.2 BACKGROUND

This section presents the relevant background material and related work.

13.2.1 UWB Physical Layer

Although all UWB transmissions comprise series of pulses, modulation may be carried out using various techniques. There are basically four methods of UWB modulation [13, 14]: (i) TH-SS; (ii) DS-SS; (iii) MB-OFDM; and (iv) delay-hopped transmitted-reference spread spectrum (DHTR-SS).

The TH-SS system is similar to traditional pulse modulation and transmits pulses at specific times in a frame. The DS-SS system is similar to a BPSK-CDMA system.

Here the information bits are multiplied by a PN chip sequence to provide channelization and spreading in frequency domain. The signal is then transmitted by phase shift keying using a Gaussian mono-pulse and shifting its phase according to pulse polarity. The reception is carried out using a rake receiver comprising a correlator for each rake finger and a maximal ratio combining of the correlator outputs. The MB-OFDM system is based on the OFDM technology implemented in 802.11 g and 802.11a [12]. Instead of using the whole spectrum as a single band as in the case of DS-SS, the MB-OFDM divides the spectrum into individual bands of around 500 MHz. The spectrum of 3–10 GHz thus contains around 13 such bands, which are grouped into four groups based on applications (implemented by Texas Instruments). With DHTR-SS, a pair of identical doublets is transmitted for each frame similar to the differential phase shift keyed (DPSK) system.

13.2.2 IEEE 802.15.3 Standards

The IEEE 802.15.3 standard is being developed for high-data-rate wireless personal area networks. The standard initially specified operation in the unlicensed frequency band between 2.4 GHz and 2.4835 GHz, and is designed to achieve data rates of 11–55 Mbps, which are required for the distribution of high-definition video and high-fidelity audio. An alternative PHY layer, based on UWB radio transmission, has been proposed as part of the IEEE TG802.15.3a.

As of November 2004, the IEEE task group TG802.15.3a has not chosen the physical layer design but is considering two proposals. The first proposal, promoted by the UWB Forum, is based on the principles of direct sequencing. DS-UWB provides support for data rates of 28, 55, 110, 220, 500, 660 and 1320 Mbps. The other proposal, developed by the MBOA [12], is based on the concept of *multiband OFDM* and supports data rates of 55, 110, 200, 400 and 480 Mbps. Multiband OFDM is a transmission technique where the available spectrum is divided into multiple bands. Information is transmitted on each band using OFDM modulation. The information bits are interleaved across all the bands to provide robustness against interference. Multiband OFDM divides the available spectrum (3.1–10.6 GHz) into 13 bands of 528 MHz each. These bands are grouped into four groups to enable multiple modes of operation for multiband OFDM devices. These are: groups A (bands 1–3), B (bands 4–5), C (bands 6–9), and D (bands 10–13). Two modes of operation have been specified. Mode 1 is mandatory and operates in frequency bands 1–3, that is, group A. Mode 2 is optional and uses seven frequency bands, three bands from group A and four bands from group C. Groups B and D are reserved for future use. Channelization in multiband OFDM is achieved using different time–frequency codes, each of which is a repetition of an ordered group of channel indexes. An example of time–frequency codes is given in Table 13.1. The beacon frames are transmitted using a predetermined time–frequency code. This facilitates reception of beacon frames by devices that have not been synchronized. There are still many technological challenges ahead, mostly around the high level of integration that UWB products require: they need to be developed at low cost and low power to meet the vision of integrated connectivity for PANs.

TABLE 13.1 Time–Frequency Codes for Multiband OFDM Devices

Channel No.	Time Frequency Codes (Mode 1)						Time Frequency Codes (Mode 2)						
1	1	2	3	1	2	3	1	2	3	4	5	6	7
2	1	3	2	1	3	2	1	7	6	5	4	3	2
3	1	1	2	2	3	3	1	4	7	3	6	2	5
4	1	1	3	3	2	2	1	3	5	7	2	4	6
5	—	—	—	—	—	—	1	5	2	6	3	7	4
6	—	—	—	—	—	—	1	6	4	2	7	5	3

13.3 MEDIUM ACCESS PROTOCOLS

The IEEE 802.15.3a group initially has selected the IEEE 802.15.3 MAC protocol specifications for channel access. This may not be efficient since the MAC protocol does not consider UWB's characteristics. The MBOA is working on its version of the MAC protocol, but has not made it publicly available yet.

We first present the details of IEEE 802.15.3 and then discuss some of the related work on UWB MAC protocols that considers the impact of channel acquisition time and the presence of multiple communication channels.

13.3.1 IEEE 802.15.3 MAC Protocol

WPANs are not created *a priori*. They are created when an application on a particular device wishes to communicate with similar applications on other devices. This network, created in an ad hoc fashion, is torn down when the communication ends.

Network Architecture The network is based on a master–slave concept, similar to the Bluetooth network formation. A piconet is a collection of devices such that one device is the master and the other devices are slaves in that piconet. The master is also referred to as the piconet controller (PNC). The master is responsible for synchronization and scheduling the communication between different slaves of its piconet.

In the 802.15.3 WPAN, there can be one master and up to 255 slaves. The master is responsible for synchronization and scheduling of data transmissions. Once the scheduling has been done, the slaves can communicate with each other on a peer-to-peer basis. This is contrary to Bluetooth PAN, where devices can only communicate with the master in a point-to-point fashion. In Bluetooth, if device d_1 wants to communicate with d_2, d_1 will send the data to the master and the master will forward the data to d_2. The two slave devices cannot communicate on peer basis.

A scatternet is a collection of one or more piconets such that they overlap each other. Thus, devices belonging to different piconets can communicate over multiple hops.

The piconet can be integrated with the wired network (802.11/Ethernet) by using a IEEE 802 LAN attachment gateway. This gateway conditions MAC data packet units to be transported over Bluetooth PAN.

Channel Access Channel access in the 802.15.3 MAC is based on superframes, where the channel time is divided into variable size superframes, as illustrated in Figure 13.1. Each superframe begins with a beacon that is sent by the PNC and is composed of three main entities: the beacon, the contention access period (CAP) and the contention free period (CFP). The beacon and the CAP are mainly used for synchronization and control information whereas the contention-free period is used for data communication. During the CAP, the devices access the channel in a distributed manner using CSMA/CA with a specified backoff procedure. The CFP is regulated by the PNC, which allocates time slots to various devices based on their demand and availability.

The beacon packet is used to send the timing information and any piconet management information that the PNC needs to send to the devices. The beacon consists of a beacon frame and any commands sent by the PNC as beacon extensions. The beacon packet contains details about the superframe duration, CAP end time, maximum transmit power level and piconet mode. The superframe duration specifies the size of the current superframe and is used along with the CAP end time to find the duration of the CAP. The resolution of superframe duration and CAP end time is 1 µs and the range is 0–65535 µs. The value of maximum transmit level is specified in dBm and may vary for each superframe. The piconet mode field describes some of the characteristics of the piconet and the superframe. It specifies whether the CAP may contain data, command or association traffic and may be used to disallow a certain type of traffic from being sent during the CAP. The piconet mode field specifies if the management time slots are being used in the current superframe. It also defines the security mode of the piconet.

All the devices in the piconet reset their superframe clock on receiving the beacon preamble. All times in the superframe are measured relative to the beacon preamble. Each device in the piconet calculates its transmission time based on the information contained in the beacon.

The CAP is used to communicate commands and asynchronous data traffic, if any. Carrier sense multiple access with collision avoidance (CSMA/CA) is the basic medium access technique used in the CAP. The type of data or commands that a device may send during the CAP is governed by the PNC by setting

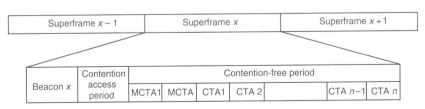

Figure 13.1 IEEE 802.15.3 superframe format.

appropriate bits in the piconet mode field of the beacon. Before transmitting each frame, the device senses the medium for a random period of time and transmits only if the medium is idle. Otherwise, it will perform a backoff procedure that is maintained across superframes and is not reset at the start of a new superframe. That is, if the backoff interval has not expired and there is not enough time left in the CAP, then the backoff interval is suspended and restarted at the begin of the next superframe's CAP. When the device gains control of the medium, it checks if there is sufficient time in the CAP for the transmission of the whole frame. CAP traffic is not allowed to intrude in the contention free period and the device must backoff until the beginning of the next superframe's CAP.

The CFP consists of channel time allocations (CTAs). CTAs are used for management commands as well as synchronous and asynchronous data streams. The PNC divides the CFP into channel time allocations that are allocated to individual devices. A device may or may not fully utilize the CTA allocated to it, with no other device being allowed to transmit during this period. The order of transmission of the frames is decided locally by the device without the knowledge of the PNC. Depending on its position in the superframe, there are two type of CTAs: dynamic CTA and pseudostatic CTA. The devices in the piconet have the choice of requesting either of the CTAs. The position of a dynamic CTA, within a superframe, can be moved on a superframe to superframe basis. This allows the PNC the flexibility to rearrange the CTAs to obtain the most efficient schedule. The scheduling mechanism for the CTAs is not specified by the draft standard and is left to the discretion of the implementer. Pseudostatic CTAs maintain their position within the superframe and are allocated for isochronous streams only. The PNC is allowed to move the location of these CTAs as long as the old location is not allocated to any other stream for a predefined constant period. The CFP may also contain management CTAs (MCTA) that are allocated just after the contention access period. MCTAs are used to send command frames that have the PNC either as the source or the destination. The PNC is responsible for determining the number of MCTAs for each superframe.

Whenever a device needs to send data to another device in the piconet, it sends a request to the PNC. The PNC allocates the CTAs based on the current outstanding requests of all the devices and the available channel time. When a source device has a frame to be sent to a destination, it may send it during any CTA allocated for that source destination pair. If such a CTA does not exist, the source may send the frame in any CTA assigned to that source as long as the source device can determine that the destination device will be receiving during that period. A device may not extend its transmission, started in the CTA, beyond the end of that CTA. The device must check whether there is enough time for transmission of the frame during the current CTA to accommodate the frame. If a device receives the beacon in error, it will not transmit during the CAP or during any management or dynamic CTA during that superframe. The device is allowed to use the pseudostatic CTAs until the number of consecutive lost beacons exceeds a constant value. Any device that misses a beacon may also listen for the entire superframe to receive frames for which it is the destination.

13.3.2 Impact of UWB Channel Acquisition Time

Unlike continuous wave technology that use sine waves to encode information, UWB technology uses very short (subnanosecond), low-power pulses (monocycles [5, 15, 16]) with a sharp rise and fall time, resulting in a waveform that occupies several GHz of bandwidth. Since the signal is spread very thinly over the entire bandwidth, the power density is very low that facilitates co-existence with existing legacy systems such as global positioning system (GPS [17]). The acquisition time for a UWB signal is thus large due to a combination of the low energy per pulse and very short pulse durations (nanoseconds or hundreds of picoseconds, typically). In a broadcast multiple access environment this can severely affect efficiency of the MAC protocol [9]. It is therefore necessary to study the impact of high acquisition time on performance metrics that include throughput, delay, and acquisition overhead, as considered in Rangnekar et al. [18]. Another approach based on aggregating multiple upper-layer packets into a larger burst frame at the MAC layer is presented in Lu et al. [10].

Timing acquisition is typically performed using a preamble in packet data systems. In high data rate applications, preamble efficiency is required so as to reduce loss of throughput. Consider a 1024 byte data payload transmitted at 100 Mbps. A 10 µs preamble amounts to an overhead of 11%, which rises to 34% for a 500 Mbps data rate. A matched filter or a correlator receiver is optimal for acquisition of a single user's preamble sequence. Analog correlators are used because digital filters are infeasible due to the excess GHz sampling rates that cannot be supported by current state-of-the-art ADC designs. The mean acquisition time depends on both the signal bandwidth and pulse duration of the UWB signal. The Dispersive nature of the multipath can also be exploited to improve acquisition performance.

The following scheduling algorithms have been considered in Rangnekar [18]:

- *Single CTA (CTA-1)*—each connection is assigned a single guaranteed time slot in each superframe. The number of CTAs allocated in each superframe, and hence the size of the superframe, depends on the number of connections in the piconet.
- *Multiple CTA (CTA-M)*—each connection is assigned multiple timeslots in the same superframe. The draft standard recommends that, if multiple CTAs are assigned to a connection, then the timeslots should be spread out through the superframe. Hence the timeslots are assigned on a round-robin basis until the maximum size of the superframe is reached or there are no more data packets in the buffer. The maximum number of timeslots assigned to each connection depends on the number of connections in the piconet at that instant.
- *Contiguous CTA (CTA-C)*—each connection is assigned multiple contiguous timeslots. The number of contiguous slots is limited by a preset constant, *maxContiguousCTA*. Depending upon the number of connections in the piconet, each connection may be assigned multiple blocks of contiguous timeslots. These blocks are assigned to each connection on a round-robin basis until the size of the superframe has reached its maximum value.

TABLE 13.2 Simulation Parameters and Values

Simulation Parameter	Value
Channel bandwidth (C)	100–500 Mbps
Number of nodes (N)	16–128
Packet size	2032 bytes
Packet generation rate (λ)	0.1–40,000 pkt/s
Maximum buffer size (B)	150 pkts
Acquisition time (T_a)	5–25 μs
T_{CAP}	1 ms
T_{SIFS}	10 μs
Guard band time (T_{Gb})	3.28 μs
Beacon time (T_B)	0.7–60 μs
CTA size	2032 bytes
Maximum superframe	65.535 ms
Maximum contiguous CTAs in CTA-C	1–100

The paper also presents performance evaluation of a network using discrete event simulation models. The system parameters varied include number of nodes in the piconet, acquisition time, channel bandwidth and packet arrival rate. The values for the simulation parameters are specified in Table 13.2 and are based on information provided in the 802.15.3 standard.

The performance metrics measured are utilization, average packet delay and acquisition overhead. The superframe utilization is given by:

$$U = \frac{T_d}{[T_B + T_{CAP} + T_d + n(T_a) + m(T_{SIFS} + T_{Gb})]} \quad (13.1)$$

where T_d = data transmission time; T_a = acquisition time; T_B = beacon transmission time, depending on the number of connections in the superframe; T_{CAP} = contention access period and is constant; T_{SIFS} = SIFS time period and is constant; T_{Gb} = guard time and is constant; n = number of times signal acquisition is needed, depending on the scheduling scheme; and m = number of CTAs allocated in the superframe.

Each superframe has just one instance of T_B and multiple instances of T_a, T_{SIFS} and T_{Gb}, where the values of T_{SIFS} and T_{Gb} are constant. Thus, larger values of T_a will result in lower per-frame utilization. For a 500 Mbps channel with $T_a = 5$ μs and superframe of maximum size 65535 μs, there can be at most 1259 CTAs per superframe. Each CTA, for a 2032 byte packet, lasts for 32.9 μs. Based on these values and those described in Table 13.2, we expect the utilization to be about 61%, but if the acquisition time is increased to 25 μs, the number of CTAs per superframe drops to 905 and the utilization falls to 44%. For a 100 Mbps channel, it can be calculated that the utilization drops from 88% to 80% as the acquisition time is increased from 5 μs to 25 μs. Thus, the adverse effect of acquisition time is more prominent in higher data rate channels.

Figure 13.2(a) presents the average packet delay of the MAC protocol vs channel acquisition time. The packet generation rate for this set of simulations was set to 4000 packets/s. The maximum service rate for the 500 Mbps channel, for the given set of parameters, was obtained as 1750 packets/s for each queue. Similarly, for the 100 Mbps channel, a maximum service rate of 523 packets/s is feasible. This decrease in service rate causes the average packet delay to increase considerably as the channel bandwidth is decreased. The average packet delay is further increased with an increase in acquisition time. This is because of the recommendations of the draft standard that cause the protocol to spend acquisition time for each packet. There is an 11% increase in average packet delay for the 100 Mbps channel and a 50% increase for the 500 Mbps channel. Thus, the increase in packet delay is more severe for the higher bandwidth channels. This handicap can be overcome by scheduling CTAs for each connection contiguously. In the following section, we will show that the contiguous scheme decreases packet delay and also reduces the impact of acquisition time.

Figure 13.2(b) and (c) presents per-packet acquisition overhead and utilization, obtained through simulations by varying acquisition time. The acquisition overhead (AO) is defined as the percentage of time in a superframe that is spent in signal acquisition relative to the time spent for data transmission: $AO = (n \cdot T_a)/T_d$, where T_a, T_d, and n are channel acquisition time, data transmission time, and number of times signal acquisition is needed, respectively. It is measured by computing the time spent in signal acquisition and data communication over the duration of the simulation. Our intention in plotting this metric is to show the amount of time lost in signal acquisition per packet transmission. Figure 13.2(b) shows that an increase in acquisition time severely affects the acquisition overhead. The 500 Mbps channel suffers the most as the acquisition overhead increases from 15% to 76% (an increase of 400%). An acquisition overhead of 76% implies that the protocol spends nearly as much time in signal acquisition as it spends in actual data transfer. Thus, there exists much scope for improvement in utilization if the acquisition overhead is controlled by efficiently scheduling the CTAs.

Figure 13.2(c) matches our expected values of U based on Equation (13.1). This plot shows the reduction in throughput as compared with the ideal throughput for each value of C. This reduction is mainly due to the the amount of time wasted in signal acquisition.

13.3.3 Multiple Channels

In a wireless network, the transmission channel has to be shared by many nodes using either a random access method such slotted Aloha or by scheduling the channel to the nodes based on user requests. As the number of nodes sharing the wireless medium increases, the amount of bandwidth available to each node drops. This effect is aggravated by the fact that the available bandwidth is already low as compared with the wired networks. One possible solution is to provide multiple wireless communication channels for simultaneous use.

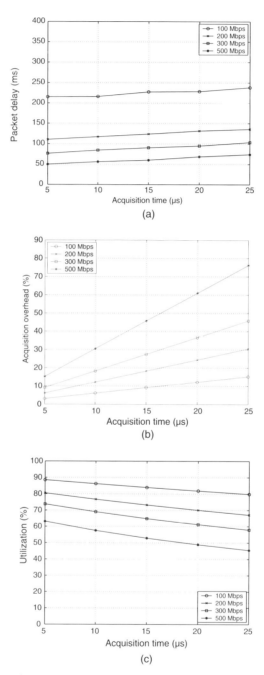

Figure 13.2 Effect of acquisition time on (a) average packet delay, (b) acquisition overhead, and (c) utilization ($\lambda = 4000$ packets/s).

Depending on the technology used for wireless transmissions, the wireless spectrum is divided into multiple simultaneous channels. For example, IEEE 802.11b communication, operating in the 2.4 GHz ISM band with 80 MHz of reusable spectrum, provides 11 channels. In this paper, we consider the IEEE 802.15.3a standard based on UWB communications being developed for wireless PANs [3, 4, 15, 19].

In this paper, we present a QoS-aware scheduling algorithm for the IEEE 802.15.3 MAC protocol that utilizes the multiple channels that are available in a UWB network. Each wireless device is equipped with a tunable transceiver that gives the node the flexibility to transmit or receive on any channel, thereby enabling sharing of the channels among the nodes. The scheduling mechanism employs a distributed dynamic channel allocation algorithm to distribute the channels among neighboring piconets based on dynamic traffic demand.

Scheduling Algorithm This section describes the mechanism used by the PNC to schedule packets over multiple channels, as studied in Rangnekar and Sivalingam [20]. The system has C distinct and nonoverlapping channels are available for use in the entire system. Each node is equipped with a half-duplex tunable transceiver. The node can transmit or receive on only one channel at a time. The transceiver is capable of tuning to different channels dynamically. The channel switch time, as defined by the multiband OFDM proposal [21], is 9 ns. All piconet nodes are synchronized to the PNC, which transmits the beacon on a predetermined default channel. All piconet nodes know the default channel and tune their transceiver to the default channel to listen to the beacon. For ease of explanation, we assume only one piconet in the system. However, it can be easily extended to multiple piconets, as presented in the next section. Let the number of nodes in a given piconet be denoted by M and the number of channels by C. The requests made by each node for transmission are stored as an $M \times M$ demand matrix. The demand matrix contains the number of packets to be transmitted by each node to every other node in the network. The objective of the scheduling algorithm is to schedule these requests on the C channels in a collision-less manner. The demand matrix is first converted into an $M \times C$ matrix to convert the scheduling problem into a time slot assignment (TSA) problem. Solving the time slot assignment problem implies finding a conflict-free assignment of requests to the channels such that the total frame size is minimized.

This scheduling problem is similar to one of the basic, well-studied problems of scheduling theory, that of nonpreemptively scheduling M independent tasks on C identical, parallel processors. The objective is to minimize the total time required to complete all the tasks. This problem is known to be NP-complete [22] and approximations to this problem such as MULTI-FIT [23] for finding near-optimal schedules have been studied. The Multi-fit [23] algorithm to convert the demand matrix into a $M \times C$ form. This matrix is then input to the interval-based scheduling (IBS) algorithm [24] that generates the transmission schedule. The details of the algorithm are presented in Rangnekar and Sivalingam [20]. The paper also considers a single piconet scenario and multiple piconet scenario. For the latter, a dynamic channel allocation, that adapts the channel allocation to the participating piconets, is presented.

Performance The performance of the scheduling algorithms has been studied in detail in Rangnekar and Sivalingam [20]. The performance metrics measured are throughput, average packet delay, and scheduling efficiency. Throughput is defined as the amount of data transmitted in the piconet per unit time. Average delay is the time between packet generation and reception. Scheduling efficiency is a measure of wastage of channel bandwidth due to the scheduling algorithm and is defined as the ratio of allotted slots to the total number of slots in a superframe. Here, we summarize the results of the single piconet system analysis.

Figure 13.3(a) presents the average packet delay of the scheduling algorithm for varying packet generation rate. A 64-node piconet with a packet size of 2032 bytes is considered. The packet generation rate (λ) is varied from 100 to 100,000 packets/s. The channel bandwidth is fixed at 500 Mbps and the number of channels is varied from 1 to 8. For a 500 Mbps channel, each superframe can accommodate up to 1394 CTAs, each of size 2032 bytes. If the superframe is filled up to its limit (65,535 μs), approximately 21,200 CTAs can be allocated per second. Hence, the service rate (μ), for the schemes with full superframe utilization, is about 2120 packets/s for each queue given that there are 10 connections in this scenario. For stable queue operation with infinite buffer capacity $\lambda \leq 2120$ packets/s. In our study, we consider buffer capacity of 300 packets/node. Thus, the packet delay values tend to be stable after the saturation load limit is reached. As the number of channels available for data transmission is increased, different nodes within the piconet can transmit simultaneously. This increases the service rate thus reducing the average packet delay.

Since we assume the packet size to be 2032 bytes, a packet generation rate of 10,000 packets/s is equivalent to 160 Mbps. Since the piconet has 10 connections, the total packet generation rate of the piconet, for $\lambda = 10,000$ packets/s, is 1600 Mbps.

As explained earlier, a 500 Mbps channel with full superframe utilization can accommodate 21,200 packets/s and hence its expected throughput is 340 Mbps. As the number of channels (C) increases, we would expect the total throughput to increase linearly with C, that is, as C is increased to 2, the throughput is expected to be 680 Mbps. However, Figure 13.3(b) shows a slight reduction in the observed throughput. This reduction is due to the decline in scheduling efficiency as C is increased. There is a 10% reduction in efficiency, as C is increased to 2, which accounts for the fall in observed throughput. As C is increased beyond 2, the fall in throughput is even more pronounced and can be explained by the corresponding drastic fall in scheduling efficiency.

Figure 13.3(c) plots the efficiency of the scheduling algorithm. For a piconet with a single channel, the efficiency is 1 as there is no slot wastage since all transmissions are sent on the same channel and ordered in time. Slot wastage is introduced when the piconet has multiple channels. Consider the example of a piconet with multiple channels, where a single node wants to transmit data to multiple nodes. Even if each

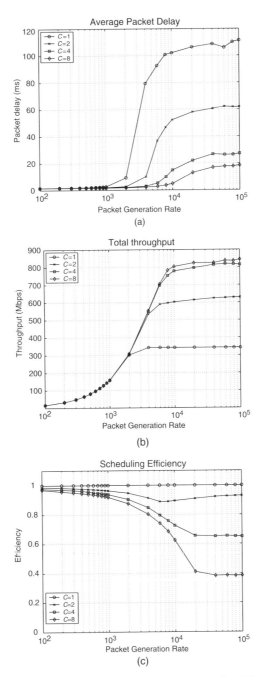

Figure 13.3 Effect of packet generation rate on (a) average packet delay, (b) total throughput, and (c) scheduling efficiency for a single piconet ($M = 64$, $B = 500$ Mbps).

of the receivers is assigned a unique channel for data reception, the sender can transmit to only one node at a time since it has only one transmitter. While the sender is transmitting to a particular receiver, timeslots on all the other channels are wasted, unless utilized by some other sender. This wastage is measured by the channel efficiency metric. It can be logically deduced that, as the number of channels increases, the slot wastage increases leading to lower scheduling efficiency.

13.4 NETWORK APPLICATIONS

UWB technology has been used in the past for inventory, locationing and ranging applications. Commercial products have been available in this domain for many years [25–27].

One example is the asset location system developed by Multispectral Solutions, Inc. Their PAL650 UWB Precision Asset Location system consists of a set of active UWB tags, UWB receivers and a central processing hub. One of the UWB tags is used as the reference beacon. The tag operates at a center frequency of 6.2 GHz with an instantaneous bandwidth of 1.25 GHz. Multilateration techniques, combined with time different of arrival (TDOA) measurements that utilize data from at least three receivers, are used by the hub to determine the location. The system range has been demonstrated to be around 600 feet with accuracies better than 1 foot [28].

An example of an UWB-based radar application is Time Domain's "Thru-Wall Sensing," a radar imaging system that can penetrate common building materials including reinforced concrete, concrete block, or sheetrock. The potential applications include tactical operations, search operations, and covert operations and urban warfare.

As mentioned earlier, UWB is now being actively considered for WPANs as part of the IEEE 802.15.3 and MBOA standards. One of the classical instances of WPAN applications is home area network-based entertainment applications, that is, networked consumer electronics applications. In this scenario, components such as high-definition television (HDTV), a DVD player and recorder, stereo speakers, and receiver component will be connected in a wireless manner using UWB. The potentially high bandwidth of UWB makes this application a reality. When combined with a home area network based on ethernet or wireless LANs, it is also possible to link multiple TVs in different rooms to receive the same streaming multimedia content. Another important WPAN application lies in connecting peripheral devices to a desktop computer. These devices can be digital cameras, digital camcorders, hard drives, printers, etc. In this peripheral connectivity application, there is an increased interest in designing a *wireless USB* interface based on UWB techniques [29, 30], with Intel Corporation being one of the leading developers.

An interesting networking example is presented in Ameti et al. [31]. In military aircraft, inter-crew communication is typically done using a wired network that connects the various crew members helmet to a central audio system. An UWB-based aircraft wireless intercommunications system (AWICS) for military aircraft has been designed and implemented by Multispectral Solutions Inc. The system takes advantage of multipath mitigation, low probability of detection, and low probability of interference features of UWB. The system has been demonstrated on CH-53E Super Stallion and CH-46E Sea Knight helicopters.

UWB is also being considered for wireless sensor networks, especially for high-data-rate applications such as those based on multimedia and video sensors. Application of UWB to sensor networks has recently gained attention. A multihop homogeneous UWB sensor system, denoted UWEN, is described in Oppermann et al. [32]. UWEN comprises low-power, low-data-rate sensors that communicate with fixed UWB nodes to transmit the sensed information. A centralized approach is employed in this scheme wherein the results from the individual sensors are fed to a sink. NanoMAC, an energy sense multiple access with collision avoidance, is implemented as the MAC technique for this scheme.

Although many of the current applications concentrate on the short range capabilities of UWB technology, UWB can be used for larger range applications [33]. Multispectral Solutions Inc. has already demonstrated the feasibility of high-data-rate (6 Mbps video link) at 2 W peak power for a line of sight range of 8 km [34] using UWB radios. A typical 802.11 device (ORiNOCO AP-2000) operates on a power of 10 W supports a data rate of 6 Mbps for a maximum range of 250 m using IEEE 802.11a technology [35]. This indicates that UWB is capable of providing higher data rates at lower power as compared to the 802.11 standards.

A summary of several possible UWB applications is presented in Fontana [36]. Future applications of UWB can take advantage of the precise locationing and ranging information to design better MAC, routing and related network protocols and for authentication protocols.

13.5 SUMMARY AND DISCUSSION

In summary, this chapter presented an overview of UWB networking and related issues such as protocols and applications. The exciting potential of UWB is in the process of being understood and we envision the future to hold much more significant promise in better integration of UWB techniques in next generation wireless products and applications.

ACKNOWLEDGMENTS

The authors are pleased to acknowledge the discussions with Prathima Agrawal, Santosh Pandey, and Minal Mishra on this chapter.

REFERENCES

1. FCC Note of Proposed Rule Making (1998). Revision of part 15 of the commission's rules regarding ultra-wideband transmission systems. ET-Docket 98–153.

2. FCC First Report and Order (2002). Revision of part 15 of the commission's rules regarding ultra-wideband transmission systems (fcc 02-48). ET-Docket 98–153.
3. Nardis, L. D., Baldi, P., and Benedetto, M.-G. D. (2002). UWB Ad hoc Networks. In *Proc. IEEE Int. Conf. on Ultra Wideband Systems and Technologies.*
4. Foerster, J., Green, E., Somayazulu, S., and Leeper, D. (2001). Ultra-wideband technology for short- and medium- range wireless communications. *Intel Technical Journal*, vol. 5, no. 2; http://developer.intel.com/technology/itj/
5. Win, M. Z., and Scholtz, R. A. (1998a). Impulse Radio: How it Works. *IEEE Communications Letters*, vol. 2, no. 2, pp. 36–38.
6. Siwiak, K. (2001). Ultra Wide Band radio: introducing a new technology. In *Proc. IEEE Vehicular Technology Conf.*, pp. 1088–1093, Rhodes.
7. Smak, K., Withington, P., and Phelan, S. (2001). UltraWide Band Radio: the emergence of a important new technology. In *Proc. IEEE Vehicular Technology Conf.*, pp. 1169–1172, Rhodes.
8. Win, M. Z., and Scholtz, R. A. (1998b). On the Robustness of Ultra-wide Bandwidth Signals in Dense Multipath Environments. *IEEE Communications Letters*, vol. 45, no. 2, pp. 10–12.
9. Ding, J., Zhao, L., Medidi, S., and Sivalingam, K. (2002). MAC Protocols for Ultra-Wide-Band (UWB) Wireless Networks: Impact of Channel Acquisition Time. In *Proc. SPIE ITCOM*, vol. 4869, Boston, MA.
10. Lu, K., Wu, D., Fang, Y., and Qiu, R. C. (2005). On medium access control for high data rate ultra-wideband ad hoc networks. In *Proc. IEEE WCNC*, New Orleans, LA.
11. IEEE (2005). IEEE 802.15 Working Group for Wireless Personal Area Networks (WPANs); http://grouper.ieee.org/groups/802/15/
12. Multiband OFDM Alliance (2005). UWB Alliance to Enable CE, PC and Mobile Communications Markets; www.multibandofdm.org
13. Saquib, M. (2004). UWB communications for military. In *3rd Annual Winter Workshop of VI U.S. Army Vetronics Institute.*
14. Welborn, M. and Shvodian, B. (2003). Ultra-wideband Technology for Wireless Personal Area Networks—the IEEE 802.15.3/3a Standards, UWBST Tutorial. In *IEEE UWBST Conference Proceedings.*
15. Scholtz, R. A. (1993). Multiple Access with Time-Hopping Impulse Modulation. In *Proceedings of IEEE MILCOM'93.*
16. Pande, D. C. (1999). Ultra Wide Band (UWB) Systems and their Implications to Electromagnetic Environment. In *Proc. International Conf. on Electromagnetic Interference and Compatibility.*
17. Multiple Access Communications Ltd (2000). An Investigation into the potential impact of ultra-wideband transmission systems. Technical Report RA0699/TDOC/99/002.
18. Rangnekar, A., Sivalingam, K., and Roy, S. (2004). Impact of Long Acquisition Times on the Performance of IEEE 802.15.3 MAC Protocol. Technical report, University of Maryland at Baltimore County (UMBC); http://dawn.cs.umbc.edu/wireless-pubs.html
19. IEEE 802.15 Working Group for WPAN (2003). Part 15.3: Wireless Medium Access Control (MAC) and Physical Layer (PHY) Specifications for High Rate Wireless Personal Area Networks (WPAN). Draft P802.15.3/D17-pre.

20. Rangnekar, A. and Sivalingam, K. M. (2004). Multiple Channel Scheduling in UWB Based IEEE 802.15.3 Networks. In *Proc. First Int. Conf. on Broadband Networks—Wireless Networking Symposium*, San Jose, CA.
21. Multi band OFDM Alliance (2003). Multi-band OFDM Physical Layer Proposal for IEEE 802.15 Task Group 3a. IEEE P802.15-03/268r2.
22. Ullman, J. D. (1976). Complexity of sequencing problems. In Coffman, E. G., editor, *Computer and Job/Shop Scheduling Theory*, Chapter 4, Wiley, New York.
23. Coffman, E., Garey, M. R., and Johnson, D. S. (1978). An application of bin-packing to multiprocessor scheduling. *SIAM Journal of Computing*, vol. 7, pp. 1–17.
24. Sivalingam, K. M., Wang, J., Mishra, M., and Wu, X. (2002). An interval-based scheduling algorithm for optical WDM star networks. *Journal of Photonic Network Communications*, vol. 4, no. 1, pp. 73–87.
25. Time Domain (2005). Time Domain: Pulse of the Future; www.timedomain.com
26. Multispectral Solutions Inc. (2005). UWB Precision Asset Location System; www.multispectral.com/products.html
27. Ultrawideband Planet (2005). The Source for Ultrawideband Business and Technology; www.ultrawidebandplanet.com/products
28. Fontana, R. J., Richley, E., and Barney, J. (2003). Commercialization of an ultra wideband precision asset location system. In *IEEE Conference on Ultra Wideband Systems and Technologies*, Reston, VA.
29. WUSB Alliance (2005). Wireless USB; www.usb.org/wusb/home
30. Staccato Communications Inc. (2005). Wireless USB: the Time is Now; www.staccatocommunications.com/products/
31. Ameti, A., Fontana, R. J., Knight, E. J., and Richley, E. (2003). Ultra Wideband Technology for Aircraft Wireless Intercommunications Systems (AWICS) Design. In *IEEE Conference on Ultra Wideband Systems and Technologies*, Reston, VA.
32. Oppermann, I., Stoica, L., Rabbachin, A., Shelby, Z., and Haapola, J. (2004). Uwb wireless sensor networks: Uwen—a practical example. *IEEE Communications Magazine*, vol. 42, pp. S27–S32.
33. Time Domain (2004). PulsON 202 UWB Reference Design.
34. Multispectral Solutions Inc. (2002). Current Trends in UWB Systems in the USA: implementation, Applications and Regulatory Issues. In *Proc. Advanced Radio Technology Symposium*.
35. Proxim Corporation (2004). ORiNOCO AP-2000 Access Point.
36. Fontana, R. (2000). Recent applications of ultra wideband radar and communications systems; www.multispectral.com/pdf/AppsVGs.pdf

CHAPTER 14

Low-Bit-Rate UWB Networks

LUCA DE NARDIS and GIAN MARIO MAGGIO

14.1 LOW DATA-RATE UWB NETWORK APPLICATIONS

UWB technology was first introduced in the context of wireless communications in the mid 1990s, driven by the demand for high-data-rate (HDR) links for multimedia traffic over short distances. A notable example is the IEEE 802.15.3a initiative for WPANs. In this context, the two main technical proposals were based upon the DS-CDMA and the MB-OFDM modulation formats, respectively.

Recently, though, there has been a growing interest in the application of the UWB technology to low-power, low-data-rate (LDR) networks, like in sensor networks, as witnessed by the creation of the IEEE 802.15.4a Task Group [1]. This trend has also been marked by the return to the "origins" of the UWB technology, deriving from radar applications, namely the use of *impulse radio*. UWB-IR systems make use of ultra short duration pulses which yield ultra wide bandwidth signals characterized by extremely low power spectral densities. These systems are particularly suited to sensor network applications as they potentially combine reduced complexity with low power consumption, immunity to multipath fading, multiaccess capabilities, resilience vs interference and support for precise ranging/localization.

In the following, we describe the recent developments of the IEEE 802.15.4a initiative for low data-rate UWB networks as well as the targeted applications.

14.1.1 802.15.4a: A Short History

In November 2002, an interest group "a" was formed to investigate a UWB alternative physical layer to the 802.15.4 WPAN standard (adopted by the Zigbee Alliance). Then, the IEEE 802.15 Low Rate Alternative PHY Task Group (TG4a) was officially formed in March 2004, with the mission amending the 802.15.4

Ultra Wideband Wireless Communication. Edited by Arslan, Chen, and Di Benedetto
Copyright © 2006 John Wiley & Sons, Inc.

standard for an alternative PHY, called 802.15.4a. By January 2005, the group had over 20 proposals to consider, mostly focused on IR or DS-UWB, but alternative technologies such as near-field ranging and "chirp" spread-spectrum (CSS) radio were also on the table. In March 2005, the baseline specification was approved. The baseline consists of two optional PHYs: (a) UWB IR, operating in unlicensed UWB spectrum; and (b) CSS, operating in unlicensed 2.4 GHz spectrum.

14.1.2 The 802.15.4a PHY

The Task Group TG4a has specified that the UWB portion of the PHY should be capable of both communications and ranging, and that it should occupy a bandwidth of at least 500 MHz. That would be centered somewhere between 3.85 GHz and 4.05 GHz. Options include two additional 500 MHz bands, an above-6 GHz band with a guaranteed bandwidth of more than 1.5 GHz and possibly a sub-gigahertz band (subject to regulations). The UWB-PHY shall be based upon impulse radio (pulse-shape independent), support different receiver architectures (coherent/ noncoherent), and multiple rates, and support simultaneously operating piconets (SOP).[1]

On the other hand, the chirp PHY will operate in the 2.45 GHz band but will not be capable of ranging. CSS uses a frequency-modulated pulse. Chirp pulses are robust and can be generated and processed without complex digital circuitry, thereby increasing battery life and reducing costs. Because it works in the 2.45 GHz band, the CSS-PHY can supposedly interoperate with ZigBee devices. On top of that, in contrast to the ZigBee radio's narrowband operation, the CSS operation should provide greater coverage and allow for lower-power operation and for mobility (fast connections) at up to 100 mph.[2]

As of today, the standard is still under evolution and many technical details of the PHY remain to be fixed. Other issues under discussion include common packets for communications and ranging, and support for multiple rates.

In the rest of the chapter, we will focus on UWB-PHY since many of the target applications within 802.15.4a require support for ranging and localization.

14.1.3 PHY: 802.15.4a vs 802.15.4

The 802.15.4a Alt-PHY standard aims to support a low-complexity, low-cost, low-power-consumption WPAN communication system with precision location, extended range, robustness, and mobility. The precision ranging capability, range, robustness, and mobility will be improved enough to satisfy an evolutionary set of industrial and consumer needs. The project will address the requirements to support sensor, control, logistic, and peripheral networks in multiple compliant co-located systems and also coexistence.

---

[1] The UWB-PHY development is sponsored by ST Microelectronics, Freescale, IBM, Mitsubishi, Philips, Renesas, Samsung, Motorola, Staccato Communications, Aetherwire & Location and others.
[2] Nanotron is the major sponsor of the CSS-PHY.

The anticipated high-level characteristics of the Alt-PHY layer, vs the 802.15.4 PHY, are summarized in the following:

- High-precision ranging/location capability (1 m accuracy and better);
- High aggregate throughput; nominal payload bit rate equal to 1 Mbps;
- Robustness and interference resistance;
- Low power consumption;
- Scalability (data rate, range, power consumption and cost);
- Reduced form factor (compatible with sensor networks or RF tags applications);
- Extended range;
- Mobility.

These additional capabilities over the existing 802.15.4 standard are expected to enable significant new applications and market opportunities. The result will enable a wide range of applications, from factory floor control, sensors, and tracking to body-area networks.

14.1.4 Technical Requirements

In this section, we summarize the fundamental requirements for UWB-PHY, as implied by the 802.15.4a applications detailed in Section 14.1.5.

The reference model used for the 802.15.4a alternate (Alt-)PHY layer is shown in Figure 14.1. The list of the Alt-PHY technical requirements follows.

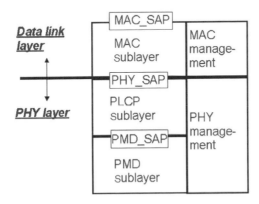

MAC_SAP: MAC service access point
PHY_SAP: PHY service access point
PLCP: PHY layer convergence protocol, contains FEC
PMD: physical medium dependent (radio)

Figure 14.1 Reference partitioning.

Topology The Alt-PHY layer shall support all types of topologies defined by IEEE 802.15.4 standard in its MAC section. This includes the capability to relay messages, coordinate cells or aggregated cells, or quasi-simultaneously concentrate data issued from multiple nodes.

The network configuration has to be highly dynamic. Thus the Alt-PHY layer must be workable without requiring complex static setup procedures and must comply with dynamic insertion and de-insertion of nodes into a network. Typical applications imply data collection by a unique or set of coordinated data collectors. Thus the corresponding Alt-PHY component may have to sustain a much higher throughput than those of the other nodes. The Alt-PHY layer must be able to maintain bidirectional links (half duplex).

Bit Rate The bit rate is categorized in the following way:

- *Individual Link Bit Rate*—this is related to a peer-to-peer link, typically between a sensor device and an information collector or between two devices (relaying of information, synchronization, mutual positioning etc.).
- *Aggregated Bit Rate*—this is typically the bit rate concentrated from many sensor devices to a data collector during a short period of time (can be during specific situations when many devices need to update their information at the same time, like alarm or emergency situations). The data collector must be capable of acquiring at least 1 Mbps of effective data.

Typical selected figures are: link bit rate, at least 1 kbps at PHY-SAP; aggregated bit rate (data collector only), at least 1 Mbps at PHY-SAP.

Location Awareness This is a mandatory function in most applications. It can be related to precise (tens of centimeters) localization in some cases, but is generally limited to about 1 m. Localization awareness may result in different applications such as precise positioning, localization aided routing, motion tracking (simple detection of an object in a determined area, or moving outside of this area). This functionality must be built into the node with basic functions embedded into the Alt-PHY and capable of being serviced in a simple and automatic way by higher layers. It is anticipated that the physical layer must be capable of providing adequate time resolution and jitter elimination to properly exercise the localization awareness functionality, for example, by providing services based on message transit time measurement.

Range The maximum distance between communicating nodes is generally from 0 m to 30 m. In some cases, mainly assets tracking, the range has to be extended to several hundreds of meters. Possibly, relay of messages could be used in such situations. In most cases the link data rate can be limited to a few kbps where the range is very large. However, if the number of nodes is very large (up to thousands), the data collector needs to absorb large aggregated data rate (in sustained mode, and particularly in burst mode).

Coexistence and Interference Resistance The alternate PHY may need to operate in an interference environment by having attributes that can be adjusted by higher layer management to deal with interference ingress (interference coming into the alternate PHY) and interference egress (interference caused by the alternate PHY).

The devices must be able to operate in high noise and high multipath environments (e.g., harsh factory environments). The Alt-PHY must be able to sustain an appropriate level of co-channel and out-of-band interference. Both indoor and outdoor applications have to be considered.

Power Consumption The device (complete communication system including Alt-PHY and MAC) must operate while supporting a battery life of months or years without intervention. Therefore very efficient power saving modes are desirable, in particular for devices that transmit sporadically. In addition, the coordination of nodes must not induce frequent wake-up of nodes. These mechanisms must be supported by the Alt-PHY layer.

Quality of Service The critical factor is the reliability of the transmission, meaning that strong error-correction methods need to be provided at PHY level. Other QoS parameters have a strong impact on PHY layer: real-time communication is required, synchronization of nodes (mainly for localization), and the capability to provide rapid reaction in emergency situations.

Complexity Complexity should be minimal to enable mass commercial adoption for a variety of cost-sensitive products. Complexity (gate count, die size) should be minimized. In a number of applications, the components are to be considered as throwaway after use.

Mobility This is a mandatory feature related to intra-cell mobility, not to roaming or handover. Nodes should be capable of reliable communication when on the move, at least for tracking. It is admitted that limited communication performance (e.g., data rate) can be tolerated in such cases. The considered applications may involve pedestrian, industrial vehicle, and optionally higher speed vehicle mobility.

14.1.5 Applications

It is anticipated that future applications will go beyond the currently defined 802.15.4 PHY capabilities, for example high-precision location capability (smart homes, asset tagging) and high aggregate throughput. The main 802.15.4a application areas, along with some practical examples, are reported below.

- *Industrial Inventory Control*—these applications specialize in location without much communication and are less time critical than others. Accurate knowledge of the state of all the items is important. Changes of state (leaving, entering the warehouse/store) are important. Examples include autonomous manifesting; retail, especially high-value items; healthcare inventory tracking; vehicle inventory for dealerships/heavy machinery dealers; and automated meter reading.

- *Home Sensing, Control and Media Delivery*—these applications are consumer-oriented and involve at least unidirectional and often bidirectional communication for support of sensing and control functions. Some of the communication elements may have higher bandwidth requirements, in contrast to applications seen above. Timescales in these applications are similar to logistics applications. Examples include sensing/tracking children/pets/assets; missing item finding/tracking; automatic appliance control (lights, heat); automatic audio sweet-spot calibration/optimization; and 3-D gaming heads-up display based on user location.

- *Logistics*—these applications generally help improve the efficiency of the operations in which they are used. Finding and tracking are essential elements for these applications, sometimes with low-rate unidirectional communication. These applications are even less time-critical and are generally more tolerant of missed communication (in other words, they can have redundancy built-in with no major impact). Examples include warehouse/supply chain management; package tracking (truck inventory, manifest, proper loading); sports tracking (NASCAR, horse, soccer); supermarket cart tracking (matching customers/advertising); and phone call forwarding/asynchronous messaging/moving maps.

- *Industrial Process Control and Maintenance*—these applications are similar to those in industrial inventory control, with the essential difference that at least unidirectional communication (uplink) is a required feature. In these applications, sensors and actuators are generally part of the item being located and the information from the sensors and information to the actuators needs to be communicated. Examples include wireless sensor networks; faulty sensor location; large structures monitoring; aircraft/ground vehicle anti-collision; and monitoring, sensing and control of industrial and environmental processes.

- *Safety/Health Monitoring*—these applications have human life at stake, are very time-critical and generally involve at least one-way communication of some sort (uplink) and may involve bidirectional communication of high-speed data. Tracking is often an important element of these applications. Examples include emergency monitoring (earthquakes, fire); preventive medicine/health monitoring/therapy; military tactical unit situational awareness (urban/rural); tracking firefighters/emergency responders; and finding avalanche victims.

- *Personnel Security*—these applications generally involve real-time location and may involve tracking and some uplink communication. A few applications require bidirectional, generally low-rate communication, often in combination with location tracking. They are generally less time-critical than others, but it is critical that the information be conveyed. These applications may have a "radius of allowed mobility" (e.g., the prisoner tracking or child tracking

applications). Examples include security and surveillance functions in public areas; workstation lock/unlock authentication; automobile auto-unlocks when owner in range; point of sale authentication/wireless ethernet authentication; and activity-based CCTV stream selection.
- *Communications*—these applications are those for which communication is primary and location is secondary. An example is body-area networks (BANs). The technology should provide data rates of 500 kbps for audio, 3–5 Mbps for video and up to 10 Mbps for server applications, but at low power and short range for BANs. UWB should provide the data rates of WLANs but at low power and with good coverage.

14.2 THE 802.15.4 MAC STANDARD

In Section 14.1 the key characteristics of the future IEEE 802.15.4a PHY were analyzed. The 802.15.4 standard released in 2003 [2], however, defined both PHY and MAC layers; as a consequence it is expected that the innovative features introduced in the 802.15.4a will impact the MAC design as well. In this section we will briefly analyze the original 802.15.4 MAC, in order to highlight how the new 802.15.4a requirements and applications may impact this MAC protocol.

14.2.1 Network Devices and Topologies

The 802.15.4 standard defines two classes of devices: full-function devices (FFD), in which all network functionalities are implemented, and reduced-function devices (RFD), that only support a reduced set of functionalities and are thus only suitable for simple applications such as sensing or executing commands.

RFD and FFD devices organize themselves in PANs. A PAN is controlled by a PAN coordinator, that is a device in charge of setting up and maintaining the PAN. The role of PAN coordinator can only be taken by an FFD device, while RFD devices can only join an existing PAN by communicating with the PAN coordinator. A PAN can adopt either of the two following network topologies:

- *Star Topology*—in this topology, devices can only exchange information with the PAN coordinator; since all communications involve the coordinator, this topology is better suited for network architectures where a device is connected to the power network, and can thus take the role of coordinator for a long time without drowning its battery power. An example of a star topology is presented in Figure 14.2.
- *Peer-to-Peer Topology*—in this topology, FFD devices can communicate directly as long as they are within physical reach, while RFD devices, due to their limitations, can only connect with the PAN coordinator. An example of a star topology is presented in Figure 14.3.

The peer-to-peer topology, thanks to its higher flexibility, potentially allows for the formation of more complex topologies, for example based on multiple clusters;

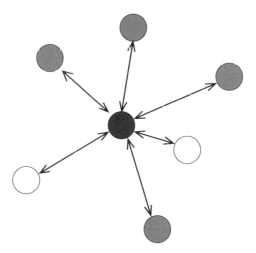

Figure 14.2 Example of star topology (dark gray circle, PAN coordinator; light gray circle, FFD device; white circle, RFD device).

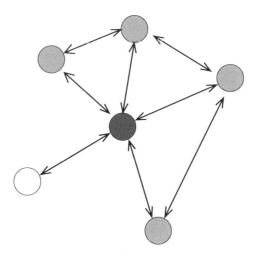

Figure 14.3 Example of peer-to-peer topology (dark gray circle, PAN coordinator; light gray circle, FFD device; white circle, RFD device).

algorithms for the creation and management of such larger network topologies are, however, not part of the 802.15.4 standard.

14.2.2 Medium Access Strategy

The medium access within a PAN is controlled by the PAN coordinator. The coordinator may choose between two different modalities: *beacon-enabled* and *nonbeacon-enabled*. In the *beacon-enabled* modality, the PAN coordinator broadcasts a periodic beacon containing information on the PAN. The period between two consecutive beacons defines a superframe structure divided into 16 slots. The first slot is always occupied by the beacon, while the other slots are used for data communication by means of random access, and form the so-called Contention Access Period (CAP). The beacon contains information related to PAN identification, synchronization, and superframe structure. The beacon-enabled modality is only adopted when the PAN is organized in a star topology. In this case, only two data transfer modes exist:

1. *Transfer from a Device to the Coordinator*—a device associated with the PAN willing to transfer data to the coordinator uses a slotted CSMA-CA protocol to access the medium; the slot to be used is selected on the basis of the information sent by the coordinator in the beacon. The coordinator may confirm the successful data reception with an optional acknowledgment message within the same slot.
2. *Transfer from the Coordinator to a Device*—when the coordinator has data pending for a device, it announces it in the beacon. The interested device selects a free slot and sends a data request to the coordinator, indicating that it is ready to receive the data. Slotted CSMA-CA is adopted to send the request. When the coordinator receives the data request message, it selects a free slot and sends data again using CSMA-CA.

In order to support low-latency applications, the PAN coordinator can reserve one or more slots that are assigned to devices running such applications without need for contention with other devices. Such slots are referred to as guaranteed time slots (GTS), and they form the contention-free period (CFP) of the superframe. An example of superframe with both CAP and CFP is shown in Figure 14.4.

In the *nonbeacon-enabled* modality there is no explicit synchronization provided by the PAN coordinator. This modality is particularly suited for PANs adopting the peer-to-peer topology, but can be adopted in a star network as well.

It should be noted that the peer-to-peer topology allows for a third transfer mode: the *peer-to-peer data transfer*, in which devices exchange data without involving the PAN coordinator, thus allowing more complex topologies and larger networks.

Since there is no superframe defined in the nonbeacon-enabled modality, no GTS can be reserved, and only random access is used. Furthermore, since no slot

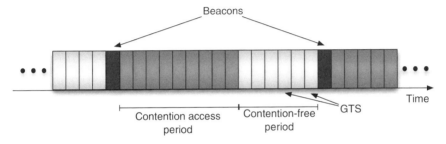

Figure 14.4 Example of superframe in beacon-enabled modality.

synchronization is available, unslotted CSMA-CA is adopted for medium sharing in all transfer modes.

14.2.3 From 802.15.4 to 802.15.4a

The new 802.15.4a standard will endorse a different transmission technology and new requirements in terms of ranging and positioning. As a consequence, the MAC will need to be re-designed in order to satisfy the new requirements by efficiently using the features of the alternative PHY. The new MAC will likely share several features with the existing 802.15.4 MAC, such as:

- *Network Topology*—the star and peer-to-peer topologies already defined in the 802.15.4 MAC provide enough flexibility to address efficiently all potential scenarios, spanning from a single link to complex mesh networking scenarios;
- *Mixed Random and Scheduled Access*—the new standard will support all applications already foreseen for 802.15.4, and will thus need the capability of dealing with both standard and more demanding, low-latency applications.

The new MAC layer will, however, need innovative solutions under two key aspects:

1. *Random Access Strategy*—the CSMA-CA approach adopted in the 802.15.4 MAC may pose serious implementation problems with the UWB radio technology; as a consequence, the 802.15.4a MAC will need a different approach in providing random access to the medium;
2. *Ranging Support*—802.15.4 does not provide any support for ranging at the MAC layer; the 802.15.4a MAC will thus require dedicated functionalities for retrieving, storing, and exchanging ranging information in order to support the new position-based applications that are the main target of 802.15.4a.

Moving from this premise, advanced MAC design issues for LDR UWB networks will be addressed in Section 14.3, and a MAC protocol that can meet the requirements posed by the new 802.15.4a standard will be described.

14.3 ADVANCED MAC DESIGN FOR LOW-BIT-RATE UWB NETWORKS

UWB technology is characterized by unique features, such as high processing gain, good robustness to multiuser interference and high synchronization latencies. The analysis of the impact of such characteristics on traditional MAC functions is thus the first step in the design of an MAC strategy specific for low data rate UWB networks. Such as analysis is carried out in [3], where available solutions for MAC functions are analyzed with respect to the requirements of UWB networks. The areas in which design can benefit from existing solutions and those which, conversely, require dedicated solutions for UWB, are identified. In particular, it is shown that issues related to admission control, packet scheduling, and power control can be addressed by adopting similar approaches to those proposed for existing wireless networks. On the other hand, medium sharing and MAC organization require specific design in order to take into account the peculiar characteristics of UWB.

In [3] it is also noted that the main innovation offered by UWB is the capability of achieving high-accuracy ranging. It should be observed, however, that this characteristic is typical of spread spectrum signals. Time of arrival estimations, for example, can be obtained in DS-CDMA systems by evaluating time shifts between the spreading code in the receiver and the same code in the received signal. The ranging precision thus depends upon the capability of determining this time shift, and is directly related to the adopted chip rate, that is, the spread signal bandwidth. A GPS system, for example, relies on this technique, and guarantees an accuracy on TOA estimation of 100 ns, corresponding to an accuracy on the order of meters in distance estimation [4].

In the case of UWB, errors in the order of centimeters can be guaranteed, much better than the precision achievable by DS-CDMA systems, thanks to a time accuracy of less than 100 ps. This precision is useful in the short-range scenarios (tens of meters) expected for UWB networks, where positioning is effective only if high accuracy can be achieved.

Ranging information can be exploited in several ways in resource management. Examples are: (a) definition of distance-related metrics for both MAC and higher layers, enabling the development of power-aware protocols, for example, [5]; (b) evaluation of initial transmission power levels, required in distributed power control protocols [6]; and (c) introduction of distributed positioning protocols in order to build a relative network map starting from ranging measurements. This map can enable location-based enhancements in several MAC and network functions, such as position-based routing, and position-aware distributed code assignment protocols in multiple channel MAC, in order to minimize MUI. The accurate ranging capability is thus the key feature of UWB enabling novel MAC functions.

The results of the above analysis formed the basis for the definition of a MAC protocol suitable for UWB systems, which is specifically designed for the special case of low data rate UWB networks: the uncoordinated, wireless, baseborn medium access for UWB communication networks, "$(UWB)^2$", originally proposed in [7].

In the following subsections a description of $(UWB)^2$ is provided, and the performance of the protocol in a typical low-data-rate scenario is evaluated.

14.3.1 $(UWB)^2$: Uncoordinated, Wireless, Baseborn Medium Access for UWB Communication Networks

$(UWB)^2$ takes advantage of data transmission of the multiple access capabilities warranted by the TH codes, and relies for access to the common channel on the high

MUI robustness provided by the processing gain of UWB. The proposed protocol also takes into account synchronization requirements.

$(UWB)^2$ is a multichannel MAC protocol. Multichannel access protocols have been widely investigated in the past, since the adoption of multiple channels may significantly increase the achievable throughput [8]. In multichannel protocols the overall available resource is partitioned into a finite number of elements. Each element of the resource partition corresponds to a channel. According to the definition of resource, a channel can therefore correspond to:

1. A time slot, as in TDMA;
2. A frequency band, as in FDMA;
3. A code, as in CDMA.

The design of an UWB MAC may adopt any of the above solutions. As described in Section 14.2, the IEEE 802.15.4 standard for example proposes a mixed TDMA/CSMA-CA MAC for low-data-rate networks [2]. TH-IR UWB, however, provides a straightforward partition of the resource in channels, each channel being associated with a TH code. The design of a multichannel CDMA MAC protocol forms, therefore, the natural basis for the design of a MAC in TH-IR UWB. Multichannel CDMA MAC algorithms, commonly referred to as multicode, have been intensively investigated for DS-CDMA networks. Among all we cite random CDMA access [9], and, more recently, multicode spread slotted aloha [10]. Note, however, that although in recent years most of the research efforts were focused on DS-CDMA, frequency hopping (FH) CDMA and TH-CDMA also provide viable solutions.

The performance of multicode MAC protocols is limited by two factors:

1. MUI, caused by the contemporary transmission of different packets from different users on different codes;
2. Collisions on the code, caused by the selection of the same code by two different transmitters within radio coverage.

Robustness of the system to MUI is determined by the cross correlation properties of the codes; the lower the cross correlation between different codes, the higher the number of possible simultaneous transmissions. The effect of code collisions can be mitigated by adopting appropriate code selection protocols. The task of assigning codes to different transmitters in the same coverage area is a challenging issue in the design of distributed networks. Within this framework, Sousa and Silvester [8] provided a thorough overview of possible code assignment solutions:

1. *Common Code*—all terminals share the same code, relying on phase shifts between different links for avoiding code collisions.
2. *Receiver Code*—each terminal has a unique code for receiving, and the transmitter tunes on the code of the intended receiver for transmitting a packet.

3. *Transmitter Code*—each terminal has an unique code for transmitting, and the receiver tunes on the code of the transmitter for receiving a packet.
4. *Hybrid*—a combination of the above schemes.

The common code scheme is a sort of limit case for a multicode protocol, since no real multicode capability is exploited. If phase shifts are too small, this solution collapses into the single Aloha channel. Note however that, in the case of very low data rate UWB networks, even the common code can be an appealing solution, since the processing gain guaranteed by the low duty cycle of UWB can provide by itself enough protection from MUI to avoid the additional complexity of multicode management.

The receiver code scheme has the main advantage of reducing receiver complexity, since a terminal must only listen to its receiving code. On the other hand, multiple transmissions involving the same receiver may result in collisions, since the same code is adopted by all transmitters.

Conversely, the transmitter code scheme avoids collisions at the receiver, since each transmitter uses its own code and thus two transmissions directed to the same receiver use different codes. On the other hand, the adoption of a transmitter solution requires in principle a receiver capable of listening to all possible codes in the network.

Hybrid schemes allow a trade-off between the above conditions. A hybrid scheme may foresee the use for signaling of either the receiver or common code schemes, over which the receiver can read the information about the code which will be used for data. A transmitter code scheme may then be used for data. When the set of codes is limited, however, the transmitter code scheme may be subject to collisions due to reassignment of the same code. In this case, a code assignment protocol is required for optimizing the use of the limited set of available codes. An example of such a protocol is presented in [11]. The solution proposed in [11] is a distributed assignment protocol for CDMA multihop networks: it guarantees that, if code C is used by terminal T, code C is never selected within a two-hops range from T, thus avoiding the occurrence of collisions.

The $(UWB)^2$ protocol applies the multicode concept to the specific case of a TH-IR UWB system. $(UWB)^2$ adopts a hybrid scheme based on the combination of a common control channel, provided by a common TH code, with dedicated data channels associated with transmitter TH codes. The adoption of a hybrid scheme can be motivated as follows:

1. It simplifies the receiver structure, since data transmissions (and corresponding TH codes) are first communicated on the control channel.
2. It provides a common channel for broadcasting; this is a key property for the operation of higher layers protocols. Broadcast messages are, for example, required for routing and distributed positioning protocols.

Note that the use of a common code at the beginning of each transmission also allows an easy transition to the adoption of a common code solution, whenever the bit rate and the offered traffic are low enough to allow the generated MUI noise to be managed in each receiver with the UWB processing gain alone. On the other hand, when high levels of MUI are expected, a correct choice of TH codes can be fundamental in meeting the application requirements. In the following we will assume that PN time hopping codes are used, but design of TH codes by itself is an open research area, and several TH-code generation algorithms have been proposed [12–16]. As regards code assignment, a unique association between MAC ID and transmitter code can be obtained by adopting, for example, the algorithm described in [17] which avoids implementing a distributed code assignment protocol.

$(UWB)^2$ is specifically designed for low data rate networks; as a consequence, it does not assume that synchronization between transmitter and receiver is available at the beginning of packet transmission, because clock drifts in each terminal may lead to complete loss of alignment between two devices in the average time between two DATA packets. As a consequence, a synchronization trailer long enough to guarantee the requested synchronization probability is added to the packet. The length of the trailer depends on current network conditions, and it is supposed to be provided to the MAC by the synchronization logic. Robust synchronization is indeed a critical issue in the deployment of TH-IR UWB networks, especially for the common code which is shared by all terminals.

$(UWB)^2$ also exploits the ranging capability offered by UWB. Distance information between transmitter and receiver is in fact collected during control packets exchange. Such information can enable optimizations of several MAC features, and allow the introduction of new functions, such as distributed positioning. Procedures adopted in $(UWB)^2$ for transmitting and receiving packets are described below. The procedures have two main objectives:

1. To exchange information such as the adopted synchronization trailer, that is, hopping sequence and length;
2. To perform ranging; since no common time reference is available, a two-way handshake is required to collect distance information by estimating the round-trip-time of signals in the air.

In the following it is assumed that, at each terminal T, MAC protocol data units (MACPDUs) resulting from the segmentation/concatenation of MAC service data units (MACSDUs) are stored in a transmit queue. The segmentation/concatenation block is also in charge of determining the amount of error protection to be added to each PDU by means of a PDU trailer.

It is also assumed that T is able to determine how many MACPDUs in the queue are directed to a given receiver R.

14.3.2 Transmission Procedure

Figure 14.5 contains the flow chart of the transmission procedure.

14.3 ADVANCED MAC DESIGN FOR LOW-BIT-RATE UWB NETWORKS

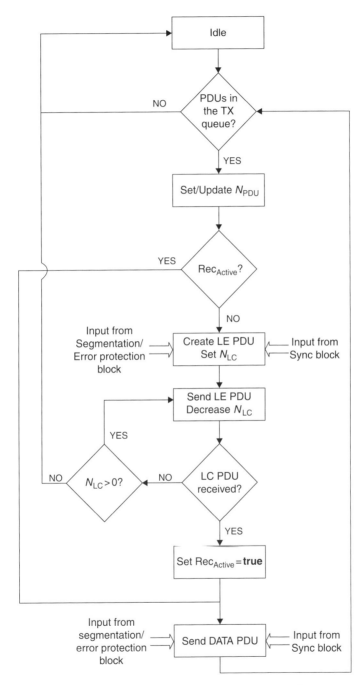

Figure 14.5 Transmission procedure in $(UWB)^2$.

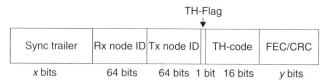

Figure 14.6 Structure of the link establish PDU in $(UWB)^2$.

Terminal T periodically checks the status of the transmit queue. Detection of one or more MACPDUs triggers the transmission procedure, which can be described as follows:

1. The ID of the intended receiver R is extracted from the first PDU in the queue.
2. T determines the number N_{PDU} of MACPDUs in the queue directed to R.
3. T checks if other MACPDUs were sent to R in the last T_{ACTIVE} s. If this is the case, T considers R as an active receiver, and moves to step 5 of the procedure.
4. If R is not an active receiver, T generates a link establish (LE) PDU. The LE PDU, shown in Figure 14.6, is composed by the following fields:
   - SyncTrailer—used for synchronization purposes;
   - TxNodeID—the MAC ID of transmitter T;
   - RxNodeID—the MAC ID of receiver R;
   - TH_{Flag}—this flag is set to true if the standard TH code associated with TxNodeID will be adopted for transmission of DATA PDUs; the flag is set to false if a different TH code is going to be adopted;
   - TH code (optional)—if the TH_{Flag} is set to false, the information on the TH-code to be adopted is provided in this field;
   - FEC/CRC—bits for error correction/revelation.
5. Terminal T sends the LE PDU and waits for a link confirm (LC) response PDU from R.
6. If the LC PDU is not received within a time T_{LC}, the LE PDU is re-transmitted for a maximum of N_{LC} times, before the transmission of the MACPDU is assumed to have failed.
7. After receiving the LC PDU, T switches to the TH code declared in the LE PDU and transmits the DATA PDU. The DATA PDU, shown in Figure 14.7, is composed of the following fields:
   - SyncTrailer—used for synchronization purposes;
   - Header, including the fields TxNodeID, RxNodeID, PDU_{Number} and N_{PDU};
   - ACK-flag—used to inform the receiver R if an ACK PDU should be sent in order to inform the transmitter T on the result of the transmission;
   - Payload—containing data information;
   - FEC/CRC—bits for error correction/revelation.

14.3 ADVANCED MAC DESIGN FOR LOW-BIT-RATE UWB NETWORKS

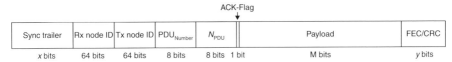

Figure 14.7 Structure of the DATA PDU in $(UWB)^2$.

8. Once the transmission is completed, T checks again the status of the data queue, and repeats the procedure until all MACPDUs in the transmit queue are served.

When the ACK-flag field is set to 1 in the DATA PDU, the transmitter expects an ACK PDU to be sent by the receiver, in order to schedule a retransmission of a packet if its reception was corrupted by noise or interference, following a predefined backoff scheme. The effect of the selected backoff scheme on performance will be analyzed in Section 14.3.4, where an evaluation of $(UWB)^2$ performance will be presented. As regards the transmission of the ACK PDU, two solutions are possible: either the receiver R transmits such PDU on the common TH code, or it transmits the ACK PDU on a receiver-specific TH code, at the price of an additional overhead required for communicating such code to the transmitter T in the case such a code cannot be derived from the MAC ID of R.

Note furthermore that when the MACSDU is constituted by a broadcast packet (e.g., a routing control packet), the MAC will adopt a simplified transmission procedure, where the DATA PDU that encapsulates the MACSDU is directly transmitted on the common TH code, without performing the LE/LC exchange. The broadcast nature of such PDU would in fact make impossible the reception of a LC PDU by all interested terminals. Furthermore, for this kind of PDUs the ACK-flag will be automatically set to 0 in order to avoid the transmission of several ACK PDUs by each neighbor of T receiving the broadcast PDU. A broadcast ID known to all terminals is set as receiver ID in these PDUs in order to inform neighbors of the broadcast nature of the transmission.

Such simplified procedure guarantees of course a lower protection of broadcast PDUs from interference; on the other hand, it makes it possible for the upper layers to have a straightforward mean to communicate broadcast information. Furthermore, the potential loss of a control broadcast packet is usually much less critical than the loss of a DATA packet since updated control information is usually retransmitted either on a periodic basis or within a short time.

14.3.3 Reception Procedure

Figure 14.8 contains the flow chart of the reception procedure. A terminal R in idle state listens to the common TH code, indicated as TH-0. When a SyncTrailer is detected, R performs the following procedure:

1. R checks the RxNodeID field. If the value in the field is neither the MAC ID of R nor the broadcast ID, the reception is aborted and the reception procedure ends.

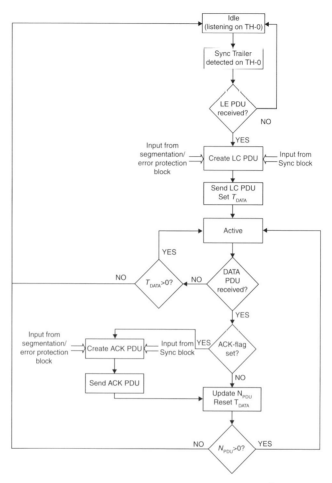

Figure 14.8 Reception procedure in $(UWB)^2$.

2. Since in the following we are not considering broadcast packets, let us assume that the RxNodeID contains the MAC ID of R. In this case, since R is assumed to be idle, MACPDUs directed to this terminal will necessarily be LE PDUs.
3. Following the reception of a LE packet, R creates an LC PDU, shown in Figure 14.9. The LC PDU is structured as follows:
   - SyncTrailer—used for synchronization purposes;
   - TxNodeID—the MAC ID of T;
   - RxNodeID—the MAC ID of R;
   - FEC/CRC—bits for error correction/revelation.

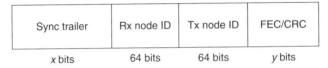

Figure 14.9 Structure of the link confirm PDU in $(UWB)^2$.

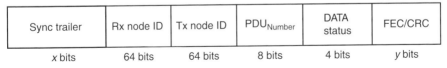

Figure 14.10 Structure of the ACK PDU in $(UWB)^2$.

4. R sends the LC PDU and moves into the active state, listening on the TH code indicated in the LE PDU. If no DATA PDU is received within a time T_{DATA}, the receiver falls back to the idle state and the procedure ends.

5. When a DATA PDU is received, R processes the payload, and extracts N_{PDU} from the header. If the ACK-flag is set to 1, R generates and sends an ACK PDU with the structure presented in Figure 14.10 reporting the status of the transmission. Next, if $N_{PDU} > 0$, R remains in the active state, since at least N_{PDU} more DATA PDUs are expected to be received from T. If $N_{PDU} = 0$, R goes back to the idle state.

It should be noted that the above procedures are related to the setup of a single link. During the reception procedure for example R also keeps on listening to the common code. It is assumed in fact that a terminal can act as a receiver on one or more links while acting as a transmitter on another link.

Finally, note that the exchange of LE/LC PDUs can also be triggered on a periodic basis for the purpose of updating distance information. This is likely to be the case, for example, if a distributed positioning protocol is adopted which relies on up-to-date distance estimations to build a network map.

14.3.4 Simulation Results

The performance of the $(UWB)^2$ was analyzed by means of simulations in order to evaluate its behavior in terms of throughput and delay. The simulation scenario consisted of N terminals, randomly located in an area of 80×80 m^2 size. Each terminal was characterized by a radio transmission range of 120 m in order to guarantee almost full connectivity between terminals. Each terminal generated MACPDUs to other terminals in the network following a Poisson process characterized by an average interarrival time T_{PDU}. The size of each MACPDU, with the format reported in Figure 14.7, was set to $L = 2000$ bits. As regards UWB physical layer parameters, the pulse rate was set to $1/T_s = 10^6$ pulses/s, $N_s = 1$, and $T_M = 1$ ns. In the simulations we assumed all terminals to adopt the same synchronization sequence of length $L_{sync} = 100$ pulses. Performance of the $(UWB)^2$ protocol was evaluated for a number of terminals N varying between 25 and 50, and for T_{PDU} values in the interval [1.25, 0.039063] s, corresponding to data rates between 1600 and 51,200 bps, respectively.

No correction capability was considered during the simulations; it was thus assumed that all bits in a packet must be correct, for a packet to be correct. As a consequence, the packet error probability was evaluated as follows:

$$PEP = 1 - \prod_{i=0}^{L-1} [1 - \text{Prob}_{\text{BitError}}(i)] \qquad (14.1)$$

where $\text{Prob}_{\text{BitError}}(i)$ is the error probability for the ith bit in the packet. Such a probability was evaluated adopting the pulse collision approach, originally proposed in [7] and further refined in [18]. In this approach, the probability of bit error is evaluated by determining the probability of collisions between pulses, and the effect of such collisions on receiver performance. Simulation results show that the pulse collision approach provides a far more accurate estimation of system performance then the one provided by the standard Gaussian approximation, especially for low-data-rate systems [18].

As already stated, two performance indicators were considered: throughput, defined as the ratio between received MACPDUs and transmitted MACPDUs; and delay. Both were evaluated in the presence of retransmissions, that is, with the ACK-flag set to 1 in all DATA PDUs.

Note that all results presented in the following take into account the control traffic consisting in the LE/LC PDUs exchanged to setup DATA PDU transfers and perform ranging.

As anticipated in Section 14.3.2, retransmissions are scheduled by a transmitter following a backoff algorithm. In evaluating the performance of the $(UWB)^2$ protocol, two different backoff algorithms were considered:

- *Immediate Retransmit*—in this algorithm retransmissions are performed as soon as the information of the transmission error is sent back by means of the ACK packet.
- *Binary Exponential Backoff (BEB)*—in this algorithm retransmissions are performed after a random delay. The average delay before attempting a retransmission for the rth time is equal to N_r times the transmission time of a DATA PDU; the value of N_r is randomly extracted in the interval $[2^0, 2^{\min(r, r_{max})}]$. In our simulations, we chose $r_{max} = 10$.

The introduction of a random element in the retransmission policy avoids the problem of systematic collisions that would occur when two devices collided and keep on re-scheduling the transmission of colliding packets at the same time.

14.3 ADVANCED MAC DESIGN FOR LOW-BIT-RATE UWB NETWORKS

It should be noted, however, that in the traffic scenarios considered for low-data-rate UWB networks, the event of collision is expected to be quite rare; furthermore, a collision between two PDUs, P1 and P2, will be destructive, that is, it will lead for example, to corruption of PDU P1, only when the power of the colliding PDU P2 is sufficiently high to overcome the MUI resiliance at the intended receiver of P1 guaranteed by the high processing gain of the UWB signal. As a consequence, in most cases PDU P2 will be received correctly, since it is characterized by a higher power level, and will not hinder the correct reception of P1 retransmission, even if it is retransmitted immediately after the reception of a negative ACK. This motivated the idea of comparing the standard BEB algorithm with the immediate retransmission of corrupted PDUs. The measured values for throughput and delay are presented in Figures 14.11 and 14.12, respectively.

Figure 14.11 shows that measured throughput was higher than 0.985 in all simulation cases. Furthermore, the two backoff schemes considered led to comparable values in all simulations, highlighting the fact that most PDU collisions are not destructive thanks to the MUI resiliance guaranteed by UWB.

This conclusion is confirmed by Figure 14.12, showing that the average delay is only slightly increased as the number of offered packets increase, and is in all cases close to the minimum value given by the transmission time of a MACPDU at the bit rate of 1 Mbps. Furthermore, the adoption of the binary exponential backoff scheme led to higher delays, since in the rare cases where a destructive collision occurs, transmitters are forced to wait on average a longer time before attempting a retransmission.

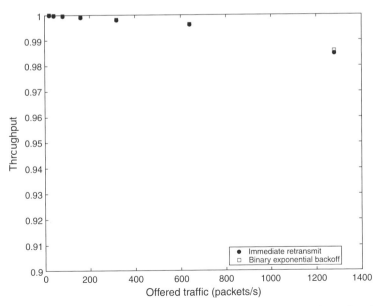

Figure 14.11 Throughput as a function of the offered traffic expressed in packets/s (open squares, binary exponential backoff scheme; solid circles, immediate retransmit scheme).

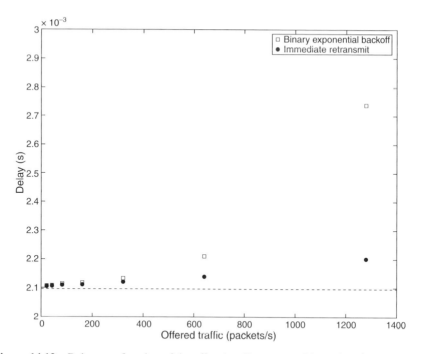

Figure 14.12 Delay as a function of the offered traffic expressed in packets/s. The dashed line shows the delay due to transmission time of a MACPDU at a bit rate of 1 Mbps (open squares, binary exponential backoff scheme; solid circles, immediate retransmit scheme).

Noticeably, the simulation results are in good agreement with theoretical results obtained in [9] and [19] for spread spectrum Aloha networks based on direct sequence. The values of throughput and delay predicted by theory for a processing gain of 30 dB and reported in [9] are in fact close to the results obtained in our simulation where the duty cycle of the signal was set to $T_M/T_s = 10^{-3}$, corresponding to approximately 30 dB of processing gain.

The $(UWB)^2$ protocol was originally conceived as a pure Aloha protocol, capable of operating without the need for a slotted time axis. In low-bit-rate application scenarios foreseeing a central controller, however, a slotted time axis could be added with low overhead, thus enabling the protocol to work in a slotted Aloha fashion. In order to highlight the impact of a slotted axis on the performance of $(UWB)^2$, a second set of simulations was performed. In these simulations the UWB channel model proposed in [20] for indoor environments was adopted, and the size of the simulated area was reduced to $40 \times 40 \, m^2$, in order to better model a typical indoor scenario.

The results of simulations comparing the performance of the proposed MAC as a function of the number of terminals are presented in Figures 14.13 and 14.14, showing throughput and delay respectively. The results were obtained considering a transmission range $R_{TX} = 70 \, m$ and a user bit rate $R = 10 \, kbps$.

14.3 ADVANCED MAC DESIGN FOR LOW-BIT-RATE UWB NETWORKS

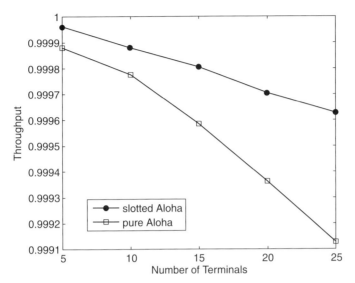

Figure 14.13 Throughput as a function of number of terminals for a full connectivity scenario ($R_{TX} = 70$ m) with user bit rate $R = 10$ kbps (circle, slotted Aloha; square, pure Aloha).

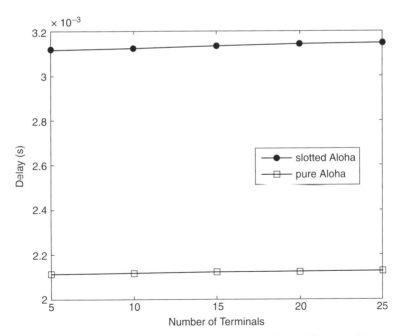

Figure 14.14 Delay as a function of number of terminals for a full connectivity scenario ($R_{TX} = 70$ m) with user bit rate $R = 10$ kbps (circle, slotted Aloha; square, pure Aloha).

Figure 14.13 shows that both slotted Aloha and pure Aloha lead to very high throughput in these conditions. Although slotted Aloha leads to a slightly higher value of throughput, the difference is quite small, of the order of 0.05%. As one would expect, however, the gap between the two strategies increases as the number of terminals (and as a consequence the offered traffic) increases.

Figure 14.14, on the other hand, shows an interesting result. In the considered low traffic scenarios, where the advantage of slotted Aloha over pure Aloha is not significant in terms of throughput, the slotted Aloha approach leads to a higher delay. This is due to the fact that in pure Aloha a packet is sent immediately, as soon as it is inserted in the queue, and thus in absence of high packet error rates, the delay is limited to the packet transmission time over the channel. Conversely, in the case of slotted Aloha the packet remains in average a time $T_{SLOT}/2$ in the queue, where T_{SLOT} is the duration of the slot, waiting for the beginning of the first slot after the insertion in the queue (the first useful for transmitting the packet). This accounts for the difference of about 1 ms in the average delay between the two strategies, remembering that we chose packets of 2000 bits, with a transmission time over the channel $T_{TRANSMIT} \simeq T_{SLOT} \simeq 2$ ms.

In conclusion, simulation results show that, in all considered scenarios, the processing gain guaranteed by UWB is high enough to manage the traffic without appreciable effects of MUI. This confirms that, thanks to the MUI robustness guaranteed by impulse radio, the $(UWB)^2$ MAC protocol is a suitable solution for low data rate UWB networks.

REFERENCES

1. "IEEE 802.15.TG4a official web page"; available at: www.ieee802.org/15/pub/TG4a.html
2. "IEEE 802.15.4 MAC standard"; available at: http://www.ieee.org/
3. L. De Nardis and M. G. Di Benedetto, "Medium Access Control design for UWB Communication Systems: review and trends," *Journal of Communications and Networks*, vol. 5, no. 4, pp. 386–393, December 2003.
4. I. A. Getting, "The global positioning system," *IEEE Spectrum*, vol. 30, no. 12, pp. 36–38, 43–47, December 1993.
5. P. Baldi, L. De Nardis, and M. G. Di Benedetto, "Modeling and Optimization of UWB communication networks through a flexible cost function," *IEEE Journal on Selected Areas in Communications*, vol. 20, no. 9, pp. 1733–1744, December 2002.
6. A. J. Goldsmith and S. B. Wicker, "Design challenges for energy-constrained ad-hoc wireless networks," *IEEE Wireless Communications*, vol. 9, no. 4, pp. 8–27, August 2002.
7. M. G. Di Benedetto, L. De Nardis, M. Junk, and G. Giancola, "$(UWB)^2$: uncoordinated, wireless, baseborn medium access control for UWB communication networks," *Journal on Mobile Networks and Applications*, vol. 10, no. 5, pp. 663–674, October 2005.
8. E. S. Sousa and J. A. Silvester, "Spreading code protocols for distributed spread-spectrum packet radio networks," *IEEE Transactions on Communications*, vol. COM-36, no. 3, pp. 272–281, March 1988.

9. D. Raychaudhuri, "Performance analysis of random access packet switched code division multiple access systems," *IEEE Transactions on Communications*, vol. COM-29, no. 6, pp. 895– 901, June 1981.
10. S. Dastangoo, B. R. Vojcic, and J. N. Daigle, "Performance analysis of multi-code spread slotted ALOHA (MCSSA) system," *IEEE Global Telecommunications Conference*, vol. 3, pp. 1839–1847, November 1998.
11. J. J. Garcia-Luna-Aceves and J. Raju, "Distributed assignment of codes for multihop packet-radio networks," in *IEEE Military Communications Conference*, vol. 1, November 1997, pp. 450–454.
12. W. Chu and C. J. Colbourn, "Sequence designs for ultra-wideband impulse radio with optimal correlation properties," *IEEE Transactions on Information Theory*, vol. 50, no. 10, pp. 2402–2407, October 2004.
13. S. Gezici, A. F. Molisch, H. V. Poor, and H. Kobayashi, "The trade-off between processing gains of impulse radio systems in the presence of timing jitter," in *IEEE International Conference on Communications*, vol. 6, June 2004, pp. 3596–3600.
14. L. Yang and G. B. Giannakis, "Ultra-wideband multiple access: unification and narrow-band interference analysis," in *IEEE Conference on Ultra Wideband Systems and Technologies*, November 2003, pp. 320–324.
15. I. Guvenc and H. Arslan, "On the modulation options for UWB systems," in *IEEE Wireless Communications and Networking Conference*, vol. 2, March 2004, pp. 914–919.
16. R. Muller, S. Zeisberg, H. Seidel, and A. Finger, "Spreading properties of time hopping codes in ultra wideband systems," in *IEEE Seventh International Symposium on Spread Spectrum Techniques and Applications*, vol. 1, 2002, pp. 64–67.
17. M. S. Iacobucci and M. G. Di Benedetto, "Computer method for pseudo-random codes generation," National Italian patent RM2001A000592, 2002, under registration for international patent.
18. G. Giancola and M. G. Di Benedetto, "A collision-based model for multi user interference in impulse radio UWB networks," in *IEEE International Symposium on Circuits and Systems*, May 2005, pp. 49–52.
19. A. Polydoros and J. Silvester, "Slotted random access spread-spectrum networks: an analytical framework," *IEEE Journal on Selected Areas in Communications*, vol. 5, no. 6, pp. 989–1002, July 1987.
20. S. S. Ghassemzadeh and V. Tarokh, "UWB path loss characterization in residential environments," in *IEEE Radio Frequency Integrated Circuits Symposium*, June 2003, pp. 501–504.

CHAPTER 15

An Overview of Routing Protocols for Mobile Ad Hoc Networks

DAVID A. SUMY, BRANIMIR VOJCIC and JINGHAO XU

15.1 INTRODUCTION

The development of wireless, multihop, mobile ad hoc networks, here simply referred to as MANETs, continues to be a topic of increasing interest to the wireless community. MANETs can be considered an emerging fourth-generation (4G) wireless system that supports anytime, anywhere and from any device communication. Currently, the MANET applications that have drawn considerable attention relate to military, emergency services and sensor networks; however, more mainstream applications relating to business, education, entertainment, and commercial settings are quite viable [1]. Indeed, the commercial viability and diverse applications for a MANET bolster the exponential growth of interest in this area. In 2005, the proliferation of individuals carrying mobile phones, palm tops, PDAs, laptops, handheld PCs and other wireless communication devices, for work or personal use, has grown considerably. As the multifunctionality of wireless devices increases, the commercial demand for ubiquitous communication services is sure to progress. In this regard, peer-to-peer (P2P) applications for MANETs are accumulating attention and discussion in the literature. Recent industry achievements have also sparked interest in UWB wireless technology as a common platform for MANETs. In view of the base-band nature of signal transmission in UWB, radio devices with a high data rate, low complexity, and low power emission become ideal components for the UWB/MANET paradigm.

While the developmental aspects of a MANET is expansive, of particular interest to network traffic engineers and protocol designers is the development of routing in MANETs. As a result, a great number of proposals for routing protocols have been presented in the literature, any of them being applicable to a UWB system. However, when considering the intrinsic advantages and constraints of UWB technology, routing protocols in ad hoc networks adopting UWB would have some notable

Ultra Wideband Wireless Communication. Edited by Arslan, Chen, and Di Benedetto
Copyright © 2006 John Wiley & Sons, Inc.

differences in their design and implementation from routing protocols used for ad hoc networks not adopting UWB. Generally speaking, two special criteria should be taken into account in routing protocols for UWB networks: (1) the exploitation of accurate positioning information potentially provided by UWB; and (2) consideration of power/energy efficiency. To further clarify the last criterion, while other routing protocols for MANETs may consider power and energy efficiency, especially in the case of sensor networks, ad hoc networks adopting UWB must adhere to specific power spectral density restrictions.

Due to the very narrow time domain radio pulses characteristic in a UWB system, UWB radios can provide much finer timing precision than other radio systems. Consequently, this implies that UWB wireless technology is capable of recovering positional information with high precision, which can be utilized in routing protocols to reduce the protocol overhead due to its directivity. Therefore, geographical information aided routing protocols, such as LAR (location aided routing) and DREAM (distance routing effect algorithm for mobility), discussed below, may be more suitable for MANETs employing UWB technology. Nevertheless, while the positioning capability of UWB is clearly a beneficial feature for the execution of routing in MANETs, it is not essential.

Since UWB signals spread over very wide bandwidths and overlap with narrowband radio systems, regulatory groups, such as the FCC in the United States, specified spectral masks on UWB power density to avoid interference with these other co-existing systems. Since UWB devices are often portable and allow user mobility, they operate on limited battery power. Therefore, power and energy efficiency become critical issues for UWB in light of the FCC constraints and limited battery capacity. Of course, these factors directly relate to the design of routing protocols. The power or energy aware routing protocols described in this chapter, for example, PARO (power-aware routing optimization) and MTRP (minimum total transmission power routing), can be efficiently applied to ad hoc networks with UWB.

This chapter provides a thorough survey of the many routing protocols currently developed, both new and old. Since some protocols have progressed through a number of versions, we have elected to discuss only the most recent version of a protocol, which includes details relating to enhancements to previous versions. We consider this to be beneficial to the reader since recent developmental facets and strategies for routing in MANETs will be illuminated. This chapter will follow the following format. First, we introduce the reader to ad hoc networks. Second, we provide the reader with the fundamentals relating to categorizing routing protocols, followed by the presentation of routing protocols that can be applied to MANETs adopting UWB. For each protocol presented, a rather detailed description of its basic routing functions is described. In addition to a fundamental discussion of each protocol, this paper will provide, for select protocols, performance evaluations, as well as subsequent modifications and extensions. We have also included examples and illustrations in order to enhance the reader's understanding and to facilitate protocol comparisons. The reader can also refer to the tables found in the Appendix, which provide more parameter details (e.g., complexities) for each protocol so that a more meticulous protocol comparison summary can be made.

Lastly, this chapter will conclude with the authors' observations relating to the development of routing protocols and suggestions for future research.

15.2 AD HOC NETWORKS

The Merriam–Webster defines ad hoc to mean: "concerned with a particular end or purpose" or "fashioned from whatever is immediately available" [2]. This definition is apropos, for ad hoc networks are formed with a purpose in mind and have no infrastructure, thus permitting unplanned, spontaneous connectivity to be implemented. However, arguably, singular purposes may not exist if one compares individual nodes of an ad hoc network belonging to a single authority (e.g., an emergency rescue team), with the individual nodes belonging to a commercial mainstream scenario, where individuals may differ in cooperation and intention [3]. While a lengthy discussion of such concerns is beyond the scope of this paper, this issue is often overlooked due to our optimistic assumptions regarding users. Nevertheless, it is highly germane to routing where unrelated users will be required to utilize scarce resources to forward packets to more distant nodes. In [3], the authors present an insightful discussion relating to incentives for MANETs.

The beginnings of ad-hoc networking can be traced back to 1968 [4, 5], where the ALOHA random access scheme, developed at the University of Hawaii for packet radio networks, provided a foundation for distributed channel-access schemes. Further development ensued and in 1973 the Defense Advanced Research Projects Agency (DARPA) developed one of the first ad hoc wireless networks, called the PRNET, a ground mobile packet radio network. It was a multihop network that allowed up to 138 nodes and used IP packets for data transport. In 1990, the committee for IEEE 802 standards formed a working group for wireless LAN (WLAN) standards. The development of the standard was completed in 7 years, and in June 1997, the first specification of the IEEE 802.11 standard for WLANs was ratified and the term *ad hoc network* came to being [6].

A MANET is formed by a confluence of self-organizing mobile devices, with wireless capabilities, that dynamically self-configure themselves to permit multihop communication. Figure 15.1 shows an illustrative example of a multihop ad hoc network (MANET). Such an autonomous network averts the need for a central infrastructure or network administrator. Rather, a MANET is self-maintained by the nodes themselves. The topology of a MANET is susceptible to frequent unpredictable changes primarily attributable to node mobility. However other types of node behavior contribute to a mutating network topology. In a MANET, mobile devices, at any time, may enter and exit the network voluntarily (e.g., by turning a device on/off or sleep mode operation) or exit involuntarily (e.g., resource depletion, such as battery capacity). The surrounding environment can also play a significant role, whether indoors or outdoors, in that mobile nodes may naturally accumulate or disperse within a particular region or be subjected to restrictions in node mobility and node dispersion [7]. Notwithstanding, the extent of topological changes as a result of node mobility (unpredictable or otherwise), includes but is

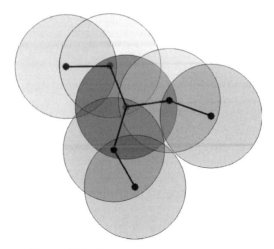

Figure 15.1 Example of a simple MANET.

not limited to the type of MANET (e.g., tactical vs commercial), its constituents (e.g., user vs robot, manner of locomotion) and the surrounding environment. Accordingly, the actually existence and impact of topological volatility, in time and space, can only be gaged on a case-by-case basis.

Mobile devices in a MANET serve as both a router and a host. Each mobile device is expected to assist in the routing of packets to other mobile devices within its radio coverage, as well as perform end terminal functionality. However, while device homogeneity may exist in certain types of MANETs, device heterogeneity may exist in other types of MANETs. While it is true that any node will have to meet minimum system performance criteria to attain viable membership in a MANET, mobile node classifications, such as weak or strong, based on disparities in resource capabilities, may be yet another characteristic of a MANET. Similarly, links between nodes may fluctuate among various states such as bidirectional, unidirectional, or broken, as well as inheriting all the traditional problems associated with wireless, mobile communications.

Providing multihop connectivity and seamless internetworking based on unpredictable route lifetimes clearly presents a difficult task. Arbitrary movement, scarce communication resources, multipath fading, interference, and fluctuations in network population are only some of the numerous considerations in the development of a routing protocol. Due to the ever-changing parameters and nuances of a MANET, available links and paths to a desired hop or destination change over time. Reducing control overhead, optimizing route discovery, route maintenance, recovery from link failures, scalability, adaptive strategies, power efficiency, and mobility are common topics discussed in proposals regarding routing protocols. Despite differences in approach to resolve these various obstacles, network developers and protocol designers continue to present insightful analysis, which provides a platform for open discourse and critique.

15.3 ROUTING IN MANETS

A majority of the literature broadly categorizes routing protocols as being either proactive, reactive or hybrid. In fact, one could say that it has nearly become a convention. However, these classifications are somewhat arbitrary in nature, since other classifications, such as topology-based or position-based, are just as valid. Our survey presents the various routing protocols under the following categories: proactive, reactive, power-aware, hybrid, and other. Admittedly, even these categories are not exhaustive. Nonetheless, these categories do provide a more expansive treatment of the various routing protocols for MANETs. Power-aware routing (PAR) is being included as a separate category since a power-aware scheme can be applied, as an extension for example, to other protocols within each of the other categories (i.e., proactive, reactive, hybrid, and other). We include the "other" category separately from hybrid in order to encompass protocols that fail to directly fall into a combinatorial archetype, such as those in the hybrid category.

While each of the above categories will be explained in greater detail later, a brief side-note at this juncture is helpful. The proactive and reactive categories stem from the criteria of when the route is computed. Of course, as hinted above, other criteria for classifying routing protocols can be made. Certainly, since the prevailing view of categorizing a routing protocol is based on some arbitrary single criteria (e.g., when the route is computed, power conservation), typically protocols of different categories share a number of common attributes. As previously mentioned, the Appendix presents parameter comparisons among the different routing protocols within each category, except the "other" category; however, some commonalities can be elicited by comparing protocols among the different categories as well.

15.4 PROACTIVE ROUTING

Proactive routing protocols, also known as table-driven routing protocols [8], require that each node maintains one or more tables containing routing information to every other node in the network or subdivision thereof. The routing information is typically updated based on some periodic route update process, but other route update policies may exist. Accordingly, such protocols are termed "proactive" because each node stores and maintains routing information to a destination before it may be actually needed. A disadvantage to proactive routing is that control messages may unnecessarily utilize network resources, such as power and link bandwidth, to maintain routing information to every node in the network or a specific region of the network. In some reports they also show that positioning information can be better exploited in the case of reactive protocols than proactive protocols [9]. An advantage associated with proactive routing is that, when a node needs routing information, the time for determining a route to a destination is minimized because up-to-date routing information is available at the time of a route request. In this regard, proactive routing protocols typify the trade-off between providing routes with less delay at the cost of excess control overhead.

Proactive routing protocols typically differ in the number and type of tables at each node as well as the methodology by which network topology changes are distributed across the network. A further distinction among proactive routing protocols is the routing algorithm implemented. There are two well-known routing algorithms, namely, distance vector and link state [10]. In distance vector (DV) routing approaches, each node v maintains for each destination w a set of distances D_{VX}^W where x ranges over the node v neighbors. Node v selects a neighbor m as the next hop for reaching destination w, if $D_{VM}^W = \min_X (D_{VX}^W)$. As this process repeats from node to node, the shortest path to destination w is selected. Each node periodically broadcasts to neighbors current estimates of its shortest distances to every other node in the network. While this algorithm is relatively easy to implement, the DV routing algorithm can produce short and long-lived loops. In link state (LS) routing methods each node maintains an image of the network topology with an associated cost for each link. A LS node periodically broadcasts link-state information of its outgoing links to its neighbors. The nodes that receive the broadcast update their network topology and apply a shortest-path algorithm to select the next hops for each destination. Naturally, nodes may differ in their view of the network because of delays in receiving link state information. These differences can result in temporary looping.

15.4.1 DSDV

Destination-sequenced distance vector (DSDV) routing protocol [10, 11] derives from the classical distance-vector distributed Bellman–Ford algorithm (DBF) [12, 13] that uses a flat addressing scheme. The two principle modifications of the DBF algorithm are that the DSDV protocol guarantees loop-free paths to a destination and provides MAC-layer support for ad hoc networks.

DSDV is a hop-by-hop routing protocol that requires mobile stations to store routing tables. Each routing table lists all reachable destination addresses, the number of hops required to reach each destination, and the sequence number assigned by the destination. Routes with a more recent sequence number are used and routes with older sequence numbers are discarded. If two routes have identical sequence numbers, then the route with the better metric (i.e., shortest route) is chosen, and the other route is either discarded or stored as a less preferable route. Each node also maintains a forwarding table with next hop information. Figure 15.2 depicts a typical MANET and Table 15.1 illustrates the forwarding table for node H.

Depending on the ad-hoc networking protocol, the addresses stored in the routing table can correspond to either layer 3 (network) or layer 2 (MAC). To resolve layer 3 network addresses into MAC addresses, and to allow operation at layer 2, each destination node includes information about the layer 3 protocol(s) it supports, along with layer 2 information. Mobile stations, in turn, include this information with their advertisements.

Mobile stations voluntarily advertise routing information to each of their neighbors by broadcasting or multicasting, or upon request to a single mobile station.

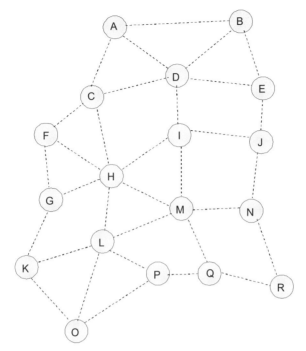

Figure 15.2 Network topology.

TABLE 15.1 Forwarding Table for Node H

Destination	Next Node	Metric	Sequence Number
A	C	2	12
B	I	3	34
C	C	1	46
D	I	2	8
E	I	3	176
F	F	1	232
G	G	1	228
H	—	—	—
I	I	1	70
J	I	2	64
K	G	2	36
L	L	1	112
M	M	1	168
N	M	2	10
O	L	2	124
P	L	2	248
Q	M	2	152
R	M	3	86

Packets are transmitted periodically (time-driven) and as topological changes are detected (event-driven). Routing table updates are sent in two different ways in order to reduce the amount of overhead, namely, a "full dump" or an "incremental." An incremental routing update should fit in one network protocol data unit (NPDU), whereas a full dump requires several NPDUs. A full dump includes all the available routing information—a full routing table. An incremental update includes only information changed since the last full dump—such as a different metric for a destination. To avoid problems caused by one-way links, mobile stations cannot insert routing information received from a neighbor unless that neighbor shows that it can receive packets from the mobile node. Thus, DSDV effectively operates using only bidirectional links.

Returning to Table 15.1, if a mobile station H determines that its route to destination node R has broken, it will advertise the destination route with an infinite metric and a sequence number one greater than the last sequence number received from the destination. This advertisement causes any mobile station routing packets through mobile station H to incorporate the infinite-metric route into its routing table. In the event a mobile station receives an infinite metric, and if it has an equal or later sequence number with a finite metric, it will trigger a route update broadcast, since routes containing any finite sequence numbers will supercede routes generated with the infinite metric.

The broadcasting of routing information by mobile stations is regarded as asynchronous and requires the mobile station to determine which route changes are significant enough to warrant the sending of an advertisement. Additionally, mobile stations use past experience to judge the settling time of routes before an advertisement is sent to ensure that the best route within a time period has been received. Thus, a node maintains data relating to the time of arrival of the first route and the best route for any destination. Mobile stations do not delay transmission if an update involves a route to a destination that was previously unreachable.

15.4.2 WRP

Wireless routing protocol (WRP) [14, 15] is a distance-vector, flat addressing protocol where each node maintains a distance table, a routing table, a link-cost table and a message transmission list. The distance table contains for each destination and neighboring node, the distance to the destination and the predecessor reported by each neighbor. The routing table contains for each destination: a destination identifier, the distance to the destination, the predecessor and successor of the chosen shortest path to the destination, and a tag used to specify whether the entry corresponds to a simple path, a loop, or an invalid.

The link-cost table lists the cost of relaying information through each neighbor and the number of periodic update periods that have transpired since the node received error-free messages from respective neighbors. The message retransmission list (MRL) consists of the sequence number of an update message, a retransmission counter (e.g., set to 3 or 4) which is decremented each time an update message is sent, an ack-required flag that specifies whether a neighbor has sent an acknowledgement (ACK), and a list of updates sent in the update message. WRP relies on the MRL to handle errors by way of retransmissions.

A node exchanges routing table update messages to its neighbors. The update message includes a sending node ID, sequence number, an update list of updates or ACKs, and a response list of nodes that need to send an ACK. An update entry specifies a destination, a distance to the destination, and a predecessor to the destination. When no updates require reporting, an update message contains an empty address, so no ACK is required. This type of update message acts as a hello message and is sent periodically for maintaining node connectivity.

A node updates its routing table after receiving an update from a neighbor or detecting a change in a link to a neighbor. A node checks the consistency of predecessor information reported by *all* its neighbors each time it processes an event involving a neighbor. This consistency checking of WRP provides a faster convergence after a single link failure and faster recovery than DBF. When processing an update of node *b*, node *c* determines if the path to the destination node through any of its other neighbors include node *b*. Thus, nodes determine whether an update received from a neighboring node affects its other distance and routing table entries.

WRP relies on a lower layer protocol for maintaining link status. Similar to DSDV, when a link fails, the corresponding distance entries in a node's distance and routing tables are marked as infinity. When a link fails or a link cost changes, the node re-computes distances and predecessors to all affected destinations, and sends an update message for all destinations whose distance and/or predecessors have changed.

When a node receives an update message error-free, it is required to send a positive ACK to the originator node. An ACK entry specifies the source and sequence number of the update message being acknowledged. Upon receiving an ACK, the node updates its MRL. To ensure connectivity between neighbors during a period of time when no routing updates or ACKs are received by a node, a node sends null update messages to its neighbors. The time interval between two null update messages is the hello interval. If a node fails to receive any type of message from a neighbor during a router dead interval (e.g., 3–4 times a hello interval), the node assumes that connectivity with that neighbor has been lost. On the other hand, if a node *c* receives an update or user message from node *b* and node *b* is not listed in the routing or distance table of node *c*, node *c* adds the entry to these tables for destination *b*.

In [15], WRP-Lite is introduced as a derivation of WRP. The main concept behind this derivation is to reduce control overhead in instances when network topology changes occur rapidly. WRP-Lite reduces control overhead by providing nonoptimal routes and triggering updates only when a path becomes invalid. While control messages contain the entire routing table, they are sent infrequently. Additionally, unlike WRP, messages are transmitted unreliably and nodes do not maintain an MRL.

15.4.3 CGSR

Clusterhead Gateway Switch Routing protocol (CGSR) [16, 17] is a modification of DSDV that uses hierarchical routing. In CGSR, mobile nodes are partitioned and form a cluster. A distributed clustering algorithm (least clusterhead change, LCC, clustering algorithm) is used within a cluster in order to elect a mobile node as a *cluster-head*. All mobile nodes within transmission range of the cluster-head belong to this cluster, so that all mobile nodes in the cluster can communicate with the cluster head and possibly with each other. *Gateway nodes* are nodes that belong to more than one cluster. Gateways select the code used to communicate within a particular cluster and are able to change their codes when they receive messages. Figure 15.3 illustrates the hierarchical routing scheme of CGSR.

Under the LCC algorithm, only two conditions cause the cluster-head to change: (1) if two cluster-heads come within range of each; and (2) if a mobile node becomes disconnected from any cluster. Minimizing the conditions that trigger the cluster-head to change provides a more stable framework in a MANET since frequent cluster-head changes adversely affect performance of other protocols such as scheduling and resource allocation that rely on it. The LCC algorithm uses either lowest-ID or highest connectivity for initialization and routine maintenance.

The LCC algorithm provides for the following scenarios. If a cluster-head moves into an existing cluster, then it may become the cluster-head, depending on ID, connectivity or some other priority. If a noncluster-head moves into an already established cluster, the node cannot challenge the current cluster-head. If a noncluster-head moves out of its cluster and does not enter any other existing cluster, then it forms a new cluster and becomes a new cluster-head.

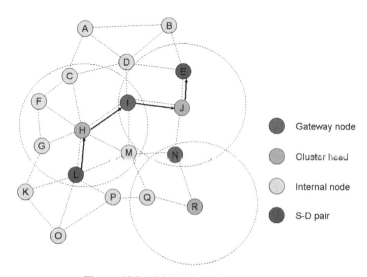

Figure 15.3 CGSR hierarchical structure.

TABLE 15.2 Routing Table

Destination Cluster	Next Hop	Metrics	Sequence Number
J	H	5	55

TABLE 15.3 Cluster Member Table

Destination	Destination Clusterhead	Sequence Number
E	J	95

As shown in Figure 15.3, the routing strategy of CGSR requires the source node L to transmit packets to the cluster-head H. The cluster-head H forwards packets to a gateway node I that connects to another cluster-head J. This cluster-head to gateway to cluster-head routing continues until the appropriate cluster-head is reached and the packets are forwarded to the destination. In Figure 15.3, node E is the destination node. Each node contains two tables: a cluster member table and a routing table. Tables 15.2 and 15.3 are illustrative examples relating to Figure 15.3.

The cluster member table is used to map a destination address to a destination cluster-head address. Destination sequence numbers are used to avoid stale routes and looping. The routing table is used to select the next node to reach the destination cluster. A node broadcasts its cluster-head member table periodically and updates it when it receives a new cluster-head member table from its neighbors.

When a node wants to transmit a packet, it selects the shortest, minimal hop, destination cluster-head according to the cluster member table and the routing table. Thereafter, the node selects the next node to transmit for that destination cluster-head according to the routing table. To improve routing efficiency, extensions to CGSR may be implemented such as priority-token-scheduling for high priority traffic (e.g., multimedia and real-time sources), gateway code scheduling between gateways and cluster-heads and path reservation until disconnect, similar to a virtual circuit.

15.4.4 STAR

Source tree adaptive routing (STAR) [18–20] is a table-driven, flat addressing scheme (but can be used with a distributed hierarchical scheme) protocol that uses link-state information to obtain efficient routing. A router reports link state information of every link it uses to reach a destination to all its neighbors. Each router maintains a *source tree* rooted at the source node, which includes the set of links used by the router in its preferred path to destinations. STAR routers also maintain a *partial topology map* (*graph*) of their network by compiling source tree information from neighbors and adjacent links to these neighbors. Of course, the links of the source tree and topology graph must be adjacent links. Each router generates its own source tree by using its topology graph. In turn, each router generates a *routing table* specifying the successor to any destination by applying a local route selection algorithm, such as Dijkstra's shortest path algorithm, on its source tree.

In STAR, two approaches to updating routing information can be used: the optimum routing approach (ORA) and the least-overhead routing approach (LORA). The ORA provides updates to provide optimum paths defined by a metric, whereas the LORA provides variable paths, which need not be optimum, thus reducing overhead. STAR was the first proactive routing protocol to use the LORA. Accordingly, our discussion relates to STAR and the LORA.

Using the LORA, a router sends source tree updates to its neighbors only when it loses all paths to one or more destinations, detects a new destination, determines local changes that can potentially create long-term looping, or changes in cost that exceed a threshold. Since source trees are exchanged among routers, new link information to a destination allows a router to infer the deletion of a link to the same destination, thereby eliminating the need to exchange deletion updates. However, if no new link exists, a router can make a deletion update when the failed link causes the router to have no paths to one or more destinations.

The basic update unit that contains source tree changes is the link-state update (LSU). The LSU includes the characteristics of the link and is time-stamped—a monotonically increasing number. A router accepts the LSU as valid if the LSU has a larger time stamp or if no entry for the link exists and the LSU is not reporting an infinite cost. One important aspect of STAR is that LSUs for operational links do not age out, so periodic flooding of LSUs is not required to validate link-state information. However, link state information for failed links is LSUs erased from the topology graph due to aging (e.g., an hour after processing the LSU). Additionally, a router will erase a link from its topology graph if the link is not found in the source trees received from its neighbors. STAR relies on an underlying protocol to insure that a router will detect, within a finite time period, the existence of a new neighbor, the disconnection between a neighbor, and the reliable transmission of LSUs.

J. J. Garcia and M. Spohn [18, 19] compared the performance of STAR (LORA) with DSR (1998 version) using the C++ Protocol Toolkit (CPT) simulator environment having 20 wireless nodes in continuous motion for 900 and 1800 s with various flows (i.e., the number of sources of data). Under all flow scenarios, STAR generated fewer update packets than DSR with comparable data delivery. Under certain scenarios, STAR was able to deliver twice or three times the number of data packets than DSR. A more recent version of DSR is discussed below.

Hong [20] introduced performance comparisons of STAR and open shortest path first (OSPF), ad hoc on-demand distance vector (AODV) and dynamic source routing (DSR) using the simulation tool PARSEC. In certain scenarios, STAR outperformed OSPF, AODV, and DSR by generating fewer routing packets while providing similar quality routing service to data delivery.

15.4.5 HSR

Hierarchical state routing (HSR) [21, 22] is a table-driven, hierarchical, link-state based routing protocol that includes, in addition to multilevel *physical* clustering (geography-based), the feature of multilevel *logical* partitioning of mobile nodes

15.4 PROACTIVE ROUTING

(i.e., a logical functional affinity, e.g., employees of the same company). The logical partitioning of nodes provides a basis for location management. This feature is noteworthy because typically hierarchical routing suffers from location management and mobility drawbacks.

As shown in Figure 15.4, the network is partitioned into physical clusters. In the physical clusters there are three kinds of nodes: a gateway node, a clusterhead node, and an internal node. The internal nodes of a physical cluster elect a clusterhead node. HSR does not adopt any specific algorithm for forming the clusters or electing clusterheads. The clusterhead node coordinates transmissions within the cluster. Within this physical cluster, each node monitors the state of a link to each neighbor and broadcasts it within the cluster. The clusterhead processes the link state (LS) information and forwards this cluster information to neighboring clusterheads via a gateway node. The shared partial topology among clusterheads forms the first level of clusters. The above procedure is performed in a recursive fashion, where cluster-heads, in turn, organize among themselves into clusters, and so on, resulting in the next higher level of clusters. Higher-level LS information is flooded down to lower level clusters so each node has a hierarchical topology image.

At level 0, we have three physical clusters, namely, C-01, C-02, and C-03. While not shown in Figure 15.4, nodes A, B, and D and nodes K, O, and P would form their other clusters, respectively. Additionally, at level 1, higher clusters C-11 and C12 are formed, as well as cluster C-21 at level 2. To implement routing at level 1, gateway node I at level 0, sends LS information for link I–J to cluster H at level 0.

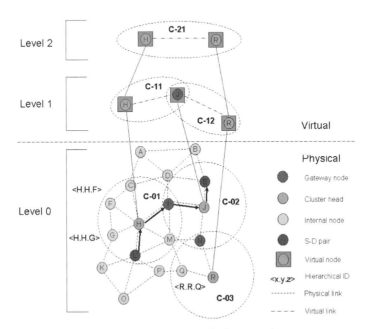

Figure 15.4 Example of HSR scenario.

Cluster H at level 0 estimates the parameters for the route H–I–J by using its own estimate of link H–I and the received estimate of link I–J. Upon calculation, an LS parameter of a virtual link between nodes H and J at level 1 is derived. These virtual links are sometimes referred to as "tunnels."

Each node has a unique identifier or node ID that is a MAC address. HSR also assigns a hierarchical ID (HID) or hierarchical address to a node, which is a concatenation of MAC addresses of nodes on a path from a top hierarchical node to the node itself. This hierarchical address is used to deliver a packet to its destination from any source node using HSR tables. So, for example, source node G has an HID(G) = <H,H,G> and node Q has an HID(Q) = <R,R,Q>. To delivery a packet from node G to node Q, node G delivers the packet to node H (level 0), upwards to node H (level 2), the top hierarchy. Node H forwards the packet to node R, which is the top hierarchy node for destination node Q. Node H has a virtual link or tunnel to node R via nodes (H, I, J, N, R). Node R ultimately delivers the packet to destination node Q along the downward hierarchical single-hop path.

Nodes at lower levels update their HIDs upon receiving routing updates from higher-level nodes. A gateway node can have multiple hierarchical addresses since a gateway node can communicate with multiple clusterheads and thus can be reached via multiple paths. Notable drawbacks exist in this hierarchical scheme, since a HID is long, and as nodes move the HID has to be continuously updated. HSR relies on logical partitioning to assist in locating mobile nodes.

In addition to a node ID (or MAC address), each mobile node is assigned a logical address of the type <subnet, host>. Each subnet corresponds to a particular group of mobile nodes that share a common characteristic. Each sub-network is associated with at least one home agent (a distributed location management server) to manage membership. All the mobile nodes of a logical sub-network know the HID of its home agent and each node registers, periodically or event-driven (e.g., a node moves to a new cluster), its current hierarchical address with the home agent. All home agents, in turn, advertise their HIDs to the top hierarchy. Since nodes are logically partitioned, HSR can define a group mobility model. Each logical sub-network has a conceptual center, whereby the center's motion represents the trajectory of the sub-network, including its location, speed, direction, etc. In this way, there is a separation of mobility management from physical hierarchy.

If a node wants to forward data packets to a destination node for which it has the MAC address, it extracts the subnet address and obtains the hierarchical address of the appropriate home agent from either its internal list or from the top hierarchy. Once the node has the hierarchical address, it forwards the packet to the home agent. The home agent delivers the data packets to the destination by eliciting the registered address from the host ID. Once a source learns a destination node's hierarchical address, home agent involvement is no longer required.

HSR offers improved scalability compared with flat, table-driven routing schemes, but with added complexity (e.g., the home agent) and routing inaccuracy. There is also a higher packet loss, since packets are dropped until a new route is established vs buffering packets until a new route is discovered.

15.4.6 OLSR

Optimized link stating routing (OLSR) [23–26] is a distributed, table-driven, link-state based protocol that performs hop-by-hop routing with the most recent route information at each node. OLSR protocol optimizes a pure link state protocol by reducing the size of a control packet and using only selected nodes called multipoint relays (MPRs) to diffuse broadcast messages in the network. This approach reduces the number of retransmissions in a flooding or broadcast procedure. Since control messages are sent periodically and include a sequence number, OLSR does not require reliable transmission or in-order delivery of its messages. OLSR is considered particularly suitable for large and dense networks, since using MPRs works well in such a context.

Each node in the network selects its MPRs among its one-hop neighbors with a bidirectional link—a unidirectional link is avoided. To accomplish this, each node periodically broadcasts hello messages containing a list of addresses of the neighbors where there exists a valid bidirectional link and a list of addresses of neighbors that are heard by the node, but for which the link has not yet been validated as bidirectional. With the hello message information, each node performs the selection of its MPRs, which are subsequently declared in a hello message sent to the selected MPRs. While OLSR fails to specify a specific algorithm for selecting MPRs, by default, the MPR set can equate to the whole neighbor set. Figure 15.5 illustrates nodes C, I, L, and M as MPRs for node H. MPRs are recalculated when there is bidirectional link failure with a neighbor; a new neighbor with a bidirectional link is added, or a change in the two-hop neighbor set with a bidirectional link is detected. The union of the neighbor sets of all MPRs contains the entire two-hop neighbor set.

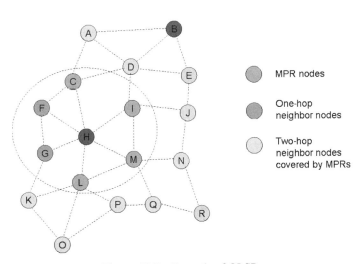

Figure 15.5 Example of OLSR.

The neighbors of a node that are not in the MPR set read and process packets but do not retransmit the broadcast packet from a node. This set of neighbors is called the MPR selectors of the node. When a node receives a hello message, it constructs its MPR selector table. The MPR selector table of a node contains the addresses of its one-hop neighbor nodes that have selected the node as its MPR, along with a corresponding MPR sequence number of that neighbor node. A node updates its MPR selector set according to hello information and increments the sequence number on each modification.

To construct a forwarding table for routing, control messages called topology control (TC) messages are periodically broadcast through the entire network by each node to declare its MPR selector set and associated sequence number. Each node calculates its topology table using received TC message information. An entry in the topology table includes an address of a destination (an MPR selector in the received TC message), an address of a last-hop node to that destination (the originator of the TC message) and the associated MPR selector set sequence number of the sender node.

Each node also maintains a routing table which allows it to route packets for other destinations in the network. The routing table is based on the information contained in the neighbor table and the topology table. The route entries in the routing table consist of destination address, next-hop address, and estimated distance to the destination. If any of these tables is changed, the routing table is recalculated to update the route information about each known destination in the network. Considering Figure 15.5, if source node H wishes to transmit packets to destination node B, a possible route would be H–I–D–B.

Wang et al. [25] proposed integrating MANETs into the Internet using Mobile IPv6 and OLSR. A test-bed was constructed that evidenced the viability of seamless handoffs between WLANs and MANETs. Benzaid et al. [24] introduced an extension of OLSR, called fast-OLSR. Fast-OLSR is designed to improve route discovery for a fast-moving node and maintain connectivity with a small number of neighbors. To achieve this, a fast-moving node establishes a small number of symmetric links refreshed at a high frequency using fast-hellos. If an MPR of a node has not received a fast-hello within an arbitrary hold time, the link is considered broken and a TC message is sent through the network. Benzaid et al. [26] provide a performance evaluation of integrating Mobile-IP and OLSR in IP networks. OLSR is used to support micromobility within an ad hoc network.

15.4.7 TBRPF

Topology dissemination based on reverse-path forwarding (TBRPF; Internet Draft, February 2004) [27] routing is a table-driven, link-state based, flat addressing scheme (but combinable with hierarchical) protocol. TBRPF was originally based on the extended reverse-path forwarding algorithm (ERPF) [28] designed for general broadcast, not for topology broadcast. However, TBRPF uses ERPF to broadcast link-state updates in the reverse direction along the spanning tree made up of the minimum hop paths from any node leading to the source of the update message. The following overview relates to version 4 of TBRPF outlined in RFC 3684 [27].

The main idea of this version of TBRPF is that each node reports only part of its source tree. TBRPF comprises two modules, namely, the neighbor discovery module and the routing module. Each module operates independently. TBRPF assumes a data link layer which supports broadcast, multicast, and unicast addressing with best-effort delivery services. Data packets are sent using UDP/IP (IPv4 or IPv6).

The TBRPF neighbor discovery (TND) protocol allows each node to detect a bidirectional link between it and a one-hop neighboring node. The routing module discovers two-hop neighbors. The TND protocol also detects when a bidirectional link breaks or becomes unidirectional. A neighbor table (NT) is maintained for each interface of a node. TBRPF accomplishes neighbor discovery using differential hello messages, which report only changes in the status of neighbors. In this way, hello message size is reduced compared with other link-state routing protocols where each hello message includes the IDs of all neighbors. Smaller hello messages permit more frequent exchanges and faster detection of topological changes. Since TND is used to sense neighbors only one hop away, the routing module is responsible for sensing more distant neighbors. For nodes with multiple interfaces, each interface must run TND separately. During each transmission of a hello, from each interface, a hello sequence number is incremented. TRBPF requires at least one hello to be sent within a hello interval. Nodes may also maintain and update link metrics, as an additional condition for changing the status of a neighbor (i.e., one-way, two-way, or lost), based on some link metric threshold.

The routing module of node maintains a source tree (ST) that provides shortest paths to all reachable nodes. Figure 15.6 illustrates the computed source tree for node H. The part of the tree that a node reports to its neighbors is called the reported sub-tree (RT). Each node computes and updates the source tree based on partial

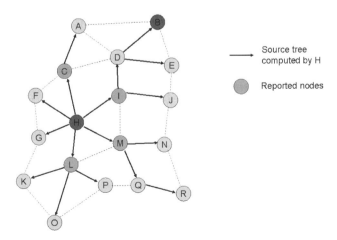

Figure 15.6 Example of TBRPF.

topology information stored in its topology table (TT), using a modification of Dijkstra's algorithm. Figure 15.6 depicts a route from node H to node B utilizing the routing module. TBRPF uses a combination of periodic topology updates (e.g., to inform new neighbors of the RT) and reports changes to the RT in more frequent differential updates. Topological updates avoid sequence numbers and are included in the same packet as a hello message. TBRPF does not use ACKs or NACKs. TBRPF allows the option to include link metrics in topology updates and to compute paths that are shortest according to the metric.

The RT consists of links (x,y) of the source tree where link x is in the reported node set (RN) based on the following. Referring back to Figure 15.6, Node H will include node M in the RN if node H determines that a neighbor will include node H as a shortest path next-hop to node M. Accordingly, neighbors in the RN are analogous to multipoint relay (MPR) selectors in OLSR. However, unlike OLSR, which does not include self-node-selection, if a node H selects neighbor node M to be in the RN, then node H is essentially selecting itself to be an MPR of node M. Each node is required to report the RT, but may report additional links up to the full network topology.

In addition to the NT and TT, each TBRPF node maintains a routing table, an interface table, a host table, and a network prefix table. The routing table comprises a list of tuples, namely, the destination IP address, the interface address of the next hop of the route, the length of the route and the ID of the local interface through which the next hop can be reached. The interface table stores information for determining the removal of an interface IP address associated with a router ID that has expired. Similarly, the host table and network prefix table contain analogous information for host IP addresses and network prefix, respectively, associated with a router ID that is used to remove expired host IP addresses and network prefixes.

15.4.8 DREAM

Distance routing effect algorithm for mobility (DREAM) [29, 30] is a fully distributed, location-aware routing protocol. Unlike the other protocols discussed thus far, a node maintains a location table (LT) that contains location information (e.g., geographic coordinates obtained by means of UWB position measurement procedures or other means such as GPS) for any other node in the network. One underlying principle of DREAM is that the greater the distance separating two nodes, the slower they appear to be moving with respect to each other. Therefore, the updating of location information becomes a function of the distance between nodes. Another underlying principle is that the frequency of sending location updates depends on the mobility of the node. In other words, slower-moving nodes require less updating than faster-moving nodes. Stationary nodes do not send control messages.

A node maintains an LT, which stores the location of each node. Using the LT information, a node can compute for every node, its direction and distance. When a source node wishes to forward packets to a destination node, it computes a forwarding zone, which is an angle whose vertex is at the source node and sides to be the outer bounds of the destination. Each neighbor subsequently performs the same operation until the data reaches the destination. The destination returns an acknowledgement packet (ACK) back to the source in the same manner.

Each node periodically broadcasts a control packet containing its coordinates. Each control packet is assigned a lifetime based on the geographical distance the packet has traversed from the sending node. In the case of a node with high mobility, short-lived packets can be disseminated frequently to inform nodes that are in close proximity and most in need of the node's location. These short-lived packets will travel a short distance from the sender and then die. On the other hand, slower moving nodes and further away destinations will not require sending control packets often. Long-lived packets may be sent to reach distant nodes. Since each control message contains only the identifier of a node and its coordinates, a control message uses only a minimal portion of the bandwidth and transmission power. With this approach, DREAM does not require route discovery.

When a node j wants to send to node k, it uses the location information for node k to obtain node k's direction and then transmits the packet to all its one-hop neighbors in the direction of node k. Each neighbor repeats this process until the destination is reached. The transmission of data is essentially loop-free. When the destination node receives a data packet, it returns an ACK packet in the same manner as the data packet. If the source node does not receive an ACK packet within a timeout period, then the source node will resort to a recovery procedure. DREAM does not provide specific implementations for the recovery procedure, but mentions that either a partial or full flooding can be used to determine a route, if any, to the destination.

Camp et al. [30] conducted performance comparisons with DREAM, location-aided routing (LAR) and dynamic source routing (DSR), and made some optimization proposals. One proposed optimization related to the transmission of control messages (i.e., location packets), which resulted in a reduction of total packets transferred in the simulation by 19%. Another proposal was considered, where instead of categorizing nearby mobile nodes based on distance, categorization was based on the number of hops. One major conclusion declared by Camp et al., was that DREAM did not appear to provide benefits over Flood protocols. However, DREAM and Flood have equivalent data packet delivery ratios for all mobile node speeds.

While geography-based routing has attracted more attention recently and can improve network scalability, reliance on location information for routing has its disadvantages. First, the inclusion of a GPS receiver in a node is an added expense, will add size to the node, does not work indoors, and may not always work outdoors depending on the surrounding environment. This is readily apparent in urban areas where there are many obstructions to prevent GPS signals from being received, along with multipath fading, etc., which could produce imprecise location information. Second, in high-mobility networks, control overhead can be considerable. Moreover, when high mobility is coupled with an urban area environment, the accuracy of location information can be degraded even further. Undeniably, UWB wireless technology may resolve many of these issues associated with GPS. The effect of mobility-induced location errors on different mobility models, including an urban setting (e.g., Manhattan model), is discussed in [31].

15.4.9 GSR

Global state routing (GSR) [32] is a link-state vector-based protocol with the advantage of no flooding (similar to DBF) of link-state packets, yet maintains full network topology. GSR provides that each node maintains a neighbor list (NL), a topology table (TT), a next hop table (NEXT), and a distance table (DT). The NL contains a list of adjacent (neighbor) nodes. The TT contains, for each destination, the link state information reported by each destination node and a timestamp indicating the time each destination node produced this link information. The NEXT contains the next hop to which the packets for the destination must be forwarded (shortest path). The DT contains the distances of the shortest path from the source to the destination. A weight function is used to compute the distance of a link, where nodes that are directly connected equal 1 for shortest path calculation. However, other type functions could be used for promoting other routing metrics.

When a node receives packets, it examines the sender field and adds this information to its NL. Then, a node processes the routing message and link state information. The node compares sequence numbers with those in local memory and uses the newest sequence number to compute the TT. A node periodically broadcasts this information to its neighbors. This procedure is similar to DSDV. The NEXT and DT are calculated in parallel with tree reconstruction using a modification of the Dijkstra's algorithm.

Each node maintains a link state table based on up-to-date information received from neighboring nodes. When a topology change occurs, nodes will not flood the network with link state packets; instead they forward the information to their local neighbors only. While this approach reduces the number of control messages transmitted through the network, the size of update messages is very large and consumes bandwidth. GSR is suitable for networks where node mobility is high.

15.4.10 FSR

Fisheye state routing [32–35] is a link-state-based protocol that originates from GSR. However, FSR reduces the overhead compared with GSR and therefore increases bandwidth and power efficiency. Unlike GSR, in FSR, the exchange of update messages among neighbors does not entail the entire TT. Instead, FSR exchanges entries of link state information about closer nodes more frequently than it does more distant nodes, thereby reducing the overall update messaging that occurs.

As the name of this protocol implies, FSR utilizes the fisheye technique for routing. The basic principal of this technique is that the level of detail and accuracy of route information decreases as one moves away from a center or focal node. Nodes partition the network into a number of scopes. A scope defines a group of destination nodes that a node can reach within a given hop interval. While route information differs among nodes, packets are routed correctly due to the ever-increasing routing accuracy of nodes as packets travel closer towards a given destination. As in GSR, FSR uses sequence numbers to make sure that the most up-to-date information is transmitted and stale information is discarded.

Since FSR does not include event-driven updating, solely relying on time-driven updating, the frequency of updating has be carefully monitored and scaled. It is implicit that, as changes in network population (topology) occur, nodes will need to appropriately modify the updating frequency and fisheye scope. Despite such adaptation, an unavoidable latency and routing inaccuracy will exist for more distant nodes. However, FSR scales well with large MANETs where there exists high mobility and bandwidth is low.

Figure 15.7 illustrates the application of the fisheye technique to a typical MANET. Node H represents the center node. The different shaded circles represent various fisheye scopes with respect to center node H. In this case, the fisheye scopes are defined by the number of hops from the center node H to neighboring nodes, which are colored dark blue, light blue and white depending on their hop distance. Depending on the size of the MANET, the specifics of a fisheye scope (e.g., radius and number) can vary.

Dimitriadis et al. [33] introduced clustered fisheye state routing (CFSR), which employs a typical hierarchical scheme comprising identifier-based ordinary, cluster-head, and gateway nodes. The primary objective of CFSR is to decrease the possibility of redundancy when routing control messages are broadcast. Clusterheads and gateways operate as a backbone and disseminate whole network population link state information, while ordinary nodes transmit their respective link state

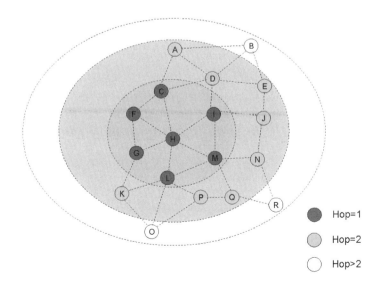

Figure 15.7 Scope of FSR—center node H.

information. Simulation results between FSR and CFSR illustrate lower bandwidth requirements for CFSR as mobility is increased.

Liang et al. [34] presented an adaptive routing table update scheme for FSR to improve routing accuracy and line overhead. The main idea of this scheme is that route table updates intervals (TUIs) should be adaptive according to a mobility metric called neighbor-changing degree (NCD). A node maintains two neighbor tables, where each table assigns a scope a_k (k = the number of hops from a focal node) value to each neighboring node. The first neighbor table assigns a scope value at the beginning of a routing update interval, and the second neighbor table assigns a scope value at the end of the routing update interval. A node ascertains the NCD by comparing the two respective scope values and modifies the routing update interval, if necessary.

A reiteration of the above was provided by Johansson et al. [35], where it was shown that higher user capacity could be achieved by adapting FSR parameters to optimum values based on the total network capacity. Further, it was shown that, if fixed FSR parameter settings are implemented, they should be based on the lowest network capacity that can occur at a given time.

15.4.11 HR

Hierarchical routing [46] is an adaptive hierarchical routing protocol that conforms to cluster-based networks adopted on the network and MAC layers. The salient features of HR are that routing changes only require an update of a forwarder ID, instead of whole path information. Each cluster updates at different frequencies and only entries that have changed. An entry is updated based on comparisons of generation/registration time values stored in a routing table of a clusterhead or central controller (CC). In this way, the count-to-infinity problem is averted.

In each cluster, a CC is selected and aware of its terminal node membership. Each CC maintains a local routing table composed of N fields (one for each member terminal) and a table update time (T-up), which provides a last-change-of-table time. Each field includes a path length (PL) measured in links between the CC and a terminal, a maximum transmission rate (MTR), which is the MTR over the PL via the weakest link, a field generation time (T-gen), which indicates the information life of this field, a field registration time (T-reg), which provides a last-change of field time, and the ID of a forwarder to the next cluster.

There is no synchronization among clusters to update routing tables. To perform an update, a CC sends an update-request (URQ) containing its T-up to its neighboring CCs. Neighboring CCs compare the T-up of the URQ to its T-reg for the entire routing table. Each CC then transmits an update response (URP) which includes all the fields where T-reg is greater than T-up. The source CC processes the URP and compares its fields to the new fields and updates those fields that meet the update criteria.

Performance evaluations between HR and an on-demand routing algorithm (ODRA) were conducted that showed HR outperformed ODRA in terms of average routing time and average routing information over large networks with moderate terminal mobility.

15.4.12 HSLS and A-HSLS

Hazy sighted link state (HSLS) [37] falls under the remit of a fuzzy sighted link state (FSLS) algorithm, similar to FSR discussed above, where the frequency of link state update transmissions relates to the distance of a neighboring node to a focal node. FSLS is based on the principle that distant node changes have little effect on local routing decisions. Nodes transmit link status updates (LSUs) at t s or multiples thereof, with a time-to-live (TTL) set to $\{s_I\}$. While the mathematical derivations are not presented herein, it is shown that FSLS is optimized by assigning $\{s_I\} = 2^I$. The generation process of LSUs can be obtained by replacing s_1, s_2, s_3, and s_4 by 2, 4, 8, 16, etc. HSLS's maximum refresh time function provides almost a linear relationship between $T(r)$, maximum refreshing time in seconds, and r, which equals distance in hops. HSLS is distinguished from FSLS by utilizing this established value, $\{s_I\} = 2^I$, in order to minimize the total overhead produced by any given node. Thus, HSLS optimizes the balance between refresh times and distances, so that the probability of making a sub-optimal next hop decision is "roughly" equal for every destination independent of the distance.

A key idea in HSLS is that, in addition to defining total overhead as the amount of bandwidth needed to construct and maintain a route, sub-optimal routing (e.g., imperfect or incorrect routing) is also considered as overhead since it reduces bandwidth and increases end-to-end delay. HSLS includes three main functions: LSU generation; LSU dissemination; and topology table maintenance.

For LSU generation, nodes operate in three different modes. Upon initialization, resetting all counters and timers, a node operates in undecided mode with LSUs assigned a TTL = 2. Nodes forward the LSU, decrementing the TTL value, until it becomes zero (0). When link change rates have been detected, a node will operate in one of two other modes. For a slow-changing local topology the node operates in SLS (standard link state) mode. In SLS mode, a node will send a global LSU (i.e., TTL = infinity) to notify neighboring nodes of a link change. Nodes that receive an LSU with a predefined value meaning infinity will not decrement this value. However, for a fast changing local topology, a node will operate in HSLS mode where LSUs are transmitted will values 2, 4, 8, etc. at times that are multiples of t_e s. If changes occur between these time instants, the node will also send LSUs with a TTL = 1.

Hello packets or level 1 LSUs (i.e., TTL = 1) are frequently exchanged for neighbor discovery and to avert short loop issues. Hello packets contain the sending-node ID and a list of neighbors. The hello packet also includes whether links are unidirectional or bidirectional, as well as the number of hellos received from a neighbor within a specified time window. Accordingly, degradation of a link can be determined by the number of hellos received within this time window.

Each node contains a TT, which is maintained by the reception of LSUs reporting incoming link state information. The TT assigns a higher cost to unidirectional links than to bidirectional ones. Level 1 LSUs are event-driven, transmitted every time a link change occurs, but could also be given a smaller time-driven periodicity. Nodes consider LSUs transmitted from local neighbors more reliable (i.e., most up-to-date), compared with other LSUs transmitted by more distant nodes. However, in situations where a node receives an LSU that a link to a distant node is down, instead of erasing a TT entry, the node will assign the highest cost to this entry. The reason for this is twofold. First, it may be some time before an alternate route to the distant node reaches the node. Second, if the node needs to forward a packet to this distant node, then it is more probable that nodes closer to the distant node will have more up-to-date information to route the packet to the distant node. Thus, the node can still utilize the un-erased topology entry information to forward the packet.

Adaptive hazy sighted link state (A-HSLS) is a derivation of HSLS to handle extremely low node mobility situations. In this type of scenario, HSLS sends a series of LSUs with TTL equal to 2, 4, and 6 until a global LSU is sent, for each link change. However, in retrospect, it would be more efficient to merely send a global LSU at the beginning, if we know that no subsequent link state changes will follow. A-HSLS uses past experience to predict near-future mobility of the network. If the time between the last global LSU being sent and the time of the last link change exceeds or is less than a threshold, then a global LSU or an LSU according to HSLS rules will be sent, respectively.

15.5 REACTIVE ROUTING

Reactive routing maintains routing information for active routes only. When a node receives a packet address to an unknown destination, a route discovery process occurs, on-demand, by flooding (partial or other) a route request packet through the network. When a node with a route to the desired destination receives the request packet, it responds with a reply packet to the source node using link reversal or piggybacking. Unfortunately, such an approach delays the actual forwarding of a packet until the route is determined. Thus, the reactive approach epitomizes the tradeoff between reducing overhead and the expense of delay due to route search. Once a route is established, it is maintained by a route maintenance procedure until the route is no longer needed or becomes inaccessible. However, route maintenance may generate a significant amount of overhead if the topology of the network changes frequently.

While there have been a number of strategies for reactive protocols, three popular approaches include source routing, hop-by-hop routing, and link reversal routing. In source routing, the source node constructs a source route in the packet header. Dynamic source routing (DSR) protocol provides an illustrative example of this approach, as described below. Source routing has many advantages, such as attributing different routes with different packets, depending on the desired metric. The main disadvantage is the amount of overhead generated by explicit routing within each packet header. However, similar to ATM virtual paths, the amount of overhead can be minimized by ascribing *flow identifiers* to each packet. Hop-by-hop routing relies on dynamically establishing route table entries at intermediate nodes. An advantage in this approach is that packet overhead is reduced, since only the destination address and next hop address are needed compared with a sequential list of nodes in source routing. Another advantage is that a route is not static and is adaptable to changes in the network environment. A disadvantage of using hop-by-hop routing is the overhead required to maintain routing tables for active routes and to prevent looping. The link reversal routing scheme decouples the dynamics of the topology of the network from long-distant control message propagation. To accomplish this, the algorithm maintains a directed acyclic graph (DAG) rooted at the destination. To clarify, a directed graph is acyclic if there are no loops and it is rooted at the destination since it is the only node that has only incoming links. Different protocols may maintain the DAG in different ways. A primary disadvantage to this approach is that some algorithms based on this scheme may be unstable and never converge or remain nonconvergent for a period of time.

15.5.1 DSR

This version of DSR [38, 39] is a pure on-demand protocol designed for MANETs of up to 200 mobile nodes and works well with very high rates of mobility. DSR uses source routing where each packet header includes a sequential list of nodes through which the packet will propagate to reach the destination. This approach allows the sender to indicate a specific route. Since the route is in the header of each packet, other nodes in the path or neighboring nodes can cache this routing information for future use. DSR provides multipath routing, unidirectional links and asymmetric route support. DSR is composed of two main mechanisms: route discovery and route maintenance.

If a source needs to send a packet, it checks its route cache. The route cache contains routes previously learned. If the destination is not found in the route cache, then the node initiates a route discovery process by transmitting a route request and buffering copies of packets to be sent in a send buffer. Figure 15.8 illustrates an example route discovery by node H to find a route to node E. A node has the option of limiting the propagation of a route request to one hop [i.e., a time-to-live (TTL) = 1, a non-propagating route request], to see if direct neighbors are the destination or have a route to the destination. Otherwise, as shown in Figure 15.8, a node may initiate an expanding ring search, where the hop limit is progressively increased if no route reply is received. In such cases, a tradeoff exists between route discovery latency and minimizing route control overhead. To avoid looping and manage route discovery requests, each node maintains a route request table that includes the above-mentioned request packet ID, a TTL field, the number of consecutive route discoveries for a destination, and the time the last route request for a destination was sent. When the destination node or an intermediate node that has a route to the destination receives the request packet, it transmits a route reply to the source, which contains the route record of the request packet minus any duplicate nodes.

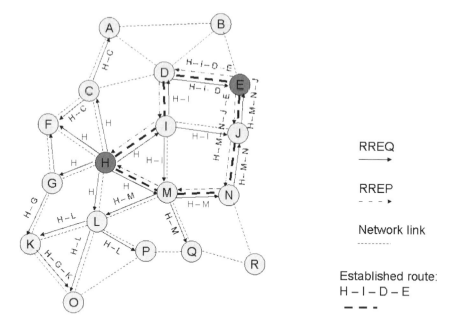

Figure 15.8 DSR route discovery from node H to E.

However, if a destination is not reachable, further route discoveries may be initiated. In Figure 15.8, node E transmits a reply packet back to node H.

Once a route is learned, a source may transmit data packets along that route. Intermediate nodes may salvage packets instead of discarding packets when links break by using another route in their route cache. Further, an intermediate node operating in promiscuous mode can support an automatic route-shortening feature, which helps eliminate unnecessary hops. In particular, if a node overhears a packet carrying the source route and it is not the intended next-hop, but it is a node designated later in the source route, the node will transmit a gratuitous route reply to the source, which provides a shorter route. To eliminate unnecessary gratuitous route replies, the node maintains a gratuitous route reply table. DSR also includes optional extensions to routing, such as flow state, to reduce source route overhead. Data packets can be forwarded along the source route by relying on the intermediate's node local knowledge of the route to the destination. This hop-by-hop forwarding of packets along a route, once the route has been discovered, eliminates the need to include an explicit route. The source/destination addresses and a flow ID are combined to identify a flow. Each node implementing a flow state maintains a flow state table, which includes such things as the previous and next-hop MAC addresses.

Route maintenance relies on some form of acknowledgment, either on the link-layer, network layer or a passive acknowledgement (e.g., node A confirms receipt to node B by overhearing node B transmit to node C), if a network interface of a node can operate in promiscuous receive mode. If a node is unable to reach a next-hop

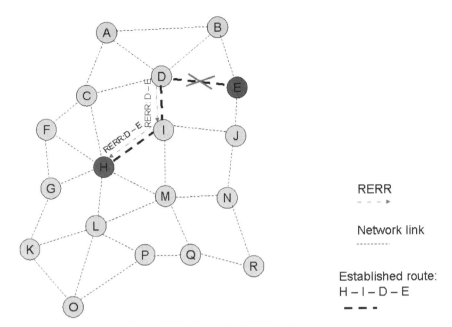

Figure 15.9 Example of DSR route error.

node after a certain number of retransmissions, the node transmits a route error to the source. Figure 15.9 illustrates the transmission of a route error by node D after the D–E link break.

Recent updates to DSR outlined in this version include rules for IP fragmentation and reassembly, nodes having multiple network interfaces, interaction of DSR with Internet's address resolution protocol (ARP), removal of optimizations for unidirectional links, based on special 127.0.0.1 and 127.0.0.2 flags in a route request and route reply and DSR options header and DSR flow state header that can share a single IP protocol number assignment. While this version of DSR specifies routing for unicast IPv4 packets, advanced options permit QoS support, multicast routing, and IPv6 support [38].

15.5.2 ARA

Ant-based routing algorithm (ARA) [40] is a protocol based on swarm intelligence, particularly mimicking ant colony behavior when seeking food. Ants use *stigmergy* methods (i.e., altering the environment as a means to communicate) to communicate by depositing *pheromones*. The concentration of pheromone in any particular space guides ants along a path towards a food source. ARA adopts this approach in its implementation of route discovery and route maintenance by utilizing ant algorithms.

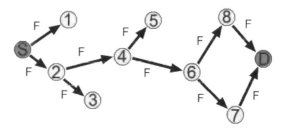

Figure 15.10 Route discovery stage of ARA.

In the route discovery stage of ARA, forward ants (FANT) are used. FANTs are small packets that include a sequence number which is broadcast by a source to its neighbors to establish a route (see Figure 15.10). Each node maintains a routing table that includes a destination address, a next hop, and a pheromone value. When a neighboring node receives a FANT, it considers the source address as a destination address, the previous node as the next hop, and calculates a pheromone value depending on the hop count, before further transmission. When a FANT arrives at the destination it processes the information and deletes it. Next, the destination node will generate a backward ant (BANT) to establish a route back to the source, where similar processing by the intermediate nodes takes place. Once the BANT is received by the source, ARA considers that a path has been established and data packets can be transmitted (see Figure 15.11).

Route maintenance also mirrors ant behavior in that data packets propagate through the network changing the pheromone value at each node. Specifically, if a data packet travels through an intermediate node, the pheromone value increases (i.e., a pheromone counter). Conversely, pheromone values are exponentially decreased over time for nodes not visited by data packets. Thus, ARA does not require any special maintenance packets to be sent for route maintenance. If a pheromone value decreases to zero, a link is deactivated. Nodes are considered to be in sleep mode if a pheromone value reaches a nominal threshold value.

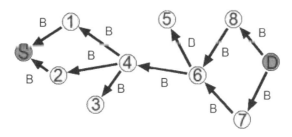

Figure 15.11 Route discovery stage of ARA.

Data packets contain source address and sequence information to avoid looping. If a node receives a duplicate packet, the node sets a duplicate error flag and transmits the duplicate data packet back to the previous node, which in turn deactivates the link. Missing acknowledgements indicate a route failure. In this case, a node sends a router error message to indicate the link failure and sets the pheromone value to zero. The node then searches for an alternative route in its routing table and, if no route is found, the node surveys its neighbors. If the neighboring nodes do not have a route, they in turn inform their neighbors, backtracking to the source. Ultimately, if no route is discovered, the source must initiate route discovery.

Since ARA does not exchange routing tables, the overhead is very small. Additionally, FANT and BANT packets do not contain much information, thus minimizing the effect of flooding. Nonetheless, reliance on flooding has scalability problems for large networks. Simulations using ns-2 were conducted and performance comparisons among ARA, AODV, DSDV, and DSR were made. The first performance metric was delivery rate. DSR and ARA had a delivery rate of >95% with low pause time and high topology changes, but with very high changes DSR outperformed ARA. The second performance metric was overhead. Again, DSR outperformed ARA in cases of high node mobility, but was on a par with DSR with low node mobility. AODV and DSDV showed poor performance in comparison to ARA in both instances.

The framework for ARA stems from the ant colony optimization routing algorithm originally designed for wired networks. Since ARA, there have been two analogous protocols proposed, probabilistic emergent routing algorithm (PERA)[41] and AntHocNet [42]. Readers intrigued with this routing approach may find these additional references interesting and informative.

15.5.3 ABR

Associativity-based routing (ABR) [43, 44] is a protocol where route selection is based on nodes having a measurable degree of associativity states (i.e., periods of connection stability between nodes). This metric is based on the observation that there exists "dormant time" or a portion of time during which a mobile user remains in a location without moving. Such a period of time translates into a period of spatial, temporal, and signal stability. Stability is represented by a threshold value of associativity ticks.

Each node periodically transmits beacons to identify it and correspondingly updates its associativity tick value for every beacon received. A high associativity value represents a low state of mobility between two nodes and the converse holds true where a high state of node mobility exists. A node will reset its associativity tick value of another node, if this other node moves out of the connectivity range of the node. Nodes in ABR maintain a routing table for existing routes. Nodes also maintain a neighboring table (NT), which is updated by the data link layer protocol through the use of beacons.

ABR comprises three phases, namely, the route discovery phase, the route reconstruction phase and the route deletion phase. The ABR route discovery phase

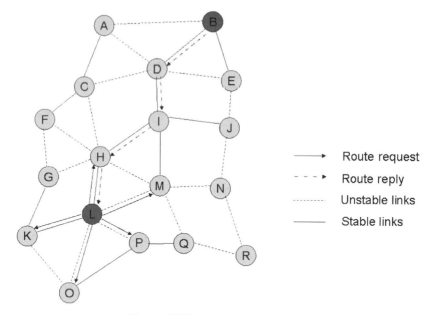

Figure 15.12 ABR route discovery.

requires a source-initiated broadcast of a query packet when a route is unknown. This process is illustrated in Figure 15.12, where node L searches for a route to destination node B by transmitting a broadcast query (BQ) packet to its neighbors. Intermediate nodes receive the query packet and discard it if it was previously processed. Otherwise, an intermediate node determines if it is the destination. If it is not the destination, the intermediate node attaches its address, associativity ticks of neighbors (if any) and any other routing metrics to the query packet before broadcast forwarding (not shown). The subsequent node will erase its upstream node neighbors' associativity tick values and maintain only the tick values of it and its upstream node. Ultimately, the query packet will reach the destination containing the route, their respective tick values, hop count, propagation delays, and route-relaying load.

The destination then selects the best route based on the accumulated information and sends a reply packet back to the source as shown in Figure 15.12. The destination will pick a route with high associativity ticks over a route with fewer hops. However, if two routes have equivalent associativity, the destination will select the route the minimum number of hops. If multiple routes still exist, the route with the least cumulative forwarding delay will be selected. Of course, ABR provides flexibility in route selection based on the QoS metric. Intermediate nodes that are in the selected route validate their route to the destination, and other routes are deemed inactive to avoid duplicate packets arriving at the destination. Under ABR, both the BQ packet and the reply packet are not of fixed length. Rather, the length of these packets depends on the number of nodes traversed. If

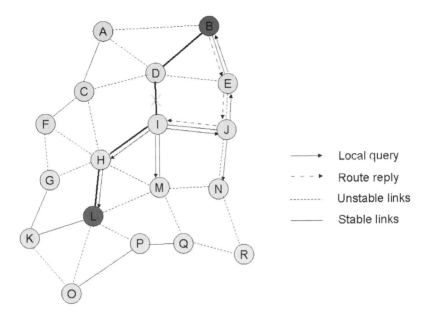

Figure 15.13 ABR route maintenance.

the selected long-lived route happens to have a node move unexpectedly, then ABR enters the route reconstruction phase.

The route maintenance phase comprises partial route discovery (route reconstruction phase), invalid route erasure, valid route update, and new route discovery, depending on the movement of respective nodes along the route. If a source node moves, then a new route initialization (query/reply) is initiated. If the destination moves, the destination's immediate upstream node erases its route. A localized query (LQ){H} process (where H signifies the hop count from the upstream node to the destination) is initiated to determine if the destination node is still reachable. If the destination node receives the LQ packet, a best partial route is selected and a reply is sent. Figure 15.13 illustrates such a scenario where a link failure occurs along the route between nodes D and I. Here, node I initiates a local query, which reaches destination node B. Destination node B sends a reply back to node I. If the destination node does not receive the LQ packet, the LQ packet times out and the initiating node backtracks to the next upstream node. This backtracking process continues until a partial route is found or the new pivot node is greater than half the hop distance between source and destination. If no partial route is found, the initiating node will send an FQ{1} packet back to the source node to begin a new route discovery or to the destination node to erase an invalid route. An FQ{1} packet, which includes a direction of propagation field, means to backtrack one hop at a time.

If a route is no longer needed, for example, the destination moves within the transmission range of the source, the source will send a FQ{1} packet to erase a route—a route delete (RD) (route deletion phase). ABR defines other situations regarding concurrent node movements, but are omitted herein. The 1996 version [43] of ABR describes being suitable for conference size MANETs. In [44] (1999 Internet Draft version), it is described that ABR uses a soft or hard state for route deletion. Hard route deletion requires a network broadcast of a route delete packet. Soft route deletion occurs when inactivity of a route reaches a threshold level, whereby the route is automatically deleted. ABR [44] also uses distinguishable, additional, routing metrics from [43], namely, relaying intermediate nodes supporting existing routes and knowledge of link capacities of selected routes.

ABR [44] requires that intermediate nodes operate in promiscuous mode to listen to packet forwarding so that a passive acknowledgment takes place. However, destination nodes will send an active acknowledgement. ABR is loop-free, avoids deadlocks and packet duplication, and maintains one route for each route request (i.e., no alternate routes are maintained) [44]. Since route selection relies on long-lived paths, throughput is likely to be high. However, route recovery time is increased if a path fails, since alternate routes are not cached [43]. In contrast, ABR [44] supports caching if the extent of node mobility does not cause frequent route invalidity. ABR does not support multicasting and has yet to cope with asymmetric links.

15.5.4 AODV

Ad hoc on-demand distance vector (AODV) [45] is another distance vector routing protocol designed for large MANETs with variable traffic and mobility rates. While many versions of the AODV algorithm exist, this discussion relates to the Internet Draft of 2003. Each AODV node maintains a routing table, where each route entry includes a destination IP address, prefix size, destination sequence number, next hop IP address, lifetime (deletion time of the route), hop count, network interface, and other state and routing flags. A salient feature of AODV is the use of destination sequence numbers for each routing table entry, which are generated by a destination node and sent to a requesting node. Destination sequence numbers ensure loop-freedom, aid in the selection of two or more routes to a given destination, and assist in updates received from control messages for route table entries. Each node has and maintains its own destination sequence number.

AODV route discovery is source-initiated using route request (RREQ)/route reply (RREP) query cycles. When a source does not have a route to a destination, the source generates an RREQ. Before the source broadcasts the RREQ, it buffers a RREQ ID and its own IP address as path discovery time information. If a route is not received after a network traversal time, the node may reinitiate a route discovery in conjunction with a binary exponential back-off algorithm up to a specified retry number. Every node that receives the RREQ, increments a hop count field of the RREQ and caches a route to the source. If a node receives a duplicate RREQ, the node discards the RREQ. If the receiving node is not the destination and does not have a current route to the destination, the node will re-broadcast the RREQ. Otherwise, if the receiving node is the destination or has a route to the destination with a corresponding destination sequence number that is greater than or equal to that of the RREQ, the node unicasts an RREP to the next hop toward the source. Additionally, if the RREQ has a G-flag set, then the node will also send a gratuitous RREP to the next hop toward the destination. Since the RREP contains a path list from the RREQ, each intermediate node may create/update a route for the source and destination.

If a RREP transmission failure occurs because the RREQ was sent over a unidirectional link, the AODV node lists the previous node in a "blacklist" for a blacklist time period. Thus, if the AODV node receives a RREQ from a blacklisted node during the blacklist period, the RREQ is ignored. In this way, if a subsequently arriving duplicate RREQ over a bidirectional link is received, the RREQ will not be discarded.

When the source receives the RREP it can begin transmission. If the source receives multiple RREPs, the source can select the route with the shortest hop count. As data packets propagate along the route, each node updates its lifetime field associated with the route. While inactive routes with expired lifetimes cannot be used to forward packets, the routing table maintains such routes for a certain period of time because they still may contain useful information when processing control messages. Additionally, AODV provides a lifetime extension to be appended to a RRQ or RREP to update the lifetime field in the routing table, which, when applied, prevents an active route timeout to occur.

An AODV node may use periodic broadcasts of hello beacons to detect and monitor links to neighbors. If a node fails to receive hello messages or other packets from a neighbor during a specified time interval, the node will assume the link to the neighbor is broken. In this case, the node will update its route information concerning that neighbor node and increase the destination sequence number by one. In addition to hello message, route maintenance can be supported by passive acknowledgements, layer 2 notifications, failure to transmit a packet after a maximum number of retransmission attempts, and Internet control message protocol (ICMP) messaging.

Any node may locally broadcast to its neighbors a route error (RERR) message (TTL = 1) for route repair, to trigger an update mechanism, and when a route break is detected. Also, if a data packet for an invalid or unreachable destination is received, the node may generate an RERR message, which includes the destination IP address and the corresponding destination sequence number.

This version of AODV omitted discussion relating to a number of facets compared with previous versions such as precursor lists, many flags, expanding ring search, multicasting, and local repair. On the other hand, some aspects were changed from previous versions, such as routing table updates, route invalidation, RERR creation, and requiring an originator sequence number in RREPs.

A recent modification to AODV is called AODV-BA (AODV with break avoidance) [46]. This protocol attempts to detect the danger of a potential link break and avoid actual link breaks. AODV-BA relies on four criteria: (1) the received radio; (2) overlap of routes; (3) battery life threshold; and (4) density. More particularly, received radio relates to the distance between nodes that is farther than the radio communication range. Overlap routes are synonymous with hotspots, that is, an intermediate node handling packets from several different sources. Lastly, density relates to the number of nodes surrounding an intermediate node sharing a wireless channel. In all four cases, intermediate nodes notify upstream nodes of a potential link break.

15.5.5 BSR

Backup source routing protocol (BSR) [47] is an offshoot of DSR, except that, in addition to establishing a primary path to a destination, a backup path is established in order to reduce the frequency of route discovery floods if the primary path link(s) break. BSR is an on-demand protocol that comprises two traditional stages: route discovery and route maintenance. However, BSR uses a new routing metric called route durability.

Similar to DSR, when a source node needs route information that it does not have in cache, it floods the entire network with a route request message (RREQ) as shown in Figure 15.14. Unlike DSR, where nodes discard duplicate packets, BSR nodes utilize duplicate packets that have traversed different paths to establish backup routes. To prevent unnecessary forwarding of RREQs, RREQs are dropped by intermediate nodes if all the following conditions are met: (1) the node is the destination node of the RREQ; (2) the node is listed in the source route; (3) the path in the duplicate packet cannot produce new backup routes with lower cost; and (4) a candidate (i.e., a threshold is defined to use cache route information) of backup routes can be obtained from cache. A node then selects primary and backup paths to a destination based on a minimal metric cost and stores them in its routing cache, as shown in Table 15.4.

When the destination receives an RREQ, it selects backup routes and returns a route reply message (RREP) to the source using the selected routes. Alternately, an intermediate node may generate an RREP, if it has a backup route in its routing cache. Allowing intermediate nodes to transmit RREPs under these circumstances reduces latency and prevents unnecessary flooding. When a source receives a RREP, it transmits data packets whose header includes the primary path and backup path.

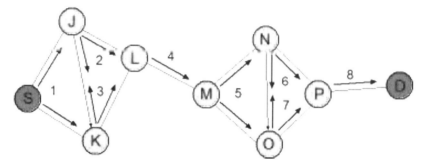

Figure 15.14 RREQ propagation in the route discovery phase of BSR.

15.5 REACTIVE ROUTING

TABLE 15.4 A Routing Cache is Constructed in All Nodes Except the Source After Route Selection Initiated by the RREQ of the Route Discovery Phase.

S	D	Primary Path ψ		Backup Path ψ'		(ψ, ψ')
J	S	S–J	(f)	S–K–J	*(f)	2
K	S	S–K	(f)	S–J–K	*(f)	2
L	S	S–J–L	(f)	S–K–L	*(f)	2
				S–J–K–L	(f)	3
M	S	S–J–L–M	(f)	S–K–L–M	*(f)	2
				S–J–K–L–M	(f)	4
N	S	S–J–L–M–N	(f)	S–K–L–M–N	*(f)	3
				S–K–L–M–O–N	(d)	6
				S–J–K–L–M–N	(f)	4
				S–J–K–L–M–O–N	(d)	7
O	S	S–J–L–M–O	(f)	S–K–L–M–O	*(f)	4
				S–K–L–M–N–O	(d)	6
				S–J–K–L–M–O	(f)	5
				S–J–K–L–M–N–O	(d)	7
P	S	S–J–L–M–O–P	(f)	S–K–L–M–O–P	*(f)	5
				S–K–L–M–N–O–P	(d)	7
				S–J–K–L–M–O–P	(f)	6
				S–J–K–L–M–N–O–P	(d)	8
D	S	S–J–L–M–O–P–D(f)		S–K–L–M–O–P–D	*(f)	6
				S–K–L–M–N–O–P–D	(d)	9
				S–J–K–L–M–O–P–D	(f)	6
				S–J–K–L–M–N–O–P	(d)	10

The paths marked (f) are forwarded, the paths marked (d) are discarded and the paths marked * are considered the most durable backup paths. Backup paths are selected based on some threshold, which in turn is based on a heuristic cost function if $C(\psi, \psi') \leq |\rho|$. The heuristic cost function of the backup routes measures the durability of the route. A back-up path provides better performance if it has a shorter length, is less link-similar to, but has more disjoint subpaths than, its primary path.

When a link failure occurs, nodes will try to forward the data packet to the destination using backup path information found in the header. Concurrently, BSR requires a route error packet (RERR) to be sent when a node is unable to deliver a data packet to the next hop of a route. The RERR includes the backup routes to the source and the direct upstream and downstream neighbors of the broken link. A node sends RERRs to upstream nodes of the route that are using the reverse backup routes in the packets. When the source receives the RERR, it updates its routing cache or, if it has insufficient information to reconstruct its backup routes, the source will initiate a new route recovery process.

A simulation model based on ns-2 was used to compare BSR with DSR based on the following performance metrics: packet delivery ratio; average end-to-end delay; and normalized control message overhead (the number of control packets transmitted per data packet). BSR outperforms DSR as to packet delivery ratio and normalized control message overhead when there is a high rate of node mobility. However, DSR outperforms BSR as to average end-to-end delay and normalized control message overhead when there is a low rate of node mobility. Notwithstanding all of the above, the current version of DSR, as discussed above, includes packet salvaging by intermediate nodes using alternate routes. Thus, one of the main favorable aspects of BSR has been nullified.

15.5.6 CHAMP

Caching and multipath routing (CHAMP) [48] protocol provides data caching and shortest multipath routing at the expense of additional storage overhead and control information. Packet salvaging can reduce energy consumption if packets otherwise dropped are forwarded to the destination. Thus, packet salvaging and multipath routing provide a high packet delivery ratio. However, when packet salvaging occurs near the source, then there may be a significant delay in packet delivery and ultimate packet reordering at the destination.

CHAMP requires three types of control packets—route request (RREQ); route reply (RREP) and router error (RERR). Each CHAMP node includes a route cache and a route request cache. The route cache contains a list of active destinations. For each destination entry, the route cache stores a destination ID, distance to the destination, a set of successor nodes to the destination, age of each successor node, and the frequency of which a successor is used. The route request cache maintains a list of every unique route request received and processed. For each request entry, the route request cache stores a source ID of the request, the destination being searched, sequence number of the request, minimum forward count, set of previous hop nodes, and the status of the request (replied or not).

Each CHAMP node also manages a send buffer and a data cache. The data cache holds copies of packets recently forwarded. Data headers include a source ID, a sequence number, as well as the previous hop, which is used to indicate where the same data is cached. The data cache is used to implement packet salvaging when routing errors occur. Before a node forwards a packet, it saves a copy of the data in its data cache and sets the previous hop address to itself. When forwarding data, the node picks the least used next hop neighbor in order to spread the data transfer over all possible routes (i.e., load balancing). If no route exists, the node will broadcast an RERR. However, if the node is the data source, it will initiate a route discovery.

In the route discovery stage, a source node broadcasts a RREQ and waits for a RREP. In the event an RREP is not received, a source will retransmit RREQs using a back-off algorithm. After a specified number of unsuccessful attempts, the upper layer is informed. When a node receives an RREQ, it assigns a route request ID and uses the fc (forward count) field of each RREQ to determine the shortest path to the destination. When the destination receives the RREQ, it sends an RREP. Intermediate nodes can also send RREPs.

For route maintenance, data acknowledgements are used to determine the state of a link and, if an acknowledgement is not received from a next-hop after forwarding a data packet, the link is deleted from route cache. A node will initiate a route repair for a limited period of time, if an alternate route does not exist. If unsuccessful, the node will broadcast a RERR with the data packet's previous hop (if the data is stored in the data cache) or any next hop (if the data is not cached).

15.5.7 DYMO

Dynamic MANET on-demand (DYMO) [49] is a current Internet Draft (July 2005) routing protocol that minimizes the use of network resources and adapts quickly to network conditions. Each node is responsible for maintaining route and link state information. All DYMO packets are transmitted via user datagram protocol (UDP). DYMO consists of three element types: a route element (RE), a route error (RERR) and an unsupported-element error (UERR). Each element type includes a portion of a fixed data structure with certain header fields. All nodes must implement an RE; however, some nodes may ignore an RERR and a UERR. Any DYMO control packet can be either unicast or MANET-cast (i.e., IP broadcast address with duplicate suppression).

DYMO is a bifurcate system comprising route discovery and maintenance. In the route discovery phase, a source node buffers its node address and node sequence number in its routing element table (RT) before it MANET-casts an RE to its neighbors within transmission range to learn a route to a destination. When the RE reaches the destination, the node replies with an RE. Interim, intermediate nodes acquire routing information to the source and destination nodes. In the event a source does not receive a route, the source may issue subsequent route discoveries according to a binary exponential back-off algorithm. During this time, data packets are buffered. However, after a specified number of failed retries, data packets are dropped and an ICMP undeliverable message is sent to the application.

DYMO nodes monitor links using several methods such as link layer feedback, hello beacons, and neighbor discovery and/or route timeouts. Some nodes in the MANET may act as a gateway to the Internet. Gateways will signify their status in an RE by setting a gateway bit. Other nodes may advertise connectivity to a node subset by setting a prefix field in an RE. If an active route is broken, a node will transmit a RERR to the source node, which in turn will initiate route discovery.

The RERR includes the unreachable node address and node sequence number. Node processing of a RERR includes setting a route timeout to the current time (the author assumes that a common time reference among nodes is necessary). If the element time-to-live (TTL) is zero and the element is the first element, the DYMO packet is dropped, if the element TTL is zero, but not the first element, the element is removed. Otherwise, if the TTL is above zero, the element is retransmitted in a packet.

Each node is assigned a sequence number and if the destination sequence number is known, it is included in the RE. When a node receives an RE, it checks its RT for the RE address. If the RT does not have a corresponding entry, then it creates one. Otherwise, the node will process the RE by checking if the route information is stale or not. If the route is not stale, then the RT is updated and any queued data packets are forwarded. If a node receives an unsupported DYMO element type, the node decides whether to send a UERR and also how to handle the unsupported element. Regardless of whereas a UERR is sent to a notify-address, the node will either skip or remove the element from the packet or set an ignored bit and skip the element.

DYMO does not include security measures but recommends authentication. The current draft specifies parameter values (e.g., network diameter, RREQ wait times, and retries, etc.) for a small, well-connected network with moderate node mobility. However, DYMO is adaptable to large networks, if the parameter values are properly adjusted. DYMO supports nodes having multiple interfaces and coverage extension to/from the Internet via a gateway. DYMO supports either IPv4 or IPv6.

15.5.8 DNVR

Presented in March 2005, dynamic Nix-vector routing (DNVR) [50] adopts the *Nix-vector* (NV) approach, originally used in wired networks, for efficient on-demand routing. DNVR provides some unique features such as route validation before use, a compressed form for source routes and reliance on MAC addresses to identify neighbors so that IP addresses do not need to be resolved using, for example, the address resolution protocol (ARP).

The main idea of the *original* NV routing method for wired networks was to reduce the number of bits to represent routing information per hop using a *neighbor index* (*Nix*). Each node represents its neighbors with an ordered set (e.g., {0, 1, 2...}) and selects its next hop from this set. An NV is created as a packet propagates through the network from the source to the destination. When a router receives a packet it will make a routing decision and inserts a *Nix* value in the packet header to add to the preexisting vector length. Once the packet reaches the destination, the complete concatenation of *Nix* values will form the NV. The destination returns the packet to the source and subsequent packets are forwarded to the destination along this NV path. As a packet passes through a router, the vector length is decremented by the appropriate number of bits (i.e., a *Nix*) and forwarded by using the *Nix* to index a table of next hop IP addresses, interface numbers and possibly layer 2 addresses.

Based on the above framework, the DNVR protocol comprises two phases: the NV creation phase and a mobility management phase. DNVR also relies on three different data structures: the NV, a neighbor table, and an NV forwarding information base (NV-FIB). An NV is a concatenation of *Nixes* with a NV length field. The NV length field signifies the number of bits of the NV minus the length field. As a packet with an NV propagates through a node, a *Nix* value is removed, thus decreasing the length of the NV. The *Nix* comprises a color field and a neighbor index field. The color field represents the number of bits for the neighbor index of the neighbor table and the neighbor index field represents the actual index of the neighbor table used to determine the next hop. To adapt the NV to a MANET with node mobility, the neighbor index field is of variable length, while the color field is fixed. This allows the *Nix* to be a variable number of bits. The color field

N_b is calculated from Equation (15.1),

$$N_b = [\log_2 \text{index}] + 1 \qquad (15.1)$$

where index $\in \{1, 2, 3\ldots\}$. The index begins from one due to the insertion of a *hidden bit* in the neighbor index field, where bit 1 always comes before the number specified in the neighbor index field. Thus, the length of the index field is $N_b - 1$ since the color N_b includes the *hidden bit* and the index value includes the prefix of the *hidden bit*.

Each node maintains a neighbor table that is indexed by the *Nix* value. Each entry includes the next hop, interface number, and lifetime of the entry. The next hop field contains the MAC address of the next hop, the interface number indicates which interface should be used for communicating with the neighbor node and the lifetime value provides the life of the entry. The NV forwarding information base (NV-FIB) is a table that stores NVs that are indexed by a *path id*. The *path id* consists of source/destination IP addresses and an NV reply number. An NV-FIB contains five parameters: a *path id*, an NV, a metric, state information, and lifetime.

In DNVR, a node initiates route discovery when it does not have a route to a destination. The node buffers data packets and locally broadcasts an NV request (NVREQ) message, which includes source/destination IP addresses, sequence number, route metric, a reverse NV and MAC address of a forwarding node's interface, to its neighbors. Intermediate nodes broadcast or unicast the NVREQ until the destination is reached. If an intermediate node has an NV to the destination, it does not generate an NV reply (NVREP) message. Rather, the intermediate node unicasts the NVREQ to the destination to *validate* the route and act as a probe to ensure route accuracy. If an NVREP is returned, the route is validated. Otherwise, a new route must be discovered. The reverse NV is copied into the NVREP header and returned to the initiating node. Upon receipt of the NVREP message, the source node can unicast the packets from its buffer. When a node receives a data packet, it reads the color field $N_b - 1$ and appends the *hidden* bit to the extracted bits to form the *Nix* value and determine the next hop.

DNVR's mobility management detects routing failures and performs mobility management functions. DNVR relies on link layer functionality for notification of link failure. Alternatively, nodes can operate in promiscuous mode to overhear packet deliveries or failure thereof. If a node detects link breakage, it generates an NV error (NVERR) message that contains the *path id* for the route that includes the broken link, forwards the message to the source, and invalidates any NV in the NV-FIB that contains the broken link. As the NVERR message propagates back to the source, NV routes are invalidated in each node's NV-FIB and the source will initiate route discovery. As for neighbor detection or neighbor management, DNVR relies on passive detection and monitoring during an NV creation phase and when a route is being used.

DNVR scales well to large networks under various degrees of mobility and traffic volume. While at first glance it may appear that the delay incurred during route validation would degrade performance, simulations presented in [51] illustrate that

DNVR is equally efficient to DSR in terms of normalized total control overhead and provides a higher packet delivery and smaller packet latency.

15.5.9 LAR

Location-aided routing (LAR) [52] is a reactive protocol that attempts to decrease overhead during route discovery by utilizing location information. Basically, this is accomplished by reducing the search space for a needed route, which in turn, reduces the number of nodes that receive route discovery packets. While this protocol suggests using the GPS to obtain location information, UWB technology is clearly a viable alternative option. It should be noted that one difference between LAR and DREAM, discussed previously, is that DREAM uses location information for data delivery, while LAR uses location information for route discovery. This distinction naturally flows from DREAM being a proactive protocol and LAR being a reactive protocol.

LAR presents two algorithms to determine a *request zone* for forwarding route requests. In LAR, a source node calculates an *expected zone* of the destination node based on previously obtained location information and estimated velocity of the destination node. In this way, the source node can determine a spatial region or *expected zone* that the destination node is situated at the time of the route request. If the source node does not have previous location information for the destination node, then the entire MANET becomes an *expected zone* and traditional flooding is used. Thereafter, the source node defines a *request zone* for the route request, which includes the *expected zone*, and may also include other regions around the *request zone*. There are a couple of reasons for additional regions being included. First, if the source node is not within the *expected zone*, then a path from source/destination must include nodes outside the *expected zone*. Second, if after a timeout period a route request does not produce a path to the destination, then the source may initiate a new route discovery with an expanded *request zone*.

The two LAR algorithms differ in how the *request zone* is defined. In the first scheme, we refer to Figure 15.15. As shown, a *request zone* is defined by the smallest rectangular that includes the source node M and the *expected zone* (i.e., a circular region) of the destination node E. In this case, node M broadcasts its route request, which specifies the route *request zone*, to all its neighbor nodes. Nodes I and N will forward the route request of node M since they are in the *request zone* of M. However, nodes H, L and Q will not forward the route request, because they are all out of M's *request zone*. Thus, neighboring nodes must determine whether they are within the *request zone* or not. When destination node E receives the route request, it sends a route reply, which includes the current location and time of node E, back to the source.

In the second algorithm of LAR, the source does not explicitly specify the *request zone*. Rather, the source node includes in its route request, the last known coordinates of the destination node (e.g., X_D, Y_D, assuming two-dimensional node movement) at $T(0)$ and an estimated coordinate of the destination node at the time of the route request $T(1)$. Nodes receiving the route request must then calculate their distance from location X_D, Y_D as DIST(i). Based on DIST(i), the neighboring

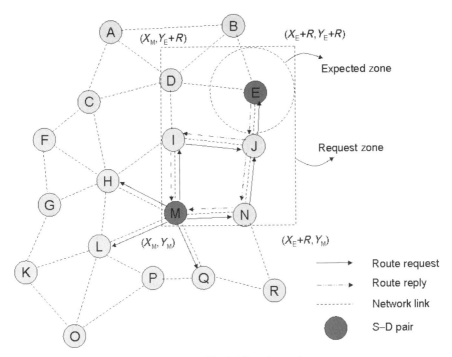

Figure 15.15 LAR: scheme 1.

node applies some parameters α and β, such that if $\alpha[\text{DIST}(S)] + \beta \geq \text{DIST}(i)$, then the neighbor node forwards the route request with $\text{DIST}(i)$ and X_D, Y_D information. That is, $\text{DIST}(i)$ replaces $\text{DIST}(S)$. Otherwise the node discards the route request. When the neighbor node (i) forwards the route request to another node (p), then the above calculation is repeated except that $\text{DIST}(i)$ replaces $\text{DIST}(S)$ that is, $\alpha[\text{DIST}(i)] + \beta \geq \text{DIST}(p)]$. An example of this process is shown in Figure 15.16. Here, nodes I and N forward M's route request since $\alpha[\text{DIST}(M)] + \beta \geq \text{DIST}(i)$ and $\alpha[\text{DIST}(M)] + \beta \geq \text{DIST}(n)$. However, node Q will not forward M's route request, because $\text{DIST}(q) > \alpha[\text{DIST}(M)] + \beta$.

Since GPS may include some error, the coordinates of a node may not be entirely reliable. While LAR addresses possible location error in scheme 1 by adding an error value e to the *expected zone* radius, scheme 2 does not account for possible location error. In either case, location error can contribute to the need by a source node to initiate multiple route discoveries. Moreover, the authors of [52] acknowledge that more work is needed to determine when the degree of location error nullifies the effectiveness of LAR schemes. Of course, UWB may alleviate these concerns.

15.5.10 LBR

Link life-based routing (LBR) [51] is another on-demand protocol based on *link life* prediction for making path selection, similar to ABR described above. The

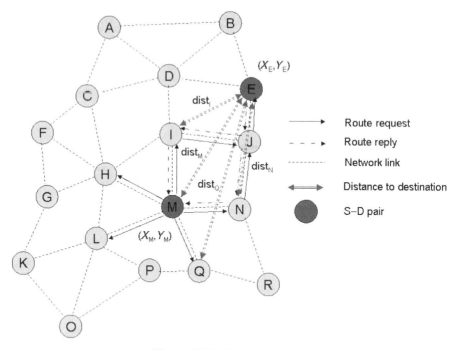

Figure 15.16 LAR: scheme 2.

worst-case expected lifetime of a link (*link life*) is derived from a linear regression method over a number of previous time-based distance samples. Another feature of LBR is that it limits the frequency of a node sending a beacon. An LBR node that has transmitted a data or control packet within a beacon-defer interval is not permitted to transmit a beacon. Under these circumstances, neighboring nodes can listen in promiscuous mode (i.e., a node can listen to a neighbor's packet even if the packet is not addressed to the node) and update their signal strength information. Accordingly, as the number of sessions increases, the numbers of beacons sent decreases.

Route discovery begins with a source node broadcasting a route request packet (RREQ), which includes source/destination addresses, to find a route to a destination. Intermediate nodes that receive the RREQ append their node numbers and link life$_{\text{diff}}$ (link life$_{\text{diff}}$ = link life − current time). To balance the load, RREQs are dropped if the number of flows through a node exceeds a flow limit. When the first RREQ reaches the destination, the RREQ is buffered and a timer is triggered. Subsequently arriving RREQs are buffered until the timer expires. The destination node then selects the best path and sends a route reply packet (RREP) back to the source, which propagates through the intermediate nodes where routing information is updated in cache. Intermediate nodes also insert a path-expunge time in cache so an entry can be removed after this time transpires.

LBR describes two methods for route maintenance. The first approach is to use beacons. If a node detects that its beacon count is not incremented after a beacon-check interval, the node can assume the link is broken. The second approach is to have the node check for packet drops during a data transmission session.

LBR also describes two methods for route reconfiguration (RREC). A reactive RREC begins only after the source node is aware of the route break. This occurs if the first hop is down or if it receives a route error packet (RERR) from an intermediate node. As the RERR propagates through the network, nodes delete route cache entries. A new RREQ is sent from the destination and buffers arriving packets until an RREP arrives at the source. The proactive RREC begins to setup a new path when the path-expunge time expires without waiting for an actual break to occur. Under simulations, packet delivery is slightly less for proactive RREC compared with reactive RREC.

15.5.11 MPABR

Multipath associativity based routing (MPABR) [52] is a recent offshoot of ABR, which establishes multipath backup paths for each communicating node pair. This overview of MPABR will not reiterate the implementation of ABR, but merely include the modifications thereto.

MPABR provides multiple loop-free node-disjoint, partial node-disjoint, or disjoint routes from the source or node-disjoint routes for intermediate nodes. This first extension to ABR is implemented in the following way. When the destination node is collecting broadcast RREQs during the *collect replies time period* it selects the shortest hop count, most stable routes based on the above-mentioned multipath criteria (e.g., disjoint, partial node-disjoint, etc.). A node-disjoint path is when a path between a source/destination pair does not share any intermediate nodes with any other path between the source/destination pair. This extension is referred to as MAPBR. A partial node-disjoint path is when pathways are node-disjoint except that they permit sharing of each intermediate node with up to one other alternative route between the same source/destination pair. This extension is referred to as MP1ABR. A disjoint path is a path formed between a source/destination pair whereby the path does not use the same neighbor nodes for any two paths of either the source or destination. This extension is referred to as MP2ABR. An intermediate node-disjoint path refers to the forming of alternative node-disjoint paths at every intermediate node in the primary route. This extension is referred to as MP3ABR. If a link failure on the primary route occurs, a RERR packet is sent back along the path to the destination. All the intermediate nodes receiving the RERR check for an alternative route in their routing tables. If an alternate route exists, then a node will use the newly found route and will not retransmit the RERR.

Another modification to ABR is aimed at reducing the path setup time, which is implemented as follows. When the destination node receives the first RREQ, it immediately sends an RREP to the source, so that, when the source receives the RREP, it can immediately start forwarding data packets along the specified route.

This first route is then deleted, if it is not optimal, once the destination selects the best path and issues another RREP.

The last proposed modification was an attempt to reduce unnecessary broadcasts. When a node receives a broadcast query it compares the neighbor list of nodes of the sender with its own list of neighbors. If it determines that its own list is merely a subset of the sending node's list, then it does not broadcast the query packet. The logic is that all of its neighbors will already have received the broadcast by the time the node transmits it. Performance comparisons between MPABR and ABR clearly evidence substantial improvement in performance when these extensions to ABR are implemented.

15.5.12 NDMR

Node-disjoint multipath routing (NDMR) [53] is a modification and extension of the AODV protocol. NDMR is designed to discover multiple node-disjoint routing paths using the request/reply technique. NDMR outlines two types of disjoint paths: link disjoint and node disjoint. Link disjoint paths do not share a common link; however, they may share a common node, whereas node-disjoint paths do not have any common node, except for the source and destination nodes, as shown in Figures 15.17 and 15.18.

It can be shown that multipath breakage is related to the number of intermediate nodes as well as the type of the disjoint path. Further, it can be shown that the probability of multipath breakage is lower for node-disjoint paths compared with link disjoint paths. Therefore, node-disjoint paths are more stable and considered a more desirable path type to be implemented.

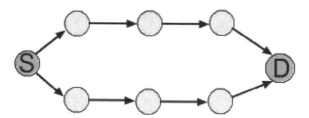

Figure 15.17 Node disjoint.

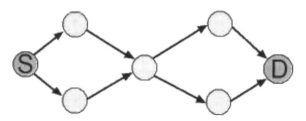

Figure 15.18 Link disjoint.

Based on the above findings, NDMR aims at building multiple node-disjoint paths. To accomplish this, the destination must judge whether a path is a node-disjoint path when it receives an RREQ during the route discovery phase. If the path is a node-disjoint path, the destination sends an RREP, which contains an sequential list of intermediate nodes, back to the source node that initiated the RREQ. The intermediate nodes, along the route, update their routing tables and reverse routing tables, respectively.

Since AODV relies on flooding RREQs during the route discovery stage, it is probable that nodes will drop duplicate RREQs, even if the RREQs come from a different path. On the other hand, to allow nodes to forward all packets would lead to control overhead saturation. Therefore, NDMR requires nodes to use a shortest hop metric to determine whether to forward duplicate RREQs. A node calculates the number of hops from the source to itself and compares that number to the shortest number of hops in the node's reverse route table entry. If the number of hops is larger than the reverse table entry, the duplicate packet is discarded. Otherwise, if it is equal to or less than the stored entry, the node attaches its address to the RREQ and broadcasts it.

Given that the destination must select and store "multiple" node-disjoint paths, it must compare whole routes when it receives duplicate RREQs. If no common node exists (except for source and destination nodes) between the received RREQ and previously stored routes in the destination's reverse route table, the RREQ route path will be recorded. Otherwise, the route in the RREQ will be discarded. Once the destination node's RREP reaches the source node, the source can record the next hop to the destination in its multiple route forward path table. The source can begin sending data packets immediately after receiving the first RREP.

NDMR nodes maintain routes by sending out periodic hello packets. If a hello packet is not received, an RERR is propagated through the network towards the source. Each node along the path to the source marks the route to the destination as invalid. The source then selects an alternate node-disjoint path or if none exists, initiates a route discovery.

15.5.13 PLBM

Although not previously discussed, multicast routing is an important facet in MANETs and desirable in a number of different situations where MANETs arise. There are two basic types of multicast protocols tree-based and mesh-based. Mesh-based protocols usually require more control overhead even though tree-based protocols rely on flooding. Nevertheless, mesh-based protocols have higher packet delivery compared with tree-based protocols. Preferred link-based multicast (PLBM) [54] is a tree-based, receiver-initiated multicast protocol that minimizes the flooding of join query packets (JQs), by permitting only certain nodes to forward them.

PLBM nodes maintain local two-hop topology and tree information in a neighbor's neighbors table (NNT) and a connect table (CT), respectively. Since PLBM is receiver initiated, it is the responsibility of each member node to acquire a connection to the multicast source. The multicast construction phase requires that a member node send a JQ only if the node is not connected to the multicast source, the NNT does include a tree node, and at least one neighbor is present in its NNT for forwarding the JQ. In the event that a tree node (i.e., multicast source, connected member, or forwarding node) is found in the member's NNT, then the member simply sends a join confirm message to the tree node, without flooding. If all the above conditions are met, then the member computes the preferred link table (PLT) using the preferred link-based algorithm (PLBA).

The PLBA assigns preference to nodes with high neighbor associativity (i.e., the number of neighbors of a node), thus reducing the number of nodes required to include all the nodes in the NNT and the number of JQ transmissions. PLBA obtains two-hop topology information from an NNT. To summarize PLBA, the current forwarding node excludes all its neighbors that have already forwarded JQ, neighbors that are shared by nodes in the preferred list, nodes that are neighbors to nodes in the traversed path of JQ, and multiple copies of a JQ reaching a node through nodes that are neighbors of nodes in the previous node's preferred list. Next, the node starts selecting neighbor nodes for the preferred list in the order of their associativity. PLBA excludes those neighbors that are neighbors of the currently included node. Thereafter, the nodes are stored in the preferred list table (PLT) whose neighbors are not covered by nodes in the preferred list. The PLBA eliminates (from the PLT) redundant nodes and nodes that are neither the destination nor have any other new outgoing links. Based on the PLBA, the number of nodes selected to forward JQ is small and is executed only when no tree node is present in the NNT or the CT.

Once the PLT has been constructed, the member node selects the first X entries to send JQs. A JQ is sent as a unicast packet only to one of the preferred nodes, while all other preferred nodes receive the JQ in promiscuous mode. Each intermediate node that receives a JQ checks its eligibility (i.e., whether the node is in the PL field of the JQ) for forwarding and, if it is not eligible, the JQ is discarded. If a node is connected to the multicast tree, it sends a join reply packet (JREP) to the source JQ node and triggers a timer to receive a join confirm packet (JCON) from it. Otherwise it forwards the JQ packet using the same conditions described above. A JREP travels a reverse path of the corresponding JQ. Intermediate nodes discard duplicate JREPs and forward (unicast) unprocessed ones. Neighbor nodes may listen using promiscuous mode. When the first JREP reaches the source, a join confirm packet (JCON) is sent back, while JREPs with the same ID and having the same or lower sequence number are discarded.

Having established a connection, data packets are unicast to minimize collisions and neighbors use promiscuous mode. PLBA uses two mechanisms for route maintenance. Link breaks are detected based on either not receiving a beacon within a specified interval or the absence of a CTS (clear to send) packet after multiple retransmissions of RTS (request to send) packets. If a tree node finds a more optimal path from the multicast source than its current uplink during a multicast session, the node sends a *prune me* message to the current uplink node and a JCON to the new member node.

15.5.14 RDMAR

Relative distance microdiscovery ad hoc routing protocol (RDMAR) [55] utilizes a relative distance microdiscovery algorithm (RDM) to minimize flooding of the entire network during the route discovery phase. RDMAR is loop-free and does not require location assistance methods (e.g., global positioning system) to limit the region of query flooding.

When no route to a destination is known, a node has two route discovery options, either to flood the entire network with route query packets (RREQ) or limit the region of flooding. A source node selects entire network flooding if no previous communication information between the source/destination pair is available. Otherwise, the source node accesses previous routing information in its routing table regarding the desired destination. Every node maintains a routing table that lists all known destinations, and respective routing information, such as next hop, an estimate of the relative distance in hops between the source and destination, last update time, route active flag, and route lifetime information. Based on the stored routing information, a node calculates a new relative distance, in terms of hops, to the destination.

In general, RDM uses the last update time of a destination node entry to calculate a *t-motion* interval. The source node then calculates an expected minimum and maximum relative distance (i.e., radius-min, radius-max) to come up with a normalized radius, in terms of hop distance, that is inserted as a time-to-live value in the header of an RREQ (see Figure 15.19). Once an RREQ is transmitted, every node that receives the RREQ creates a reverse route to the source and discards copies

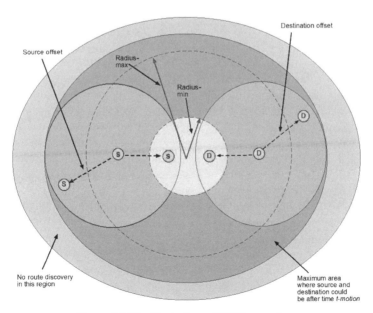

Figure 15.19 Illustration of RDM procedure.

of the RREQ to minimize control overhead. A route reply (RREP) can be sent only by the destination node of the RREQ in order to avoid stale routes. If an RREP is not received, the source is governed by an exponential back-off algorithm and limited to a specified number of retransmissions. RDMAR also includes an urgent route reply packet (U_RREP) in the instance where a node becomes completely disconnected from the rest of the network, and during re-entrance, a pending RREQ exists. When a connected node realizes its new neighbor, it transmits a U_RREP to all nodes that need a route to the entering node.

During data packet transmission, intermediate nodes forward the data and also send messages to previous nodes in order to establish bidirectional connectivity and to secure route information for future acknowledgements back to the source. Upon load failure and retransmission failure, a node initiates the RDM algorithm for local repair or may notify the source depending on the intermediate node's proximity to the source/destination. When the intermediate node that experiences link failure is close to the source, it enters a failure notification phase (FN). An FN packet is transmitted and received only by other nodes that rely on the intermediate node to reach the destination. This is accomplished by nodes maintaining a dependent list (DL), which lists all the neighbors that use the node as a router to reach a destination. When an FN packet is received, a node deletes the associated route to the destination. If a node that has an empty DL receives an FN packet, the node uses its routing cache to forward the FN packet to its destination. Finally, to reduce error messaging, nodes maintain a copy of a data packet to allow retransmission through an alternate path, if one exists.

15.5.15 SOAR

Source-tree-on-demand adaptive routing (SOAR) [56] is a protocol that depends on the exchange of minimal source trees (MSTs) containing link state information to neighbors for routing to active destinations, as shown in Figure 15.20. Sequence numbers are used for updating and validating minimal source trees, which reduces overhead. SOAR depends on lower (link) layer notification or unsuccessful transmissions as a measure of link connectivity to neighbors and does not rely on periodic link state beacons. SOAR can be implemented on top of UDP and IP and has access to data packets from the network and upper layers.

Figure 15.20 shows the minimal source tree of router I advertised to its neighboring nodes. Router I is aware of the links to nodes J–U. However, router I only has active flows to destinations L, N, Q, and U. In this case, router I only advertises those links within the curved boundary. For example, router does not report links K–M, L–O, Q–S, and R–T.

When a router receives a data packet from the application layer, if it has a route in its route table to the destination, the data is immediately forwarded to the next hop. However, a router will initiate a route discovery when it does not have next-hop information to the destination. In this case, the router buffers the data and transmits a *nonpropagating* query packet to its neighbors requesting link state information to

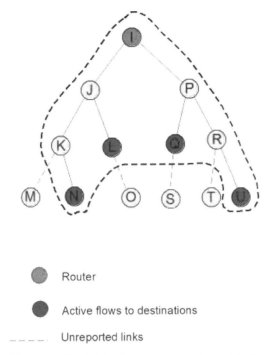

Figure 15.20 Minimal source tree exchanged by SOAR.

generate a route to the destination. If no reply packets are received, then the router sends a *propagating* query packet with a max hop limit. If need be, this process oscillates between nonpropagating/propagating queries, continually increasing the timeout period, until a specified number of attempts have been made. Each router manages a query table to manage the queries sent for each destination.

When a router receives a data packet from the network and a path to the destination is not known or it determines forwarding could lead to a loop, the router will discard the data and send an update packet to its neighbors for an alternate path. To limit the number of update packet transmissions based on incoming data packets, the router adheres to a minimum update time interval. A neighbor sends a reply packet in response to a query if it has complete path information to the destination based on its MST. Otherwise, the neighbor will forward the query to its neighbors, if the max hop and timeout period permits. A node will forward a reply packet if it is in the path of the source/destination pair, has a path to the query router and has a new route to the reply router. Otherwise, the node will send an update packet if there is an increase of distance to an important destination.

Each control packet (e.g., query, reply, and update) includes multiple link state information updates and is sent unreliably in a limited broadcast. A router uses LSU information to update a partial topology table (PT) and the MST. A modified

Bellman–Ford or Dijkstra's SPF is used as a path selection algorithm on the PT to calculate a source tree (ST) and update the RT. SOAR designates *important* nodes as nodes that act as a router/relay or are sending data packets. A router determines the *important* nodes by a path traversal through the ST. Routers also maintain a distance table (D), where each entry specifies a destination sequence number and a last heard time. Each router is assigned a sequence number and each link is assigned a cost, where ∞ equates to link failure. A router will increment its sequence number when an adjacent link is broken or a new link is formed. The outgoing links of a router have the same sequence number. If a head node in the PT has links with different sequence numbers, the router will send packets along the links with the higher sequence number and delete the lower sequence link.

Figure 15.21 illustrates some examples relating to minimal source trees and partial topology tables. These examples assume that the network has converged to the same sequence number for each node and that all nodes are *important* to each other. As shown, when link (A–B) breaks, node A increments its sequence number to 31. Node A also reports this break to node C; however, node D's update has not reached node C, so the minimal tree of node D remains the same. Example 8 in Figure 15.21 illustrates the partial topology table of node D that reflects the link (A–B) break and assigns and advertises a sequence number of 31 with an infinite cost.

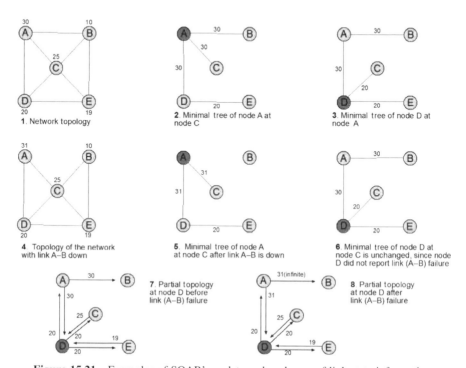

Figure 15.21 Examples of SOAR's update and exchange of link state information.

SOAR is considered bandwidth-efficient because it produces minimal overhead by reducing flooding and capitalizing on the redundancy of MSTs at the expense of suboptimal routes. Additionally, SOAR does not require periodic link state beaconing when there are no network changes. Rather, a SOAR node selectively advertises link states of links that only have active flows to its neighbors.

15.5.16 TORA

Temporally ordered routing algorithm (TORA) [11, 57, 58] is a protocol based on a "link reversal" algorithm that builds a directed acyclic graph (DAG) rooted at the destination. Multiple routes to a source–destination pair are created to avoid overhead when a topological change arises and to provide a quick reaction to such changes. When a reaction to a topological change is necessary, control overhead is minimized to the locality of the change. TORA is loop-free and scalable since nodes only maintain information about adjacent nodes. TORA is a ternary system that provides three basic functions: route creation, route maintenance, and route erasure.

Figure 15.22 illustrates route discovery for TORA. If a node does not have a route to a desired destination, the node floods the network with a query packet (QRY), which includes a destination ID. When a node with a route to the destination or the destination itself receives the query packet, it broadcasts an update packet (UPD), which includes the destination ID and a value (analogous to height) with respect to the destination. As the UPD propagates towards the source node, each node assigns itself a height value greater than the height value of the node sending the UPD. Also, each node, other than the destination, maintains a link-state array (LS) for each link. The height values provide the state of a link and its

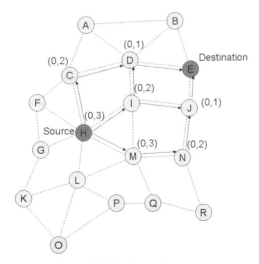

Figure 15.22 Route discovery.

392 AN OVERVIEW OF ROUTING PROTOCOLS FOR MOBILE AD HOC NETWORKS

direction. If a node is higher than its neighbor, the link is marked "up," and if the reverse is true, then the link is marked "down." The height of the destination is always zero. Links that are undirected are assigned a null value.

Nodes with a height value other than null perform route maintenance. TORA outlines a number of scenarios for route maintenance regarding a node, other than the destination node, that has no downstream links but needs to modify its height. Figure 15.23 depicts a logical flow chart for five situations relating to route maintenance. In situation 1, if a node does not have a downstream link due to link failure, the node will define a new reference level if it has at least one upstream node, otherwise, it will set its height value to null. In the second situation, a node sends the reference level of its highest neighbor and selects a height that is lower than all neighbors with that reference level. In the third situation, a node receives the same height level from all its neighbors and the node reflects back a higher sub-level by setting an r bit.

In the fourth situation, if the node defines a reference level and the level propagates back to the node as a higher sub level from its neighbors, then a network partition exists. The node then must initiate a route erasure. In situation 5, a link failure occurs between the time the node propagates a reference level and the return of a

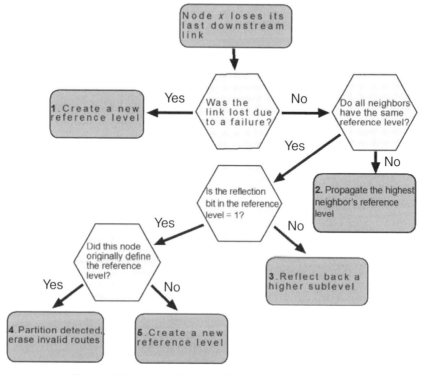

Figure 15.23 Logic flow for TORA's route maintenance.

higher sublevel from its neighbors. In this situation, a partition may not exist, so the node defines a new reference level. In any scenario, once a node determines its new height, it broadcasts a UPD to all its neighbors.

If a network partition is detected, a node sets its height and all its neighbor's height to NULL, updates its LS and broadcasts a clear (CLR) packet. The CLR packet includes a destination ID and a reference level. Nodes that receive the CLR packet and have a matching reference level set their height value to null, unless the node is the destination, in which case its height value is set to zero.

TORA assumes that packets are received sequentially and correctly. TORA also assumes that the lower level protocol provides for neighbor awareness. In [11], it is discussed that TORA is implemented on top of the Internet MANET encapsulation protocol (IMEP) to support these requirements. TORA is considered best for large, dynamic, bandwidth-constrained networks. Performance comparisons in [11] revealed that overhead in TORA is significant compared with DSR, particularly when there is high mobility. The packet delivery ratio as a function of node mobility rate and network load (i.e., number of sources) was also measured for TORA. TORA performed better in lower network load scenarios, regardless of node mobility, whereas in higher network load situations, packet delivery dropped.

15.6 POWER-AWARE ROUTING

A sector of the research community is focused on the development of routing protocols whose aim is to reduce energy expenditure for packet delivery to the destination. A further extension of this idea is to prolong the network lifetime and prevent network partitions from developing due to node failure. Thus, saving energy in an ad hoc network falls under two primary categories: power control and power management. In light of the above criteria, PAR is a special type of protocol and interest in it has grown since the application of power control and energy conservation is far-reaching and germane in MANETs with scarce power resources.

The centerpiece of power control is the selection of the appropriate transmission power, which dictates the range of the signal and the number of nodes that will receive it. It has been shown that reducing the transmission power can improve network capacity (i.e., simultaneous transmissions within a region of the network), yet may reduce connectivity and create network partitions [59]. Krunz et al. [59] discuss various approaches for transmission power control from both a network and an MAC layer standpoint.

While transmission power is a major consumer of energy, a wireless node also consumes energy in idle mode, when for example it performs computations or data collection. A wireless interface also consumes power even when idle and a transceiver consumes power in both receive and transmit modes. Thus, depending on the network traffic and the computational complexity of the routing approach, this additional consumption of power can be considered.

Power management for MANETs can span and utilize all layers of the protocol stack. Table 15.5 illustrates some well-known techniques that address power

TABLE 15.5

Protocol Layer	Power Management Approaches
Data link	1. Turn radio off or operate in sleep mode 2. Avoid retransmissions 3. Avoid collisions in channel access 4. Use contiguous slots for transmission and reception 5. Have receiver in standby mode whenever possible
Network	1. Consider battery life in route selection 2. Reduce control overhead 3. Minimize control packet size 4. Consider route relaying load 5. Efficient route reconfiguration techniques
Transport	1. Manage packet loss locally 2. Use power-efficient error control 3. Avoid retransmissions
Application	1. Utilize base stations for power-intensive computations and avoid using mobile nodes 2. Use an adaptive mobile QoS scheme 3. Use proxies for mobile clients

conservation in MANETs [60]. One power saving technique for nodes of a MANET is operation in sleep mode. The author is aware of three approaches to sleep mode operation: connected active subset, asynchronous wakeup and synchronous wake up [61]. Interestingly, similar to MANET routing protocols, power management approaches range from proactive (nodes active all the time) to reactive (all nodes in power saving mode by default). Examples of power management schemes will not be presented.

While PAR protocols offer identifiable benefits, typically these benefits are juxtaposed by a decrease in network throughput and an increase in packet delay. Accordingly, PAR may not be suitable for all MANET applications and network types.

15.6.1 BEE

Battery energy efficient (BEE) [62] is a routing protocol that manages battery consumption in order to maximize the lifetime of the network. Nodes are assumed to be battery-powered. The basis of BEE stems from two principles of battery operation: *recovery effect* and *rate capacity effect*. The *recovery effect* relates to the fact that the performance of a battery is enhanced when it is discharged in short time intervals with idle periods in between vs continuous discharge, because during the idle periods the battery partially recovers its lost capacity due to a diffusion process. The *rate capacity effect* states that a battery delivers less energy when the requested current exceeds the rated current value, which depends on the chemistry of the battery.

BEE nodes are limited to a predetermined transmission range, which is a function of the distance between source–destination pairs. For nodes within this predetermined transmission range, the transmission energy is divided into a few levels from e-minimum to e-maximum. The *recovery effect* and its impact on battery status (an energy increase) are calculated based on the transmission rate and a mean energy value for a node to transit a packet. Conversely, the battery status is reduced, in light of the *rate capacity effect*, when the energy level to transmit is greater than e-minimum. Under this framework, the routing algorithm of BEE assigns a cost to each route based on battery behavior and the energy transmission amount.

15.6.2 EADSR

Energy aware dynamic source routing (EADSR) [63] is an extension to DSR (version 9) for selecting minimum energy routes. EADSR nodes are able to measure signal strength received and use dynamic power control on each packet transmitted. The measure of energy consumption is loosely proportional to transmit power. EADSR requires that a node know its min/max transmit power, and the receiver sensitivity of their interfaces, and is able to operate in promiscuous mode. EADSR relies on the MAC layer to support bidirectional links. EADSR is gaged for MANETs with static or low mobility.

When a source needs to transmit a packet, the packet header will include, in addition to route information, the minimum power that the packet is to be transmitted for each hop of the route. If the node does not have a route to the destination, a route request is broadcast at maximum power. The route request packet will include link energy information (LEI) for every link in the route (at this point only the source exists) as well as the transmit power set to maximum. Intermediate nodes compute the minimum recommended transmit power (MRTP) based on the nodes' interface transmit power and receiver sensitivity. When the request packet reaches the destination, a reply packet is transmitted back, and the nodes in route use the computed MRTP to transmit the packet.

For route maintenance, when a data packet propagates through the network, nodes compute new estimates for MRTP and compare these values to the MRTP values in the header. If a predefined difference exists, a link flag is set in the header and when the data packet arrives at the destination it sends a gratuitous reply to the source with the corrected MRTPs. In cases where an MRTP is set below a minimum or above a maximum level of a transmitter, the MRTP is adjusted to the maximum or minimum level of the transmitter.

15.6.3 MTPR/MBCR/MMBCR/CMMBCR

In [64], there are four PAR approaches discussed. The first PAR is minimum total transmission power routing (MTPR), which depends on obtaining a route with minimum total power. The route with the minimum total power can be determined from a hop-by-hop transmission power metric from source to destination. Noncommunication-related power consumption is not considered. This approach can be applied to a standard shortest path algorithm such as Dijkstra and Bellman–Ford. However, the Dijkstra algorithm can result in selecting a path with more hops, concomitant to which an increase in end-to-end delay and greater instability (node movement).

The second PAR approach is minimum battery cost routing (MBCR). MBCR considers the total transmission power as an important routing metric; however, it considers more directly the lifetime of each node. In this regard, the battery capacity of a node is inversely proportional to cost—as battery capacity decreases the cost for that node increases. The MBCR approach tries to minimize the overuse of select nodes, analogous to load balancing. However, this approach is limited toward achieving this objective, since the selection of a route is based on the *summation* of battery capacities in a route. Therefore, it is possible to select a route with a smaller total battery capacity cost, but containing one or more nodes close to battery depletion.

The third PAR approach is min–max battery cost routing (MMBCR). MMBCR addresses the above issue by avoiding the selection of a route that contains nodes having minimal or least battery capacity comparatively speaking to the other nodes in alternative routes. Disappointingly, such a criteria could lead to selecting a route that has a higher total battery cost. Lastly, conditional max–min capacity routing (CMMBCR) is a further modification of the above by setting a battery capacity threshold value. Therefore, all nodes within a particular route must meet the threshold level. If this is true, then the route with the minimum total transmission power is selected.

While CMMBCR appears to present the more efficient PAR approach, there has been a recent modification to MTPR [65]. In [65], the authors propose a new routing strategy called Q-MTPR that couples the energy issues of MTPR with a QoS routing strategy. Briefly, Q-MTPR does not only use energy state information, but also resource state information such as bandwidth, delay, and jitter to select a preferable path.

15.6.4 PARO

Power-aware routing optimization (PARO) [66] uses a minimum transmission power routing metric, without consideration of the cost of listening and processing overhead. The primary construct of PARO is designed for static environments (e.g., sensor networks); however, application to mobile environments is permitted with additional enhancements, such as route maintenance.

PARO implementation is divided into three main areas: overhearing, redirecting, and route-maintenance. The PARO model requires that nodes be able to dynamically adjust their transmission power on a per-packet basis. Nodes must also be capable of overhearing and measuring the SNR of neighbor transmissions that are above a threshold level. Lastly, PARO assumes transmission power reciprocity between two nodes on any given link. To minimize interference, PARO assumes a MAC, such as CSMA.

Route discovery is on a per-node on-demand basis. Nodes listen to transmissions of neighbors and measure SNRs to learn MTP toward neighborhood nodes. Each node maintains an overhear table and runs an overhearing algorithm that compiles this information in order to create or refresh an entry in an overhead table. The overhead table contains the ID of the overhead node, time of the overhearing and the MPT to communicate with the neighboring node. The MTP information can be placed in headers of outgoing packets.

In conjunction with the overhearing algorithm, a redirecting algorithm is used to perform route optimization and to select new routes that require less transmission power to forward packets. The redirecting algorithm consists of two basic functions: a compute–redirect and a transmit–redirect. The compute–redirect decides whether route optimization is feasible and the transmit–redirect decides when to send route–redirect messages. Each node maintains a redirect table and entries are created when route optimization is feasible. Each entry includes source/destination IDs, creation time of the entry, previous/next-hop IDs and the total transmission power for route traversal.

Once the node finds a path that requires less transmit power, the node becomes a redirector and transmits a redirect message to the sender. The redirect message includes a new energy efficient path with an optimization percentage value (OPV). However, only one redirector between two communicating nodes can be added to a path at a time. Accordingly, PARO may require multiple iterations to attain a fully optimized route. In the event of redirector node contention, the OPV can be used by the source. Additionally, potentially contending redirecting nodes can overhear redirect messages (OPV) and refrain from sending redirect messages, thus reducing overhead.

PARO avoids transmitting signaling packets for route maintenance and primarily relies on data traffic. However, when data traffic is low between nodes, then PARO requires enhancements to the overhearing and redirecting algorithms. Overhearing is modified in that a source node will transmit route-maintenance packets toward a destination whenever there are no data packets to transmit within a route-timeout period. Redirecting is modified since mobility can affect the position of redirector nodes along a route in such a way that transmission power is not optimized. Therefore, nodes transmit route–redirect messages to any given source to overcome this problem. In this process, a node may overhear a transmission to another node that has moved. A route-redirect message informs a source to re-route its data packets toward a new next-hop node along a designated route to reach the same destination.

A recent publication in 2005 [67] concludes that PAR with BASIC-like power control (e.g., PARO) is less energy-efficient than a conventional shortest-hop routing protocol with BASIC-like power control. This finding runs contrary to previous studies. For those unfamiliar with BASIC protocol, it is similar to IEEE 802.11 where request-to-send (RTS) and clear-to-send (CTS) packets are sent at maximum transmit power. When a destination receives a RTS packet, it determines a minimum transmit power level (MTPL) and includes this information with the CTS packet before it transmits the CLS packet back to the source. Following this RTS/CTS exchange, the source and destination transmit data and acknowledgements at MTPLs.

The basic idea behind the above conclusion is that PAR will route a packet through more hops with minimum transmit power than a shortest path route to the destination. In addition, BASIC-like power control does not provide spatial reuse, thus packets travel longer hops, which create more overhead, more contention and decreased throughput. Accordingly, PARO with BASIC power control may provide less energy savings than shortest-path routing with BASIC-like power control.

15.6.5 PAWF

Power-aware weighted forwarding function (PAWF) [68] is a position-based routing protocol with power-aware capabilities for hop-by-hop routing with energy efficiency and network lifetime improvement. The power-aware forwarding function requires position and energy information of nearby neighbors and the position of the destination. PAWF assumes that nodes use a GPS device for position information. PAWF relies on beacons for the exchange of neighbor information and updates are based primarily on periodic beacons, but piggybacking and MAC layer failure feedback are used to increase accuracy. The periodic beacon includes position information, an energy consumption value and an energy residual value of the sender. Nodes also maintain a neighbor table, which includes neighbor ID, position and energy information based on the periodic beacons received. PAWF relies on two modes of operation: greedy forwarding and recovery. The forwarding function utilizes the neighbor information to select the next hop node. Figure 15.24 illustrates a simple model of PAWF.

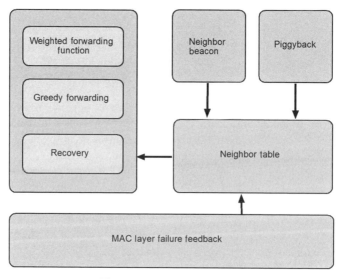

Figure 15.24 PAWF model.

15.6 POWER-AWARE ROUTING

Unlike some other PAR protocols, PAWF claims that it does not induce additional routing overhead. The energy cost added by a GPS device is considered negligible and omitted. Additionally, the energy consumed by nodes when in sleep-mode or during idling time is disregarded. PAWF nodes use only one fixed power level for transmission of packets.

PAWF ordinarily applies a greedy mode with an asymmetric exponentially weighted forwarding function, where the next hop node is chosen from a candidate set N'_c based on weighted values. Asymmetry means that the forwarding goal (i.e., equalizing the energy usage of nodes and network lifetime) and the energy consumption metric are calculated under different weight values in the forwarding function. The greedy mode operates as the candidate set is nonempty and at least one neighbor is closer to the destination than the current node. However, if the current node does not have a neighbor that can offer positive forwarding achievement, then PAWF applies a recovery strategy called the planar graph face-method used in greedy perimeter-stateless routing (GPSR). This recovery strategy forwards packets by a right-hand rule on the faces of a planar graph, which is distributed among the nodes in the network. However, the recovery strategy does not yield the shortest path nor is it power-aware. PAWF does not switch back to the greedy mode until a packet reaches a node that is closer to the destination than the node at the entrance of the recovery process.

PAWF defines two aspects of the forwarding function: forwarding achievement and candidate set of the next hop node. The forwarding achievement $Gain(n)$ of node n is defined as: $Gain(n) = Distance(c, D) - Distance(n, D)$, where c represents the current node and D represents the destination. The candidate set N'_c of the next hop node is the set of all neighbors of the current node c, which have a shorter distance (i.e., positive forwarding achievement) to D than c.

PAWF uses two energy consumption metrics that are applied to the weighted function, namely, node residual energy and node energy consumed. In the greedy forwarding mode, only nodes belonging to the candidate set N'_c qualify to be selected as the next hop. The weighted forwarding function is defined as:

$$F_C = (n_i) = \exp\left(\frac{Gain(n_i)}{MaxGain}\right)^2 + \exp\left(1 - \frac{E_c(n_i)}{MaxE_c}\right) + \exp\left(\frac{E_l(n_i)}{MaxE_l}\right) \quad (15.2)$$

where for every candidate node n_i we have $Gain(n_i)$ (previously defined); $E_c(n_i)$ denotes consumed energy; $E_l(n_i)$ denotes residual energy; and the maximum forwarding achievement, the maximum energy consumed and the maximum energy residual values of all nodes in the candidate set N'_c are denoted as MaxGain, $MaxE_c$, and $MaxE_l$, respectively. The weighted forwarding function selects the next hop node for the current node c, such that $F_C(n_{next})$ is the maximum over all n_i that are members of the candidate set N'_c.

The underlying theme of this weighted forwarding function is that energy consumption metrics and the forwarding achievement value are not applied linearly (applied linearly meaning that a small value of forwarding achievement is counterbalanced by a large residual energy value, which may result in more hops, increased delay and relay operations). Rather, a node with a larger forwarding achievement and low battery level may be selected as the next hop compared with a node with a smaller forwarding achievement and larger battery level. Thus, PAWF obtains better results in forwarding achievement while maintaining power-awareness.

15.6.6 MFP/MIP/MFP$_{energy}$/MIP$_{energy}$

Conventional routing protocols focused on power/energy conservation did not take into account other network performance metrics such as the throughput and the end-to-end delay. For ad hoc networks adopting UWB, several routing schemes with energy-aware and link-adaptive routing metrics were proposed in [69]. The ranging capability offered by UWB is utilized and adaptive modulation is applied to take advantage of favorable link conditions.

The main idea behind the energy-aware and link-adaptive routing metrics is that, by considering the link quality in the corresponding routing metrics, the inherent spatial diversity of the multihop network is efficiently exploited. Additionally, taking into account the nodes residual battery capacity results in extended network lifetime. Therefore, based on the availability of a node's location, link quality and next hop battery capacity information, the routing metrics in [69] integrate the measure of next hop remaining battery capacity with the throughput performance measures, that is, maximum forward progress (MFP) or maximum information progress (MIP).

There are four routing metrics discussed in [69], that is, MFP, MIP, MFP$_{energy}$, and MIP$_{energy}$. MFP measures the one hop throughput in terms of forward progress in the direction to the final destination with the aim of minimizing the total number of hops to the destination. MFP$_{energy}$ combines the neighbor node's remaining battery capacity with the forward progress of that node, so it can avoid the selection of nodes with lower residual battery capacities and prolong the network lifetime. MIP adapts the number of transmitted packets to the link quality and then balances the achievable next hop transmission distance and spectral efficiency. MIP$_{energy}$ additionally considers network lifetime in addition to other criteria used by MIP.

15.7 HYBRID ROUTING

A notable and perhaps natural transition from the development of purely proactive or reactive protocols to the development of hybrid protocols has taken place. Protocol designers are beginning to cherry-pick favorable attributes of one or more routing approaches, including but not limited to proactive, reactive, and geography-based, and combine them to form a hybrid routing approach. Undeniably, hybrid protocols provide greater scalability and adaptability to varying network states. This is accomplished by adjusting the contribution of each component of the hybrid protocol according to measurable and predictable network characteristics and desired performance metric(s). While it is evident that trade-offs still exist, hybrid routing attempts to manage these trade-offs in an optimal way, both in time and space.

There are two primary schemes found in hybrid protocols, namely cluster-centric and node-centric [70]. In cluster-centric schemes, clusters are formed with a static or variable periphery that serves as a partition between routing strategies. Alternately, in node-centric schemes, each node acts as a central node to an arbitrary number of surrounding nodes (i.e., an *implicit* cluster) having its own routing design.

15.7.1 MultiWARP

Multihop wireless ad hoc routing protocol (MultiWARP) [71] is a distributed protocol that tries to minimize the number of route request (RREQ) packets sent during route discovery. To minimize flooding of RREQs, MultiWARP applies the NP-complete set covering problem combined with a hybrid routing approach. The hybrid routing approach uses a proactive routing algorithm with a region awareness limiter in terms of hop-count distance.

Each node maintains a routing table and, when data needs to be transmitted, the source node searches the table for the desired destination. If the destination is present, the packets are forwarded with source-routing headers. However, if the routing table does not contain a route to the desired destination, a reactive request packet (RREQ) is generated and a limited broadcast with a TTL of 1 takes place.

To minimize the number of nodes that the RREQ propagates through to reach the destination, a set of candidate nodes that provide set cover must be determined. Based on the topology already known, nodes are selected that are one hop less than the awareness region. Any nodes that are terminating are discounted and nodes that can be reached using fewer hops are selected. These sets of candidate nodes have knowledge of all nodes within MAX-HOPs from themselves, which in turn, can determine their candidate nodes, so that the RREQ packets can be covercast to find a route to the destination. Using the set covering problem, matrices are constructed to find the optimum solution(s) or minimum subsets to completely cover the candidate nodes. Since multiple solutions are probable, for each solution the overlap of covered nodes is calculated. In this way, the selected solution is robust and will have alternate paths in place, in case links should break during the route discovery process.

RREQs include the addresses of previously visited nodes so that these nodes are not considered as candidate nodes at the next RREQ initiating node. Alternatively, to reduce packet overhead, this information can be calculated from a subsequent RREQ initiator's routing table.

For route maintenance, each node broadcasts a routing update (RUPDT) packet to its direct neighbors (i.e., one hop). Each RUPDT contains all the routes that are less than or equal to a MAX-HOP value (region awareness). Further, each route contained in the RUPDT includes an expiry counter to remove stale nodes. The expiry counter decrements by one after it is transmitted until it reaches zero and the route is expunged.

15.7.2 SHARP

Sharp hybrid adaptive routing protocol (SHARP) [72] is a highly adaptive hybrid that employs both a proactive and reactive protocol to conduct routing. A unique feature of SHARP is that each node can dynamically tailor its operation toward one of three application-specific performance metrics at the routing layer, specifically, minimizing packet overhead, bounding loss rate, and controlling delay jitter. SHARP relies on data and network characteristic measurements to dynamically adjust zone size and the amount of control overhead generated in a local region of the network.

The proactive routing component of SHARP is based on DSDV and TORA. SHARP defines a *proactive zone* around one or more nodes that are the centerpiece of data activity (hotspots). These nodes are designated as center nodes within a *proactive* zone and serve to establish a *zone radius*. As shown in Figure 15.25, nodes D and L are designated as center nodes for their respective zones. SHARP dynamically adjusts the *zone radius* of a *proactive zone* based on the degree of data activity and mobility of the network (i.e., designating a center node having high data activity with a large zone radius). All other nodes within a *proactive zone* use node-specific proactive routing to maintain a route only to the center node. Nodes that are not within a *proactive zone*, typically nodes with little or no data traffic, rely on reactive routing, to be discussed below.

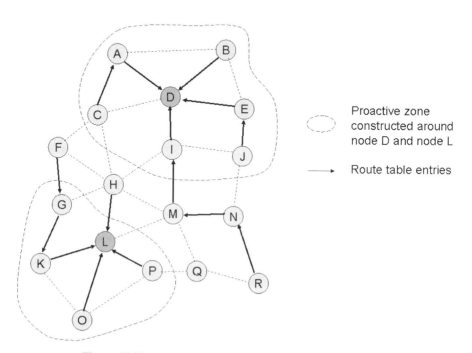

Figure 15.25 Proactive zones constructed around destinations.

SHARP employs a DAG rooted at the destination for route creation and maintenance, similar to TORA. The DAG is constructed using a *construction protocol* and maintained using an *update protocol* instituted at periodic intervals. The center node broadcasts a *construction packet*, which contains a DAG ID, zone radius, and TTL, within its *proactive zone*. Each node within the *proactive zone* is assigned a height value for routing and such nodes forward the *construction packet* once after waiting a random time interval. The *update protocol* relies on periodic beacons with a TTL of 1 for route maintenance. Lost update packets or the inability to transmit data to a neighbor serves as a detection mechanism for link failure. Failure recovery is based on TORA; however, unlike TORA, the *construction protocol* periodically reconstructs the DAG to ensure a path exists to the destination. Event-triggered control packets are used for link-failures; however, if a high periodicity for update packet transmissions exists, event-driven control overhead may be superfluous.

The reactive component of SHARP is based on AODV routing protocol, but with route caching enhancement. Nodes that reside outside a *proactive zone* rely on AODV route discovery, which requires broadcasting a route request. If a response node is within a *proactive zone*, it acts as an intermediary to the destination, setting the destination height to zero. When the responding node receives the data packets, it forwards them to the destination using proactive routing.

SHARP nodes monitor network characteristics, such as average link lifetime, average node-degree and forward this information to the center node. The center node processes this statistical information, along with data characteristics, such as the number of sources and distances thereof, to compile a new *zone radius*. The new *zone radius* is broadcast to the nodes of the new *proactive zone* before DAG reconstruction. Concurrently, to attain different application-specific goals and fulfill an optimal balance between proactive and reactive routing, SHARP utilizes different mechanisms. SHARP-PO protocol performs quantitative analysis to determine the optimum radius for *proactive zones* so as to minimize packet overhead in the network. SHARP-LR protocol is used to minimize loss rate and SHARP-DJ is used to provide an application-specific delay jitter.

15.7.3 SLURP

Scalable location update-based routing protocol (SLURP) [73] is a location-aware routing method where each node has GPS capabilities. In SLURP, the MANET is considered rectangular in shape such that it is divided into multiple home regions, each having a unique ID. Each node also has a unique address and is aware of its neighbors (e.g., MAC layer information). SLURP also requires a *static mapping f* that maps a node's ID to a particular *home region*, as in Equation (15.1):

$$f(\text{NodeID}) \longrightarrow \text{Region ID} \qquad (15.3)$$

The function f provides: (a) an even distribution of nodes within every region; (b) scalability to different geographic sizes or shapes; and (c) that the exit and entry

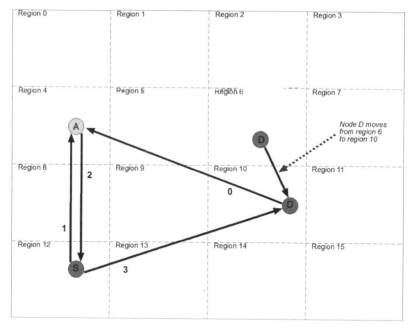

Figure 15.26 Location management. 0 = location update message sent by node D; 1 = location query message sent by node S; 2 = location reply message sent by node A; 3 = data packet sent to D's current location.

of nodes in the network are transparent. All SLURP nodes of the MANET have this mapping information. As shown below, Figure 15.26 illustrates an example network using the above-described location management of SLURP. Each SLURP node maintains a table containing every home region addresses. In this regard, SLURP is based on approximate geographical routing and a simple static mapping procedure.

SLURP requires that approximate location information about a node be maintained. To accomplish this, each node broadcasts a location update message to its fellow nodes in its home region. Each node maintains a node list of all other nodes within their region. When a node enters a new region or exits, location packets are sent. When a region is empty, the entering or exiting node broadcasts to nodes in the surrounding regions.

When the destination home region is known (calculated from GPS information), but a source node does not have topological routing tables, a most forward with fixed radius (MFR) algorithm without backwards progression is used. An example of the MFR algorithm is shown in Figure 15.27, where source node S transmits its packets to node 3, since node 3 is closest in physical distance to the destination node D.

Since the destination home region is known, a source node transmits a packet using the MFR algorithm to the center of the destination's home region. Once the

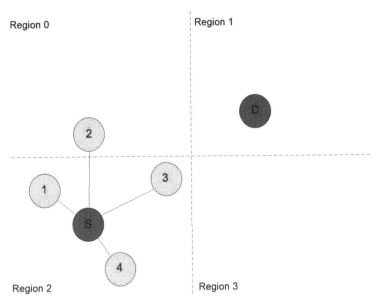

Figure 15.27 Example of MFR.

packet reaches the center node and if it has a cached route to the destination, the query packet is routed to the destination node. Otherwise, the center node broadcasts a location discovery packet (LDP) to neighboring nodes, which eventually reaches the destination. The destination generates a location reply packet (LRP) that propagates through the network and ultimately contains the necessary route information.

For location discovery, every node maintains a location cache that stores node ID, x–y coordinates, current region ID, and best neighbor for routing to a destination. When a source node wishes to send data to a destination it must transmit a LDP, which contains the destination/source IDs, current location, sequence number and level of discovery. Level of discovery relates to the extent of the broadcast. A level 1 query is designated when there is at least one reachable node in a region, whereas a level 2 query is designated when the region is empty, so all surrounding regions are considered the destination's home region. As the LDP propagates through the network toward some node in the destination's home region using MFR, each node updates its location cache with the source's location. Once a node in the destination's home region is reached, it sends a LRP that includes the destination's current region ID.

If a source node of home region A wishes to send data to destination node of home region B, the source node searches its static mapping and forwards a message to home region B to query the destination node's current location. A reply packet is sent back to the source with the destination node's location. MFR is used to forward data packets towards the destination, after which local delivery is performed with a DSR-like approach. Error messages are sent to the source node in case of delivery failures.

15.7.4 ZRP

As the name implies, zone routing protocol (ZRP) [74] divides the network into different zones on a node-by-node basis and employs a flat addressing scheme, which allows zone overlap. Synonymous with SHARP, ZRP employs proactive routing when the destination is within a zone and reactive routing when the destination is outside a zone. ZRP proposes two zone sizing schemes and includes an extension for unidirectional routing. ZRP also introduces a concept called *bordercasting*, implemented by a bordercast resolution protocol (BRP), to forward route queries to border nodes of a zone.

Each zone radius is measured in hops from a center node. ZRP categorizes nodes of a zone as either interior or peripheral (i.e., border nodes). ZRP accommodates any proactive link-state routing protocol termed intra-zone routing protocol (IARP), and any reactive protocol termed inter-zone routing protocol (IERP), for network connectivity. This is an important attribute of ZRP, since it does not specify a particular protocol, but allows for flexible implementation. However, there must be a certain level of commonality between the IERP and IARP. For example, IERP must support IARP routing table lookup. Also, IARP link state metrics should be compatible with the IERP. Regardless of the selected protocols employed, the IARP maintains routes within a zone and the IERP provides route discovery and maintenance outside a zone.

When a source has data to send it first uses IARP information. However, if a route does not exist, then the source uses BRP to *bordercast* route request packets to the border nodes of the zone. This can be done as a multicast to avoid unnecessary flooding and reduce control overhead. If a border node knows a route, it responds with a reply packet. If a route is unknown, the border node acts as a center node and *bordercasts* route queries to its border nodes. This process continues until a node with destination route information is found. The node responds by sending a reply back to the source. Route information in the reply packet can either by next-hop addresses or a complete route (e.g., source routing). An example of bordercasting is shown in Figure 15.28.

BRP controls route queries initiated by IERP to peripheral nodes during a *bordercast*. The source node can compute a multicast (bordercast) tree or this can be done at each node within a zone provided it knows the topology as known by the source. Since there may be situations where queries could backtrack to previously visited nodes or covered routing zones (zones that already received the query), BRP provides two levels of query detection to remove redundancy and provide maximum efficiency. The first-level query detection is QD1 and the second-level query detection is QD2. QD1 relates to multiple channel networks. Thus, nodes actually relay a query and are able to store source address and query ID in a query detection table. QD2 relates to single-channel networks where nonrelaying query nodes can listen to

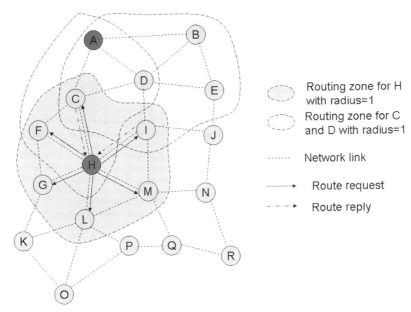

Figure 15.28 Route finding from node H to node A.

the query traffic of its relaying neighbor nodes that are within its radio transmission range. ZRP also supports something called early termination (ET), which allows a node to prevent a route request from entering a covered region.

ZRP relies on any neighbor discovery protocol (NDP) supported by the MAC layer to detect new neighborhood nodes and link failures. Periodic hello beacons can be used. If the MAC layer does not support an NDP, then the IARP must manage this function. ZRP relies on its knowledge of local topology within a zone for route maintenance. If a link failure or suboptimal path exists, data packets can be redirected, thus providing more robustness.

As mentioned above, ZRP proposes two traffic adaptive zone-resizing schemes: min searching and IARP:IERP ratio. The periodicity of zone-resizing is not mentioned in the literature. However, either scheme presents a rather computationally intensive framework since so many factors can be considered in zone resizing. Nevertheless, providing a dynamic rather than a static zone radius for each node can be beneficial in striking a balance between proactive/reactive routing. Under min searching, the routing zone is incremented or decremented by 1 depending on the total minimum ZRP traffic (i.e., IARP and IERP traffic). The IARP/IERP ratio scheme simply compares the ratio of traffic to a threshold and depending on the outcome of the comparison the zone size is increased or decreased. To prevent frequent zone radius adaptation, a multiplicative hysteresis value is used to improve network stability.

15.7.5 AZRP

Adaptive zone routing protocol (AZRP) [75] is an extension of an older 2002 Internet Draft version of ZRP and not the newer version of ZRP discussed above. However, AZRP provides an adaptive means for zone resizing that is distinct from the two zone-resizing schemes mentioned above. In AZRP, the zone radius varies based on a rate of packet loss metric reported by the IERP and the density and number of nodes inside the zone. The number of nodes inside the zone is used to predict a predetermined number of nodes outside the zone boundaries (e.g., three hops away). This information is then used to predict a future route failure rate. The algorithm generates a hops-weighted table where a lesser weight value is assigned to a route failure that is farther from the zone boundary than a route failure closer to the zone boundary. AZRP takes into account the past route failure rate within a variable period of time T and soft thresholds to determine whether the zone radius is increased or decreased by 1. The periodicity of zone resizing is variable depending on network traffic conditions and node mobility.

15.7.6 IZR

Independent zone routing (IZP) [76] adopts many aspects of ZRP with one significant modification. While ZRP [74] briefly mentions combining the two resizing schemes, IZP describes the approach in more detail. IZP combines the min searching and adaptive traffic estimation schemes (as previously described in ZRP) for zone resizing. Upon initialization, a zone radius is set to 1 and min searching is applied to determine the optimal zone radius that results in the least control overhead. A threshold is then set by the node that corresponds to the ratio of both the reactive and proactive components at the optimal zone radius. Once the threshold is established, the adaptive traffic estimation scheme is used to adjust the radius in relation to the threshold. The adaptive traffic estimation scheme monitors the varying network characteristics to determine the proper adjustment. However, when the adaptive traffic algorithm is used, it is possible if the network is static and a high call rate exists or a high rate of node mobility in relation to the call rate exists that extreme zone radii may be selected, resulting in purely proactive or reactive routing. To prevent this from happening, the adaptive algorithm should hand over to the min searching algorithm either periodically or on an event-driven basis (e.g., when an extreme change in the network characteristics takes place).

15.7.7 TZRP

Two-zone routing protocol (TZRP) [70] is an extension to ZRP that aims to reduce inaccurate topology information, especially during times of high mobility. Each TZRP node is a center node for two different type zones: a *crisp zone* and a *fuzzy zone*. The radius of the *fuzzy zone* is always larger, in terms of hops, than the radius of the *crisp zone*. This two-zone system allows TZRP the ability to decouple traffic characteristics from node mobility by adjusting independently the size of each zone. As a result, TZRP can reduce total control overhead. Specifically, the *crisp zone* can be used to manage routing overhead due to mobility, while the *fuzzy zone* can be used to manage routing overhead due to traffic pattern, as explained below.

A node maintains the topology of the *crisp zone* proactively, while the *fuzzy zone* is managed using a *fuzzy-sighted* proactive approach (e.g., hazy sighted link state, HSLS). Since the *crisp zone* is maintained proactively, TZRP provides effective bordercasting. The perimeter of the *crisp zone* divides proactive routing and fuzzy proactive routing, while the perimeter of the *Fuzzy zone* divides proactive routing and reactive routing. When the network experiences low-mobility, the *crisp zone* can be proactively maintained with relatively little control overhead. However, during high-mobility periods, the amount of control overhead to maintain the *crisp zone* is greatly increased. In ZRP, this leads to reducing the radius of the *proactive zone* and increasing the number of nodes for *reactive routing*. In TZRP, the *fuzzy zone* is kept large and used to provide fuzzy proactive routing, since for example, HSLS is basically independent of node mobility patterns and is long-timer-based compared with the short-timer-based implementation of the *crisp zone*. Thus, TZRP uses a short-timer based implementation as a crisp IARP and an HSLS-based implementation as a fuzzy IARP, to provide the decoupling of traffic pattern from node mobility and to provide event-driven IARP and timer-based IARP, respectively.

TZRP nodes can generate *crisp* LSUs and *fuzzy* LSUs. Each node x counts the current time in T s and wakes up every t_S s, and calculates the largest positive integer i such that $T \bmod(2^{i-1} \times t_e) = 0.1$. If a positive integer i exists and there was a link change during the last $(2^{i-1} \times t_e)$ s, then node x is sending a *fuzzy* LSU with a TTL $= L$. If $L < ZR_C - 1$, then L is set to $ZR_C - 1$ and if $L > ZR_F - 1$, then L is set to $ZR_F - 1$. Otherwise, if a positive integer i does not exist, node x checks if a link change occurred during the last t_S s. If so, node x is sending a *crisp* LSU with a TTL $= ZR_C - 1$.

The *crisp zone* is maintained when LSUs are sent. When, for example, $ZR_C \geq 2$, each node will send LSUs with TTL $= 1$ to maintain shortest paths to every two-hop neighbor. Nodes select the minimum number of one-hop neighbors to forward the LSU from a *forwarding set*, which is included in the LSU. When a node receives the LSU, it updates its link state table, and forwards the LSU, if the TTL is greater than 0 and it appears in the *forwarding set*, otherwise it discards the LSU. The intrazone routing table uses link state table information for computing and updating shortest paths to known neighbors. If a node's shortest paths equal ZR_C, then these paths constitute the node's *bordercast tree*.

When a source node wants to send data to a destination node, it initially relies on a crisp IARP route or a fuzzy IARP. However, if the destination node is unreachable, then a *reactive bordercasting* procedure is invoked. Specifically, a node will broadcast a route request packet (RREQ) to all its one-hop neighbors. A neighbor node in the *forwarding set* will determine if it has a route to the destination, and if it does then it will unicast the RREQ to the destination and the destination will respond with a route reply packet (RREP) to the source node. Otherwise the node will forward the RREQ packet based on its own bordercast tree, marking all nodes that are ZR_C or fewer hops from the source node as *covered*, and including its own *forwarding set* in the RREQ packet.

TZRP primarily depends on periodic HELLO beacons to detect new and broken links. However, the MAC layer may also provide like functions. If a link breakage occurs on an active route, then the upstream node checks whether the destination can be reached using a crisp or fuzzy IARP route. If so, then the route is repaired locally. Otherwise, a route error message (RERR) is sent back to the source node.

Performance comparisons between TZRP and ZRP show that TZRP significantly reduces reactive control overhead and provides a much-improved balance between proactive/reactive control overhead under many (ZR_C, ZR_F) settings. However, the authors of [70] indicate that more research needs to be done to implement efficient means to dynamically adapt the crisp/fuzzy zone radii.

15.8 OTHER

In addition to the routing protocols described and categorized above, there are also many other protocols proposed to meet the challenges of routing in ad hoc networks. By taking into account the different issues of ad hoc networks, such as time-variant channel properties, limited bandwidth, mobility of nodes and potential multiple paths among nodes, routing protocols can be designed to further improve network performance. Accordingly, we present a brief introduction to this type of routing protocol.

The quality of the channels among the mobile nodes is inevitably time-variant due to mobility of nodes and signal propagation effects such as shadowing and small-scale fading. Therefore, the link quality in an identified route may become worse and even no longer usable due to this time-variant channel property. Exploitation of link quality information in routing protocols can allow the protocols to dynamically adapt to the change of channel conditions, resulting in improved performance. In [77], two channel adaptive protocols termed RICA (receiver-initiated channel adaptive) and BGCA (bandwidth guarded channel adaptive) have been proposed. The main idea of BGCA is that, when a link is in deep fading, the upstream node (node nearer to the source node than to the destination node) will execute a local search to find a partial route to the destination with better link quality. BGCA is designed based on AODV, and also incorporates some ingredients of ABR, for example, the local search for a partial route and selection of the shortest route. The major feature of RICA is to make use of the time-variant channel property and let the routing between a pair of source and destination nodes adapt to the channel state information of the whole route. The primary difference between BGCA and RICA relating to channel adaptation is that in BGCA only a few links with worse qualities will be changed, but in RICA the entire route between the source and destination nodes will be reconstructed when suffering worse channel conditions.

BGCA and RICA are both end-to-end routing protocols, which implies that they may not be able to adapt to rapid changes in link qualities. In [78] a next-hop channel adaptive routing protocol was proposed by considering both the spatial and temporal diversity of ad hoc networks. Multiple next hop alternatives during routing provide spatial diversity and utilization of channel state information in routing provides temporal diversity.

The multihop transmission nature of ad hoc networks leads to possible multipaths existing between the source and destination nodes. Therefore, routing data over multiple disjoint paths seems the logical choice because it minimizes the diminishing effects of unreliable links and a constantly changing topology [79]. Admittedly, some protocols supporting multipath routing have already been described and explained above, such as DSR, TORA, CHAMP, NDMR, and MPABR. However, there are still many other multipath routing protocols proposed for various reasons, such as load-balancing, congestion avoidance, lower frequency of route inquiries, and to achieve a lower overall routing overhead [80, 81].

15.9 CONCLUSION

While this chapter presents an overview of new and old routing protocols for MANETs, there are many other protocols that have been omitted. Indeed, the wireless community is continuing to be overwhelmed with literature regarding the development of routing protocols for MANETs. Notably, there are certain topics that are being studied and appear in the literature with greater frequency than other topics. Based on our own personal observations, there has been a steady increase in the number of publications that relate to multipath routing [80], multicasting [82], location-aware routing [83, 84] and channel adaptive routing [77, 78]. However, it is our opinion, that there are more fundamental issues relating to protocol design that should be discussed and brought to the forefront in the literature. The remainder of this chapter will briefly broach these issues.

UWB has received increasing attention for its broader applicability to telecommunication systems. It is a promising field to create small, low-cost and high-data-rate transceivers that could be connected to each other for information sharing and exchange. In this context, the UWB devices will form distributed and self-organizing multihop UWB ad hoc networks. Design of routing protocols for efficient data forwarding and reliable transmission in such networks becomes necessary and important. This chapter provides an overview of routing protocols for MANETs and explains the main concept of each protocol in detail with illustration. While it is understood that all of the routing protocols discussed can be directly applied to ad hoc networks adopting UWB technology, greater optimizations between the lower layers and the routing protocol may be achieved. Clearly, two aspects of a routing protocol's design and optimization in an ad hoc network employing UWB should be given special attention, that is, location information and energy efficiency. In this regard, the cross-layer design of routing protocols may present more optimal path selection and an accretion in overall network performance. Additionally, there are still some questions about routing protocols that remain unanswered, such as security and quality of service, even after so many answers are suggested by the work represented in this chapter. We believe that future work is needed to resolve these other issues.

Another issue relates to the need for protocol designers to incorporate real testbeds and field tests in the process of protocol development and performance evaluation. The authors have uncovered recent exemplary articles that vocalize the notion that, while computer simulations and emulators have a utility, their application in the ultimate development and measure of performance is limited [85–87]. Undoubtedly, the MANET research community is cognizant of the limitations associated with simulation testing (e.g., unrealistic movement scenarios, simplified models). Nevertheless, simulation testing has almost become the only practice, resulting in an unfortunate situation where scientific credibility is being compromised by convenience. While the authors recognize that there are a multitude of reasons for this practice, it is important that routing protocol designers strive to gradually distance themselves from this practice as an end-all, so that more accurate and useful performance data can be gleaned. This can only occur when a transition towards ad hoc testbeds occurs. The authors believe that more research and development needs to be initiated so that protocol designers have at their disposal the necessary tools that more closely mirror actual implementation. In the meantime, protocol designers should be cognizant that the majority of researchers rely on suboptimal testing methods and that published results are as credible as the weakest link, which happens to be current modeling, simulation, and emulation tools.

Another issue that, to the authors' knowledge, has not been previously discussed in the literature is for protocol designers to adopt new benchmarks for performance comparisons. Having surveyed over 100 publications relating to routing protocol proposals, it is peculiar that there are a select number of protocols that have attained the level of benchmarks, despite the fact that most performance comparisons made with these select protocols are left wanting to some degree. If protocol designers are aiming to establish dominance in the wireless community by running simulations and publishing performance comparison results, it would seem scientifically pragmatic not to select protocols whose performance have already been continually surpassed by other protocols. Clearly some protocols have gained some notoriety in the literature, and perhaps deservedly so based on performance issues. Nevertheless, as new protocols are being developed and heldup for comparison, we should never lose sight of the bottom line: quantitative analysis.

The development of routing protocols for MANETs is burgeoning. It is evident that different demands placed on an ad hoc protocol make it nearly impossible to develop a single protocol that could be applied to all applications and scenarios. However, any ad hoc protocol needs to be adaptive and self-configurative to a certain degree. Since routing in MANETs is an important aspect in this new paradigm shift in wireless communications, we are obliged to advance our scientific research with a new evolution of experimental tools for testing and evaluating routing protocols.

REFERENCES

1. J. Hoebeke, I. Moerman, B. Dhoedt, and P. Demeester "An overview of mobile ad hoc networks: Applications and challenges," *The Journal of the Communications Network*, vol. 3, no. 3, pp. 60–66.
2. Merrian–Webster Online Dictionary; www.m-w.com (accessed 5 February 2005).
3. Huang, Crowcroft, and Wassell, "Rethinking incentives for mobile ad hoc networks," *SIGCOMM'04 Workshops*, pp. 191–196.
4. N. Abramsson "The ALOHA system—another alternative for computer communications," in *AFIPS Conf. Proc.*, vol. 37, FJCC, 1970, pp. 695–702.
5. M. Frodigh, P. Johansson, and P. Larsson, "Wireless ad hoc networking—the art of networking without a network," *Ericsson Review*, vol. 4, 2000, pp. 248–263.
6. S. Korotygin "Development of wireless network technologies: IEEE 802.11 standard"; www.digit-life.com/articles/wlan/index.html (accessed 18 February 2005).
7. Shiflet, Belding-Royer, and Perkins, "Address aggregation in mobile ad hoc networks," *IEEE Communications Society*, 2004, pp. 3734–3738.
8. P. Samar and Z. Haas, "Strategies for broadcasting updates by proactive routing protocols in mobile ad hoc networks," *MILCOM 2002 Proceedings*, vol. 2, 2002, pp. 873–878.
9. L. De Nardis, G. Giancola, M.-G. Di Benedetto, "A power-efficient routing metric for UWB wireless mobile networks," *Vehicular Technology Conference*, vol. 5, October 2003, pp. 3105–3109.
10. C. E. Perkins and P. Bhagwat, "Highly dynamic destination sequenced distance-vector routing (DSDV) for mobile computers," *Proceedings of the SIGCOMM '94 Conference on Communications Architectures, Protocols and Applications*," August 1994, pp. 234–244; available at http://people.nokia.net/~charliep/ (accessed 19 February 2005).
11. J. Broch, D. A. Maltz, D. B. Johnson, Y.-C. Hu, and J. Jetcheva, "A performance comparison of multi-hop wireless ad hoc network routing protocols," *Proceedings of the 4th Annual ACM/IEEE International Conference on Mobile Computing and Networking*, October 1998, pp. 85–97.
12. L. R. Ford Jr and D. R. Fulkerson, *Flows in Networks*, Princeton University Press, Princeton, NJ, 1962.
13. R. E. Bellman, *Dynamic Programming*, Princeton University Press, Princeton, NJ, 1957.
14. S. Murthy and J. J. Garcia-Luna-Aceves, "An efficient routing protocol for wireless networks," *ACM Mobile Networks and Applications Journal*, Special Issue on Routing in Mobile Communication Networks, October 1996, pp. 183–197.
15. J. Raju and J. J. Garcia-Luna-Aceves, "A comparison of on-demand and table driven routing for ad-hoc wireless networks," *IEEE International Conference on Communications*, vol. 3, 2000, pp. 1702–1706.
16. C.-C. Chiang and M. Gerla, "Routing and multicast in multihop, mobile wireless networks," *IEEE 6th International Conference on Universal Personal Communications Record*, 1997, pp. 546–551.
17. C.-C. Chiang, H.-K. Wu, W. Liu and M. Gerla, "Routing in clustered multihop, mobile wireless networks with fading channel," *Proceedings of IEEE SICON*, April 1997, pp. 197–211.
18. J. J. Garcia-Luna Aceves and M. Spohn, "Source-tree routing in wireless networks," *Seventh IEEE International Conference on Network Protocols*, 1999, pp. 273–282.

19. J. J. Garcia-Luna-Aceves and M. Spohn, "Efficient routing in packet-radio networks using link-state information," *IEEE Wireless Communications and Networking Conference*, vol. 3, 1999, pp. 1308–1312.
20. Hong Jiang "Simulation of source tree adaptive routing protocol (STAR)", 2000, pp. 1–9; http://citeseer.ist.psu.edu/cs (accessed 22 February 2005).
21. G. Pei, M. Gerla, X. Hong and C.-C. Chiang, "A wireless hierarchical routing protocol with group mobility," *IEEE Wireless Communications and Networking Conference*, vol. 3, 1999, pp. 1538–1542.
22. A. Iwata, C.-C. Chiang, G. Pei, M. Gerla, and T.-W. Chen, "Scalable routing strategies for ad hoc wireless networks," *IEEE Journal on Selected Areas in Communications*, vol. 17, no. 8, 1999, pp. 1369–1379.
23. P. Jacquet, P. Muhlethaler, T. Clausen, A. Laouiti, A. Qayyum, and L. Viennot, "Optimized link state routing protocol for ad hoc networks," *Technology for the 21st Century Proceedings, IEEE International*, 2001, pp. 62–68.
24. M. Benzaid, P. Minet, and K. A. Agha "Analysis and simulation of fast-OLSR," *The 57th IEEE Semiannual Vehicular Technology Conference*, 2003, vol. 3, pp. 1788–1792.
25. L. Lamont, M. Wang, L. Villasenor, T. Randhawa, R. Hardy, and P. McConnel, "An Ipv6 and OLSR based architecture for integrating WLANs and MANETs to the Internet," *The 5th International Symposium on Wireless Personal Multimedia Communications*, 2002, vol. 2, pp. 816–820.
26. M. Benzaid, P. Minet, and K. A. Agha, "Performance evaluation of the implementation integrating mobile-IP and OLSR in full-IP networks," *IEEE Wireless Communications and Networking Conference*, vol. 3, 2004, pp. 1697–1702.
27. Request For Comments 3684, February 2004 at www.ietf.org/rfc/rfc3684.txt (accessed 24 February 2005).
28. B. Bellur and R. G. Ogier, "A reliable, efficient topology broadcast protocol for dynamic networks," *Eighteenth Annual Joint Conference of the IEEE Computer and Communication Societies*, vol. 1, 1999, pp. 178–186.
29. S. Basagni, I. Chlamtac, V. R. Syrotiuk, and B. A. Woodward, "A distance routing effect algorithm for mobility (DREAM)," *Proceedings of the Fourth Annual ACM/IEEE International Conference on Mobile Computing and Networking*, 1998, pp. 76–84.
30. T. Camp, J. Boleng, B. Williams, L. Wilcox, and W. Navidi, "Performance comparison of two location based routing protocols for ad hoc networks," *Twenty-First Annual Joint Conference of the IEEE Computer and Communications Societies*, vol. 3, 2002, pp. 1678–1687.
31. D. Son, A. Helmy, and B. Krishnamachari, "The effect of mobility-induced location errors on geographic routing in mobile ad hoc and sensor networks: analysis and improvement using mobility prediction," *IEEE Transactions on Mobile Computing*, vol. 3, no. 3, 2004, pp. 233–245.
32. G. Pei, M. Gerla, and T.-W. Chen, "Fisheye state routing: a routing scheme for ad hoc wireless networks," *2000 IEEE International Conference on Communications*, vol. 1, pp. 70–74.
33. G. Dimitriadis and F. N. Pavlidou "Clustered fisheye state routing for ad hoc wireless networks," *4th International Workshop on Mobile and Wireless Communications Network*, 2002, pp. 207–211.
34. W. Liang and Z. Nai-tong, "Adaptive routing table update scheme of the scalable routing in large ad hoc networks," *9th Asia-Pacific Conference on Communications*, vol. 2, 2003, pp. 595–599.

35. E. Johansson, K. Persson, M. Skold, and U. Sterner, "An analysis of the fisheye routing technique in highly mobile ad hoc networks," *59th IEEE Vehicular Technology Conference*, 2004, pp. 2166–2170.
36. N. Prabagarane, C. A. Navin, B. Partibane, V. Nagarajan, and R. Krishnakiran, "Hierarchical routing algorithm for cluster-based multihop mobile adhoc network," *IEEE Wireless Communications and Networking Conference*, vol. 2, 2004, pp. 1116–1120.
37. C. A. Santivanez and R. Ramanathan, "Hazy sighted link state (HSLS) routing: a scalable link state algorithm," Internetwork Research Deptartment, 31 August 2001 (revised March 2003); www.cuwireless.net/OSI/progress_report.html (accessed 1 March 2005).
38. D. B. Johnson, D. A. Maltz, and Y.-C. Hu, "The dynamic source routing protocol for mobile ad hoc networks"; www.ietf.org/internet-drafts/draft-ietf-manet-dsr-10.txt> IETF MANET Working Group, Internet Draft, 19 July 2004, expired 19 January 2005 (visited 18 February 2005).
39. D. A. Maltz, J. Broch, J. Jetcheva, and D. B. Johnson, "The effects of on-demand behavior in routing protocol for multihop wireless Ad hoc network," *IEEE Journal on Selected Areas in Communication*," vol. 17, no. 8, August 1999, pp. 1439–1453.
40. M. Gunes, U. Sorges, and I. Bouazizi, "ARA—the ant-colony based routing algorithm for MANETs," *ICPP Workshop on Ad hoc Networks*, 2002, pp. 79–85.
41. J. S. Baras and H. Mehta, "A probabilistic emergent routing algorithm for mobile ad hoc networks," *Proceedings at WiOPT03: Modeling and Optimization in Mobile, ad hoc and Wireless Networks*, 2003.
42. F. Ducatelle et al., "Ant agents for hybrid multipath routing in mobile ad hoc networks," *Proceedings of the Second Annual Conference on Wireless On-demand Network Systems and Services*, IEEE, New York, 2005.
43. C. Toh "A novel distributed routing protocol to support ad-hoc mobile computing," *IEEE 15th Annual International Phoenix Conference*, 1996, pp. 480–486.
44. C.-K. Toh, "Long-lived ad hoc routing based on the concept of associativity," Internet-Draft, March 1999 (expired November 1999), pp. 1–15; www.ietf.org/proceedings/99nov/I-D/draft-ietf-manet-longlived-adhoc-routing-00.txt. (See also http://cktoh.1accesshost.com).
45. S. Das, C. Perkins, and E. Royer, Ad hoc on demand distance vector (AODV) routing, Internet Draft (expired 19 April 2003); www.draft-eitf-manet-aodv-11.txt
46. M. Tauchi, T. Ideguchi, and T. Okuda, "Ad-hoc routing protocol avoiding route breaks based on AODV," *Proceedings of the 38th IEEE Hawaii International Conference on System Sciences*, 2005, pp. 1–7.
47. S. Guo and O. W. Yang "Performance of backup source routing in mobile ad hoc networks," *Wireless Communications and Networking Conference*, vol. 1, 2002, pp. 440–444.
48. A. Valera, W. K. G. Seah, and S. V. Rao, "CHAMP: a highly resilient and energy-efficient routing protocol for mobile ad-hoc networks," *4th International Workshop on Mobile and Wireless Communications Network*, 2002, pp. 43–47.
49. I. Chakeres et al., "Dynamic MANET on-demand routing protocol (DYMO)," draft-ietf-manet-dymo-00 (work in progress, expired 5 July 2005); www.ietf.org/internet-drafts/draft-ietf-manet-dymo-00.txt (accessed 24 February 2005).
50. Dynamic Nix-Vector Routing (DNVR); 13–17 March 2005 (pending publication in IEEE).

51. B. S. Manoj, R. Ananthapadmanabha, and C. S. R. Murthy, "Link life based routing protocol for ad hoc wireless networks," *10th International Conference on Computer Communications and Networks*, 2001, pp. 573–576
52. (a) P. McCarthy and D. Grigoras, "Multipath associativity based routing," *Proceedings of the Second Annual Conference on Wireless On-demand Network Systems and Services*, 2005.
52. (b) Y.-B. Ko and N. H. Vaidya, "Location-aided routing (LAR) in mobile ad hoc networks," *Wireless Networks*, vol. 6, 2000, pp. 307–321.
53. X. Li and L. Cuthbert, "Stable node-disjoint multipath routing with low overhead in mobile ad hoc networks," *The IEEE Computer Society's 12th Annual International Symposium on MASCOTS*, 2004, pp. 184–191.
54. R. S. Sisodia, I. Karthigeyan, B. S. Manoj, and C. S. R. Murthy, "A preferred link based multicast protocol for wireless mobile ad hoc networks, *IEEE International Conference on Communications*, 2003, vol. 3, pp. 2213–2217.
55. G. Aggelou and R. Tafazolli, "RDMAR: a bandwidth-efficient routing protocol for mobile ad hoc networks," *ACM International Workshop on Wireless Mobile Multimedia*, 1999, pp. 26–33.
56. S. Roy and J. J. Garcia-Luna-Aceves, "Using minimal source trees for on-demand routing in ad hoc networks," *IEEE INFOCOM 2001*, pp. 1172–1181.
57. V. D. Park and M. S. Corson, "A performance comparison of the temporally-ordered routing algorithm and ideal link-state routing," *Proceedings of Third IEEE Symposium on Computers and Communication*, 1998, pp. 592–598.
58. V. D. Park and M. S. Corson, "A highly adaptive distributed routing algorithm for mobile wireless networks," *IEEE Sixteenth Annual Joint Conference of the IEEE Computer and Communication Societies, INFOCOM'97*, April 1997, vol. 3, pp. 1405–1413.
59. M. Krunz, A. Muqattash, and S.-J. Lee, "Transmission power control in wireless ad hoc networks: challenges, solutions, and open issues," *IEEE Network*, vol. 18, no. 5, 2004, pp. 8–14.
60. J. Nie and Z. Zhou, "An energy based power-aware routing protocol in ad hoc networks," *IEEE International Symposium on Communications and Information Technology*, vol. 1, 2004, pp. 280–285.
61. B. Awerbuch, D. Holmer, and H. Rubens, "The pulse protocol: mobile ad hoc network performance evaluation," *Proceedings of the Second Annual Conference on Wireless On-demand Network Systems and Services*, 2005.
62. C. F. Chiasserini and R. R. Rao, "Routing protocols to maximize battery efficiency," *21st Century Military Communications Conference*, vol. 1, 2000, pp. 496–500.
63. T. X. Brown, S. Bhandare, and S. Doshi, The energy aware dynamic source routing protocol, Internet Draft (expired December 2003); http://ftp.ist.utl.pt/pub/drafts/draft-brown-eadsr-00.txt
64. C.-K. Toh, "Maximum battery life routing to support ubiquitous mobile computing wireless ad hoc networks," *IEEE Communications Magazine*, vol. 39, no. 6, pp. 138–147.
65. Z. Zhou and Y. Mao, *A New QOS Routing Scheme In Mobile ad hoc Network—Q-MTPR*. IEEE, New York, 2004, pp. 389–393.
66. J. Gomez et al., "PARO: supporting dynamic power controlled routing in wireless ad hoc networks," *Wireless Networks*, vol. 9, 2003, pp. 443–460; http://comet.ctr.columbia.edu/~campbell/papers/winetparo.pdf

67. E.-S. Jung and N. H. Vaidya, "Power aware routing using power control in ad hoc networks," February 2005; www.crhc.uiuc.edu/wireless/papers/pcmr-tech.pdf
68. J. Shen and J. Harms, "Position-based routing with a power-aware weighted forwarding function in MANETs," *IEEE International Conference on Performance, Computing and Communications*, 2004, pp. 347–355.
69. J. Xu, B. Peric, and B. Vojcic, "Energy-aware and link-adaptive routing metrics for ultra wideband sensor networks," *Second International Workshop on Networking with Ultra Wide Band Workshop on Ultra Wide Band for Sensor Networks*, Rome, July 2005, pp. 1–8.
70. L. Wang and S. Olariu, "A two-zone hybrid routing protocol for mobile ad hoc networks," *IEEE Transactions on Parallel and Distributed Systems*, vol. 15, no. 12, 2004, pp. 1105–1116.
71. S. M. Van Der Werf and K.-S. Chung, "Multi-hop wireless ad-hoc routing protocol (MultiWARP)," *5th International Symposium on Multi-Dimensional Mobile Communications*, vol. 2, 2004, pp. 961–965.
72. V. Ramasubramanian, Z. J. Hass, and E. G. Sirer, "SHARP: a hybrid adaptive routing protocol for mobile ad hoc networks," *Proceedings of the 4th ACM International Symposium on Mobile ad hoc Networking and Computing*, 2003, pp. 303–314.
73. S.-C. M. Woo and S. Singh, "Scalable routing protocol for ad hoc networks," *Wireless Networks*, vol. 7, 2001, pp. 513–529.
74. Z. J. Haas, M. R. Pearlman, and P. Samar, "The zone routing protocol (ZRP) for ad hoc networks," Internet Draft (expired January 2003); www.ietf.org/proceedings/02nov/I-D/draft-ietf-manet-zone-zrp-04.txt
75. S. Giannoulis et al., "A hybrid adaptive routing protocol for ad hoc wireless networks," *IEEE International Workshop on Factory Communication Systems*, 2004, pp. 287–290.
76. P. Samar, M. R. Pearlman, and Z. J. Haas, "Independent zone routing: an adaptive hybrid routing framework for ad hoc wireless networks," *IEEE/ACM Transactions on Networking*, vol. 12, no. 4, 2004, pp. 595–608.
77. X.-H. Lin, Y.-K. Kwok, and V. K. N. Lau, "A quantitative comparison of ad hoc routing protocols with and without channel adaptation," *IEEE Transactions on Mobile Computing*, vol. 4, no. 2, March/April 2005, pp. 111–128.
78. M. R. Souryal, B. R. Vojcic, and R. L. Pickholtz, "Information efficiency of multihop packet radio networks with channel-adaptive routing," *IEEE Journal on Selected Areas in Communications*, vol. 23, no. 1, January 2005, pp. 40–50.
79. A. Tsirigos and Z. J. Haas, "Analysis of multipath routing—Part I: the effect on the packet delivery ratio," *Wireless Communications, IEEE Transactions*, vol. 3, no. 1, January 2004, pp. 138–146.
80. S. Mueller and D. Ghosal, "Analysis of a distributed algorithm to determine multiple routes with path diversity in ad hoc networks," *3rd International Symposium on Modeling and Optimization in Mobile, Ad hoc, and Wireless Networks*, April 2005, pp. 277–285.
81. Z. Ye, S. V. Krishnamurthy, and S. K. Tripathi, "A framework for reliable routing in mobile ad hoc networks," *INFOCOM 2003. Twenty-Second Annual Joint Conference of the IEEE Computer and Communications Societies*, vol. 1, 30 March to 3 April 2003, pp. 270–280.
82. Y. Sasson, D. Cavin, and A. Schiper, "A location service mechanism for position-based multicasting in wireless mobile ad hoc networks," *Proceedings of the 38th Annual Hawaii International Conference on System Sciences*, January 2005.

83. L. Blazevic, J.-Y. L. Boudec, and S. Giordano, "A location-based routing method for mobile ad hoc networks," *IEEE Transactions on Mobile Computing*, vol. 4, no. 2, March/April 2005, pp. 97–110.
84. T. Park and K. G. Shin, "Optimal tradeoffs for location-based routing in large scale ad hoc networks," *IEEE/ACM Transactions on Networking*, vol. 13, no. 2, April 2005, pp. 398–410.
85. D. Kotz, C. Newport, R. S. Gray, J. Liu, Y. Yuan, and C. Elliott, "Experimental evaluation of wireless simulation assumptions," Dartmouth Computer Science Technical Report TR2004-507, June 2004, pp. 1–20.
86. E. Nordström, P. Gunningberg, and H. Lundgren, "A testbed and methodology for experimental evaluation of wireless mobile ad hoc networks," *Proceedings of the First International Conference on Testbeds and Research Infrastructures of the Development of Networks and Communities (TRIDENTCOM)* 2005, pp. 1–10.
87. I. D. Chakers and E. M. Belding-Royer, "AODV routing protocol implementation design," *Proceedings of the 24th International Conference on Distributed Computing Systems Workshop (ICDCSW'04)*.
88. P. Trakadas, Th. Zahariadis, S. Voliotis, and Ch. Manasis "Efficient routing in PAN and sensor networks," Special Issue on Wireless Pan and Sensor Networks, *ACM SIGMOBILE Mobile Computing and Communication Review*, vol. 8, no. 1, 2004, pp. 10–17.
89. M. Abolhasan, T. Wysocki, and E. Dutkiewicz "A review of routing protocols for mobile ad hoc networks," *Ad hoc Networks*, vol. 2, 2004, pp. 1–22.
90. L. A. Latiff and N. Fisal "Routing protocols in wireless mobile ad hoc network—a review," *APCC 2003. The 9th Asia-Pacific Conference on Communications*, vol. 2, 2003, pp. 600–604.
91. E. M. Royer and C.-K. Toh, "A review of current routing protocols for ad hoc mobile wireless networks," *IEEE Personal Communications*, April 1999, pp. 46–55.
92. Y. Chun, L. Qin, L. Yong, and S. MelLin, "Routing protocols overview and design issues for self-organizing network," *Communication Technology Proceedings, WCC-ICCT 2000, International Conference*, vol. 2, 2000, pp. 1298–1303.
93. A. Boukerche "Performance evaluation of routing protocols for ad hoc wireless networks," *Mobile Networks and Applications*, vol. 9, no. 4, 2004, pp. 333–342.
94. X. Hong, K. Xu, and M. Gerla, "Scalable routing protocols for mobile ad hoc networks," *IEEE Networks*, July/August 2002, pp. 11–21.
95. T.-W. Chen and M. Gerla, "Global state routing: a new routing scheme for ad-hoc wireless networks," *Proceedings of the IEEE International Conference on Communications*, 1998, pp. 171–175.

APPENDIX

This appendix contains parameter comparisons between the various protocols within each category discussed above, except the "other" category. Some parameters of a protocol include more than one value since their value is dependent upon on a given scenario, such as worst case, best case, etc.

Appendix A: Proactive Routing Protocol Comparisons

TABLE 15.A1

Proactive Protocol	Routing Info	Routing Metric	RS	Number/Data Structure of RI at Node	U	MC	MP	HM	Updating Policy	EA
DSDV	DV	Shortest path	F	2, tables	No	No	No	Yes	Periodic and as needed	No
WRP	DV	Shortest path	F	3, tables 1, list	No	No	No	Yes	Periodic and as needed	No
CGSR	DV	Shortest path (to destination clusterhead)	H	2, tables	No	No	No	No	Periodic	No
STAR	LS (source tree)	Variable	F or H[a]	1, table[b] x, graph[b] x, source tree[b] 1, list[b]	Yes	No	No	No	As needed[c]	No
HSR	LS	Variable	H	1, table and location management[d]	No	No	No	No	Periodic and event-driven[e]	No
OLSR	LS (MPRS)	Variable	F	3, tables	Yes[f]	No	Yes	Yes	Periodic	No
TBRPF	LS	Variable	F or H	1, source tree	No	Yes	Yes	Yes	Periodic and differential	No
DREAM	Location	Forwarding zones based on destination location	F	6, tables 1, table	No	No	Yes	No	Periodic as a function of node mobility	No

(*continued*)

TABLE 15.A1 Continued

Proactive Protocol	Routing Info	Routing Metric	RS	Number/Data Structure of RI at Node	U	MC	MP	HM	Updating Policy	EA
GSR	LS—fisheye	Variable	F	3, tables 1, list	Yes	No	Yes	No	Periodic	No
FSR	LS—fisheye	Variable	F	3, tables 1, list	Yes	No	Yes	No	Periodic	No
HR	HTU	Variable	H	1, table	No	No	No	No	Periodic for each cluster	No
HSLS	LS	Variable	F	1, table	Yes	No	No	Yes	Hybrid	No
A-HSLR	LS	Variable	F	1, table	Yes	No	No	Yes	Hybrid	No

DV, distance vector; EA, energy-aware, F, flat; H, hierarchical; HM, hello messages; HTU, hierarchical time-update; LS, link state; MC, multicasting; MP, multipath routing; MPRS, multipoint relay selectors; RS, routing structure; U, unidirectional link support.

[a] In [18], it is noted at the top of p. 274 that STAR could be used with a hierarchical scheme.
[b] This is information stored in a router; x-graph and x-source tree signify that each router stores its own source tree and topology graph, as well as those reported by its neighbors, thus x is a variable number.
[c] The specific updating policy depends on whether ORA or LORA is being used.
[d] The number of logical levels determines the number of additional tables.
[e] The updating policy refers to node registration of its hierarchical address to its home agent of the logical subnetwork.
[f] It should be noted that links to MPRs must be bidirectional.

TABLE 15.A2

Proactive Protocol	Computation Complexity	Communication Complexity	CT	TC	MC	DC	PC	Extensions
DSDV	$O(N)$	$O(N)$	$O(D*I)$	$O(D)$	$O(N)$	$O(N)/I$	$O(1)$	No
WRP	$O(N)$	$O(N)$	$O(N*I)$	$O(H)$	$O(N)$	$O(N)/I$	$O(1)$	WRP-Lite is a streamlined version with reduced overhead
CGSR	N/A	$O(N)$	$O(D)$	$O(D)$	$O(2N)$	N/A	$O(N)$	CGSR + PTS, offers priority token scheduling; CGSR + PTS + GCS, adds gateway code scheduling; CGSR + PTS + GCS + PR, adds path reservation until disconnect (similar to a virtual circuit)
STAR	$O(N^2)$	$O(N)$?	$O(D)$	$O(H)$	$O(N^2)$	N/A	N/A	No
HSR	N/A	$O(N)$?	$O(D)$	$O(D)$	$O(N^2*M) + O(V) + O(N/V) + O(N/N_{CL})$	N/A	$O(N_C)$	No
OLSR	$O(N^2)$	$O(N)$	$O(D)$	$O(D)$	$O(N^2)$	$O(d)/I$	$O(N)$	Fast-OLSR, improves route discovery for fast moving nodes
TBRPF	$O(N^2)$	$O(N)$	$O(D*I)$	$O(H)$	$O(N^2)$	$O(N)/I$	$O(N)$	No
DREAM		$O(N)$	$O(N*I)$	N/A	$O(N)$	N/A	N/A	No
GSR	$O(N^2)$	$O(N)$	$O(D*I)$	$O(D)$	$O(N^2)$	$O(d) + O(N-d)/I$	$O(1)$	No
FSR	$O(N^2)$	$O(N)$	$O(D*I)$	$O(D)$	$O(N^2)$	$O(d) + O(N-d)/I$	$O(1)$	CFSR, hierarchical routing scheme [34, 35], adaptive routing table update schemes
HR	N/A	$O(N)$	$O(D)$	$O(D)$	$O(N^2)$	$O(N)/I$		No
HSLR	$O(N^2)$	N/A	$(d-1)*I$	N/A	$O(N)$	N/A	N/A	A-HSLR
A-HSLR	$O(N^2)$	N/A	$(d-1)*I$	N/A	$O(N)$			No

CT, convergence time, D, diameter of network; d, degree of node connectivity; DC, data complexity; H, height of routing tree; I, routing update interval; M, number of hierarchical levels; MC, memory complexity; N, number of nodes in the network; N_C, number of nodes in a cluster; N_{CL}, number of logical nodes in a logical cluster; PC, packet complexity; TC, time complexity (link addition/failure); V, number of virtual IP subnets.

Computation complexity, the number of computation steps for a node to perform a routing operation. Communication complexity, the number of messages needed to perform a protocol operation. Convergence time, the time required to detect a link change. Data complexity, aggregate size of control packets exchanged by a node in a time unit. Memory complexity, the memory space required to store the routing information. Packet complexity, average number of routing packets exchanged by a node in a time unit. Time complexity, the number of steps needed to perform a protocol operation.

Appendix B: Reactive Routing Protocol Comparisons

TABLE 15.B1

Reactive Protocol	RSTGY	RM	RS	Route Recovery Methodology	PRR	U	MP	MC	Beacons
ARA	Hop-by-hop	SP[a]	F	Erase route (link), localized query (hop-by-hop)	Yes	No	Yes	No	No
ABR	Hop-by-hop	Associativity, SP, load and delay	F	Erase route; localized query	Yes	No	No	No	Yes
AODV	Hop-by-hop	Freshest and SP	F	Erase route; notify source	No	No	Yes	Yes	No
BSR	Source routing	PP = SP or shortest delay path BP = durability[b]	F	Erase route; notify source with backup routes and forward packet using backup routes	Yes	Yes	Yes	No	No
CHAMP	Hop-by-hop	Shortest multipath and load balancing	F	Erase route; forward with alternate path or previous hop notification	Yes	No	Yes	No	No
DNVR	Nix-vector	Variable	F	Erase route, notify source	No	No	No	No	No

Protocol	Routing Strategy	Routing Metric		Route Maintenance					
DSR	Source routing	SP	F	Erase route; notify source	No	Yes	Yes	Yes	No
DYMO	Hop-by-hop	Variable	F	Erase route; notify source	No	No	No	Yes	Yes[c]
LAR	Hop-by-hop	SP within a request zone	F	Erase route; notify source	No	Yes	No	No	No
LBR	Hop-by-hop	Stability-based	F	Erase route; notify source	No	Yes	No	No	Yes
MPABR	Hop-by-hop	Associativity, SP, multi-path criteria	F	Erase route; forward with alternate path	Yes	No	Yes	No	Yes
NDMR	Hop-by-hop	SP and node-disjoint path	F	Erase route; notify source	No	No	Yes	No	Yes
PLBM	Hop-by-hop	SP and Preferred Link (neighbor degrees)	F	Erase route; route repair or local tree construction	Yes	No	No	Yes	Yes
RDMAR	Hop-by-hop	Relative distance, SP, load balancing	F	Erase route; local repair or notify source[d]	Yes	No	Yes	No	No
SOAR	Hop-by-hop	SP	F	Route repair	Yes	Yes	No	Yes	No
TORA	Link Reversal	SP	F	Link reversal; route repair	Yes	No	Yes	No	No

BP, backup path; MC, multicasting; MP, multipath routing; PP, primary path; PRR, partial route recovery; RM, routing metric; RS, routing structure; RSTGY, routing strategy; SP, shortest (hop-count) path; U, unidirectional link support.

[a] A forward ant and a backward ant establish pheromone track values depending on the number of hops from source to destination, respectively.

[b] BSR's backup path assigns a heuristic cost that essentially consists of the following parameters: shortest distance and "less-link similar" to the primary path.

[c] Hello messages can be used, but other methods are noted such as route timeouts and link layer feedback.

[d] Depends on the relative distance of the node from the source and destination.

TABLE 15.B2

Reactive Protocol	TC (Initialization)	TC (Post-Failure)	CC (Initialization)	CC (Post-Failure)
ARA[a]	$O(2D)$	0 or $O(I+D)$ or $[O(Z)$ or $TO]+O(2D)$	$O(2N)$	0 or $O(X+N)$ or $O(2N)$
ABR	$O(D+WT+Z)$	$O(I+Z)$	$O(N+Y)$	$O(X+Y)$
		$[O(Z)$ or $TO]+O(D+WT+Z)$		$O(Y)+O(N+Y)$
AODV	$O(D+Z)$	$O(D)$	$O(N+Y)$	$O(N)$
		$[O(Z)$ or $TO]+O(D+Z)$		$O(Y)+O(N-Y)$
BSR[2]	$O(D+WT+Z)$	$O(D+WT+Z)$ or 0	$O(N+Y)$	$O(Z)+O(N+Z)$ or 0
CHAMP	$O(D+Z)$	$O(S+Z)$ or $O(D+Z)$	$O(N+Y)$	$O(M+Z)$ or $O(I+Y)$
DNVR	$O(D+Z)$	$O(Z)+O(D+Z)$	$O(N+Y)$	$O(Y)+O(N+Y)$
DSR[3]	$O(2DD)$ or $O(DD+Z)$	$O(Z)+[O(2DD)$ or $O(DD+Z)]$ or 0	$O(2NN)$ or $O(NN+Y)$	$O(X)+[O(2NN)$ or $O(NN+Y)]$ or 0
DYMO	$O(D+Z)$	$O(Z)+O(D+Z)$	$O(N+Y)$	$O(Y)+O(N+Y)$
LAR	Scheme 1: $O(R+Z)$	$O(R+Z)$	$O(E+Z)$	$O(E+Z)$
	Scheme 2: $O(S+Z)$	$O(S+Z)$	$O(M+Z)$	$O(M+Z)$
LBR[d]	$O(D+WT+Z)$	$O(Z)+O(D+WT+Z)$	$O(N+Y)$	$O(Y)+O(N+Y)$
MPABR	$O(D+WT+Z)$ or $O(D+Z)$	$O(Z)$ or 0	$O(N+Y)$	$O(Y)$ or 0
NDMR	$O(D+WT+Z)$	$O(Z)+[0$ or $O(D+WT+Z)]$	$O(N+Y)$	$O(Y)+[0$ or $O(N+Y)]$
PLBM	$O(S+WT+Z)$	$O(S+Z)$	$O(M+Y)$	$O(M+Y)$
RDMAR	$O(S+Z)$	$O(I+Z)$ or $O(I+Z)+O(S+Z)$ or 0	$O(M+Y)$	$O(M+Y)$ or $O(2M+2Y)$ or 0
SOAR	$O(DD+Z)$	$O(2I)$	$O(NN+Y)$	$O(2X)$
TORA	$O(2D)$	$O(2I)$	$O(2N)$	$O(2X)$

CC, communication complexity (route discovery phase); D, diameter of the network; DD, diameter of a subset of the network up to the full network (e.g., expanding ring search); E, number of nodes in the rectangular request zone; I, diameter of the affected network segment; M, number of nodes in the localized region; N, number of nodes in the network; NN, number of nodes in a subset of the network up to the full network (e.g., expanding ring search); 0, cache hit; R, length × width of rectangular request zone; S, diameter of the nodes in the localized region; TC, time complexity (route discovery phase); TO, time out period; WT, weight time (usually at the destination for route selection based on one or more criterion); X, number of nodes affected by a topological change; Y, total number of nodes forming the directed path where the REPLY or RERR packet transits; Z, diameter of the directed path where the REPLY or RERR packet transits.

[a]We describe three post-failure situations: (1) a route error occurs at node z and node z has an alternate route; (2) node z does not have an alternate route but neighboring nodes do. The packet travels along the affect path until a valid path is found. The destination subsequently sends a BANT; (3) the source initiates a new route discovery after a timeout or receiving a RERR packet.

[b]BSR requires intermediate nodes to select backup paths, which requires wait-times, as well as at the destination.

[c]DSR provides that the RREQ may be a full flood or an expanding ring-search. Also, a RREP may follow a reverse sequence path (bidirectional links) or node may initiate a route discovery back to the source if a path is not available in its cache.

[d]We include only the reactive route reconfiguration approach and do not include the complexities utilizing the alternate proactive route reconfiguration discussed in [51].

Appendix C: Power-Aware Routing Protocols Comparisons

TABLE 15.C1

Power-Aware Protocol	Power Saving Technique(s)	Routing Metric	Routing Strategy	Other Considerations
BEE	Power control / Power management	Battery behavior / Minimize TTP	Hybrid hop-by-hop[a]	Battery behavior includes non transmission power costs (e.g., node processing)
EADSR	Power control	Minimize TTP	Source routing	Tracking energy costs of a route (route maintenance)
MTPR	Power control	Minimize transceiver power and SP	Hop-by-hop	None
MBCR	Power control / Power management	Total battery cost, transceiver power, SP	Hop-by-hop	None
MMBCR	Power control / Power management	Avoid least battery capacity, transceiver power, SP	Hop-by-hop	None
CMMBCR	Power control / Power management	Battery capacity above threshold, minimize transceiver power, SP	Hop-by-hop	None
PARO	Power control	Minimize TTP	Hop-by-hop	Multiple iterations to attain a fully optimized route
PAWF	Power management	Weighted forwarding function[b]	Hop-by-hop	Requires position information
MFP/MIP/ MFP$_{energy}$/MIP$_{energy}$	Power management	Forward progress Spectral efficiency Battery remaining capacity	Hop-by-hop	Requires position information of destination and neighbor nodes

RI, routing information; SP, shortest path; TTP, total transmission power.

[a] BEE presents two different schemes for selecting a path. In the first scheme, the source selects from a set of routes. In the first scheme may become unacceptable when the network topology changes frequently. In such instances, a second scheme could be applied where the source or destination selects x routes "independently and uniformly at random" from all available routes. In both schemes the cost function is used to select the appropriate path among the set of routes (in the first scheme) or among the x-selected routes (in the second scheme).

[b] PAWF, the weighted forwarding function is described above with sufficient detail. However, the routing metric may be characterized as emphasizing forwarding achievement over energy.

425

Appendix D: Hybrid Routing Protocol Comparisons

TABLE 15.D1

Hybrid Protocol	RM	RRM	RS	RMI	MP	Beacons	U	MC
MultiWARP	SP	Route repair at place of failure	F	Local routing table and cover matrix	Yes	Yes	No	No
SHARP	Variable[a]	Local route repair; Link reversal to repair DAG	F[b]	DAG rooted at the destination	Yes	Yes	No	No
SLURP	MFR for interzone forwarding; DSR for intrazone routing	Notify source; local query	H	Location table, region table, location cache, node list, hash table	Yes	No	No	No
ZRP	SP	Route repair at place of failure; notify source	F	Intrazone and interzone tables, bordercast tree	No	Yes[c]	Yes[d]	No
AZRP	SP	Route repair at place of failure; notify source	F	Intrazone and interzone tables, bordercast tree	No	Yes	Yes	No
IZR	SP	Route repair at place of failure; notify source	F	Intrazone and interzone tables; expected node list; peripheral node list	No	Yes	Yes	No
TZRP	SP	Route repair at place of failure; notify source	F	Intrazone and interzone tables, link state table, bordercast tree	No	Yes[e]	Yes	No

DAG, directed acyclic graph; F, flat; H, hierarchical; MC, multicasting; MP, multipath; RM, route metric; RMI, route maintained in; RRM, route reconfiguration methodology; RS, routing structure; SP, shortest hop count; U, unidirectional link support.

The reader should keep in mind that the hybrid protocols utilize an underlying proactive and reactive protocol. Accordingly, the parameters reflected above may differ depending on what proactive and reactive protocol is used, as well as its respective version. The table above presents the parameters according to the implementation discussed in the reference.

[a] Best or optimal path depending on application-specific performance metric of the node.
[b] There are central nodes in each proactive zone, but unlike typical hierarchical schemes, they are not used as gateways to other zones.
[c] ZRP relies on a NDP at the MAC layer for hellos. However, if the MAC layer does not include an NDP, the IARP must provide it.
[d] Provided by IARP and works only when source/destination are in the same zone.

TABLE 15.D2

Hybrid Protocol	TC(RD)	TC(RM)	CC(RD)	CC(RM)	Adaptive Zone/Region Size
MultiWARP	Intra HR: $O(I)$ Inter HR: $O(2D)$	$O(I)$ or $O(2D_{HR})$ $O(2D)$	$O(N_{HR})$ $O(N_{HRL}) + O(X+Y)$	$O(N_{HR})$ $O(E+N_{HRL}) + O(X+Y)$	No
SHARP	Intra-Zone: $O(I)$ Inter-Zone: $O(2D)$	$O(I)$ $O(2D)$	$O(N_Z)$ $O(N_Z) + O(X+Y)$	$O(N_Z)$ $O(E+N_Z) + O(X+Y)$	Yes
SLURP	Intra HR: $O(2D_{HR})$ Inter HR: $O(2D)$	$O(2D_{HR})$ $O(D+2D)$	$O(2N_{HR})$ $O(X) + O(N_{HR}+Y)$	$O(2N_{HR})$ $O(E+X) + O(N_{HR}+Y)$	No
ZRP	Intra-Zone: $O(I)$ Inter-Zone: $O(2L)$	$O(I)$ $O(2D)$	$O(N_Z)$ $O(N+Y)$	$O(N_Z)$ $O(N+Y)$	No
AZRP	Intra-Zone: $O(I)$ Inter-Zone: $O(2D)$	$O(I)$ $O(2D)$	$O(N_Z)$ $O(N+Y)$	$O(N_Z)$ $O(N+Y)$	Yes
IZR	Intra-Zone: $O(I)$ Inter-Zone: $O(2D)$	$O(I)$ $O(2D)$	$O(N_Z)$ $O(N+Y)$	$O(N_Z)$ $O(N+Y)$	Yes
TZRP	Intra-Zone: $O(I)$ Inter-Zone: $O(2D)$	$O(I)$ $O(2D)$	$O(N_Z)$ $O(N+Y)$	$O(N_Z)$ $O(N+Y)$	Yes

CC, communication complexity (route discovery/route maintenance); D, diameter of the network; D_Z, diameter of a home region; E, number of nodes an error packet traverses to reach the source; HR, home region; I, periodic update interval; N, number of nodes in the network; N_{HR}, number of nodes in a home region; N_{HRL}, number of nodes of the candidate set in a home region, which can be less than N_{HR}; N_Z, number of nodes in a zone; TC, time complexity (route discovery/route maintenance); X, number of nodes a location query packet traverses to reach the home region of the destination; Y, number of nodes a route reply packet traverses to reach the source.

CHAPTER 16

Adaptive UWB Systems

FRANCESCA CUOMO and CRISTINA MARTELLO

16.1 INTRODUCTION

Recent trends demonstrate the great attention directed to the development of UWB products in view of an extensive use of this technology in the WPAN sector [1]. UWB offers great potentialities due to the low-cost devices, simple RF circuitry and high-bandwidth support. As for this latter point, it should be noted that other WPAN technologies (e.g., Bluetooth) are not suitable for high-data-rate connection (even in the short range). This is the reason why UWB is the candidate technology for the IEEE 802.15.3 standard to support QoS for multimedia streams [2]. Recently UWB has been indicated to be a leading candidate for enabling the digital home, where people are expected to share photos, music, video, data, and voice among networked consumer electronics, PCs, and mobile devices throughout the home and even remotely. In this context users will be able to stream video content from a PC or consumer electronics device—such as a camcorder, DVD player or personal video recorder—to a high-definition television display without the use of any wires.

Recent research and industry achievements have shown the great potentialities of a UWB physical layer (PHY) as well as a MAC layer in supporting high data throughput with low power consumption for distances of less than 10 m, which is very applicable to the digital home requirements. The fastest data rate publicly shown over UWB is now an impressive 252 Mbps, and a data rate of 480 Mbps is expected to be shown in the not-too-distant future. Another key aspect driving the UWB world into the market is that, in the United States, the FCC has mandated that UWB radio transmissions can legally operate in the range from 3.1 GHz to 10.6 GHz, at a transmit power of −41 dBm/MHz. Japanese regulators have issued the first UWB experimental license allowing the operation of a UWB transmitter in Japan.

Ultra Wideband Wireless Communication. Edited by Arslan, Chen, and Di Benedetto
Copyright © 2006 John Wiley & Sons, Inc.

Besides the high data rates achievable with UWB, a key potential is in the capability to meet QoS constraints and to adapt to the environmental conditions. By designing suitable MAC and routing protocols a large number of transmitters can operate simultaneously in the same area, yielding increased spectral reuse and achieving high capacity per area. Some theoretical works [3] show that the most promising approach to improving the capacity bound in power-constrained wireless ad-hoc networks is to employ unlimited resources, such as UWB.

The analysis of potentialities of UWB as a networking paradigm has been carried out in [4] and [5]. In the paper by Cuomo [4], a general framework for radio resource sharing in UWB is provided by considering two main classes of traffic: (i) elastic-dynamic (also known as best effort); and (ii) guaranteed QoS (reserved bandwidth). A joint power and rate assignment is presented as an optimization problem that is solved for the elastic-dynamic traffic as well as for the guaranteed QoS traffic. In the latter case, a target rate value comes as a requirement and the proposed algorithm checks whether feasible power levels can be set in all transmitters, so that the required bit rates are supported. Coexistency of the two classes of traffic is studied in [6]. The goal of [5] is to define design objectives for multihop UWB best-effort ad hoc networks. The authors give guidelines on how to organize access to the medium as well as routing in order to optimize the use of the UWB system and the performance perceived by users.

Finally, some works explicitly concentrate on the MAC layer design and analysis [7–9]. The work in [7] proposes an ALOHA-like approach for the design of the UWB MAC layer. The algorithm exploits typical features of impulse radio such as large processing gain, and is conceived in conjunction with a synchronization strategy which foresees the presence of a synchronization sequence in each transmitted packet. Performance analysis of the synchronization tracking mechanism shows that, under the preliminary simplistic hypothesis of an AWGN channel, and for a sufficient number of pulses in the synchronization sequence, a fairly high probability of successful synchronization can be achieved, even in the presence of several users and multiuser interference. Also, [8] addresses the critical aspect of time acquisition in UWB systems, while [9] proposes and evaluates a number of MAC protocols for UWB.

The focus of this chapter is on adaptivity in UWB systems. More specifically, we describe how to exploit the UWB adaptability to support wireless links in ad hoc networks based on UWB. We show how to dynamically set up wireless communications among devices distributed in a given area, without the support of a centralized infrastructure. Each wireless link should be characterized by two main QoS parameters: (i) a given PHY transmission rate (that typically comes as a requirement from the upper layers); (ii) a given SINR able to insure a target bit error rate at the receiver side. We design an admission control (AC) scheme that guarantees that the active wireless links in the system maintain their QoS parameters. Adaptivity is achieved by exploiting interference measurements and inter-device signaling.

16.1.1 Related Work on Adaptive UWB Systems

Adaptability is one of the main features exploited in UWB systems. Adaptive assignment of PHY and MAC parameters can be used to:

- Set-up the desired quality for a given communication;
- Achieve reliable transmissions;
- Mitigate the effects of the environmental and mutual UWB interference at the receiver;
- Support the desired throughput;
- Achieve data rate assignment granularity.

Several works have appeared in the literature dedicated to the analysis of the UWB adaptability to achieve some or a combination of the aforementioned features. A dynamic channel coding is used in [10] to constantly adapt the rate to the level of interference experienced at the destination. The proposed protocol is fully distributed and is based on the adoption of a threshold demodulator that at the receiver detects when the received energy of a pulse is larger than a threshold. This means that this pulse has collided with other interfering pulses. In such a case the chip is skipped and an erasure is declared. The rate is then adapted to the highest value that allows successful reception of a data packet at the receiver. A variable encoding rate is achieved by puncturing the data to be sent.

The authors of the paper in [11] propose a UWB physical layer that adapts its modulation scheme to efficiently meet QoS requirements. The system employs M-ary PPM and adapts its pulse repetition interval (PRI) and/or the number of bits per symbol ($\log 2m$). To efficiently meet QoS requirements, the authors propose to dynamically configure a UWB system with a three-step resource allocation procedure that in the first step examines the current QoS requirements, the resources of each node and the current environmental conditions. The second step identifies a suitable system configuration to meet the current requirements, resources, and environmental conditions through a cost function. The third step allocates available resources to achieve the desired QoS. In the paper, the QoS requirements include data rate, BER, and energy dissipation. The environmental conditions include link distance and level of interference, and the resources are the possible values of PRI and m. Simulations of the adaptive system show that it improves performance significantly as compared with a conventional nonadaptive system under variable environmental and QoS requirements. The adaptive system improves BER by 50%, data rate by 260%, or energy by 60% without sacrificing the performance of any other parameter.

A rate adaptive MAC protocol for high-rate PANs is proposed in [12]. The data rate is selected on the basis of the channel condition estimated from the received data frame at PHY layer. The adopted MAC layer scheme is the one proposed in the IEEE 802.15.3 Task Group. The selected rate information is delivered via rate-adaptive ACK (RA-ACK). When the piconet controller receives this RA-ACK it updates the data rate of the communication link. Using this rate adaptation scheme, the WPAN system efficiently copes with the time-varying channel. To obtain better performance, a constant PHY frame length mechanism is proposed so that the channel efficiency is increased due to reduced overheads. Simulation results show that the proposed rate-adaptive MAC protocol gives a 58% throughput gain over the nonrate-adaptive MAC protocol in IEEE 802.15.3.

Finally, UWB adaptability is also exploited when dealing with the application of UWB as transmission technology of a wireless sensor network (WSN). Güverc et al. [13] analyze the adaptation of multiple access parameters in cluster-based UWB-impulse radio WSNs in both synchronous and asynchronous communication scenarios. For synchronous communications, an orthogonal sequence construction is presented, which assigns variable processing gains to the sensors, and acquires the desired BER requirement at each sensor. For asynchronous communication systems, Gaussian approximation methods are used to adapt the transmission powers and processing gains of the sensors. Computer simulations results demonstrate the data rate and power savings improvements with the proposed approach.

16.2 A DISTRIBUTED POWER-REGULATED ADMISSION CONTROL SCHEME FOR UWB

As stated in Section 16.1, the focus of this chapter is in the support of UWB wireless communications in an ad hoc network composed of a number of distributed UWB terminals (named simply *terminals* in the following). The considered scenario is typical of an office or a home where different terminals exchange data in an ad hoc fashion (i.e., without a supporting centralized infrastructure). Since the exchanged data may require guaranteed performance (in terms of data rate or delay), the focus is on the admission control of the communications in the system. Each communication between two terminals in the system is named in the following *link*.

The UWB IR considered in this section works by transmitting extremely short pulses (of duration 0.1–1.5 ns), named monocycles, on a time axis structured in time frames of duration T_f (typically about 100 ns), as illustrated in Figure 16.1. A monocycle is characterized by an energy level E_m. Each time frame is divided into N_h short time periods of duration T_c. A symbol is transmitted by N_s pulses according to a PPM scheme. The multiple access is based on the adoption of pseudorandom TH codes whose elements are chosen among N_h possible T_c-shifts within the period T_f.

One of the appealing features of UWB based on IR regards the possibility of supporting distributed flexible radio resource management schemes. System efficiency is also achievable when different wireless links are mutually asynchronous. In accordance with the multiple access scheme considered in this chapter, different transmissions use different TH codes. The AC decisions are taken at terminals in a distributed fashion, on the basis of information regarding neighboring wireless links and obtained by measurements and signaling. The distributed nature of these operations insures that the transmission parameters of a link are adapted to the

16.2 ADMISSION CONTROL SCHEME FOR UWB

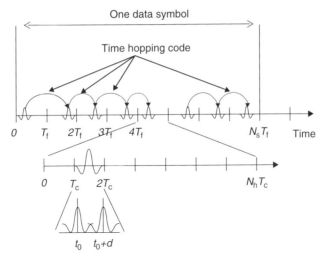

Figure 16.1 Organization of the time axis of UWB IR multiple access.

status of the neighboring links and to the conditions of the system. It is worth noting that in this latter aspect resides one of the key potentials of UWB transmission. As indicated in [1], "the novel and unconventional approach underlying the use of UWB is based on sharing optimally the existing radio spectrum resources rather than looking for still available but possibly unsuitable new bands." This is the reason why a UWB radio resource control should take care of the environment in terms of *introduced* interference on one side and of *perceived* interference on the other.

The distributed AC scheme presented in this chapter jointly assigns powers and rates to links and operates in an *incremental* way: The decision whether a new link can be established is based on the current interference conditions used to coordinate the power selection process of the entering new link. The already active links do not change the transmission powers selected during their AC phase. Power is configured by tuning the monocycle energy E_m or by changing the parameters T_c or N_h. Transmission rates are varied by controlling suitably the parameters T_c, N_h, or N_s.

The SINR value, denoted by γ, for a UWB multiple access scheme based on TH can be computed under the hypothesis of Gaussian approximation of the multiuser interference [14]. For the i-th link the SINR is determined according to the following formula:

$$\gamma_i = \frac{P_i \cdot g_{ii}}{R_i \cdot (\eta_i + \sigma^2 T_f \sum_{j=1, j \neq i}^{N} P_j \cdot g_{ij})} \quad (16.1)$$

where N is the number of the active links in the system; P_i is the average power emitted by the ith link's transmitter ($P_i = E_{m,i}/\{T_{c,i} \cdot N_{h,i}\}$); g_{ij} is the path gain from the jth link's transmitter to the ith link's receiver; R_i denotes the transmission rate of the ith link (in a TH scheme transmitting one bit per symbol, the transmission rate is $R_i = 1/\{T_{f,i} \cdot N_{s,i}\} = 1/\{T_{c,i} \cdot N_{h,i} \cdot N_{s,i}\}$); η_i is the noise spectral density power at the ith link's receiver; and σ^2 is an adimensional parameter depending on the pulse shape.

As it can be noticed, different UWB parameters have an impact on the SINR that, besides increasing with power as in a generic wireless system, depends on the pulse shape via σ, on the TH period via T_f and on the number of pulses per symbol via R.

In the following the link's QoS requirements are expressed in terms of transmission rate, R, and target SINR, γ^T.

16.2.1 Problem Formalization

The evaluation of admissibility of a configuration of N links in the wireless ad hoc network consists of finding out if there exists a proper set of transmission powers P_i, $i = 1, \ldots, N$, which satisfies the requirement on the rate R_i and insures $\gamma_i \geq \gamma_i^T$ for each link $i = 1, \ldots, N$.

More precisely, this problem can be formalized according to the following matrix form which identifies a well-known condition for the existence of a feasible solution:

$$\begin{cases} (\mathbf{I} - \mathbf{F}) \cdot \mathbf{P} \geq \mathbf{u} \\ \mathbf{P} \geq \mathbf{0} \end{cases} \tag{16.2}$$

where $\mathbf{I}$ is the $N \times N$ identity matrix; $\mathbf{P}$ is the column vector of the N transmission powers; $\mathbf{F}$ is an $N \times N$ matrix whose elements depend on the system topology (e.g., terminals reciprocal distances); in particular $F_{ii} = 0$ and

$$F_{ij} = \frac{\gamma_i^T \cdot R_i \cdot \sigma^2 T_f \cdot g_{ij}}{g_{ii}}$$

with $i \neq j$; and $\mathbf{u}$ is an N-dimensional column vector related essentially to noise powers:

$$u_i = \frac{\gamma_i^T \cdot R_i \cdot \eta_i}{g_{ii}}.$$

Both $\mathbf{F}$ and $\mathbf{u}$ depend on the the desired transmission rates and target SINRs. In Equation (16.2) the inequalities between vectors have to be taken as inequalities component by component.

The condition of existence of a feasible solution of the problem (16.2) consists of a constraint for the maximum modulus eigenvalue of $\mathbf{F}$, ρ_F, and is $\rho_F < 1$. If a solution of the problem (16.2) exists, the minimum power configuration is called the *Pareto-optimal* solution ([15]) and is provided by the following expression:

$$\mathbf{P}^* = (\mathbf{I} - \mathbf{F})^{-1} \cdot \mathbf{u} \tag{16.3}$$

which has the property that every other power configuration sets transmission powers at values that are not lower than their corresponding Pareto-optimal ones. In other words, any other solution $\mathbf{P}$ can be expressed as $\mathbf{P} = (\mathbf{I} - \mathbf{F})^{-1} \cdot (\mathbf{u} + \Delta\mathbf{u})$ where $\Delta\mathbf{u}$ is a column vector of N real positive values.

We name $\mathcal{D}$ the domain of the feasible solutions $\mathbf{P} \geq \mathbf{P}^*$. Since a typical real scenario is power-constrained, solutions are selected without exceeding the maximum level of transmission power, named P_{bound}, that a device can emit. Therefore, an admissible topology is characterized by a nonempty domain $\mathcal{D}$ of solutions which is composed of: (i) the Pareto-optimal solution $\mathbf{P}^*$ allowing the desired SINR levels at the minimum transmission powers to be matched exactly; (ii) the set of solutions $\mathbf{P}^-$ such that $\mathbf{P}^* < \mathbf{P}^- \leq \mathbf{P}_{\text{bound}}$ (denoted by $\mathcal{D}^-$); these solutions do not exceed the maximum powers and assure SINR levels greater than the target; and (iii) the set of solutions $\mathbf{P}^+$ (denoted by $\mathcal{D}^+$) such that $P_i^+ > P_{\text{bound}}$ for some $1 \leq i \leq N$; in this case the bound on the maximum power is exceeded. When the topology is admissible but the activation of all links requires to set some transmission power above the maximum P_{bound}, it happens that the set $\mathcal{D}^-$ is empty and the set $\mathcal{D}^+$ includes the Pareto-optimal solution $\mathbf{P}^*$.

16.2.2 Power Selection in UWB

To support QoS in terms of the bit rate and target SINR and to apply, in an incremental way, a power adaptability, receivers maintain a nonnegative parameter (named maximum extra interference, MEI). The MEI of a receiver (or equivalently of a link) is defined as the amount of interference that can be tolerated by the receiver itself without endangering the QoS level of the communication link.

When a link has its MEI equal to zero, no other interfering emissions can be tolerated; when the MEI is positive, other links can be activated, provided that the overall interference they produce does not reduce one or more MEIs below zero. In other terms, a positive MEI means that the link is maintaining its negotiated QoS, while a MEI ≤ 0 means that the QoS is not assured any more.

The aim of our AC mechanism is primarily to guarantee that the MEIs of all active links in the considered area are never negative. In addition, for efficiency reasons, the AC procedure tends to balance all MEIs within the system, so as to avoid bottleneck regions and regions characterized by terminals with high MEIs. The link block probability is related to the MEI values. This probability is high if just a single MEI is low and, conversely, the probability is low if the MEIs are all high.

The MEI level perceived by the ith link, denoted by M_i, depends on the link's QoS parameters, transmission power and current interference conditions according to the following expression:

$$\gamma_i^{\text{T}} = \frac{P_i \cdot g_{ii}}{R_i \cdot (\eta_i + \sigma^2 T_{\text{f}} \sum_{j=1, j \neq i}^{N} P_j \cdot g_{ij} + \sigma^2 T_{\text{f}} M_i)} \tag{16.4}$$

where γ_i^{T} and R_i denote the desired SINR and data rate, respectively.

The AC scheme proceeds in an incremental way: given a set of active links, the two entities (transmitter and receiver) willing to establish a new link take the access decision by measuring the system. Once the admissibility of the new link has been verified, the links' power levels will be maintained at a power configuration, $\mathbf{P}^-$, included in the domain $\mathcal{D}^-$, thus insuring that the transmission powers are within P_{bound}. To guarantee that UWB terminals can operate in unlicensed mode, P_{bound} is the maximum power value imposed by the regulatory bodies and obtained by the average EIRP value of 0.566 mW derived from a power spectral density of -41.3 dBm/MHz.

Power levels are computed on the basis of the current MEI values $M_i, i = 1, \ldots, N$, according to:

$$\begin{cases} \mathbf{P}^- = (\mathbf{I} - \mathbf{F})^{-1} \cdot (\mathbf{u} + \Delta\mathbf{u}), \\ \Delta u_i = \dfrac{\gamma_i^\mathrm{T} \cdot R_i \cdot \sigma^2 T_\mathrm{f} M_i}{g_{ii}}, \quad i = 1, \ldots, N. \end{cases} \quad (16.5)$$

As stated in Section 16.2.1, the power configuration $\mathbf{P}^-$ provides for power levels greater than the corresponding Pareto-optimal ones and can be expressed as $\mathbf{P}^- = \mathbf{P}^* + \Delta\mathbf{P}$ where $\Delta\mathbf{P} = (\mathbf{I} - \mathbf{F})^{-1} \cdot \Delta\mathbf{u}$ is a vector of positive elements. MEI levels can be expressed as functions of the additional powers $\Delta P_i, i = 1, \ldots, N$, employed by the N links. In particular, it is derived the following expression of the MEI of the ith link, M_i, by exploiting Equation (16.4) and substituting $P_i = P_i^* + \Delta P_i, i = 1, \ldots, N$:

$$M_i = \frac{g_{ii} \Delta P_i}{\sigma^2 T_\mathrm{f} \gamma_i^\mathrm{T} R_i} - \sum_{j=1, j \neq i}^{N} g_{ij} \Delta P_j. \quad (16.6)$$

In Equation (16.6) the contribution of the terms depending on P_i^*, for $i = 1, \ldots, N$, is null since the Pareto-optimal solution entails null MEIs.

Equation (16.6) highlights a tradeoff: the additional power ΔP_i used by the ith link increases the relevant MEI, M_i, while the terms $\Delta P_j, j = 1, \ldots, N, j \neq i$, of the other active links, reduce M_i. Furthermore, MEI is inversely proportional to the QoS parameters; as an example, the support of a high transmission rate leads to a reduced MEI level if the transmission power is kept constant.

The AC rule for an $(N + 1)$th link, given N active ones, consists of the comparison between the minimum power, $P_{\min, N+1}$, needed to satisfy the link's QoS requirements (γ_{N+1}^T and R_{N+1}) on the basis of the current interference level measured at the receiver $I_{N+1} = \sum_{j=1}^{N} P_j \cdot g_{N+1 j}$, and the maximum power, $P_{\max, N+1}$, bounded by P_{bound} and satisfying the constraints imposed by the MEI levels of the N active links.

The minimum and maximum power are derived according to the two following equations:

$$P_{\min, N+1} = \frac{\gamma_{N+1}^\mathrm{T} \cdot R_{N+1} \cdot (\sigma^2 T_\mathrm{f} I_{N+1} + \eta_{N+1})}{g_{N+1 N+1}}, \quad (16.7)$$

$$P_{\max, N+1} = \min\left\{P_{\text{bound}}, \min_{1 \leq j \leq N}\left\{\frac{M_j}{g_{jN+1}}\right\}\right\}. \quad (16.8)$$

The access can take place if:

$$P_{\min, N+1} \leq P_{\max, N+1}. \quad (16.9)$$

We select a suitable transmission power level, within the range $[P_{\min, N+1}, P_{\max, N+1}]$. As stated before, the considered criterion for power selection is to keep balanced the MEI values in the system. In fact, it is to be noticed that the access probability for the $(N+1)$th link, defined as $\text{Prob}\{P_{\min, N+1} \leq P_{\max, N+1}\}$, is as high as the MEI values M_i, $i = 1, \ldots, N$. In Equation (16.8) the lowest MEI constitutes a bottleneck for further accesses.

At the access of the $(N+1)$th link, the optimal working point $\mathbf{P}^-$ [see Equation (16.5)] can be set by choosing suitable P_{N+1}. In particular, the power P_{N+1} that balances MEIs will be the one that maximizes the minimum MEI. An example of the potential impact of a new link's transmission in terms of MEIs is shown in Figure 16.2 for the case $N = 2$ where also the selected power P_{N+1} is indicated (in the example P_3). In the figure the MEIs of the already active links decrease as a function of the power of the new entering link. On the other side, the MEI of the new link increases as P_3 increases. The transmission power value is computed according to the following equation which provides the minimum value among the abscissas of intersection between the new link's MEI and the active links' ones:

$$P_{N+1} = \min_{1 \leq i \leq N} \left\{ \frac{M_i^- + I_{N+1} + \eta_{N+1}/\sigma^2 T_f}{g_{iN+1} + \frac{g_{N+1 N+1}}{\sigma^2 T_f \gamma_{N+1}^T R_{N+1}}} \right\} \quad (16.10)$$

where M_i^- denotes the value of the MEI for the generic ith link before the $(N+1)$th link's access at power P_{N+1}.

The selected power P_{N+1} actually represents just a suboptimal choice with respect to the Pareto-optimal solution since the access is managed in an incremental

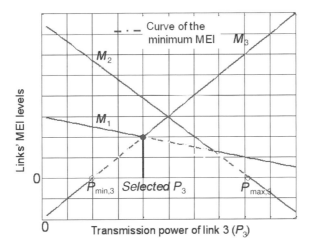

Figure 16.2 Example of P_{N+1} selection in case of two already active links.

438 ADAPTIVE UWB SYSTEMS

way, that is without reconfiguring transmission powers of the active links. On the other hand, this power selection is *adaptive* to the current system conditions (represented by M_i^- and I); for this reason the proposed approach is named in the following adaptive MEI (A-MEI).

16.2.3 Steps of the Access Scheme

This subsection describes the A-MEI operations performed at the transmitter (TX) and the receiver (RX) of a link that should be activated in the system. In Figure 16.3 we report the messages exchanged among terminals. The proposed scheme is based on the assumption that each terminal that should initiate a communication as TX acquires the current MEI values of its neighboring receivers (named in the following n_RXs). The acquisition of MEIs allows the TX to compute the maximum power it can emit so that the n_RXs still maintain their negotiated QoS, even if further interference will be introduced in the system by the new transmission. As a consequence, the implementation of the access scheme requires an explicit interlink signaling: each terminal advertises on a common channel its current MEI level (MEI message). The common channel is constituted by a

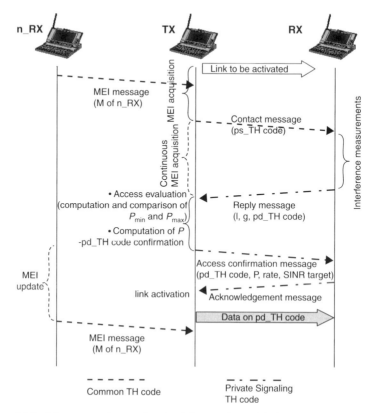

Figure 16.3 Message exchanges for the AC procedure with the adaptive MEI approach.

common TH-code (c_TH code) shared by all terminals with a random access procedure. In Equation (16.8) the MEIs of all the N links in the system are considered. However, since the impact that the transmission could have on a generic terminal is inversely proportional to the distance of this terminal from the TX, it is sufficient that only MEIs of the neighboring terminals, n_RXs—which are less than or equal to N—are acquired.

The TX estimates the reciprocal path gains between itself and the n_RXs; this estimate is derived by the MEI messages by comparing the relevant transmitted and received powers: The first one can be known *a priori* while the second one can be measured.

Besides the MEI advertisements, the AC of the new link requires a signaling exchange between the TX and the RX. This is obtained by the exchange of a contact message in the TX → RX direction and a reply message in the RX → TX direction. The contact message is sent on the c_TH code and signals a private signaling TH code (ps_TH) to be used between the TX and the RX for the next steps of the procedure. The reception of a contact message at the RX triggers the measurement of the perceived interference, I. The RX answers the TX by sending the reply message on the ps_TH code, and communicates the values of the measured interference path gain in the TX → RX direction and the TH code that could be used for the data transmission (private data TH code—pd_TH code).

The admission rule is checked by the TX by:

- Computing P_{min} on the basis of the QoS parameters, the interference measured at the receiver and the path gain;
- Computing P_{max} on the basis of the received MEIs and P_{bound}; the TX continues to update the MEIs from n_RXs;
- Comparing P_{min} and P_{max} and checking if condition in Equation (16.9) is satisfied.

If the access is possible, the computation of the transmission power is performed by the TX in accordance to Equation (16.10). If the access is denied (since it results in $P_{min} > P_{max}$), it may be possible to reconfigure the QoS request. For example, the desired rate R could be reduced to decrease P_{min}.

In the case of access success, the TX also checks the pd_TH code indicated by the RX for transmission and sends an access confirmation message notifying to the RX of the access decision and the parameters selected for transmission (pd_TH code, P, rate and target SINR). After the acknowledgment sent back by the RX, the TX can transmit data on the active link.

16.3 PERFORMANCE ANALYSIS

This section is dedicated to present quantitative results derived by simulating the distributed power-regulated AC scheme. We simulated topologies composed of pairs of terminals each representing a link trying to enter the system.

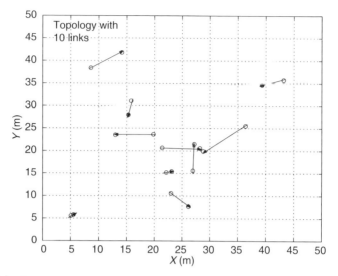

Figure 16.4 Example of a topology with 10 links to establish: The arrows are in the direction TX → RX.

Specifically, UWB terminals are placed within a square area (50 × 50 m). Transmitters are randomly placed in the area and receivers are randomly located within a circle of radius 10 m centered in the corresponding transmitter. Figures 16.4 and 16.5 illustrate two examples of topology composed of 10 and

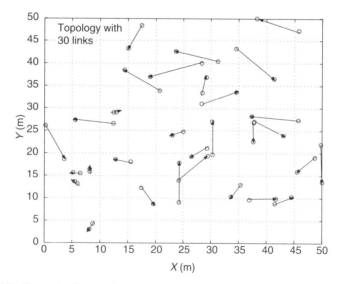

Figure 16.5 Example of a topology with 30 links to establish: The arrows are in the direction TX → RX.

TABLE 16.1 QoS Requirements Parameters for the Three Classes of Traffic: Multimedia, Voice, and Data

Class of Traffic	Data Rate	Target SINR (dB)
Multimedia	1 Mbps	15
Voice	400 kbps	15
Data	5 Mbps	8

30 links to be activated. As stated, in the considered topologies, a receiver is forced to be within a maximum distance from the corresponding transmitter; nevertheless, this receiver may be interfered with by disturbing transmitters that in the topology are very close to it.

As for traffic, we consider three classes: multimedia, voice, and data. Each class is charaterized by a different pair of the QoS parameters R and γ^T (see Table 16.1). We simulate only homogeneous scenarios, that is scenarios where all links are supposed to have the same QoS parameters.

The PHY layer parameters are set according to the values reported in Table 16.2. In particular, we assume a simplified path-loss model that leads to the following expression of the path gain g between two terminals, as a function of their distance d:

$$g = \frac{G_0}{d^\rho} \quad (16.11)$$

where ρ is the path-loss exponent and G_0 is a constant term. The value considered for ρ in simulations, $\rho = 3.5$, is typical of an indoor scenario with NLOS propagation.

The two performance metrics studied in this simulation campaign are (i) the achieved throughput measured as number of links successfully activated (throughput expressed in bps can be derived by multiplying the number of links and their respective data rate); (ii) the average power employed by terminals.

TABLE 16.2 Values Adopted for the Transmission Parameters

Parameter	Symbol	Value
Maximum power	P_{bound}	0.556 mW
UWB parameters depending on the pulse shape form	$\sigma^2 T_f$	1.9230×10^{-10} s
Exponent of distance in the path gain expression	ρ	3.5
Path gain at 1 m distance	$\frac{G_0}{1m^{3.5}}$	2.1023×10^{-6}
Spectral density of thermal noise	η_0	-196.0871 dB W/Hz

The performance analysis is oriented to compare the presented power-regulated AC scheme, based on adaptation of MEIs (A-MEI), to:

1. A similar scheme based on adoption of MEIs; the difference is that power level selected for transmission at the link activation is computed in order to achieve an initial constant MEI level equal for all links and compatible with the maximum power, P_{bound} (this scheme is named constant MEI, C-MEI);
2. A theoretical optimum strategy which is supposed to have the capability of re-allocating all power levels at each new link entrance according to the Pareto-optimal power solution (this scheme is named in the following MIN-POW).

16.3.1 Impact of the Initial MEI on Performance of MEI-Based Power Regulation Schemes

In this section we investigate the impact on performance of the *initial* MEI in the two schemes based on adoption of MEIs, A-MEI and C-MEI. Initial MEI (denoted in the following by M_0) has a different meaning in the two schemes: in A-MEI it is the value selected by the *first* link activated in the system; in C-MEI it is the value that *each* link acquires when it is admitted in the system. The presented results are derived in scenarios with 5, 25, or 50 links to be activated.

The average transmission powers as a function of M_0, when MEIs are either adapted to the environment (A-MEI) or initially configured at a constant level (C-MEI), are shown in Figures 16.6–16.8, respectively, for 5, 25, and 50 links.

Figure 16.9 is the plot of the throughput achieved by A-MEI and C-MEI vs M_0. The employed transmission powers increase as M_0 grows for both A-MEI and C-MEI and tend to saturate. In C-MEI, the saturation floor is closer to the upper bound, P_{bound}, than the saturation level gained by A-MEI. Irrespective of the actual network conditions in terms of active links and current interference, with C-MEI terminals are forced to transmit at the power level allowing the constant initial MEI, M_0, to be achieved. Conversely, with A-MEI the transmission power selected at the link activation is adapted to the current network conditions and in particular the MEI is chosen on the basis of the MEIs currently perceived by receivers, close to the the new link.

The saturation of transmission powers reflects the saturation of the number of established links, as illustrated in Figure 16.9. For both A-MEI and C-MEI, initially, throughput, in terms of number of links established with success, increases as M_0 increases. For high values of M_0, the behaviors of A-MEI and C-MEI slightly differ and the more the number of links in the area the more significant this difference, as in the case of 50 links. As M_0 grows, the curves relevant to A-MEI saturate due to the maximum power constraint that limits also the actually achievable MEIs. Instead, as for C-MEI, the curves reach a peak and then decrease, tending to a

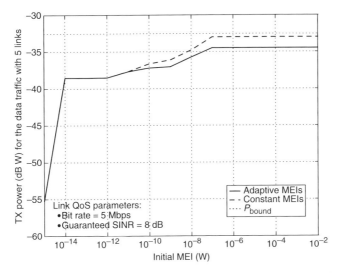

Figure 16.6 Average transmission power of the data traffic versus initial MEI for five links: comparison between adaptive and constant selection of MEIs.

constant value. In this case, since M_0 represents the initial MEI that each new link tries to achieve, the entering links generate higher interference that future links will have to overcome with higher transmission powers, thus quickly saturating the maximum power constraint. For low number of links (e.g., in the case of five links), the two access schemes perform almost the same since the adaptation of

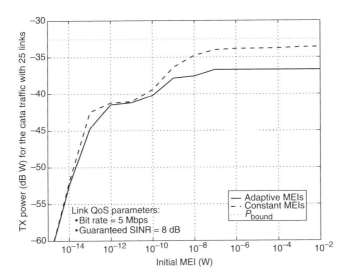

Figure 16.7 Average transmission power of the data traffic versus initial MEI for 25 links: comparison between adaptive and constant selection of MEIs.

444 ADAPTIVE UWB SYSTEMS

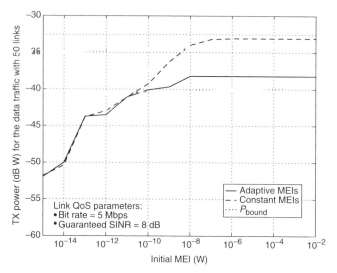

Figure 16.8 Average transmission power of the data traffic vs initial MEI for 50 links: comparison between adaptive and constant selection of MEIs.

MEIs poorly impacts the activation of the subsequent links, whose number is low. In general, a strategy with adaptive selection of MEIs becomes important as the number of links increases. Another interesting observation concerns the behavior for very low values of M_0 ($M_0 < 10^{-11}$ W in Figure 16.9). In this case, C-MEI outperforms A-MEI in terms of throughput. This effect is due to the fact that C-MEI

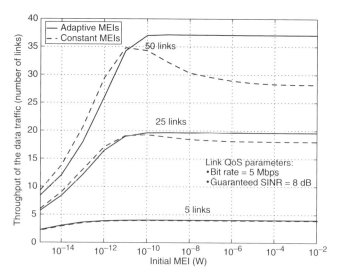

Figure 16.9 Achieved throughput of the data traffic versus initial MEI for 5, 25, and 50 links: comparison between adaptive and constant selection of MEIs.

provides higher MEIs with respect to A-MEI and, when the probabilities of access success are low, this results in better performance.

16.3.2 Performance Behavior as a Function of the Offered Load

This section investigates system performance as function of the offered load. In particular, the offered load is tuned by varying the number of links that must be activated in a given topology. We compare the MEI-based strategies—A-MEI and C-MEI—to the theoretical reference represented by the MIN-POW scheme. We consider the three traffic classes introduced above: multimedia, voice, and data.

Figures 16.10–16.12 represent the achieved throughput as a function of the number of links to be activated for the three traffic classes, respectively. It is worth remarking that, in these plots, the abscissa indicates the number of links trying to enter the UWB system while the ordinate reports the number of links that actually succeed in the AC procedure, and thus that are activated.

The throughput achieved by the two MEI based strategies (A-MEI and C-MEI) is lower than the theoretical maximum (represented by the MIN-POW curve) due to the adoption of a suboptimal access strategy. Such an effect is due to the trade-off between system performance and simplicity of the adopted access scheme, which operates in a distributed fashion based on partial information locally gathered. An additional loss in performance is due to the necessity to protect the QoS negotiated by already active links. As the offered load increases, the loss in performance, measured as the network throughput, increases too.

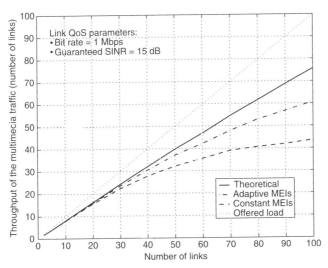

Figure 16.10 Achieved throughput of the multimedia traffic versus number of links to be activated: comparison among adaptive selection of MEIs, constant selection of MEIs and theoretical maximum throughput.

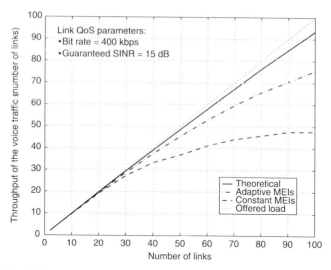

Figure 16.11 Achieved throughput of the voice traffic versus number of links to be activated: comparison among adaptive selection of MEIs, constant selection of MEIs, and theoretical maximum throughput.

The comparison among the throughput achieved with each of the three traffic classes highlights an interesting behavior of the considered AC schemes. In terms of number of activated links, all access strategies perform better in case of voice traffic than in case of multimedia and data traffic. This behavior is due to

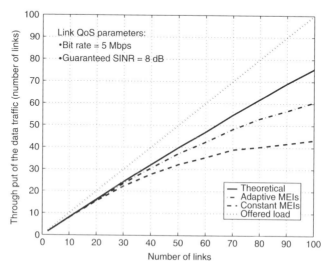

Figure 16.12 Achieved throughput of the data traffic versus number of links to be activated: comparison among adaptive selection of MEIs, constant selection of MEIs, and theoretical maximum throughput.

the lower QoS required by voice traffic, expressed by both lower rate and target SIR. In addition, performance relevant to the achieved throughput as number of activated links is almost the same for the multimedia and the data traffic. This effect is due to the fact that the required QoS impacts the admissibility of a link by means of the product of rate and target SIR and not by the two parameters separately. In particular, the QoS parameters adopted for the multimedia and data traffic (see Table 16.1) are equivalent in terms of product of required rate and SINR. As a consequence, the number of links that can be admitted with a given AC strategy is the same in case of multimedia or data links. However, since the data rate supported by each link is different (1 Mbps for multimedia traffic and 5 Mbps for data traffic), the overall throughput measured in Mbps is different too. The throughput in Mbps achieved with multimedia traffic is lower than that relevant to the data traffic due to the more stringent QoS request in terms of SINR.

Figures 16.13–16.15 report the results relevant to the average transmission powers vs the number of links to be activated, for each of the considered traffic class. In these figures the considered level of the upper bound, P_{bound}, is also reported. Thanks to the local check of the power upper bound constraint [see Equation (16.8)] during the AC procedure, the obtained transmission power levels are always below P_{bound}. The adoption of margins (specifically, the maintenance of a positive tolerable extra interference, MEI, at the receivers) implies that allocated powers are always higher than theoretical minimum values (see the MIN-POW curves). Nevertheless, such a gap is in the order of few dB (generally,

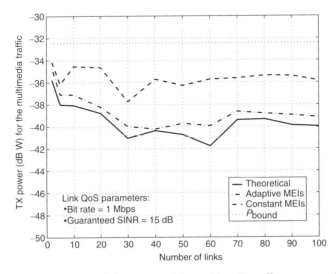

Figure 16.13 Average transmission power of the multimedia traffic versus number of links to be activated: comparison among adaptive selection of MEIs, constant selection of MEIs, and theoretical minimum power.

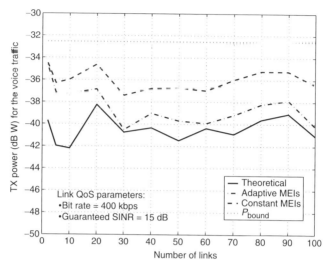

Figure 16.14 Average transmission power of the voice traffic versus number of links to be activated: comparison among adaptive selection of MEIs, constant selection of MEIs, and theoretical minimum power.

about 1–2 dB). The same loss in performance, however, guarantees the possibility of handling new accesses with little effort in terms of AC complexity. With respect to C-MEI, the adaptation of MEIs results in lower transmission powers (this difference is about 3–4 dB).

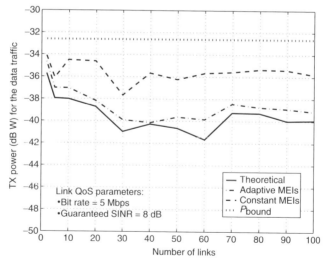

Figure 16.15 Average transmission power of the data traffic versus number of links to be activated: comparison among adaptive selection of MEIs, constant selection of MEIs, and theoretical minimum power.

16.4 SUMMARY

This chapter deals with adaptive UWB systems and describes access schemes and protocols, which exploit adaptivity in UWB and provide efficient and high performing networking in wireless UWB networks.

The chapter describes a distributed admission control scheme based on the maintenance of a positive maximum extra interference level at each receiver. This guarantees a suitable control on the number of links that can be activated; also, it guarantees that links already active can always provide the negotiated QoS characteristics, in terms of data rate and SINR. During the activation of a wireless UWB link, adaptability is achieved by measuring the current interference at the candidate receiver and by gathering information on the MEI values at receivers, which are near to the candidate transmitter.

Simulations results show that, in case of data traffic (data rate of 5 Mbps and target SINR of 8 dB) and of quite big networks (e.g., 100 links), the adaptive MEI scheme achieves a throughput 20% less than the theoretical one. On the contrary, a gain of 40% with respect to the nonadaptive schemes (denoted above as "constant MEI") is obtained. By using adaptive UWB mechanisms, the power used by terminals can be lower than the upper bound fixed by FCC and very close to the theoretical minimum.

REFERENCES

1. Porcino D. and Hirt W., "Ultra-wideband radio technology: potential and challenges ahead," *IEEE Communications Magazine*, pp. 66–74, July 2003.
2. IEEE P802.15.3 Draft Standard for Telecommunications and Information Exchange Between Systems—LAN/MAN Specific Requirements—Part 15.3: Wireless Medium Access Control (MAC) and Physical Layer (PHY) Specs for High Rate Wireless Personal Area Networks (WPAN); www.ieee802.org/15/pub/TG3.html, 2003.
3. Zhang H. and Hou J. C., "Capacity of Wireless ad Hoc Networks under Ultra Wide Band with Power Constraint," *Proc. of IEEE Infocom 2005*, March 2005.
4. Cuomo F., Martello C., Baiocchi A. and Capriotti F., "Radio resource sharing for ad hoc networking with UWB," *IEEE Journal on Selected Areas in Communications*, vol. 20, no. 9, pp. 1722–1732, December 2002.
5. Radunovic B. and Le Boudec J. Y., "Optimal power control, scheduling and routing in UWB networks," *IEEE Journal on Selected Areas in Communications*, vol. 22, no. 7, pp. 1252–1270, September 2004.
6. Erseghe T., Laurenti N., Nicoletti P. and Sivieri A., "An algorithm for radio resource management in UWB ad hoc networks with concurrent guaranteed service and best effort traffic," *Proc. of WPMC 2004*, Italy, September 2004.
7. Di Benedetto M. G., De Nardis L., Junk M. and Giancola G.,"(UWB)^2: uncoordinated, wireless, baseborn medium access for UWB communication networks," *Mobile Networks and Applications* special issue on WLAN Optimization at the MAC and Network Levels, vol. 10, no. 5, pp. 663–674, October 2005.

8. Lu K., Wu D., Fang Y. and Qiu, R., "Performance analysis of a burst-frame-based MAC protocol for ultra-wideband ad hoc networks," *Proc. of IEEE International Conference on Communications (ICC 2005)*, Seoul, 16–20 May 2005.

9. Chu, Y. and Ganz A., "MAC protocols for multimedia support in UWB-based wireless networks," *International Workshop on Broadband Wireless Multimedia (BroadWim 2004)*, San Jose, CA, October 2004.

10. Merz R., Le Boudec J-Y., Widmer J. and Radunovic B., "A rate-adaptive MAC protocol for low-power ultra-wide band ad-hoc networks," in *Proc. Ad-Hoc Now 2004*, 22–24 July 2004.

11. August N. J., Thirugnanam R. and Ha, D. S., "An adaptive UWB modulation scheme for optimization of energy, BER, and data rate," *IEEE Conference on UWB System and Technologies*, pp. 182–186, May 2004.

12. Kim B. S., Fang Y. and Wong T. F., "Rate-adaptive MAC protocol in high-rate personal area networks," *Proc. of IEEE WCNC 2004*, pp. 1394–1399, March 2004.

13. Güvenc I., Arslan H., Gezici S. and Kobayashi H., "Adaptation of multiple access parameters in time hopping UWB cluster based wireless sensor networks," *Proc. of IEEE International Conference on Mobile Ad-hoc and Sensor Systems (MASS)*, pp. 235–244, October 2004.

14. Win, M. Z. and Scholtz, R. A. "Ultra-wide bandwidth time-hopping spread-spectrum impulse radio for wireless multiple-access communication," *IEEE Transactions on Communications*, vol. 48, no. 4, pp. 679–690, 2000.

15. Cuomo, F. and Martello C., "A distributed power regulated algorithm based on SIR margins for adaptive QoS support in wireless networks," *Proc. of Personal Wireless Communications 2003*, vol. 2775 of Lecture Notes on Computer Science. Springer, Heidelberg, pp. 114–127, 2003.

CHAPTER 17

UWB Location and Tracking—A Practical Example of a UWB-based Sensor Network

IAN OPPERMANN, KEGEN YU, ALBERTO RABBACHIN, LUCIAN STOICA, PAUL CHEONG, JEAN-PHILIPPE MONTILLET, and SAKARI TIURANIEMI

17.1 INTRODUCTION

Sensor networks are typified by devices with low complexity that have limitations on processing power and memory, and severe restrictions on power consumption. By the very nature of the application, traffic in sensor networks is often bursty with long periods of no activity. A device may remain idle for long periods of time sending only periodic information, then suddenly be required to send significant amounts of information when an event occurs. For devices deployed in the field, this has significant implications for the design of efficient medium access protocols, radio communications technology and the reliability of information transfer. For devices involved in continuous monitoring, the flow of traffic will be more stable. However efficient multiple access, reliability and battery life are still major considerations.

Since the US FCC released the First Report and Order in 2002 covering commercial use of Ultra Wideband (UWB) [1], there has been greatly increased interest in UWB-based applications. This in turn has ignited interest in the use of UWB for sensor networks and fueled research in the area. Impulse radio-based UWB technology has a number of inherent properties which are well suited to sensor network applications. In particular, impulse radio-based UWB systems have potentially low complexity and low cost, have noise-like signals, are resistant to severe multipath and jamming, and have very good time domain resolution allowing for location and tracking applications [1].

The low complexity and low cost of impulse radio UWB systems arises from the essentially baseband nature of the signal transmission. Unlike conventional radio

---

Ultra Wideband Wireless Communication. Edited by Arslan, Chen, and Di Benedetto
Copyright © 2006 John Wiley & Sons, Inc.

systems, the UWB transmitter produces a very short time domain pulse which is able to propagate without the need for an additional RF mixing stage. The RF mixing stage takes a baseband signal and "injects" a carrier frequency or translates the signal to a frequency which has desirable propagation characteristics. The very wideband nature of the UWB signal means that it spans frequencies commonly used as carrier frequencies. The signal will propagate well without the need for additional up-conversion and amplification. The UWB receiver also does not require the reverse process of down-conversion. Again, this means that a local oscillator in the receiver can be omitted, which means the removal of associated complex delay and phase tracking loops. High achievable burst data rates for UWB systems mean that sensors can transfer their payload data quickly and spend much of the rest of the time "asleep" or in a low-power state.

To realize the benefits of UWB in sensor networks, careful consideration must be given to the design of the MAC, conservation of power and efficient radio technology. The solutions developed depend very much on the application examined. This chapter illustrates an example of a low complexity UWB system (UWEN) which supports low data rate communications with location and tracking for various applications. The system concept is targeted at recreational activities such as cross country skiing, athletics, and running. It may equally well be applied to asset tracking and inventory control. The concept includes the development of small, low power UWB devices which are carried by the user. Information from the UWB devices is collected by *fixed nodes* in the network, which also exchange information with a central position server to determine the location of the UWB devices. This position information is stored in the network along with sensor data from the UWB devices for later retrieval.

17.2 MULTIPLE ACCESS IN UWB SENSOR SYSTEMS

The very wide bandwidth of UWB systems means that many potential solutions exist to the issue of bandwidth usage. Devices may use all or only a fraction of the bandwidth available in the 3.1–10.6 GHz band. These devices will still be classed as UWB provided they use at least 500 MHz. Major candidates for the physical layer signal structure of UWB systems include impulse radio, OFDM, multicarrier and hybrid techniques. All of these possible techniques mean that different UWB devices may or may not be able to detect the presence of other devices. The main issues to be addressed by an UWB MAC include coexistence, inter operability and support for location/tracking.

The potential proliferation of UWB devices of widely varying data rates and complexities will require co-existence strategies to be developed. Strategies for ignoring or working around other devices of the same or different type based on physical layer properties will reflect up to the MAC layer. Optimization of the UWB physical layer should lead to the highest-efficiency, lowest-BER, lowest-complexity transceivers. The assumptions of the physical layer will, however, have implications for MAC issues such as initial search and acquisition process, channel access protocols, interference avoidance/minimization protocols, and power adaptation protocols. The quality of the achieved "channel" will have implications on the link level, which may necessitate active searching by a device for better conditions, which is what happens with other radio systems.

The most common requirement of MAC protocols is to support interworking with other devices of the same type. With the potentially wide range of device types, the MAC design challenge is to be able to ensure cooperation and information exchange between devices of different data rate, QoS class or complexity. In particular, emphasis must be placed on how low-complexity, low-data-rate devices can successfully produce limited QoS networks with higher complexity, HDR devices.

17.2.1 Location/Ranging Support

Location/ranging support is integrally linked to the MAC. This includes strategies for improving signal timing accuracy and for exchanging timing information to produce estimates of the device position. It is possible for any single device to estimate the arrival time of a signal from another device based on its own time reference. This single data point in relative time needs to be combined with other measurements to produce a 3-D position estimate relative to some system reference. Exchange of timing information requires cooperation between devices. Being able to locate all devices in a system presents a variation of the "hidden node" problem. The problem is further complicated for location because multiple receivers need to detect the signal of each node to allow a position in three dimensions to be determined.

Tracking requires that each device is able to be sensed/measured at a suitable rate to allow a reasonable update rate. This is relatively easy for a small number of devices, but difficult for an arbitrarily large number of devices. Information exchange between devices of timing and position estimates of neighbours (ad hoc modes) requires coordination, and calculation of position needs to be done somewhere (centralized or distributed) and the results fed to the information sink.

Finally, it is important to have the received signal as unencumbered by multiple access interference as possible in order to allow the best estimation of time of arrival. Every 3.3 ns error in delay estimation translates to a minimum 1 m extra error in position estimation.

All of these issues—information exchange, device sampling rate, node visibility, signal conditioning—require MAC support. They are significant obstacles to existing WLAN and other radio systems offering reliable location/tracking when added on to the MAC post-design.

17.2.2 Constraints and Implications of UWB Technologies on MAC Design

Some qualities of UWB signals are unique and may be used to produce additional benefit. For example, the accurate ranging capabilities with UWB signals may be exploited by upper layers for location-aware services. Conversely, some aspects of UWB pose problems that must be solved by the MAC design. For example, using a carrier-less impulse radio system, it is cumbersome to implement the carrier sensing capability needed in popular approaches such as carrier-sense, multiple access/collision avoidance (CSMA/CA) MAC protocols.

Another aspect that affects MAC design is the relatively long synchronization and channel acquisition time in UWB systems. In [2], the performance of the CSMA/CA protocol is evaluated for a UWB physical layer. CSMA/CA is used in a number of distributed MAC protocols and it is also adopted in the IEEE 802.15.3 MAC.

The time to achieve bit synchronization in UWB systems is typically high, of the order of few milliseconds [2]. Considering that the transmission time of a 10,000 bit packet on a 100 Mbps rate link is only 0.1 ms, it is easy to understand the impact of synchronization acquisition on CSMA/CA based protocols. The efficiency loss due to acquisition time can be minimized by using very long packets. However this may impact performance in other ways.

Acquisition preambles are typically sent with higher transmit power than data packets [3]. This impacts both the interference level and the energy consumption in highly burst traffic. This effect must be taken into account when determining the efficiency of the system.

The adoption of CSMA/CA as a distributed protocol must be jointly evaluated with the performance of the underlying UWB physical layer. In general it may not be a suitable choice for an UWB MAC unless proper synchronization techniques are developed. One solution to this problem is the exploitation of the very low duty cycle of impulse radio. Synchronization can be maintained during silent periods by sending low power preambles for synchronization tracking [3]. This approach is feasible only for communications between a single pair of nodes, which is not the case in peer-to-peer networks.

17.3 UWB SENSOR NETWORK CASE STUDY

The UWEN (UWB wireless embedded networks) project is developing a system offering low rate communications with location and tracking for various applications. The system concept is targeted at recreational activities such as cross country skiing, athletics, and running. The concept includes the development of small, low-power UWB devices that are carried by the user. Data from the UWB devices is collected by *fixed nodes* in the network that also exchange signal time-of-arrival information to determine the location of the UWB devices. This location information is stored in the network along with sensor data from the UWB devices for later retrieval.

The low complexity and low cost of impulse radio UWB systems arise from the essentially baseband nature of the signal transmission. Unlike conventional radio systems, the UWB transmitter produces a very short time domain pulse which is able to propagate without the need for an additional RF mixing stage. A typical RF mixing stage takes a baseband signal and "injects" a carrier frequency or translates the signal to a frequency that has desirable propagation characteristics. The very wideband nature of the UWB signal means it spans frequencies commonly used as carrier frequencies. The UWB signal will propagate well without the need for additional up-conversion. The UWB receiver also does not require the reverse process of down-conversion. This means a local oscillator in the receiver can be omitted as well as the removal of complex delay and phase-tracking loops with their associated power consumption. High achievable burst data rates for UWB systems means that sensors can transfer their payload data quickly and spend much of the rest of the time "asleep" or in a low-power state.

Figure 17.1 shows an example usage of the system in the initial target application of snow sport/recreation. Each user carries an UWB tag with a sensor for gathering biometric data, in this case heart rate readings. Communication takes places in a master–slave manner with the low-cost, mobile UWB devices sending and receiving information and the fixed nodes controlling the operation of the devices. The fixed nodes exchange information about the perceived position of each sensor in the network. The estimated position of the user, along with biometric information, is stored in the network for later retrieval. The UWB devices relay information on the speed, direction, and heart rate data from the mobile user. The UWB signal itself is used to position the tag based on calculations performed by the fixed nodes.

In this chapter, we also investigate the performance of a number of position estimation methods which make use of TOA estimation for this low-cost/low-complexity UWB system. The performance evaluation is performed in terms of the root-mean-square error (RMSE) of the position coordinates estimation and the

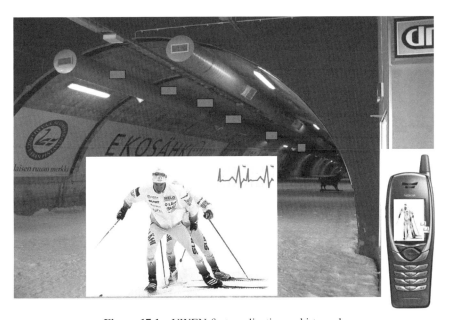

Figure 17.1 UWEN first application—ski tunnel.

position estimation failure rate. A simple and reliable two-step timing acquisition technique is proposed to obtain the TOA estimate. We then present the formula for direct calculation of position coordinates using either TOA or time difference of arrival (TDOA). Kalman filtering is examined as a means to improve the location accuracy when tracking moving UWB devices. Simulation results are presented for performance evaluation and comparison.

The focus is on the investigation/development of practical and feasible location algorithms using UWB for position localization and tracking suitable for implementation in an ASIC device. Some basic architecture information is given about the UWB devices.

17.4 SYSTEM DESCRIPTION—UWEN

17.4.1 Communications System

A major objective of the UWEN project is to develop a low-complexity, low-power-consumption transceiver and system architecture. For this reason, a noncoherent modulation scheme has been adopted utilising binary pulse position modulation (BPPM) combined with direct sequence (DS). The use of BPPM allows very simple, noncoherent "energy collection"-based receiver architectures to be adopted. The DS "overlay" sequence is used to randomize the spectrum of the transmitted signal so as to avoid strong spectral lines associated with simple pulse repetition.

The MAC solution must also be low complexity and support an environment with a minimum of interuser interference. This lead to the choice of TDMA. The system also uses time division duplexing TDD. There is an "up-link" time frame when the UWB devices can send information to the fixed nodes, and there is a "down-link" time frame where the UWB devices receive commands and information. Each of these two time frames is introduced by a beacon that carries information on the availability and the structure of the network. The TDMA system has an aggregate data rate of 5 Mbps which may be divided amongst the numerous devices. Considering a target device data rate target of several kbps, there may be many hundreds of devices in the system.

17.4.2 Transmitted Signal

The UWB signal used for this study is based on a train of short pulses randomized in phase by a scrambling sequence. The bit interval is divided into two time slots (binary modulation). As the detection procedure is based on energy collection, the separation of different users can only be done in the time domain. The transmitted signal for single user is given by

$$s(t) = \sum_{k=-\infty}^{\infty} \sum_{f=1}^{N} w_{\text{tr}}(t - kT_{\text{b}} - jT_{\text{c}} - \delta d_k)(c_{\text{p}})_j$$

17.4 SYSTEM DESCRIPTION—UWEN

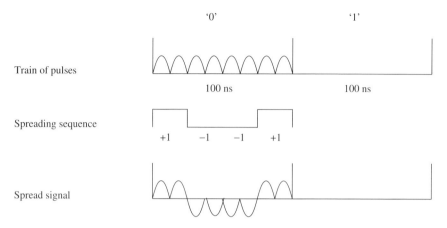

Figure 17.2 Bit position modulation.

where $w_{tr}(t)$ is the transmitted pulse with pulse width T_p. T_b is the symbol interval, $\delta = 100$ ns is the delay used to distinguish different symbols and $d_k \in [0,1]$ is the transmitted symbol. $T_c = NT_p$ where N is an integer, T_c is the chip interval, and $(c_p)_j$ is the jth chip of the pseudorandom (PR) code. The PR code is bipolar with values $\{-1, +1\}$. The data rate R_d is defined by $1/T_b = 1/(2\delta)$.

Randomization of the transmitted pulses is required in order to smooth the spectrum. Since the system is TDMA-based, the randomization does not serve any channelization role, so all devices will use the same scrambling code sequence with a reasonable chip frequency. This means that more than one UWB pulse will be affected by the same scrambling sequence chip. The modulation technique used for BPPM as shown in Figure 17.2.

The receiver in the UWB tag collects the energy in the two possible bit windows and the bit decision is based on the comparison between the energies, as shown in Figure 17.3.

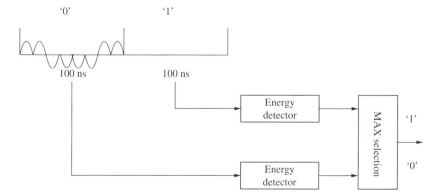

Figure 17.3 Bit decision process.

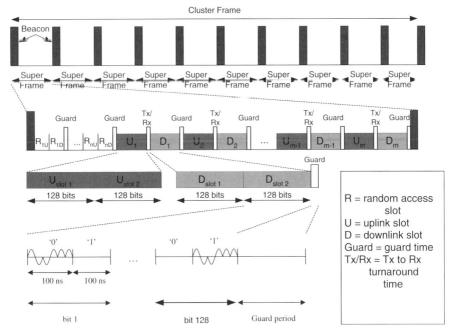

Figure 17.4 Frame structure.

17.4.3 Framing Structure

The requirement to handle a large and flexible number of users with fixed nodes exchanging information to calculate position, and subsequently supplying the information to user, implies the transmission of a large amount of data signaling data. The very low power of UWB signals requires that, during periods of TOA measurement, only one user is transmitting at a time. A frame structure is shown in Figure 17.4.

When a tag (slave) enters the network, a registration process begins. The tag detects the beginning of the frame by detecting a beacon and randomly selects a time slot inside the registration window. This slot is selected randomly from R_{xU}, random access slots (shown in white in Figure 17.4) to minimize collision with other devices registering. In the "downlink" slot, the controlling fixed node (master) replies to successfully registered devices in the same slot position of the registration window.

The reply carriers the number of a new slot position for the tag to use for communications. The tag will use this time slot for the time it is registered with the network.

17.4.4 Location Approach

In the system, there is a limited set of fixed nodes which are closely time synchronized by sharing the same local clock through cable connections. The fixed nodes are positioned at known coordinates in the area being monitored. Since each tag transmits data in different, preassigned time slots, multiple access interference is greatly reduced. Due to the drift in the clock of the mobile devices as well as the fixed nodes, synchronization between the fixed nodes and the mobile devices is performed once every second. This is achieved by broadcasting a beacon from one of the fixed nodes. The TOA of the beacon is used as the reference clock for the mobile devices to transmit data according to the preassigned time slots.

At any given time, the TOA measurements from a specific group of fixed nodes are collected for position estimation by the system. Since each tag is moving, the group of fixed nodes will change over time. In general, the fixed nodes with the strongest received signal powers are selected to provide the TOA estimates and their position coordinates are employed for the mobile tag position estimation.

There are many position estimation techniques using radio signals. Signal strength, angle of arrival, TOA measurements, time of flight/round trip time, and TDOA can all be exploited for position estimation. The most straightforward way to estimate position is to directly solve a set of simultaneous equations [4] based on the TOA/TDOA measurements. Exact solutions can be obtained for 2-D location with two fixed nodes using two TOA measurements (with known transmit time) or with three fixed nodes using three TDOA measurements. For a 3-D location, four fixed nodes are needed to obtain exact solutions using TDOA measurements.

For an over-determined system (with redundant fixed nodes), several different approaches have been proposed such as spherical interpolation [5–8], the two-stage maximum likelihood method [9], and the linear-correction least square approach [10]. Also several iterative approaches have been investigated for position estimation. Taylor series expansion can be used to iteratively produce a linear least-square solution [11, 12]. However, to maintain good convergence, the Taylor series method may require a close estimate of the actual location as a starting point which may be difficult to obtain in some practical applications. A different iterative method for location comes from nonlinear optimization theory. The gradient-based algorithms may be employed for position estimation [13, 14]. One is the quasi-Newton algorithm [15], which has been used in the UWB precision assets location system [16]. The other is the Gauss–Newton type Levenberg–Marquardt method [17].

In this chapter, several location algorithms have been investigated based on noniterative techniques utilizing TOA information. Results will be presented in later sections.

17.5 SYSTEM IMPLEMENTATION

17.5.1 Transceiver Overview

The UWB transceiver architecture for the UWB devices is based on a noncoherent structure utilizing bit position modulation (BPPM) [18–20]. While sacrificing some

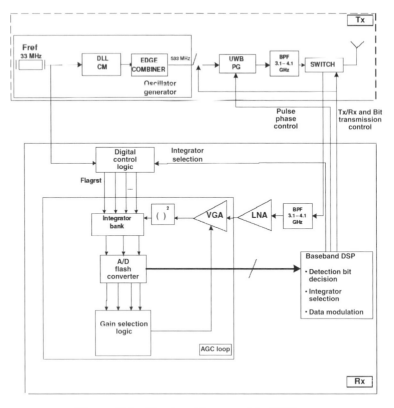

Figure 17.5 Circuit architecture of the UWB tag.

performance in terms of spectral efficiency, it greatly simplifies the implementation and decreases the size and cost of the circuit. The architecture for the UWB tag transceiver is presented in Figure 17.5. There is no signal correlation process which occurs in the receiver, so the modulation scheme employed must be orthogonal in the time domain. The receiver simply collects the signal energy in different time windows and determines the transmitted bit based on the detected maximum energy. The bit synchronization process for the receiver makes use of a short preamble which reduces the receiver complexity while still achieving good performance in the presence of multipath fading [18–20].

17.5.2 Transmitter

The transmitter is based on a delay locked loop (DLL) and UWB monocycle pulse generator. The transmitter module contains a clock generator, a UWB pulse generator and a UWB antenna. The clock generator is composed of a 33 MHz quartz oscillator, a DLL and a digital edge combiner used as for clock multiplication to produce a 533 MHz timing signal that drives the pulse generation. The UWB pulse generator generates a monocycle of 350 ps typical width once every 1.87 ns.

For UWB systems, accurate timing is very important. Subnanosecond pulses need to be transmitted using an accurate reference clock. Clocking at high speed, however, leads to significant power consumption and so must be done sparingly. The choice of the 33 MHz crystal reference clock to drive the UWB transceiver and digital components was based on the desire to keep the overall system clock as low as possible.

The higher speed clock is generated only as required for pulse transmission. The use of a DLL coupled with an edge combiner avoids classic frequency synthesis techniques using a PLL and N divider combination. The approach used is presented in Figure 17.6. The high-speed clock used for pulse triggering is generated by taking each edge of the crystal oscillator output and generating a burst of well-controlled, evenly spaced edges that span one period of the crystal oscillator. These evenly-spaced edges are then combined by an AND–OR based edge combiner. The phase detector generates UP/DOWN signals according to the phase difference between the reference signal and the final output stage of the voltage controlled delay line (VCDL). The phase detector architecture was presented in [21]. The DLL is in a "locked" state if the input and output of the voltage controlled delay line are in phase. One delay cell contains two cascaded CMOS inverters. The delay is adjusted by the charge pump and loop filter through control voltage designed to minimize the phase error. A classification of charge pump architectures

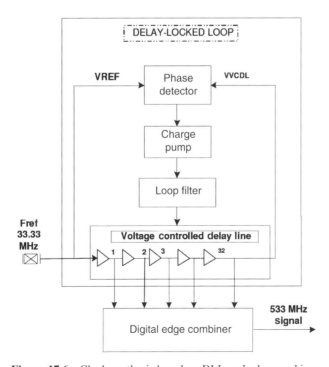

Figure 17.6 Clock synthesis based on DLL and edge combiner.

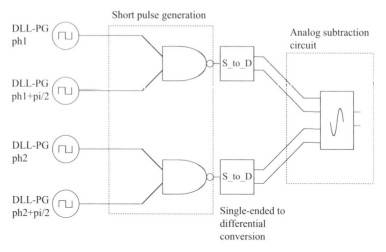

Figure 17.7 UWB pulse generator schematic.

is given in [22]. The charge pump architecture used is based on a single ended architecture with an active amplifier on the output, as found in [22].

17.5.3 UWB Pulse Generator

The UWB pulse generator used in the transmitter is presented in Figure 17.7 [22]. It creates two short pulses from two trigger signals which have a fixed delay between

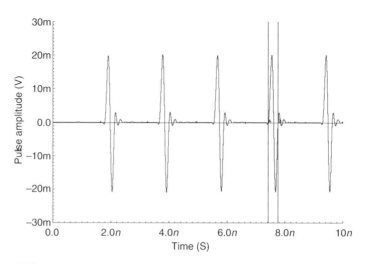

Figure 17.8 Generated pulse train (~350 ps pulses), without randomizing sequence.

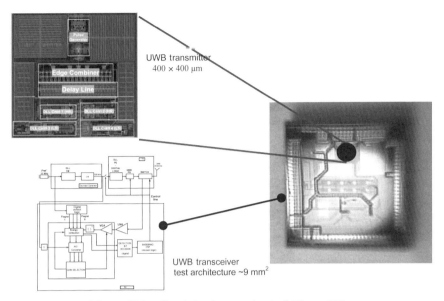

Figure 17.9 Circuit implementation in 0.35 μm SiGe.

them. The short pulses generated by the NAND-gates are converted into a differential signals by a micromixer [23]. Two of these differential pulses with a fixed time delay are subtracted from each other in an analog linear subtraction circuit. The output is a differential monocycle pulse with a typical pulse width of 350 ps. An example of a generated pulse train (without the randomizing pulse train is shown in Figure 17.8. Figure 17.9 shows the implemented UWB transmitter circuit in 0.35 μm SiGe.

17.6 LOCATION SYSTEM

To achieve accurate position estimation, we must first acquire accurate TOA measurements. There are numerous TOA estimation algorithms in the literature. A comprehensive literature review on code acquisition and delay estimation for direct-sequence spread spectrum signals can be found in [24, 25].

The extremely short, very low-duty cycle UWB pulses with very low power spectral density, pose a challenge for synchronization in UWB systems. One method proposed in the literature for UWB timing recovery employs an ML approach [26, 27]. A second method applies correlators in the traditional way, but makes use of techniques to obtain rapid timing acquisition. For example, a look-and-jump search and a bit reversal search approach have been proposed in [28]. Special code design has been employed in [29]. Chip-level post-detection integration (CLPDI) has been proposed in [30] and applied to UWB in [31]. Another method is the frequency-domain treatment of UWB synchronization using spectral estimation [32]. For low-cost and low-complexity applications, energy collection-based timing acquisition [20] is a promising approach. This technique is particularly suitable for indoor communications where dense multipath exists.

In this project, we employ a two-stage approach for fast timing acquisition to obtain the time-of-arrival of the desired signal. In the first stage, a bank of integrators is employed. The received signal is first squared. Each integrator then integrates the squared signal for a period of time, T_{int}, usually a fraction of one symbol duration. A search is performed over one symbol duration. The first integrator starts integration at a chosen time point. Each of the other integrators begins integration after a delay of T_{int} compared with its preceding integrator. The start time point of the integrator whose output is the maximum among all the integrators provides a coarse TOA estimate. With a probability dependant on the SNR, the coarse TOA estimate will indicate the region containing the first received pulses. If the first stage search is successful, the coarse TOA estimate will satisfy

$$\tau_0 - T_{int} \leq \hat{\tau}_0 \leq \tau_0 + T_m$$

where τ_0 and $\hat{\tau}_0$ are the true and the coarse TOA estimate respectively and T_m is the multipath spread. In the second stage, a refined search is constrained to this uncertain region. This may be achieved though the use of a backward and forward search using the same bank of integrators. The difference between the start time points of two adjacent integrators for the fine search can be as small as the clock period. Figures 17.10 and 17.11 show two examples of the time sequence and outputs of the integrators at the second stage. For simplicity, integration spans the interval $\tau_2 - \tau_0 + T_w$, as shown in Figures 17.10 and 17.11, where the true TOA is τ_0,

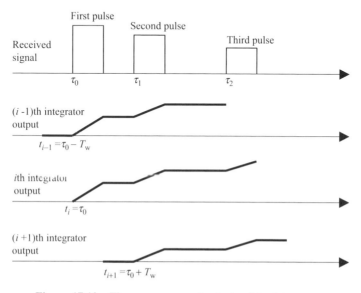

Figure 17.10 Time sequence and outputs of the integrators.

and T_w is the pulse width. In practice, the integration interval for the second stage can be chosen based on the predicted/estimated channel parameters. The maximum selection criterion is also applied at the second stage. The process may continue over a sequence of symbols to produce multiple TOA estimates which can be further processed to obtain more accurate estimates.

It is worth noting that NLOS propagation conditions introduce additional errors in TOA measurements as the received pulses are delayed as a result of propagating through material which is denser than air. Although it is difficult to accurately predict the additional delay, techniques exist to partially compensate the NLOS impact on the accuracy of TOA estimation. Statistical information of the NLOS error, if available, can be employed to reduce the NLOS effect [33]. Also, a variety of techniques in NLOS identification and LOS reconstruction have been proposed to mitigate the NLOS effect [34–37]. In the absence of any information about NLOS error, and when LOS propagation results are not available, nonparametric techniques may be applied to produce TOA measurements [38].

To examine the accuracy of the proposed two-stage TOA estimation approach, we consider an outdoor environment with dimensions $400 \times 100 \times 100$ m. The data rate is 5 Mbps and each bit consists of only one pulse [16]. A four-path channel model is employed to approximate a snow-covered environment. Whilst being a considerable simplification, this channel model serves to demonstrate the merits of the techniques employed and simplifies the evaluation process. The first channel path signal has constant amplitude (corresponding to a LOS signal) while the other three paths have Nakagami fading amplitudes with a fading value of $m = 1.5$ [39]. The fading amplitudes can be either positive or negative with equal probability [40]. The delay of the second path is 2 ns and the fourth path is 12 ns. The power ratio (Rician factor) of the direct path to the fading paths is equal to one (i.e., 0 dB). The receiver sampling rate in the fixed nodes is assumed to be 2 GHz. In Figure 17.10, the ith integrator switches on at $t_i = \tau_0$ where τ_0 is the true TOA. In Figure 17.11, the ith integrator switches on at $t_i = \tau_0 + T_w/2$ where T_w is the pulse width. These two time intervals correspond to best and worse scenarios (on-time and offset by half an integration window).

It is assumed that the first-stage (coarse) search is successful and the TOA errors are limited to the integrating period of one integrator after the search. In this case, the TOA errors are limited to ± 12.5 ns. A TOA estimate is produced over each symbol interval. In the simulation results, a total of 50,000 symbols are examined to produce 50,000 TOA estimates for an SNR of 8 dB. Figure 17.12 shows the amplitude distribution of the TOA estimation errors when the time instants for switching on the integrators are as shown in Figure 17.10. Figure 17.13 shows the corresponding results when the time instants of switching on the integrators are as shown in Figure 17.11.

Figure 17.14 shows the RMS error of the TOA estimation with respect to signal-to-noise ratio for two different Rice factors for the channel model used. In this figure, *syn* denotes results when triggering of the integrators are as shown in Figure 17.10 (best case). The other results are obtained when time instants of the integrators are as shown in Figure 17.11 (worst case).

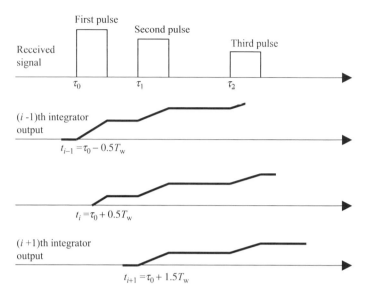

Figure 17.11 Time sequence and outputs of the integrators.

The results indicate that, to achieve RMSE values of below 1 ns, an SNR of 15 dB or more is required. RMSE improvements below 1 ns require significant increases in SNR. An interesting observation is that higher Rician factor results in worse performance for some high SNR values. As may be expected, below about

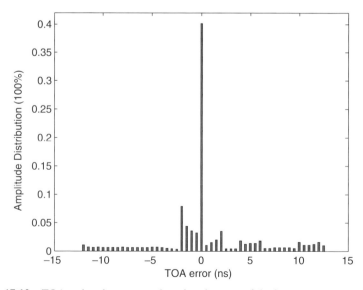

Figure 17.12 TOA estimation errors when time instants of the integrators are as shown in Figure 17.10.

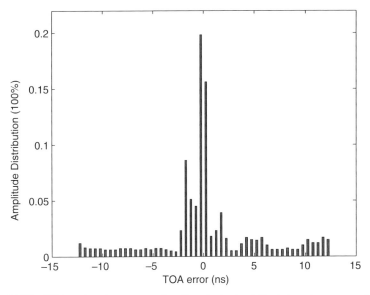

Figure 17.13 TOA estimation errors when time instants of the integrators are as shown in Figure 17.11.

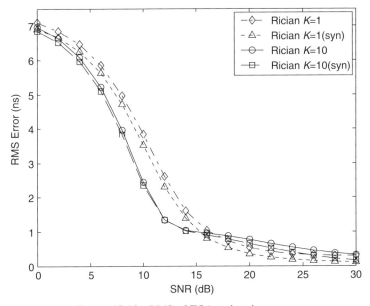

Figure 17.14 RMS of TOA estimation errors.

15 dB, the RMSE results for the low Rician value are considerably worse than the high Rician value results. The integration approach does however produce useable TOA estimates suitable for use in accurate location techniques.

17.7 POSITION CALCULATION METHODS

There exist many iterative and noniterative position estimation algorithms. The choice of algorithm depends somewhat on the computation budget per tag and the desired accuracy of the position result. The accuracy of the position calculation in turn depends on the number and quality of the TOA estimates which can be included in the calculation for a given tag. This section will focus on lower-complexity, noniterative position techniques. The choice is the result of the requirement to locate and track many hundreds or thousands of devices with moderate position accuracy.

Non-iterative techniques include direct calculation [4, 41] and the least-square techniques [7, 9]. The noniterative algorithms are simple and easy to implement. The device position may be determined as follows.

The distance between fixed node i and the mobile tag is given by

$$\sqrt{(x-x_i)^2 + (y-y_i)^2 + (z-z_i)^2} = c(t_i - t_0) \quad i = 1, 2, 3, 4, \quad (17.1)$$

where (x, y, z) and (x_i, y_i, z_i) are the coordinates of the tag and fixed node respectively, c is the speed of light, t_i is the signal TOA at fixed node i, and t_0 is the unknown transmit time at the mobile tag. In the development of the expressions, we ignore the difference between the true and the measured TOA for simplicity. Squaring both sides of Equation (17.1) gives

$$(x-x_i)^2 + (y-y_i)^2 + (z-z_i)^2 = c^2(t_i - t_0)^2 \quad i = 1, 2, 3, 4. \quad (17.2)$$

Subtracting Equation (17.2) for $i = 1$ from Equation (17.2) for $i = 2, 3, 4$ produces

$$ct_0 = \frac{1}{2}c(t_1 - t_i) + \frac{1}{2c(t_1 - t_i)}(\beta_{i1} - 2x_{i1}x - 2y_{i1}y - 2z_{i1}z) \quad i = 2, 3, 4 \quad (17.3)$$

where

$$x_{i1} = x_i - x_1, \; y_{i1} = y_i - y_1, \; z_{i1} = z_i - z_1, \; \beta_{i1} = x_i^2 + y_i^2 + z_i^2 - (x_1^2 + y_1^2 + z_1^2).$$

Define the TDOA between sensors i and j as

$$\Delta t_{ij} = t_i - t_j$$

17.7 POSITION CALCULATION METHODS

Eliminating t_0 from Equation (17.3) yields

$$a_1 x + b_1 y + c_1 z = g_1 \qquad (17.4)$$

where

$$a_1 = \Delta t_{12} x_{31} - \Delta t_{13} x_{21}, \quad b_1 = \Delta t_{12} y_{31} - \Delta t_{13} y_{21}, \quad c_1 = \Delta t_{12} z_{31} - \Delta t_{13} z_{21}$$

$$g_1 = \frac{1}{2}(c^2 \Delta t_{12} \Delta t_{13} \Delta t_{32} + \Delta t_{12} \beta_{31} - \Delta t_{13} \beta_{21})$$

and

$$a_2 x + b_2 y + c_2 z = g_2, \qquad (17.5)$$

where

$$a_2 = \Delta t_{12} x_{41} - \Delta t_{14} x_{21}, \quad b_2 = \Delta t_{12} y_{41} - \Delta t_{14} y_{21}, \quad c_2 = \Delta t_{12} z_{41} - \Delta t_{14} z_{21}$$

$$g_2 = \frac{1}{2}(c^2 \Delta t_{12} \Delta t_{14} \Delta t_{42} + \Delta t_{12} \beta_{41} - \Delta t_{14} \beta_{21}).$$

Combining Equations (17.4) and (17.5) yields

$$x = Az + B \qquad (17.6)$$

where

$$A = \frac{b_1 c_2 - b_2 c_1}{a_1 b_2 - a_2 b_1}, \quad B = \frac{b_2 g_1 - b_1 g_2}{a_1 b_2 - a_2 b_1}$$

and

$$y = Cz + D \qquad (17.7)$$

where

$$C = \frac{a_2 c_1 - a_1 c_2}{a_1 b_2 - a_2 b_1}, \quad D = \frac{a_1 g_2 - a_2 g_1}{a_1 b_2 - a_2 b_1}.$$

Then, substitution of Equations (17.6) and (17.7) into Equation (17.3) with $i = 2$ produces

$$c(t_1 - t_0) = Ez + F \qquad (17.8)$$

where

$$E = \frac{1}{c\Delta t_{12}}(x_{21}A + y_{21}C + z_{21})$$

$$F = \frac{c\Delta t_{12}}{2} + \frac{1}{2c\Delta t_{12}}(2(x_{21}B + y_{21}D) - \beta_{21}).$$

Substituting Equations (17.6–17.8) into Equation (17.1) for $i = 1$ followed by squaring gives

$$Gz^2 + Hz + I = 0 \qquad (17.9)$$

where

$$G = A^2 + C^2 - E^2 + 1$$
$$H = 2[A(B - x_1) + C(D - y_1) - z_1 - EF]$$
$$I = (B - x_1)^2 + (D - y_1)^2 + z_1^2 - F^2.$$

The solutions to Equation (17.9) are

$$z = -\frac{H}{2G} \pm \sqrt{\left(\frac{H}{2G}\right)^2 - \frac{I}{G}}$$

If both estimated z values are reasonable, they are substituted into Equations (17.6) and (17.7) to produce the coordinates x and y, respectively. Since there is only one desirable solution, we remove the solution with either no physical meaning or which is beyond the monitored area. If both solutions are reasonable and they are very close, we may choose the average as the position estimate. Cases where there are no acceptable results include cases with two complex solutions, or when both solutions are beyond the area being examined.

When the transmit time t_0 is available, only three fixed nodes are required to determine the position variables. It is straightforward to derive the solution of the position coordinates in this case. The device position may be determined as follows:

Define

$$f_{1i} = \frac{1}{2}\{c^2[(t_1 - t_0)^2 - (t_i - t_0)^2] + \beta_{i1}\}, \quad i = 2, 3.$$

Also define

$$A_1 = \frac{x_{21}z_{31} - x_{31}z_{21}}{x_{31}y_{21} - x_{21}y_{31}}, \qquad B_1 = \frac{x_{31}f_{12} - x_{21}f_{13}}{x_{31}y_{21} - x_{21}y_{31}},$$

$$C_1 = \frac{y_{31}z_{21} - y_{21}z_{31}}{x_{31}y_{21} - x_{21}y_{31}}, \qquad D_1 = \frac{y_{21}f_{13} - y_{31}f_{12}}{x_{31}y_{21} - x_{21}y_{31}}.$$

Then we have

$$\hat{z} = \frac{F_1}{E_1} \pm \sqrt{\left(\frac{F_1}{E_1}\right)^2 - \frac{G_1}{E_1}}, \tag{17.10}$$

where

$$E_1 = A_1^2 + C_1^2 + 1, \quad F_1 = A_1(y_1 - B_1) + C_1(x_1 - D_1) + z_1$$
$$G_1 = (x_1 - D_1)^2 + (y_1 - B_1)^2 + z_1^2 - c^2(t_1 - t_0)^2$$

and

$$\hat{x} = C_1\hat{z} + D_1 \tag{17.11}$$
$$\hat{y} = A_1\hat{z} + B_1. \tag{17.12}$$

Different formulae may be derived for direct position calculation; however, the method presented has the desirable property that it does not involve matrix operations.

Evaluation of the performance of the different techniques is performed in terms of the RMSE of the coordinate estimation results and the failure rate. The RMSE is defined as

$$\sqrt{\frac{1}{3N_pN_s} \sum_{i=1}^{N_p} \sum_{j=1}^{N_s} \left[(x^{(i)} - \hat{x}^{(ij)})^2 + (y^{(i)} - \hat{y}^{(ij)})^2 + (z^{(i)} - \hat{z}^{(ij)})^2\right]}$$

where N_p is the number of different position combinations of the fixed nodes and the mobile tag and N_s is the number of TOA samples at each SNR for each position combination. $(x^{(i)}, y^{(i)}, z^{(i)})$, $(\hat{x}^{(ij)}, \hat{y}^{(ij)}, \hat{z}^{(ij)})$ are the true and estimated position coordinates of the mobile tag of interest, respectively.

The failure rate includes cases for which there is no solution or the solution is unreasonable, including cases when the solutions are beyond the monitored area, both solutions are complex-valued, the two solutions are reasonable but not close to each other, or inversion of a singular matrix is involved.

In the simulation results, the monitored area examined has dimensions of 90 (l) × 90 (w) × 10 m (h). The positions of the fixed nodes and the mobile tag are randomly generated. For each point, the results of 1000 runs are averaged. New random positions of the fixed nodes and the mobile tag are generated for each run. Whilst this approach provides a general insight to the performance of a particular location technique, the location of the fixed nodes can have a substantial impact on the location performance. Therefore, the performance of one selected fixed node configuration is also examined. This is the configuration from the 1000 random fixed node configurations which produces the best results in terms of RMSE.

The TOA estimation errors are produced using the synchronization technique described earlier. Throughout the rest of the chapter, the results corresponding to the time instants of the integrators as shown in Figure 17.11 are employed.

Figures 17.15 and 17.16 show the accuracy and failure rate of the direct method and the spherical interpolation algorithm [7]. The TOA-based approach (results denoted "DM3") is seen to be sensitive to the accuracy of transmit time. "DM3(inacc t0)" presents results with a transmit time error of 4 ns. When the transmit time error is in the order of tens of nanoseconds, the TOA-based method performs very poorly. Only when nearly error-free transmit time information is available (compared with the time reference at fixed nodes) is the approach described by Equations (17.10)–(17.12) suitable.

Also shown are results for the spherical-interpolation method with five, six and eight fixed nodes (denoted "SI5," "SI6," and "SI8"). The technique does not work well in the case of four fixed nodes, as indicated in [7], so the corresponding results are not presented. At relatively high SNR, the spherical-interpolation method performs well when at least five fixed nodes are employed.

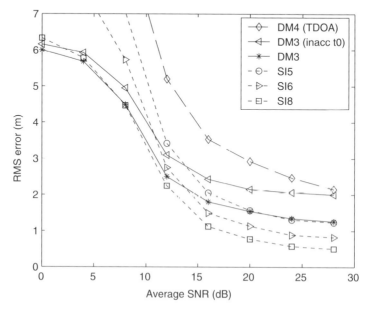

Figure 17.15 RMSE of DM and SI position estimation techniques.

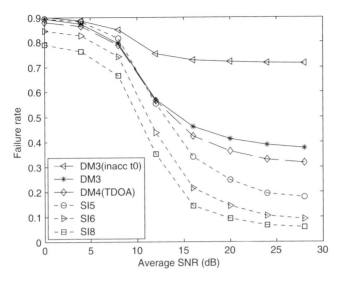

Figure 17.16 Failure rate of DM and SI position estimation techniques.

17.8 TRACKING MOVING OBJECTS

When dealing with mobile devices, tracking should be included in the algorithms considered. The system should be able to update the position estimate at a reasonable rate to follow the moving devices. At each time instant, a number of TOA measurements are collected from a specific set of fixed nodes. Members of the set of fixed nodes will change as the tag moves. Usually the fixed nodes closest to the tag are employed to provide the time measurements since in general shorter distance means higher signal power so better performance can be obtained.

Tracking performance can be improved by smoothing the individual position results using techniques such as Kalman filtering which have been widely used in modern control systems, tracking, and navigation systems [42]. References illustrating the use of Kalman filtering for smoothing of position or velocity estimates can be found in [43–47]. Other filtering approaches can be used for smoothing position estimates. A linear least squares approach is used in [48] to simultaneously obtain smoothed estimate of the position and the speed.

In this section, we are interested in applying the well known Kalman filtering for position smoothing. The Kalman filtering algorithm is given in [49]. Note that Kalman filtering can be implemented in two different ways. One technique uses an individual Kalman filter for smoothing each position coordinate so that three filters are used in total. The other uses only one filter to smooth the three coordinates. Figure 17.17 shows the block diagram of the proposed location and tracking system.

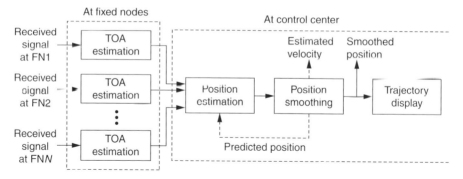

Figure 17.17 Block diagram of position location and tracking.

17.8.1 Simulation Results

In this section, we examine the performance of the proposed position location and tracking system (as shown in Figure 17.17). We use one of the realistic field structures, a snow covered slope of dimensions about $400 \times 100 \times 100$ m shown in Figure 17.18. The fixed nodes will be deployed along both sides of the slope and mounted on poles of varying height. The skier moves from A to B (120 m) at a speed of 8 m/s. The skier moves from B to C (160 m) at a speed of 10 m/s and finally from C to D (120 m) at a speed of 8 m/s.

First we examine the performance of the different position estimation algorithms under the more realistic circumstance. Two hundred different combinations of fixed node positions are tested and then the results are averaged. Tables 17.1 and 17.2 compare the averaged results of the two algorithms at SNR of 12 dB while Tables 17.3 and 17.4 show the results at SNR of 16 dB. In the tables, FNs refers to fixed nodes, SI is the spherical-interpolation method while DM is the direct method.

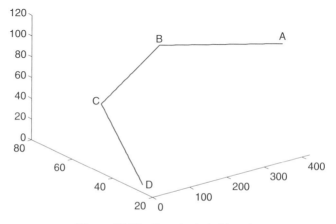

Figure 17.18 Hypothetical ski track.

17.8 TRACKING MOVING OBJECTS

TABLE 17.1 Averaged RMSE (m) of SI and DI Algorithms at SNR of 12 dB

No. of FNs	4	5	6	8
SI		3.0	1.14	0.40
DM	11.91			

TABLE 17.2 Averaged Failure Rate (%) of SI and DI Algorithms at SNR of 12 dB

No. of FNs	4	5	6	8
SI		8.4	3.1	1.9
DM	48.0			

For the parameters examined, the spherical-interpolation method provides the best trade-off between performance and complexity when there are at least five fixed nodes. To achieve submeter accuracy, at least six fixed nodes are needed with SNR of up to 16 dB.

Table 17.5 shows the averaged RMSE before and after position smoothing using Kalman filtering. The three estimated tracks (before smoothing) are produced using the SI algorithm with five fixed nodes under three different sets of fixed node configurations. Since the one-Kalman-filter scheme and the three-Kalman-filter scheme produce the same results, only results from one of them are listed.

Figure 17.19 shows the original, estimated and smoothed tracks using Kalman filtering. Dimensions are in meters. The RMSE before smoothing is 2.70 m. The effectiveness of smoothing is clearly demonstrated. Note that the speed of the moving object can also be provided by the smoothing techniques. The

TABLE 17.3 Averaged RMSE (m) of SI and DI Algorithms at SNR of 16 dB

No. of FNs	4	5	6	8
SI		1.55	0.33	0.09
DM	7.54			

TABLE 17.4 Averaged Failure Rates (%) of SI and DI Algorithms at SNR of 16 dB

No. of FNs	4	5	6	8
SI		3.4	1.3	0.9
DM	46.6			

TABLE 17.5 Averaged RMSE Before and After Smoothing

Before Smoothing (m)	After Smoothing (m)
4.13	2.30
2.70	1.30
1.40	0.99

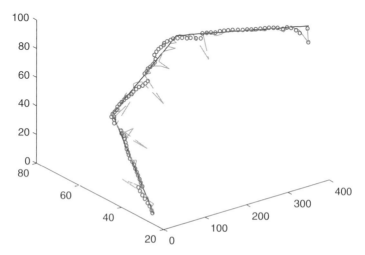

Figure 17.19 Original (—), estimated (— —) and Kalman-filtering smoothed (○) tracks.

discontinuities in the position estimate in Figure 17.19 are the result of poorly selected positions for the fixed nodes. The large "instantaneous" position errors lead to relatively large errors in the smoothed path estimate after Kalman filtering. Careful fixed node selection will minimize these errors.

17.9 CONCLUSION

In this chapter we have explored some of the features of an UWB sensor system with location and tracking capabilities which is in the process of deployment. The system is based on the use of fixed and mobile UWB devices arranged in a master–slave configuration with a TDMA framing structure. The emphasis is on developing relatively high accuracy location techniques for low-cost UWB devices. To support this objective, the core of the UWB devices has been implemented in 0.35 μm SiGe technology resulting in a solution capable of producing approximately 500 million pulses per second with an average duration of 350 ps and a burst data rate of 5 Mbps.

The communications capability of the system is complemented by location and tracking capabilities. Several position estimation approaches were examined which employ UWB technology for outdoor recreational activities. To maintain the low complexity requirement, the emphasis was placed on the noniterative methods (i.e., spherical interpolation and direct calculation). Performance comparisons of the two methods were performed under different scenarios. With the proposed TOA estimation technique and certain number of fixed nodes, accurate position estimates can be obtained even under a realistic field structure.

One of the algorithms examined will ultimately be employed in the location system utilizing the ASIC devices being manufactured for the project.

ACKNOWLEDGMENTS

The authors would like to acknowledge the support of the Finnish national agency TEKES and the industrial partners in the UWEN project, Elektrobit Inc., Softentix, and Polar Electric. Partial support from the EU funded project PULSERS is also acknowledged.

REFERENCES

1. FCC, First Report and Order, 22 April 2002, FCC 02-48.
2. J. Ding, L. Zhao, S. R. Medidi, and K. M. Sivalingam, "MAC protocols for ultra-wideband (UWB) wireless networks: impact of channel acquisition time", *SPIE ITCOM Conf.*, Boston, MA, July 2002.
3. S. S. Kolenchery, J. K. Townsend, and J. A. Freebersyser, "A novel impulse radio network for tactical military wireless communications," *Proc. MILCOM'98*, 1998.
4. B. T. Fang, "Simple solutions for hyperbolic and related position fixes," *IEEE Trans. Aerosp. Elecctron. Syst.*, vol. 26, pp. 748–753, September 1990.
5. B. Friedlander, "A passive localization algorithm and its accuracy analysis," *IEEE J. Ocean Engng.*, vol. 12, pp. 234–245, January 1987.
6. II. C. Schau and A. Z. Robinson, "Passive source location employing intersecting spherical surfaces from time-of-arrival differences," *IEEE Trans. Acoust. Speech, Signal Process.*, vol. 35, pp. 1223–1225, August 1987.
7. J. O. Smith and J. S. Abel, "Closed-form least squares source location estimation from range difference measurements," *IEEE Trans. Acoust. Speech, Signal Process.*, vol. 35, pp. 1661–1669, December 1987.
8. K. W. Cheung, H. C. So, W. K. Ma, and Y. T. Chan, "Least squares algorithms for time-of-arrival-based mobile location," *IEEE Trans. Signal Process.*, vol. 52, pp. 1121–1128, April 2004.
9. Y. T. Chan and K. C. Ho, "A simple and efficient estimator for hyperbolic location," *IEEE Trans. Signal Process.*, vol. 42, pp. 1905–1915, August 1994.

10. Y. Huang, J. Benesty, G. W. Elko, and R. M. Mersereau, "Real-time passive source localization: a practical linear-correction least-squares approach," *IEEE Trans. Speech Audio Process.*, vol. 9, pp. 943–956, November 2001.
11. W. H. Foy, "Position-location solutions by Taylor-series estimation," *IEEE Trans. Aerosp. Elecctron. Syst.*, vol. 12, pp. 187–194, March 1976.
12. D. J. Torieri, "Statistical theory of passive location systems," *IEEE Trans. Aerosp. Elecctron. Syst.*, vol. 20, pp. 183–198, March 1984.
13. P. E. Gill, W. Murray, and M. H. Wright, *Practical Optimization*. London: Academic Press, 1981.
14. R. Fletcher, *Practical Methods of Optimization*. Chichester: J Wiley, 1987.
15. R. Fletcher and M. J. D. Powell, "A rapidly convergent descent method for minimization," *Comput. J.*, vol. 6, pp. 163–168, 1963.
16. R. J. Fontana, E. Richley, and J. Barney, "Commercialization of an ultra wideband precision asset location system," *Proc. IEEE Conf. UWB Systems and Technologies*, pp. 369–373, 2003.
17. D. Marquardt, "Algorithm for least-squares estimation of nonlinear parameters," *SIAM J. Appl. Math.*, vol. 11, pp. 431–441, 1963.
18. A. Rabbachin, R. Tesi, and I. Oppermann, *Bit Error Rate Analysis for UWB systems with a Low Complexity, Non-Coherent Energy Collection Receiver*, Lyon: IST, 2004.
19. A. Rabbachin and I. Oppermann, "Synchronisation analysis for UWB systems with a low-complexity energy collection receiver," *2004 Joint UWBST&IWUWBS*, Kyoto, May 2004.
20. A. Rabbachin and I. Oppermann, "Synchronization analysis for UWB systems with a low-complexity energy collection receiver," *Proc. IEEE Joint UWBST and IWUWBS*, pp. 288–292, 2004.
21. B. Razavi, *Design of Analogue CMOS Integrated Circuits*. New York: McGraw-Hill Higher Education, 2001.
22. W. Rhee, "Design of high-performance CMOS charge pumps in phase-locked loops," *Proc. 1999 IEEE International Symposium on Circuits and Systems ISCAS*, vol. II, pp. 545–548, 30 May to 2 June 1999.
23. S. Tiuraniemi, "CMOS pulse generator design for TH-PPM UWB system," *Proc. Finnish Wireless Communications Workshop (FWCW'03)*, Oulu, pp. 136–139, October 2003.
24. K. Yu, Code acquisition and detection for spread-spectrum mobile communications, PhD thesis, School of Electrical and Information Engineering, University of Sydney, 2002.
25. K. Yu and I. B. Collings, "Performance of low complexity code acquisition for direct-sequence spread spectrum systems," *Proc. IEE Commun.*, vol. 150, no. 6, pp. 453–460, 2003.
26. V. Lottici, A. D'Andrea, and U. Mengali, "Channel estimation for ultra-wideband communications," *IEEE J. Select. Areas Commun.*, vol. 20, pp. 1638–1645, December 2002.
27. J.-Y. Lee and R. Scholtz, "Ranging in a dense multipath environment using an UWB radio link," *IEEE J. Select. Areas Commun.*, vol. 20, pp. 1677–1683, December 2002.
28. E. A. Homier and R. A. Scholtz, "Rapid acquisition of ultra-wideband signals in the dense multipath channel," *Proc. IEEE Conf. UWB Systems and Technologies*, pp. 105–109, 2002.

29. R. Fleming, C. Kushner, G. Roberts, and U. Nandiwada, "Rapid acquisition for ultra-wideband localizers," *Proc. IEEE Conf. UWB Systems and Technologies*, pp. 245–249, 2002.
30. J. Iinatti and M. Latva-aho, "A modified CLPDI for code acquisition in multipath channel," *Proc. IEEE PIMRC*, pp. F6–F10, 2001.
31. S. Soderi, J. Iinatti, and M. Hamalainen, "CLPDI algorithm in UWB synchronization," *Proc. International Workshop on UWB Systems*, June 2003.
32. I. Maravic, M. Vetterli, and K. Ramchandran, "Channel estimation and synchronization with sub-Nyquist sampling and application to ultra-wideband systems," *Proc. International Workshop on UWB Systems*, June 2003.
33. S. Al-Jazzar and J. Caffery Jr, "ML and Bayesian TOA location estimation for NLOS environments," in *Proc. IEEE VTC Fall*, pp. 1178–1181, 2002.
34. M. P. Wylie and J. Holtzmann, "The non-line of sight problem in mobile location estimation," in *Proc. IEEE Conf. Universal Personal Communications*, pp. 827–831, 1996.
35. P.-C. Chen, "A non-line-of-sight error mitigation algorithm in location estimation," in *Proc. IEEE WCNC*, pp. 316–320, 1999.
36. S.-S. Woo, H.-R. You, and J.-S. Koh, "The NLOS mitigation technique for position location using IS-95 CDMA networks," in *Proc. IEEE VTC Fall*, pp. 2556–2560, 2000.
37. L. Cong and W. Zhuang, "Non-line-of-sight error mitigation in TDOA mobile location," in *Proc. IEEE GLOBECOM*, pp. 680–684, 2001.
38. M. McGuire, K. N. Plataniotis, and A. N. Venetsanopoulos, "Location of mobile terminals using time measurements and survey points," *IEEE Trans. Veh. Technol.*, vol. 52, pp. 999–1011, July 2003.
39. M. D. Yacoub, J. E. Vargas, and L. G. R. Guedes, "On higher order statistics of the Nakagami-m distribution," *IEEE Trans. Veh. Technol.*, vol. 48, pp. 790–794, May 1999.
40. J. Forester, "Channel modeling sub-committee Report (Final)," IEEE 802.15 SG3a task group, December 2002.
41. K. Yu and I. Oppermann, "Performance of UWB position estimation based on TOA measurements," *Proc. IEEE Joint UWBST and IWUWBS*, pp. 400–404, 2004.
42. M. S. Grewal and A. P. Andrews, *Kalman Filtering Theory and Practice*. Englewood Cliffs, NJ: Pretice Hall, 1993.
43. R. Doraiswami, "A novel Kalman filter-based navigation using beacons," *IEEE Trans. Aerosp. Elecctron. Syst.*, vol. 32, pp. 830–840, April 1996.
44. M Hellebrandt and R. Mathar, "Location tracking of mobiles in cellular radio networks," *IEEE Trans. Veh. Technol.*, vol. 52, pp. 1558–1562, September 1999.
45. C.-D. Wann, Y.-M. Chen, and M.-S. Lee, "Mobile location tracking with NLOS error mitigation," *Proc. IEEE GLOBECOM*, pp. 1688–1692, 2002.
46. M. McGuire and K. N. Plataniotis, "Dynamic model-based filtering for mobile terminal location estimation," *IEEE Trans. Veh. Technol.*, vol. 52, pp. 1012–1031, July 2003.
47. M. Najar and J. Vidal, "Kalman tracking for mobile location in NLOS situations," *Proc. IEEE PIMRC*, pp. 2203–2207, 2003.
48. M. Hellebrandt, R. Mathar, and M. Scheibenbogen, "Estimating position and velocity in a cellular radio network," *IEEE Trans. Veh. Technol.*, vol. 46, pp. 65–71, February 1997.

49. K. Yu, J. P. Montillet, A. Rabbachin, P. Cheong, and I. Oppermann, "UWB location and tracking for wireless embedded networks", *EURASIP J. Signal Process.*, special issue (in press).
50. G. R. Aiello, "Challenges for ultra-wideband (UWB) CMOS integration," *Microwave Symposium Digest, 2003 IEEE MTT-S International*, vol. I, pp. 8–13, June 2003.
51. D. Dickson and P. Jett, "An application specific integrated circuit implementation of a multiple correlator for UWB radio applications," *Proc. Military Communications Conf., MILCOM 1999. IEEE*, vol. II, pp. 1207–1210, October 1999.
52. C. L. Bennett and G. F. Ross, "Time-domain electromagnetics and its applications," *Proc. IEEE* vol. 66, pp. 299–318, March 1978.
53. J. Foersters and Q. Li, "UWB channel modelling contribution from Intel," IEEE P.802.15-02/279-SG3a.
54. S. M. Kay, *Fundamentals of Statistical Signal Processing: Estimation Theory.* Upper Saddle River, NJ: Prentice Hall, 1993.
55. G. F. Ross, "The transient analysis of certain TEM mode four-port networks," *IEEE Trans. Microwave Theory Technol.*, vol. 14, no. 11, pp. 528–547, 1966.
56. A. Rabbachin, L. Stoica, S. Tiuraniemi and I. Oppermann, "A low cost, low power UWB based sensor network", IWUWBS/UWB-ST, Kyoto, Japan, May 2004.
57. I. O'Donnell, M. Chen, S. Wang, and B. Brodersen, "Ultra-wideband hardware design," *An Ultra-Wideband Technology Workshop: From Research to Reality*, Los Angeles, CA, 3–4 October, 2002.

INDEX

ABR. *See* Associativity-based routing (ABR)
AC complexity, 448
Achieved throughput
   data traffic, 446f
   multimedia traffic, 445f
   voice traffic, 446f
Acknowledgement (ACK), 349
AC mechanism, 435
AC phase, 433
AC procedure, 435, 438f, 445, 447
Acquisition algorithms
   comparison, 38t
Acquisition time, 306f
AC rule, 436
ACS. *See* Add–compare–select (ACS) unit
AC scheme, 430, 433, 436, 439, 442, 446
AC signaling change, 439
AC strategy, 447
Adaptation of multiple access parameters, 432
Adaptive hazy sighted link state (A-HSLS), 364
   proactive routing, 363
      protocol comparisons, 420t, 421t
Adaptive UWB system, 429–448
   distributed power-regulated admission control scheme, 432–433
   performance analysis, 439–448
   power selection, 435–437
   problem formalization, 434
   related work, 431
Adaptive zone routing protocol (AZRP), 408
   hybrid routing, 408
      protocols comparisons, 426t, 427t
ADC. *See* Analog-to-digital-converter (ADC)
Add–compare–select (ACS) unit, 293
Additive white Gaussian noise (AWGN), 148, 149, 160
   channels, 51, 231, 241
      BER, 89f

double-sided power spectral density, 174
Address resolution protocol (ARP)
   DNVR, 378
   DSR, 367
Ad hoc on-demand distance vector (AODV), 352, 372
   reactive routing, 372–373
      protocol comparisons, 422t, 424t
   routing protocol
      SHARP, 403
Admissible topology, 435
Advanced MAC design for low-bit-rate UWB networks
   reception procedure, 331
   simulation results, 331
   transmission procedure, 328–330
   Uncoordinated, Wireless, Baseborn Medium access for UWB Communication Networks, 325–327
Aggregated bit rate
   Alt-PHY layer, 318
A-HSLS. *See* Adaptive hazy sighted link state (A-HSLS)
Aircraft wireless intercommunications system (AWICS), 310
Alamouti space-time coding, 216, 217f
Algorithm outline
   UWB channel estimation with frequency-dependent distortion, 29–31
All-rake (ARake) receivers, 93, 94, 95
ALOHA random access scheme, 343
Alt-PHY layer
   complexity, 319
   mobility, 319
   vs. 802.15.4 PHY, 317
   quality of service, 319
   range, 318

Ultra Wideband Wireless Communication. Edited by Arslan, Chen, and Di Benedetto
Copyright © 2006 John Wiley & Sons, Inc.

Alt-PHY technical requirements, 317
A-MEI operations, 438
Analog delay lines
    mismatch, 173f
Analog-to-digital-converter (ADC), 161, 278
Analog UWB space–time coding, 209
Analysis of noise sensitivity
    UWB channel estimation and synchronization, 25
Analyzing MIMO scenarios, 212
Angle of arrival (AoA), 44
    measurements, 45f
        CRLB, 47
        UWB systems, 48–49
    modeling, 46
    UWB geolocation, 45–48
Annihilating filter approach, 18
Ant-based routing algorithm (ARA), 367
    reactive routing, 367–368
        protocol comparisons, 422t, 424t
Antenna, 131–156
    design, 132–140
        single-band and multiband schemes, 132–135
    source pulses, 136
    transmit and PDS, 136–140
    transmit-receive, 141–147
Antenna and pulse
    vs. BER performance, 148–156
Antenna array geometries, 46f
Antenna systems, 132–133
    FD, 150
    pulsed UWB system, 151
    return loss and gain, 152f
    template pulses, 154
    transmission efficiency, 153f
    transmit and receive capability, 146
AoA. See Angle of arrival (AoA)
AODV. See Ad hoc on-demand distance vector (AODV)
ARA. See Ant-based routing algorithm (ARA)
ARake. See All-rake (ARake) receivers
ARP. See Address resolution protocol (ARP)
Array gain, 206
Associativity-based routing (ABR), 369
    reactive routing, 369–371
        protocol comparisons, 422t, 424t
    routing, 369–371
Asymptotically optimum receiver structure, 57f
Asynchronous multi-user channels
    modulation schemes, 97f
Autocorrelation functions, 98f

Average angle, 214
Averaged failure rate
    SI and DI, 475t
Averaged root mean square error
    SI and DI, 475t
    smoothing, 476t
Average power, 106, 260, 433, 441.
    See also Peak to average power ratio (PAPR)
    clusters, 195
Average transmission power, 442, 443f, 444f, 447f, 448, 448f
AWGN. See Additive white Gaussian noise (AWGN)
AWICS. See Aircraft wireless intercommunications system (AWICS)
AZRP. See Adaptive zone routing protocol (AZRP)

Backup source routing protocol (BSR), 374
    reactive routing, 374–375
        protocol comparisons, 422t, 424t
Backward ant (BANT)
    ARA, 368
Bandpass filter (BPF), 166
Bandwidth, 213f
Bandwidth-antenna spacing, 206
Bandwidth guarded channel adaptive (BGCA), 410
BANT. See Backward ant (BANT)
Battery energy efficient (BEE)
    power-aware routing, 394
        protocol comparisons, 426t
Beacon-enabled modality, 322
Beacon packet, 301
Beamforming, 205
BEB. See Binary exponential backoff (BEB)
BEE. See Battery energy efficient (BEE)
Bellman–Ford shortest path first
    SOAR, 390
BEP. See Bit error probability (BEP)
BER. See Bit error rate (BER)
Binary data mapping schemes, 82f, 87–88
Binary exponential backoff (BEB), 334
Binary orthogonal phase vectors
    Hadamard partition, 119
Binary pulse position modulation (BPPM), 456
    time-hopping UWB signal, 259
Bit decision process, 457f
Bit error probability (BEP), 233
Bit error rate (BER), 2
    vs. distance for pulsed and nonpulsed-OFDM systems, 294f

performance, 127f
  binary data mapping formats, 81f
  incident power, 154f
  integration interval, 177
  template pulses, 154f
  timing misalignment, 97t
Bit position modulation, 457f, 459
Bit rates
  Alt-PHY layer, 318
  MB-OFDM system, 281t
Blind (non-data aided) algorithm, 244
Blind timing offset estimation, 68
Block diagram of the transmitter
  PCTH system, 247f
Bluetooth network formation, 300
Bluetooth PAN, 300
Bordercast
  TZRP, 409
  ZRP, 406
BPF. *See* Bandpass filter (BPF)
BPPM. *See* Binary pulse position modulation
  (BPPM)
BPSK, 88, 89, 90, 92t, 94f, 95, 96, 97t
  CDMA system, 298
BQ. *See* Broadcast query (BQ) packet
Broadband antenna systems
  system transmission efficiency, 147f
Broadband planar dipole system, 145
Broadcast query (BQ) packet
  ABR, 370f
Broadercast resolution protocol (BRP)
  ZRP, 406
BRP. *See* Broadercast resolution
  protocol (BRP)
BSR. *See* Backup source routing
  protocol (BSR)

Caching and multipath routing (CHAMP),
  376
  reactive routing, 376
  protocol comparisons, 422t, 424t
Calculated fidelity
  single-band scheme, 147t
Call for proposal (CFP), 294
CAP. *See* Contention Access Period (CAP)
Carrier frequency offset
  (CFO), 123
Carrier sense multiple access with collision
  avoidance (CSMA/CA), 301, 454
CC. *See* Central controller (CC)
CCDF. *See* Complementary cumulative
  distribution function (CCDF)
CDMA. *See* Code division multiple accessing
  (CDMA)

Central controller (CC), 362
CFO. *See* Carrier frequency offset
  (CFO)
CFP. *See* Call for proposal (CFP); Content
  ion-free period (CFP)
CFSR. *See* Clustered fisheye state
  routing (CFSR)
CGSR. *See* Clusterhead Gateway Switch
  Routing protocol (CGSR)
CHAMP. *See* Caching and multipath routing
  (CHAMP)
Channel impulse response (CIR), 187,
  188, 189
Channel models, 187
Channel parameters, 13, 14, 16, 24, 32, 49,
  50, 127, 161, 171, 172, 178, 186,
  189, 465
Channel state information (CSI), 217
Channel time allocations (CTAs), 302
Channel transfer function (CTF), 188
Cheby shev approximation problem, 111
Chip-level post-detection integration
  (CLPDI), 463
Chip spaced sampling, 161
Chirp pulses, 316
Chi-square, 197
Chong, Chia-Chin, 187
Circuit implementation, 463f
Clear (CLR) packet
  TORA, 393
Clear-to-send (CTS)
  packets
    PARO, 396
    PLBM, 386
Closed form, 215
CLPDI. *See* Chip-level post-detection
  integration (CLPDI)
CLR packet
  TORA, 393
Cluster, 193–196
  arrival times, 193
  average power, 195
  beamforming, 210
  decaying phenomenon, 195
  member table, 351t
  power decaying phenomenon, 195
Clustered fisheye state routing
  (CFSR), 361
Clusterhead, 350
Clusterhead Gateway Switch Routing protocol
  (CGSR), 350
  hierarchical structure, 350f
  proactive routing, 350
    protocol comparisons, 419t, 421t

Clustering
  algorithm, 350, 352
  channel model, 189–190
  MPCs, 187
  multilevel physical, 353
  phenomenon, 187, 193, 207, 213, 214, 220
Cluttered way, 209
CMMBCR. *See* Conditional max-min capacity routing (CMMBCR)
CMOS inverters, 461
Coarse synchronization, 38t
Coarse ToA estimation
  RSS measurements, 60
Coded Beacon sequences, 68
Coded IR-UBW
  iterative interference cancellation and decoding, 241
Code division multiple accessing (CDMA)
  BPSK, 298
  direct sequence, 12, 20, 169, 326
    spread spectrum, 256
  frequency hopping (FH), 326
  TH, 326
Coexistence and interference resistance
  Alt-PHY layer, 319
Collect replies time period
  MPABR, 383
Common channel, 200, 325, 327, 438
Common code, 326
Communications, 321
Complementary cumulative distribution function (CCDF), 194
  logarithmic, 194f
Complex orthogonal phase vectors
  FFT, 118
Complications, 39f
Computational complexity and alternative solutions
  UWB channel estimation and synchronization, 27
Conditional max-min capacity routing (CMMBCR)
  MTPR, 396
  power-aware routing, 395
  protocol comparisons, 426t
Connect table (CT)
  PLBM, 385
Constrained frequency design with linear phase filters, 113–114
Constrained frequency response approximation, 113
Construction packet
  SHARP, 403

Contention Access Period (CAP), 301, 323
Content ion-free period (CFP), 323
Contiguous CTA (CTA-C), 303
Continuous orthogonal frequency division multiplexing signals, xv
Conventional correlation-based approaches, 58–59
Convolutional encoder, 281
Coordinator
  transfer from device, 323
Correlation-based ToA estimation receiver, 58f
CP. *See* Cyclic prefix (CP)
C++ Protocol Toolkit (CPT), 352
CPT. *See* C++ Protocol Toolkit (CPT)
Cramer–Rao lower bounds (CRLBs), 43, 51
Crisp zone
  TZRP, 409
CRLB. *See* Cramer–Rao lower bounds (CRLBs)
Crossing bands, 119
CSI. *See* Channel state information (CSI)
CSMA/CA. *See* Carrier sense multiple access with collision avoidance (CSMA/CA)
CT. *See* Connect table (CT)
CTA. *See* Channel time allocations (CTAs)
CTA-1. *See* Single CTA (CTA-1)
CTA-C. *See* Contiguous CTA (CTA-C)
CTA-M. *See* Multiple CTA (CTA-M)
CTF. *See* Channel transfer function (CTF)
CTS. *See* Clear-to-send (CTS)
Cyclic prefix (CP), 280

DAC. *See* Digital-to-analog converters (DACs)
DAG. *See* Directed acyclic graph (DAG)
DARPA. *See* Defense Advanced Research Projects Agency (DARPA)
Data mapping, 78
Data traffic
  achieved throughput, 446f
  average transmission power, 448f
DBF. *See* Distributed Bellman–Ford algorithm (DBF)
DCM. *See* Dual-carrier modulation (DCM)
Defense Advanced Research Projects Agency (DARPA), 343
Delay and sum beamformer, 221
Delay-hopped transmitted-reference (DHTR) scheme, 172f, 175
Delay-hopped transmitted-reference spread spectrum (DHTR-SS), 298
Delay locked loop (DLL), 460
  clock synthesis, 461f
Delta-K model, 187

INDEX   **485**

Demapping, 78
Demodulation, 78
Design receiver algorithms, 183
Destination-sequenced distance vector
    (DSDV), 346
  proactive routing, 345–347
    protocol comparisons, 419t, 421t
  routing protocol, 345
  SHARP, 402
Deterministic models, 185–186, 189, 214
Device
  transfer from coordinator, 323
DFT. See Discrete Fourier transform (DFT)
    coefficients
DHTR. See Delay-hopped transmitted-reference
    (DHTR) scheme
Differential phase shift keyed (DPSK)
    system, 299
Diffraction, 184
Digital communication, 78
Digital equivalent models
  for pulsed OFDM
    transmitter and channel, 287f
Digital-signal processors (DSP), 161
  based UWB pulse design, 109f
Digital-to-analog converters (DACs), 278
Dijkstra's algorithm, 358
  MTPR, 396
Dijkstra's shortest path first
  SOAR, 390
Dirac delta function, 189
Directed acyclic graph (DAG), 365
  TORA, 391
Direction of arrival (DoA), 209
Direct matrix pencil algorithm, 16
Direct sequence (DS), xv, 456
Direct sequence-code division multiple
  accessing (DS-CDMA), 12, 20, 169, 326
Direct sequence spread spectrum–code division
  multiple accessing (DSSS-CDMA), 256
Dirty template, 13
Discrete Fourier transform (DFT)
  coefficients, 18, 24, 25, 30–32
  transmitted pulse, 36
Discrete wavelet transform (DWT), 269
DIST(i)
  LAR, 380, 381
Distance dependence path
  loss, 189–190
Distance routing effect algorithm for mobility
  (DREAM), 342, 358
  proactive routing, 358–359
    protocol comparisons, 419t, 421t
Distance table (DT), 360

Distance-vector distributed Bellman–Ford
    algorithm, 346
Distributed, 433
Distributed AC scheme, 433
Distributed Bellman–Ford algorithm (DBF)
  distance-vector, 346
Diversity gain, 206
Diversity order of UWB-MAS, 209
D-K model, 187
DLL. See Delay locked loop (DLL)
DM
  failure rate, 473f
DNVR. See Dynamic Nix-vect or routing
    (DNVR)
DoA. See Direction of arrival (DoA)
Double-dB gain, 210, 222
DPSK. See Differential phase shift keyed
    (DPSK) system
DREAM. See Distance routing effect
    algorithm for mobility (DREAM)
DS. See Direct sequence (DS)
DSDV. See Destination-sequenced distance
    vector (DSDV)
DSP. See Digital-signal processors (DSP)
DSR. See Dynamic source routing (DSR)
DS-UWB-IR signaling structure, 80f
DT. See Distance table (DT)
Dual-carrier modulation (DCM), 282
DWT. See Discrete wavelet transform (DWT)
Dynamic MANET on-demand (DYMO), 377
  reactive routing, 377
    protocol comparisons, 423t, 424t
Dynamic Nix-vect or routing (DNVR), 378
  reactive routing, 378
    protocol comparisons, 422t, 424t
Dynamic source routing (DSR), 352
  reactive routing, 364–366
  route discovery, 366f
  route error, 367f

EADSR. See Energy aware dynamic source
    routing (EADSR)
Effective isotropic radiated power
    (EIRP), 131
Efficient spectral utilization, 103
EGC. See Equal gain combining (EGC)
802.15.3 or 802.15.4. See IEEE 802.15.3
Energy aware dynamic source routing
    (EADSR)
  power-aware routing, 395
    protocol comparisons, 426t
Energy detector receiver
  block diagram, 176f
Energy dissipation, 431

Equal gain combining (EGC), 94, 238, 289
Ergodic channel capacity of UWB communication, 216f
ESPRIT. *See* Estimation of signal parameters via rotational invariance techniques (ESPRIT) algorithm
Estimation. *See also* Time of arrival (ToA) estimation
  blind timing offset, 68
  closely spaced paths, 24–25
  coarse ToA
    RSS measurements, 60
  error, 96–98
  fine ToA
    low-rate correlation outputs, 60–61
  low-complexity timing offset
    dirty templates, 65–66
  multiple bands
    receiver block diagram, 34f
  nonadjacent bands, 32–34
  received pulse shape, 163–164
  receiver
    correlation-based ToA, 58f
  two-step, 36–38, 38t
  two-step delay, 39f
  two-step ToA
    algorithm, 62f
    low-rate samples, 59
Estimation of signal parameters via rotational invariance techniques (ESPRIT) algorithm, 16
Expected zone
  LAR, 380, 381
Extremely high data rates, 205

Failure notification phase (FN)
  RDMAR, 388
False alarm rate, 97
Fast Fourier transform (FFT), 85
  algorithms, 277
FCC spectral mask, 104, 106, 108, 120f, 121, 123, 127
  indoor communications, 107f
FD. *See* Frequency domain (FD)
FDM. *See* Frequency division multiplexing (FDM)
FEC. *See* Forward error correction (FEC) coding
FFD. *See* Full-function devices (FFD)
FFT. *See* Fast Fourier transform (FFT)
FH. *See* Frequency hopping (FH)
Fidelity
  single-band scheme, 147t
Field generation time, 362

Field programmable gate arrays (FPGA), 161
Field registration time, 362
Filter bank approach
  UWB channel estimation, 32
FIM. *See* Fisher information matrix (FIM)
Fine synchronization, 38t
Fine ToA estimation
  low-rate correlation outputs, 60–61
Finger estimation error, 96–98
Finite impulse response (FIR), 269
  filter, 18
Finite rate of innovation, 15
FIR. *See* Finite impulse response (FIR)
Fisher information matrix (FIM), 48
Fisheye state routing (FSR), 360, 361f
  proactive routing, 360–361
    protocol comparisons, 420t, 421t
Flexible interference avoidance, 103
Flow identifiers, 365
FN. *See* Failure notification phase (FN)
Forward error correction (FEC) coding, 243
Forwarding set
  TZRP, 409
Forwarding table
  node h, 347f
FPGA. *See* Field programmable gate arrays (FPGA)
Frame spaced sampling, 161
Frame structure, 458f
Frequency bands, 111
Frequency decaying factor, 192
Frequency dependence path loss, 191
Frequency division multiplexing (FDM), 83
Frequency domain (FD), 135f, 137f
  canceling NBI, 268
  channel estimation, 15–16, 68
  measurement technique
    frequency sweeping, 188f
  technique
    channel sounding, 188
Frequency hopping (FH), xv, 104
  CDMA, 326
  multiband, 124f
FSLS. *See* Fuzzy sighted link state (FSLS) algorithm
FSR. *See* Fisheye state routing (FSR)
Full-function devices (FFD), 321
Fully distributed, 358, 431
Fuzzy sighted link state (FSLS) algorithm, 363
Fuzzy zone
  TZRP, 409

Gateway nodes, 350
Gaussian doublets, 265f

Gaussian monocycle, 120
Gaussian monocycle pulse, 104
Gaussian pulse, 80, 265f
　power spectrum, 108f
Generalized maximum likelihood scheme
　first-path detection, 63–65
Generalized spreading sequence, 228
Generated pulse train, 462f
Generation time (T-gen), 362
Generic UWB receiver structure., 166f
Gibbs phenomenon, 31
Global positioning system (GPS), 70, 303
　device
　　PAWF, 398
　　LAR, 381
Global state routing (GSR), 360
　proactive routing, 360
　　protocol comparisons, 420t, 421t
GPS. *See* Global positioning system (GPS)
GPSR. *See* Greedy perimeter-stateless routing (GPSR)
Grating lobes, 222
Greedy perimeter-stateless routing (GPSR)
　PAWF, 398
GSR. *See* Global state routing (GSR)
GTS. *See* Guaranteed time slots (GTS)
Guaranteed time slots (GTS), 323

Hadamard codewords, 119
Hankel matrix, 20
Hazy sighted link state (HSLS), 363
　proactive routing, 363
　　protocol comparisons, 420t, 421t
HDR. *See* High-data-rate (HDR)
HDTV. *See* High-definition television (HDTV)
Hermite orthogonal polynomials, 104
HID. *See* Hierarchical ID (HID)
Hidden bit
　DNVR, 379
Hidden Markov model (HMM), 271
Hidden node, 453
Hierarchical ID (HID), 354
Hierarchical routing (HR), 362
　proactive routing, 362
　　protocol comparisons, 420t, 421t
Hierarchical state routing (HSR), 353
　proactive routing, 352–354
　　protocol comparisons, 419t, 421t
High-data-rate (HDR)
　links, 315
　systems, 164
High-definition television (HDTV), 310
Higher-rank channel models, 29f

Highpass filtering
　antennas, 141
High time resolution
　UWB geolocation, 54
HMM. *See* Hidden-Markov model (HMM)
Home region
　SLURP, 403
Hop-by-hop routing protocol, 346
HR. *See* Hierarchical routing (HR)
HSLS. *See* Hazy sighted link state (HSLS)
HSR. *See* Hierarchical state routing (HSR)
Hybrid, 327
Hybrid routing, 400–408
Hypothetical ski track, 474f

IARP. *See* Intra-zone routing protocol (IARP)
ICMP. *See* Internet control message protocol (ICMP) messaging
Identifier or node ID, 354
IEEE 802.15.3
　standards
　　UWB networks and applications, 299
　superframe format, 301f
IEEE 802.15.3a, 184
IEEE 802.15.3 MAC
　channel access, 301
　protocol
　　UWB networks and applications, 300–302
IEEE 802.15.4
　vs. PHY:802.15.4a, 316
IEEE 802.15.4 MAC
　standard, 321–324
　　low-bit-rate UWB networks, 321–324
　　medium access strategy, 322–323
　　network devices and topologies, 321–322
IEEE 802.15.4 PHY
　vs. Alt-PHY layer, 317
　communications, 321
　home sensing, control and media delivery, 320
　industrial inventory control, 319
　industrial process control and maintenance, 320
　logistics, 320
　personnel security, 320
　safety/health monitoring, 320
IEEE 802.15.4a, 184, 315
　standards, 70
IEEE 802.15.4a PHY, 316
IERP. *See* Inter-zone routing protocol (IERP)
IFFT. *See* Inverse fast Fourier transform (IFFT)
IIR. *See* Infinite impulse response (IIR)

IMEP. *See* Internet MANET encapsulation protocol (IMEP)
Immediate retransmit, 334
Impulse radio, 315, 452
Impulse radio (IR), 2
Impulse radio ultra wideband (IR-UWB), xv
IMST, 212
   database, 211
   GmbH, 211
Independent zone routing (IZR), 408
   hybrid routing, 408
      protocols comparisons, 426t, 427t
Individual angles, 214
Individual link bit rate
   Alt-PHY layer, 318
Infinite impulse response (IIR), 269
Integrate and dump circuitry
   output, 178
Intel database, 212
Inter-cluster parameters, 190
Interference rejection, 207, 222
Interference sources
   IFI, 164, 170
   ISI, 164, 170
   MAI, 165
   NBI, 164, 170
Internet control message protocol (ICMP) messaging
   AODV, 373
Internet MANET encapsulation protocol (IMEP)
   TORA, 393
Intersymbol interference (ISI), 2
Inter-zone routing protocol (IERP)
   ZRP, 406
Intra-cluster parameters, 190
Intra-zone routing protocol (IARP)
   TZRP, 409
   ZRP, 406
Inverse fast Fourier transform (IFFT), 85
   algorithms, 277
IP
   SOAR, 388
IR. *See* Impulse radio (IR)
ISI. *See* Intersymbol interference (ISI)
Iterative (turbo) algorithms
   MAI, 240–242
Iterative MUD algorithms, 240
IZR. *See* Independent zone routing (IZR)

JCON. *See* Join confirm packet (JCON)
Join confirm packet (JCON)
   PLBM, 386

Join query packets (JQs)
   PLBM, 385
   transmissions
      PLBM, 386
Join reply packet (JREP)
   PLBM, 386
JQ. *See* Join query packets (JQs)
JREP. *See* Join reply packet (JREP)

Kalman–Bucy filter, 269
Kalman filtering, 473
Kalman filtering smoothed, 476f
Kolmogorov-Smirnov (K-S), 197
   test, 215
K-S. *See* Kolmogorov-Smirnov (K-S)

Laplace-distributed random variables, 214
LAR. *See* Location-aided routing (LAR)
Large-scale fading, 189
LBR. *See* Link life-based routing (LBR)
LCC. *See* Least clusterhead change (LCC)
LDP. *See* Location discovery packet (LDP)
Least clusterhead change (LCC), 350
Least-mean-squares (LMS) algorithm, 269
Least-overhead routing approach (LORA), 352
Least-squares (LS) procedure, 14
LEI. *See* Link energy information (LEI)
Lifetime of a link (link life)
   LBR, 382
Linear minimum mean square error combining, 235
Linear receivers
   MAI, 232
Line-of-sight (LOS), 52
Link disjoint
   NDMR, 384f
Link energy information (LEI)
   EADSR, 395
Link life
   LBR, 382
Link life-based routing (LBR), 381
   reactive routing, 381–382
      protocol comparisons, 423t
Link-state array (LS)
   TORA, 391
Link-state update (LSU), 352, 353, 363
   TZRP, 409
LLR. *See* Log-likelihood ratio (LLR)
LMS. *See* Least-mean-squares (LMS) algorithm
Localized query (LQ)
   ARB, 371
Location-aided routing (LAR), 380
   reactive routing, 380
      protocol comparisons, 423t, 424t

scheme 1, 381f
scheme 2, 382f
Location awareness
  Alt-PHY layer, 318
Location discovery packet (LDP)
  SLURP, 405
Location reply packet (LRP)
  SLURP, 405
Location table (LT), 358
Logic flow
  TORA, 392f
Log-likelihood ratio (LLR), 61, 242
Log-normal, 197
  distribution, 187, 190, 197
LORA. *See* Least-overhead routing approach (LORA)
LOS. *See* Line-of-sight (LOS)
Low-bit-rate UWB networks, 315–340
  advanced MAC design, 324–340
  applications, 315–319
  802.15.4 MAC Standard, 321–324
Low-complexity timing offset estimation
  dirty templates, 65–66
Low data-rate UWB network applications
  technical requirements, 317–318
LQ. *See* Localized query (LQ)
LRP. *See* Location reply packet (LRP)
LS. *See* Least-squares (LS) procedure;
    Link-state array (LS)
LSU. *See* Link-state update (LSU)
LT. *See* Location table (LT)

MAC
  mixed random and scheduled access, 324
  network topology, 324
  random access strategy, 324
  ranging support, 324
MAC address, 354
MACPDU. *See* MAC protocol data units (MACPDU)
MAC protocol data units (MACPDU), 328
  transmission procedure, 330
  transmission time, 335
MAI. *See* Multiple access interference (MAI)
Mainlobe width, 222
Management channel time allocations (MCTAs), 302
MANET. *See* Mobile ad hoc networks (MANETs)
Mapping, 78
Maravic, Irena, 11
M-ary data mapping schemes, 82f, 89–90
  SER, 90f, 91t
MAS. *See* Multiantenna systems (MAS)

Master-slave concept, 300
Matched filter, 38t, 167
Matched-filtering, sampling and despreading
  received signal, 229f
MaxContiguous CTA, 303
Maximal ratio combining (MRC), 93, 167
Maximum extra interference (MEI), 435, 437
  transmission power, 443f, 444f
Maximum forward progress (MFP), 400
  power-aware routing, 400
    protocols comparisons, 426t
Maximum information progress (MIP), 400
  power-aware routing, 400
    protocols comparisons, 426t
Maximum likelihood (ML)
  criterion, 14
  estimation approach, 13
  estimator, 16
  sequence detection
    MAI, 232
Maximum transmission rate (MTR), 362
Max-min capacity routing (MMBCR)
  power-aware routing, 395
    protocol comparisons, 426t
  routing, 395
MBCR. *See* Minimum battery cost routing (MBCR)
MBOA. *See* Multiband OFDM Alliance (MBOA)
MB-OFDM. *See* Multiband OFDM (MB-OFDM)
MCTA. *See* Management channel time allocations (MCTAs)
Mean excess delay, 197, 198
  average values, 199f
Mean square error (MSE), 25
Measurement campaigns, 50, 189, 197
MEI. *See* Maximum extra interference (MEI)
Message exchanges for the AC procedure
  with the adaptive MEI approach, 438f
Message retransmission list (MRL), 348
MFP. *See* Maximum forward progress (MFP)
MFR. *See* Most forward with fixed
    radius (MFR)
MIMO and UWB, 205–226
  benefits, 206–208
  multiantenna techniques literature review, 208–210
    beamforming, 209
    spatial diversity, 209
    spatial multiplexing, 208
  spatial channel measurements and modeling, 211–214
  spatial diversity, 216–219
  spatial multiplexing, 215

490 INDEX

Minimum battery cost routing (MBCR)
  MTPR, 396
  power-aware routing, 395
    protocol comparisons, 426t
Minimum mean square error (MMSE), 93, 234
  combining
    canceling NBI, 268
    linear, 235
  receivers, 170
Minimum pulse-to-pulse duration, 164
Minimum recommended transmit power (MRTP)
  EADSR, 395
Minimum total transmission power routing (MTPR), 342, 395
  power-aware routing, 395
    protocol comparisons, 426t
Minimum transmit power level (MTPL)
  PARO, 396
MIP. *See* Maximum information progress (MIP)
Missed detection rate, 97
ML. *See* Maximum likelihood (ML)
MMBCR. *See* Max-min capacity routing (MMBCR)
MMSE. *See* Minimum mean square error (MMSE)
Mobile ad hoc networks (MANETs), 341, 344f, 361
  DNVR, 378
  routing, 345
  routing protocols, 341–428
Modulated Gaussian pulse, 153
Modulated pulses, 92, 145, 171, 284, 316
Modulation, 78
  signal constellation, 86f
Modulation scheme, 78, 86–96, 281, 289, 431, 456, 460
Modulation techniques, 78
Most forward with fixed radius (MFR), 405f
  algorithm
    SLURP, 404
MPABR. *See* Multipath associativity based routing (MPABR)
MPC. *See* Multipath components (MPCs)
MPR. *See* Multipoint relay (MPR)
MRC. *See* Maximal ratio combining (MRC)
MRL. *See* Message retransmission list (MRL)
MRTP. *See* Minimum recommended transmit power (MRTP)
MSE. *See* Mean square error (MSE)
MST
  SOAR, 389
MSWF. *See* Multistage Wiener filter (MSWF)
MTPL. *See* Minimum transmit power level (MTPL)

MTPR. *See* Minimum total transmission power routing (MTPR)
MTR. *See* Maximum transmission rate (MTR)
Multiaccess code design, 3
Multiantenna systems (MAS), 206
Multiband OFDM (MB-OFDM), xv, 263f, 277, 278–283
  band planning, 278, 279f
  coding, 281
  frequency repetition spreading, 280
  improvement to MB-OFDM, 283
  MB-OFDM transceiver, 282
  vs. MB-pulsed-OFDM systems
    chip area comparison, 291–292
    complexity comparison, 290
    performance comparison, 293–294
    power consumption comparison, 290
    system parameters, 290
  modulation, 280
  sub-band hopping, 278–279
  supported bit rates, 281
  time-frequency codes, 300t
  time repetition spreading, 280
  transmitter and receiver structures system, 282f
Multiband OFDM Alliance (MBOA), 298
Multiband pulse design, 122
Multiband pulsed OFDM UWB system
  pulsed-OFDM
    digital equivalent model and diversity, 286–287
    receiver, 288
    signal spectrum, 284
    transmitter, 284
  selecting up-sampling factor, 289
Multiband schemes
  avoiding NBI, 263
  pulses and spectra, 135f
  waveforms, 143f
Multiband UWB, 83–84
  signaling (pulse-based), 84, 85f
Multicarrier approach
  avoiding NBI, 261–262
Multicarrier UWB, 85
Multihop wireless ad hoc routing protocol (MultiWARP), 401
  hybrid routing, 401
  protocols comparisons, 426t, 427t
Multilevel logical partitioning, 353
Multilevel physical clustering, 353
Multimedia traffic
  achieved throughput, 445f
  average transmission power, 447f

INDEX **491**

Multipath associativity based routing (MPABR), 383
  reactive routing, 383
    protocol comparisons, 423t, 424t
Multipath coefficients
  estimation, 163
Multipath components (MPCs), 184
  clustering, 187
  phenomenon, 193
Multipath effects
  propagation mechanisms, 185f
Multipath propagation, 93–94
  UWB geolocation, 52
Multiple access interference (MAI), 4, 14, 95
  mitigation in UWS, 227–248
    at receiver side, 231–243
    signal model, 228–229
    at transmitter side, 244–248
  UWB geolocation, 53
Multiple ad hoc networks. *See* Mobile ad hoc networks (MANETs)
Multiple CTA (CTA-M), 303
Multiple orthogonal pulse, 104
  design, 123–124
Multipoint relay (MPR), 355
  selectors, 358
Multiresolution approach, 36
Multistage block-spreading UWB access
  MAI mitigation, 247
Multistage Wiener filter (MSWF), 244
MultiWARP. *See* Multihop wireless ad hoc routing protocol (MultiWARP)

NACF. *See* Normalized autocorrelation function (NACF)
Nakagami distributions, 197, 217
Nakagami frequency selective fading, 216f
Nakagami-m factor, 197
Narrowband antenna systems
  system transmission efficiency, 147f
Narrowband channel models, 6, 183, 196
Narrowband interference (NBI), 4, 255–271, 256
  avoiding, 261–266
  canceling, 267–271
  jamming resistance
    DSSS systems, 258
  robustness, 126f
  scenario for multicarrier modulation systems, 262f
NBI. *See* Narrowband interference (NBI)
NCD. *See* Neighbor-changing degree (NCD)
NDMR. *See* Node-disjoint multipath routing (NDMR)

NDP. *See* Neighbor discovery protocol (NDP)
Neighbor-changing degree (NCD), 362
Neighbor discovery protocol (NDP)
  ZRP, 407
Neighbor index (Nix)
  DNVR, 378
Neighboring table (NT), 357
  ABR, 369
Neighbor list (NL), 360
Neighbor's neighbors table (NNT)
  PLBM, 385, 386
NEIP. *See* Normalized effective interference power (NEIP)
NESP, 122
  direct maximization, 111–112
Network architecture, 300
Network topology, 347f
NEXT. *See* Next hop table (NEXT)
Next hop table (NEXT), 360
Ning Chen, Zhi, 131
Nix. *See* Neighbor index (Nix)
Nix-vector forwarding information base (NV-FIB)
  DNVR, 378
Nix-vector reply (NVREP) message
  DNVR, 379
Nix-vector request (NVREQ) message
  DNVR, 379
NL. *See* Neighbor list (NL)
NLOS. *See* Nonline-of-sight (NLOS)
NNT. *See* Neighbor's neighbors table (NNT)
Node(s)
  ID, 354
  TORA, 392
  updates, 349
Node-disjoint multipath routing (NDMR), 384, 384f
  reactive routing, 384
    protocol comparisons, 423t, 424t
Noise sensitivity
  UWB channel estimation and synchronization, 25
Noisy template, 13
Nonadjacent bands
  UWB channel estimation, 32–33
Nonbeacon-enabled modality, 322, 323
Non-data aided algorithm, 244
Non-ideal channels
  timing recovery, 37f
Nonline-of-sight (NLOS), 52, 55
  propagation
    UWB geolocation, 53
Nonpropagating query packet
  SOAR, 388

Nonpulsed OFDM
 clock rates, 292t
Normalized autocorrelation function (NACF), 81
Normalized beampattern BP, 221f
Normalized effective interference power (NEIP), 126
Normalized ray relative power vs. relative delay, 196f
Notch filtering, 266f
NT. See Neighboring table (NT)
NULL
 TORA, 393
Numerical example
 UWB channel estimation and synchronization, 28
NV-FIB. See Nix-vector forwarding information base (NV-FIB)
NVREP. See Nix-vector reply (NVREP) message
NVREQ. See Nix-vector request (NVREQ) message
Nyquist sampling rate, 14, 20, 160

ODRA. See On-demand routing algorithm (ODRA)
OFC. See Optimal frame combining (OFC)
OFDM. See Orthogonal frequency division multiplexing (OFDM)
Offered traffic
 throughput as function, 335f
Off-time
 connective pulses, 159
OLSR. See Optimized link stating routing (OLSR)
OMC. See Optimal multipath combining (OMC) scheme
On-demand routing algorithm (ODRA), 362
On-off keying (OOK), 2, 176
OOK. See On-off keying (OOK)
Open shortest path first (OSPF), 352
Optimal frame combining (OFC), 237, 237f, 240
Optimally synthesized pulse
 power spectrum, 120f
Optimal multipath combining (OMC) scheme, 238, 239f, 240
Optimal positioning algorithms
 UWB geolocation, 55–57
Optimal pulse combining schemes, 234
Optimal receiver
 extension to colored noise and interference scenarios, 169
Optimization percentage value (OPV)
 PARO, 396

Optimized link stating routing (OLSR), 355, 355f
 proactive routing, 355
 protocol comparisons, 419t, 421t
Optimum routing approach (ORA), 352
OPV. See Optimization percentage value (OPV)
ORA. See Optimum routing approach (ORA)
Orientation-dependent transfer function, 150
Orthogonal frequency division multiplexing (OFDM), 2, 85–86, 86f, 261, 277–295, 452
 continuous signals, xv
 digital equivalent models for pulsed transmitter and channel, 287f
 MB-OFDM vs. MB-pulsed-OFDM, 290–295
 MD band planning, 279f
 multiband, 278–283
 transmitter and receiver structures, 282f
 nonpulsed
  clock rates, 292t
 pulsed
  clock rates, 292t
  scheme, 283
  transmitter structure, 285f
 and pulsed-OFDM UWB system, 284–289
 receiver, 262
 research, 277
Orthogonal iteration, 28
Orthogonality formulation, 115–116
Orthogonal pulse design, 116, 117
OSPF. See Open shortest path first (OSPF)

Packet generation rate, 308
Packet generation time, 309f
Packet size, 308
PAM. See Pulse-amplitude-modulation (PAM)
PAN, 321
PAPR. See Peak to average power ratio (PAPR)
PAR. See Power-aware routing (PAR)
Parallel interference cancellation (PIC) scheme, 244
Pareto-optimal solution, 434
Parks–McClellan algorithm, 110
Parks–McClellan filter, 122
PARO. See Power-aware routing optimization (PARO)
Partial-rake (PRake) receivers, 93
Partial topology map, 351
Partial topology table (PT)
 SOAR, 389

INDEX **493**

Path id
  DNVR, 379
Path length, 362
Path loss, 189
  as a function of distance, 191f
  as a function of frequency, 192f
PAWF. *See* Power-aware weighted forwarding function (PAWF)
PDP. *See* Power delay profile (PDP)
PDS. *See* Power density spectrum (PDS)
PDU
  UWB, 330f
Peak gain, 222
Peak to average power ratio (PAPR), 87
Peer-to-peer data transfer, 323
Peer-to-peer topology, 321, 322f
PERA. *See* Probabilistic emergent routing algorithm (PERA)
Perceived interference, 433
Phase-shift-keying (PSK), 2
PHY:802.15.4a *vs.* 802.15.4, 316
PIC. *See* Parallel interference cancellation (PIC) scheme
Piconet controller (PNC), 300
Planar square dipole, 140
Plane wave incident, 47f
PLBA. *See* Preferred link-based algorithm (PLBA)
PLBM. *See* Preferred link-based multicast (PLBM)
PM algorithm based pulse
  power spectrum, 121f
PN. *See* Pseudo random (PN)
PNC. *See* Piconet controller (PNC)
Poisson distribution, 193, 195
Polarity code, 228
Polynomial realization of model based
  methods, 16–20
Positioning, 205
Positioning algorithms
  ranging and optimal
    UWB geolocation, 55–57
Position location and tracking
  block diagram, 474f
Power
  consumption
    Alt-PHY layer, 319
  decaying phenomenon
    clusters and MPCs, 195
  management
    protocol layer, 394t
  method, 27, 36, 37
Power-aware routing (PAR), 345, 393–400
  approaches, 395

Power-aware routing optimization (PARO), 342, 396
  power-aware routing, 396–397
  protocols comparisons, 426t
Power-aware weighted forwarding function (PAWF), 398
  power-aware routing, 398–399
  protocols comparisons, 426t
Power delay profile (PDP)
  S-V channel, 195, 195f
Power density spectrum (PDS), 132
Power spectral density (PSD), 37f, 91, 132
  UWB modulations, 92t
PPM. *See* Pulse position modulation (PPM)
PRake receivers, 93
Preferred link-based algorithm (PLBA)
  PLBM, 386
Preferred link-based multicast (PLBM), 385
  reactive routing, 385–386
  protocol comparisons, 423t, 424t
PRI. *See* Pulse repetition interval (PRI)
Private signaling TH code, 439
Proactive routing, 345–363
  protocol comparisons, 419t–421t
Proactive zone
  SHARP, 402, 403
Probabilistic emergent routing
  algorithm (PERA)
  ARA, 369
Probability density functions, 88f
Pro-ESPRIT, 16
Prolate-spheroidal (PS) pulse, 121
  power spectrum, 121f
Propagating query packet
  SOAR, 389
PS. *See* Prolate-spheroidal (PS) pulse
PSD. *See* Power spectral density (PSD)
Pseudochaotic time hopping
  MAI mitigation, 246
Pseudo random (PN), 2
PSK. *See* Phase shift-keying (PSK)
PT. *See* Partial topology table (PT)
Pulse(s)
  designs
    narrowband interference avoidance, 125
    transceiver power efficiency, 126–127
  detectors, 242
  discarding receivers, 233
  NESP, 121t
  shaper, 6, 103–128
  shaping
    avoiding NBI, 264–265
  symbol iterative detectors, 242
  width, 68, 79, 80, 120–121, 457, 463, 465, 480

Pulse-amplitude-modulation (PAM), 2, 105, 228
   TH-IR signal, 45f
Pulsed OFDM
   clock rates, 292t
   digital equivalent models
      transmitter and channel, 287f
   scheme, 283
   transmitter structure, 285f
Pulsed UWB systems, 148f, 148–150
Pulse position modulation (PPM), 2, 95f, 176
   modulation parameter, 265
Pulse repetition interval (PRI), 431
Puncturing, 281

Q-MTPR, 396
QoS-aware scheduling algorithm, 307
QoS parameters, 447
QPSK. See Quadrature phase shift
   keying (QPSK)
QRY. See Query packet (QRY)
Quadrature phase shift keying (QPSK), 282
   modulation, 280
Quasi-decorrelator, 233
Quasi-minimum mean square error, 234
Query packet (QRY)
   TORA, 391

RA-ACK. See Rate-adaptive ACK (RA-ACK)
Radiated electric fields
   waveforms, 140
Radiation transfer function, 142f
Radio frequency (RF) carriers, 297
Radix-4 multipath delay commutator (R4MDC)
   structure, 291
Rake finger, 268
Rake receivers, 93
   with M branches, 235
   structure, 168f
Rake reception, 166
RAKERX, 200
Range
   Alt-PHY layer, 318
Ranging and optimal positioning algorithms
   UWB geolocation, 55–57
Ranging problem, 56f
Rate-adaptive ACK (RA-ACK), 431
Rate adaptive MAC protocol, 431–432
Rate capacity effect
   BEE, 394
Ray arrival times, 187, 193–194
Rayleigh distribution, 196, 215
Rayleigh fading channels, 89
   BER, 89f
Rayleigh pulse, 153

Rayleigh tap delay line model, 187
Ray tracing, 185–186
RD. See Route delete (RD)
RDM. See Relative distance microdiscovery
   algorithm (RDM)
RDMAR. See Relative distance
   microdiscovery ad hoc routing protocol
   (RDMAR)
RE. See Route element (RE)
Reactive routing, 364–392
   protocol comparisons, 422t–424t
   TZRP, 409
Receive antenna, 132–134, 137, 141, 146, 148,
   206–209, 215–220, 223
Received pulse shape estimation, 163–164
Received signal strength (RSS), 44
   based positioning algorithm, 50
   MAI, 229
   modeling, 49–50
   UWB geolocation, 49–50
Receiver (RX), 438
   algorithms, 183
   antenna gain, 189
   block diagram, 17f
   code, 326
   structure
      for pulsed OFDM system, 289f
   structures
      multiband OFDM system, 282f
Recovery effect
   BEE, 394
Rectangular lattice, 45
Reduced-function devices (RFD), 321
Reference partitioning, 317f
Reference pulses
   averaged, 172f
Reflection, 184
Registration time (T-reg), 362
Relative distance microdiscovery ad hoc routing
   protocol (RDMAR), 387
   reactive routing, 387
   protocol comparisons, 423t, 424t
Relative distance microdiscovery algorithm
   (RDM)
   RDMAR, 387, 387f
Reported node set (RN), 358
Reported sub-tree (RT), 357
Request-to-send (RTS)
   PARO, 396
   PLBM, 386
Request zone
   LAR, 380
RERR. See Route error (RERR)
RF. See Radio frequency (RF) carriers

INDEX   **495**

RFD. *See* Reduced-function devices (RFD)
Ricean finding, 215
Rice factor, 197
R4MDC. *See* Radix-4 multipath delay commutator (R4MDC) structure
R4MPC FFT implementation structure 64-point, 293f
RMS. *See* Root mean square (RMS)
RMSE. *See* Root mean square error (RMSE)
RN. *See* Reported node set (RN)
Root mean square (RMS)
  delay spread, 197, 198
    average values, 199f
  TOA estimation errors, 467f
Root mean square error (RMSE), 26, 187
  averaged
    SI and DI, 475t
    smoothing, 476f
  DM, 472f
  TOA estimation, 465
  values, 466
Route delete (RD)
  ARB, 372
Route deletion phase
  ARB, 372
Route discovery
  ABR, 370f
  ARA, 368f
  TORA, 391f
Route element (RE)
  DYMO, 377
Route error (RERR)
  AODV, 373
  BSR, 375
  CHAMP, 376
  DYMO, 377
  LBR, 383
  MPABR, 383
  NDMR, 385
  TZRP, 410
Route maintenance
  ABR, 371f
Route reconfiguration (RREC)
  LBR, 383
Route reply packet (RREP), 374
  AODV, 372
  CHAMP, 376
  LBR, 382
  NDMR, 385
  RDMAR, 388
  TZRP, 409
Route request (RREQ)
  AODV, 372
  BSR, 374, 374f

CHAMP, 376
LBR, 382
MPABR, 383
MultiWARP, 401
NDMR, 385
RDMAR, 387, 388
TZRP, 409
Routing cache
  BSR, 375t
Routing table, 351, 351t
Routing update (RUPDT) packet
  MultiWARP, 401
RREC. *See* Route reconfiguration (RREC)
RREP. *See* Route reply packet (RREP)
RREQ. *See* Route request (RREQ)
RSS. *See* Received signal strength (RSS)
RT. *See* Reported sub-tree (RT)
RTS. *See* Request-to-send (RTS)
RUPDT. *See* Routing update (RUPDT) packet
RX. *See* Receiver (RX)

Saleh-Valenzuela (S-V) channel model, 187, 189
SC. *See* Selection Combining (SC)
Scalable location update-based routing protocol (SLURP), 403
  hybrid routing, 403–405
    protocols comparisons, 426t, 427t
Scattering, 184
Scatternet, 300
SC-FDE. *See* Single carrier transmission with frequency domain equalization (SC-FDE)
Scheduling algorithm, 307
  efficiency, 308
  performance, 308
Scheduling problems, 307
SDR. *See* Software defined radio (SDR)
Search-based methods, 36
Selection Combining (SC), 289f
Selective-rake (SRake) receivers, 93, 94
Semi-infinite linear program (SILP), 112
Sensing, 205
Sensor-CLEAN algorithm, 13
Sequential (SEQ) strategy
  orthogonal pulse, 125f
  pulse design, 118
Sequential UWB pulse design, 117–119
  linear phase filters, 118–119
Set of transmission power, 434
Shadowing, 192
SHARP. *See* Sharp hybrid adaptive routing protocol (SHARP)

Sharp hybrid adaptive routing protocol (SHARP), 402
hybrid routing, 402
protocols comparisons, 426t, 427t
Shift
frequency, 145f
Shortest path first (SPF)
SOAR, 390
SIC. *See* Successive interference cancellation (SIC)
SICLC. *See* Soft interference canceller-likelihood calculators (SICLCs)
Sidelobe level, 222
Signaling exchange, 439
Signaling scheme, 35f
Signal-to-interference-plus-noise ratio (SINR), 96, 210
target, 434
Signal-to-noise ratio (SNR), 26, 35f, 81
SILP. *See* Semi-infinite linear program (SILP)
Simulation parameters and values, 304t
Simulation tool PARSEC, 352
Single-band scheme
calculated fidelity, 147t
pulses and spectra, 137f
waveforms, 143f
Single carrier transmission with frequency domain equalization (SC-FDE), 205
Single CTA (CTA-1), 303
Single-pulse designs
spectral utilization efficiency, 120–121
Singular value decomposition (SVD), 22
SINR. *See* Signal-to-interference-plus-noise ratio (SINR)
SISO. *See* Soft-input soft-output (SISO) channel decoders
SLURP. *See* Scalable location update-based routing protocol (SLURP)
Small-scale amplitude fading statistics, 196
Small-scale fading, 189
SNR. *See* Signal-to-noise ratio (SNR)
SOAR. *See* Source-tree-on-demand adaptive routing (SOAR)
Soft-input soft-output (SISO) channel decoders, 241
Soft interference canceller-likelihood calculators (SICLCs), 241
Software defined radio (SDR), 160
Source pulses
waveforms, 140f
Source tree (ST), 351, 357
Source tree adaptive routing (STAR), 351
proactive routing, 351
protocol comparisons, 419t, 421t

Source-tree-on-demand adaptive routing (SOAR), 388, 390f
minimal source tree exchanged, 389t
reactive routing, 388–390
protocol comparisons, 423t, 424t
Space time coding (STC), 205
Alamouti, 216, 217f
analog UWB, 209
Spatial correlation, 213f
Spatial diversity, 205
Spatial multiplexing, 205
Spatial multiplexing MIMO&UWB systems, 215
Spectral density shaping
radiated electrical fields, 138f
Spectral masks, 79
Spectral re-growth
transmitter nonlinearities, 107
Spectrum crossover of the narrowband interferers, 256f
Spectrum of a pulse train, 286
Speech processing
vs. UWB, 211t
SPF. *See* Shortest path first (SPF)
Spread spectrum (SS) systems, 258
SRake receivers, 93, 94
SS. *See* Spread spectrum (SS) systems
ST. *See* Source tree (ST)
STAR. *See* Source tree adaptive routing (STAR)
Star topology, 321, 322f
Statistical-based channel mode, 189
Statistical models, 186–187
STC. *See* Space-time coding (STC)
Stigmergy methods
ARA, 367
Stop bands, 111
Sub-bands
mode 1 device, 279f
Subnanosecond low-power pulses, 303
Suboptimal pulse combining schemes, 234
Sub-pulse rate, 14
Subspace-based algorithm, 23
Subspace based approach, 16, 20–24
Subspace iteration, 28
Subspace method, 38t
Successive interference cancellation (SIC), 243
Superframe
beacon-enabled modality, 323f
S-V. *See* Saleh-Valenzuela (S-V) channel model
SVD. *See* Singular value decomposition (SVD)
Symbol detectors, 242
Symbol spaced sampling, 161
Synchronization, 162–163

INDEX 497

Synthesized pulse
    power spectrum, 109
System transfer function
    magnitude, 142f

Table updates intervals (TUIs), 362
Target SINR, 434
Task Group TG4a, 316
TBRPF. *See* Topology dissemination
    based on reverse-path forwarding
    (TBRPF)
TC. *See* Topology control (TC) messages
TD. *See* Time domain (TD)
TDD. *See* Time division duplexing (TDD)
TDMA, 456
TDoA. *See* Time difference of arrival (TDoA)
TDT. *See* Timing with dirty templates (TDT)
    algorithm
Telemedicine, 1
Template pulses
    parameters, 153t
Temporal correlation coefficient, 197, 198
Temporal dispersion, 198
Temporally ordered routing algorithm
    (TORA), 391
    reactive routing, 391–392
        protocol comparisons, 423t, 424t
    SHARP, 402
Terminals
    delay as function of, 337f
    throughput as function, 337f
T-gen. *See* Generation time (T-gen)
TH. *See* Time-hopping (TH)
Thin-wire straight dipole, 136
Threshold bandwidth, 216
Threshold region, 209
Throughput
    data traffic, 446f
    multimedia traffic, 445f
    offered traffic, 335f
    terminals, 337f
    voice traffic, 446f
Thru-Wall Sensing
    TD, 310
Time axis of UWB IR multiple
    access, 433f
Time-based approaches, 52
    UWB geolocation, 51
Time difference of arrival (TDoA), 44
    measurements, 310
Time division duplexing (TDD), 456
Time domain (TD), 135f, 137f, 141
    canceling NBI, 271
    measurement technique, 187–188, 271

technique
    channel sounding, 187
    pulse transmission, 188f
Thru-Wall Sensing, 310
UTD, 186
Time frequency codes
    associated preamble patterns, 279t
Time-frequency domain techniques
    canceling NBI, 269
Time-hopping (TH)
    CDMA, 326
    code, 105
    code construction algorithm
        synchronous IR-UWB, 246f
    IR-UWB receiver, 165–177, 169
        differential detector, 175
        energy detector, 176–177
        optimal matched filter, 167–170
        TR-based scheme, 171–174
    IR-UWB signal, 228
        with pulse-based polarity randomization,
        229f
        structure, 80f, 159f
    sequence design
        MAI mitigation, 245
    UWB, 2, 80f, 81f
        time gating pulses, 259f
Time of arrival (ToA) estimation, 44
    algorithms
        UWB geolocation, 58–68
    approach, 465
    errors, 466f, 467f
    low-rate correlation outputs, 60–61
    two-step
        low-rate samples, 59
    two-step algorithm, 62f
Time sequence and outputs
    integrators, 464, 466f
Time slot assignment (TS) problems, 307
Time-to-live (TTL), 363
    DSR, 365
    DYMO, 377
    SHARP, 403
Timing acquisition, 303
Timing jitter, 96–98
    BER, 98f
Timing with dirty templates (TDT)
    algorithm, 67, 68
TLS-ESPRIT algorithm, 16
TND. *See* Topology dissemination
    based on reverse-path forwarding
    (TBRPF),
    neighbor discovery protocol
ToA. *See* Time of arrival (ToA) estimation

498 INDEX

Topology, 440f
  Alt-PHY layer, 318
Topology control (TC) messages, 356
Topology dissemination based on
    reverse-path forwarding (TBRPF),
    356, 357f
  neighbor discovery protocol, 357
  proactive routing, 356–357
  protocol comparisons, 419t, 421t
Topology table (TT), 358, 360
TORA. See Temporally ordered routing
  algorithm (TORA)
TR. See Transmitted-reference (TR)
Traffic
  delay as function of, 336f
  QoS requirements parameters, 441t
Transition bands, 111
Transmission parameters
  values adopted, 441t
Transmission rate, 290, 362, 395, 430,
    433–436
Transmit antenna
  efficiency, 140f
  multipath model, 141f
Transmit–receive antenna system, 132, 134f
Transmitted-reference (TR)
  receiver structure, 171f
  scheme
    integrator output, 174
  signaling, 14
Transmitted signal
  MAI, 228
Transmitter (TX), 438
  and RX antenna gain, 189
  structures
    multiband OFDM system, 282f
Transmitter code, 327
T-reg. See Registration time (T-reg)
Triangulation method, 49f
TS. See Time slot assignment (TS) problems
TT. See Topology table (TT)
TTL. See Time-to-live (TTL)
TUI. See Table updates intervals (TUIs)
Turbo algorithms
  MAI, 240–242
Two-step delay estimation, 39f
Two-step estimation, 38t
  UWB localizers low-complexity
    rapid acquisition, 36–38
Two-step ToA estimation
  algorithm, 62f
  low-rate samples, 59
Two-way ranging protocols, 69f
  UWB geolocation, 69

Two-zone routing protocol (TZRP), 408
  hybrid routing, 408
  protocols comparisons, 426t, 427t
TX. See Transmitter (TX)
TZRP. See Two-zone routing protocol (TZRP)

UCA. See Uniform circular array (UCA)
UDP. See User datagram protocol (UDP)
UERR. See Unsupported-element error (UERR)
ULA. See Uniform linear array (ULA)
Ultra wideband (UWB)
  advantages, 298
  applications, 3
  benefits, 2, 206–208
  challenges, 3
  channel estimation and synchronization
    frequency-dependent distortion, 29–30
    low-complexity rapid acquisition in UWB
      localizers, 34–35
    multiple bands, 32–33
    performance evaluation, 25–28
    at SubNyquist sampling rate, 11–42
  channel estimation at SubNyquist sampling rate,
    14
    closely spaced path estimation, 24
    frequency-domain channel estimation, 15
    polynomial realization of model-based
      methods, 16–19
    subspace-based approach, 20–23
  channel modeling, 183–204
    channel sounding techniques, 187–189
    classification, 185–186
    statistical-based, 189–199
    UWB multipath propagation channel
      modeling, 184–186
  channel Nakagami fading, 215
  communication networks
    uncoordinated, wireless, baseborn medium
      access, 325–327
  definition, 1
  geolocation, 43–70
    location-aware applications, 70
    positioning techniques, 44–51
    ranging and positioning, 55–69
    signal model, 44
    time-based positioning error sources,
      52–54
  location and tracking, 451–480
    case study, 454–455
    communications system, 456
    framing structure, 458
    location approach, 458
    location system, 463–467
    multiple access, 452–453

position calculation methods, 468–472
pulse generator, 462
simulation, 474
system description, 456–458
system implementation, 459
tracking moving objects, 473
transceiver, 459
transmitted signal, 456–457
transmitter, 460–461
MAS
diversity order, 209
and MMO, 205–226
modulation options, 77–102
data mapping, 87–90
data mapping and transceiver complexity, 92
modulation performances, 93–98
signaling techniques, 78–86
spectral characteristics, 91
multiantenna techniques
literature review, 208–210
multiple-access interference mitigation, 227–247
Nakagami fading channels, 206
vs. narrowband systems, 297
networks and applications, 297–311
channel acquisition time, 303
IEEE 802.15.3 MAC protocol, 300–302
IEEE 802.15.3 standards, 299
medium access protocols, 300
multiple channels, 305–309
network applications, 310
physical layer, 298
PDU, 331f, 333f
performance, 333
potential proliferation, 452
protocol
backoff algorithms, 334
multicode concept, 327
pulse generator schematic, 462f
pulse shaper design, 103–127
examples and comparisons, 120–127
FIR digital pulse design, 108–109
optimal orthogonal, 115–119
optimal single, 110–113
transmit spectrum and pulse shaper, 105–107
receiver architectures, 157–182
channel estimation, 161–163
interference, 164
sampling, 160
system model, 158–159
reception procedure, 332f
RSS approach, 50–51

signaling techniques, 78
SIMO wireless systems
wideband (WB), 217f, 218f, 219f
statistical-based channel modeling
large-scale characterization, 190–192
philosophy and mathematical framework, 189
small-scale characterization, 193–196
system design, 199
temporal dispersion and correlation properties, 197–198
tag
circuit architecture, 459f
transmission procedure, 329f
wireless embedded networks, 454
ski tunnel, 454
Ultra wideband (UWB)-IR
signaling, 79–82
Ultra wideband (UWB)-PHY
requirements, 317
Unequal prefiltering, 222
Uniform circular array (UCA), 45
Uniform linear array (ULA), 45
Uniform theory of diffraction (UTD), 186
time domain, 186
Unsupported-element error (UERR)
DYMO, 377, 378
UPD. See Update packet (UPD)
Update packet (UPD)
TORA, 391
Update protocol
SHARP, 403
Update-request (URQ), 362
Update response (URP), 362
Upper physical limit
maximum indoor data rate, 209
URP. See Update response (URP)
URQ. See Update-request (URQ)
User datagram protocol (UDP)
DYMO, 377
SOAR, 388
UTD. See Uniform theory of diffraction (UTD)
UWB. See Ultra wideband (UWB)
UWEN. See Ultra wideband (UWB), wireless embedded networks

VBLAST. See Vertical Bell Laboratory layered space-time (VBLAST) algorithm
VCDL. See Voltage controlled delay line (VCDL)
Vector network analyzer (VNA), 134f, 188
Vector transfer function, 133

Vertical Bell Laboratory layered space-time (VBLAST) algorithm, 209
Vetterli, Martin, 11
Virtual transmitters
  generic room, 214
Viterbi algorithm, 281
VNA. *See* Vector network analyzer (VNA)
Voice traffic
  achieved throughput, 446f
  average transmission power, 448f
Voltage controlled delay line (VCDL), 461

Waveforms
  multiband schemes, 143f
  radiated electric fields, 140
  single-band scheme, 143f
  source pulses, 140f
Wavelet, 269
Weibull distributions, 197
Wideband models, 183
Wireless ad hoc networks. *See* Mobile ad hoc networks (MANETs)
Wireless personal area networks (WPANs), xv, 1, 122
  multiband approaches, 263f

Wireless routing protocol (WRP), 348
  derivation, 349
  proactive routing, 348–349
    protocol comparisons, 419t, 421t
Wireless routing protocol (WRP)-Lite, 349
Wireless sensor network (WSN), 432, 454
Wireless sensors, 1
Wireless telemetry, 1
Wireless USB interface, 310
WLAN, 453
WPAN. *See* Wireless personal area networks (WPANs)
WRP. *See* Wireless routing protocol (WRP)
WSN. *See* Wireless sensor network (WSN)

Xtreme Spectrum-Motorola proposal of a dual-band approach, 263f

Zone radius
  SHARP, 402
Zone routing protocol (ZRP), 406
  hybrid routing, 406–407
    protocols comparisons, 426t, 427t
  TZRP, 408
ZRP. *See* Zone routing protocol (ZRP)